Protecting Biodiversity, Culture, and Communities Through Forest Conservation

Md. Shafiqul Islam
https://orcid.org/0000-0002-5547-046X
International University of Business Agriculture and Technology, Bangladesh

IGI Global Scientific Publishing
Publishing Tomorrow's Research Today

Vice President of Editorial Melissa Wagner
Director of Acquisitions Mikaela Felty
Director of Book Development Jocelynn Hessler
Production Manager Mike Brehm
Cover Design Jose Rosado

Published in the United States of America by
IGI Global Scientific Publishing
701 East Chocolate Avenue
Hershey, PA, 17033, USA
Tel: 717-533-8845 | Fax: 717-533-7115
Website: https://www.igi-global.com E-mail: cust@igi-global.com

Library of Congress Cataloging-in-Publication Data

Names: Islam, Md Shafiqul, 1970- editor
Title: Protecting biodiversity, culture, and communities through forest conservation / edited by Md Shafiqul Islam.
Description: Hershey, PA : IGI Global Scientific Publishing, [2026] | Includes bibliographical references and index. | Summary: "This book aims to explore community knowledge for the protection of biodiversity, and their culture through forest conservation as an essential component of global sustainability efforts"-- Provided by publisher.
Identifiers: LCCN 2025019853 (print) | LCCN 2025019854 (ebook) | ISBN 9798337339924 hardback | ISBN 9798337339948 ebook
Subjects: LCSH: Biodiversity conservation | Conservation of natural resources | Sustainable living | Forest management ‡x Citizen participation
Classification: LCC QH75 .P785 2026 (print) | LCC QH75 (ebook)
LC record available at https://lccn.loc.gov/2025019853
LC ebook record available at https://lccn.loc.gov/2025019854

British Cataloguing in Publication Data
A Cataloguing in Publication record for this book is available from the British Library.

Table of Contents

Detailed Table of Contents

Biodiversity is the count of both species (plants and animals) that exist in a given area over a period of time. Biodiversity is invisible and just like a "feel word", no one can see it or touch it or hold it, only feel it by the heart. Biodiversity is also counted both economic and non-economic benefits and losses. Again it covers both direct benefits and indirect benefits to the environment and human. The developing countries, especially the tropical countries including Bangladesh, India and Indonesia was rich in biodiversity in the world. Biodiversity conservation is directly linked with culture and communities, forest types and forest management. It is important in understanding the cultural context of forests and community engagement. Community involvement and their cultural efforts may ensure protection of biodiversity. All these efforts are the key drivers to make success of conservation of biological resources of the countries. It is important to raise the voices of the forest communities with their cultural practices. Therefore, the need of review the biodiversity and forest conservation strategies with the communities' cultural participation that have effectively contribute in protecting biodiversity. The study used qualitative data with document analysis and literature review method. Results showed that local people participation and their cultural practices helpful in protection of biodiversity and forest conservation. The ethnic communities are losing their occupation and migrating to the urban areas for their wellbeing or livelihood activities. They are sourcing new employment opportunities due to loss of biodiversity and lost forest. Many threats and reasons were identified by the respondents as the reasons of biodiversity loss. Even though the lost forest could not able to accommodate their existing culture which are useful to protect biodiversity and forest conservation. Some myths, beliefs and taboos were prevailed in the society that may hindrance the conservation efforts. The results concluded that effective biodiversity policies must reflect the cultural practices of the forest communities since they change makers of conservation. It is important to intricate local communities and their cultural practices in protecting biodiversity and forest conservation. It gives us important message to take care about the people and take care of the earth to maximize the benefits of biodiversity. Effective and applicable biodiversity policy and forest conservation policy can save the last forest and the lost people of the communities.

Ferdous Ahmed, International University of Business Agriculture and Technology, Bangladesh
Md. Sazid Hasan, International University of Business Agriculture and Technology, Bangladesh

Forests stand alone for maintaining all vital functions of biodiversity of wildlife and climate regulation as well as carbon sequestration and water filtration. This chapter investigates the effects of climate change on forest degradation specifically intended for ecosystems and communities as well as indigenous populations whose forests cost both culturally and economically. However, the aims of this chapter papers are: (a) To analyze the extent and drivers of forest loss; (b) examine the ecological consequences of forest degradation on biodiversity; (c) assessing the socio-economic and cultural impacts of deforestation on local communities; (d) to propose sustainable strategies and policy recommendations. In deforestation alone (excluding regrowth), tropical primary forests are still declining at rate of several million hectares/year. However, in 2023, primary tropical forest lost 3.7 million hectares. Simultaneously, in tropical regions especially (Amazon, Congo, Southeast Asia), in some cases 50-90% of logging may be illegal or driven by weak governance.

Niamjit Das, Center for Natural Resource Studies, Dhaka, Bangladesh & Department of Forestry and Environmental Science, School of Agriculture and Mineral Sciences, Shahjalal University of Science and Technology, Sylhet, Bangladesh

Forest ecosystems are facing an unprecedented convergence of threats from deforestation, climate change, and socio-cultural disruption. This chapter investigates key drivers of forest loss—agriculture, logging, and infrastructure—and quantifies their impacts on fragmentation and biodiversity decline. It analyzes how climate change amplifies these pressures via altered climates, intensified wildfires, and pest outbreaks, disrupting species composition and ecosystem function. Socio-cultural impacts are explored, including displacement of Indigenous communities, erosion

The "Wood Wide Web" is an underground network created by mycorrhizal fungi that connects trees and plants. This hidden system enables the sharing of nutrients, information, and warnings, showing that forests function as a unified, cooperative community rather than isolated trees. Trees use this network to help each other, adapt to environmental changes, and promote overall forest health. Although much remains unknown, this chapter explores the biological, ecological, and philosophical aspects of forest communication, discusses current research and tools, and highlights its role in supporting biodiversity and climate resilience. It invites a deeper appreciation of nature's intelligence and interconnectedness.

This chapter discusses governance and innovation issues related to forest conservation and biodiversity in developing countries, focusing on institutional, political, and technological aspects. It does not address biology or culture, but rather focuses on policy disjunction, poorly coordinated inter-sectoral collaboration, and institutional fragmentation. Challenges include scarcity of information and technological resources, complexity of regulatory frameworks, political myopia, and data scarcity. This chapter also discusses emerging governance innovations, including adaptive technologies, open-source monitoring, cross-border conservation diplomacy, and others. This multidisciplinary approach, complemented by case studies, informs the chapter's recommendations for building resilient, inclusive conservation systems that are strategically supported by data. Ilsa emphasizes the need for active participation by policymakers, academics, and practitioners in the Global South working toward sustainable conservation strategies.

This abstract offers a comprehensive analysis of Human-Elephant Conflict (HEC), a critical socio-ecological challenge in India. Using the compelling case of Chinnathambi, a translocated wild elephant, it critiques the limitations of reactive conservation strategies. The chapter contextualizes HEC within global and Indian frameworks, integrating quantitative data on conflict impacts with qualitative insights from diverse stakeholders. It examines the indispensable ecological roles of elephants as keystone species and scrutinizes existing policy frameworks, highlighting deficiencies in corridor protection, compensation mechanisms, and inter-sectoral coordination. A key focus is the justice dimension, addressing community perceptions of inequitable conservation policies. The chapter concludes by outlining forward-looking solutions, including corridor restoration, advanced early warning systems, community-based initiatives, and integrated land-use planning.

Biodiversity conservation is a global imperative in the face of accelerating habitat degradation, climate change, and unsustainable development. This chapter delves into comprehensive strategies to safeguard biodiversity—from local grassroots actions to global frameworks, while integrating scientific innovation, traditional ecological knowledge, policy, and community engagement. The chapter explores in situ and ex-situ conservation methods, ecosystem-based approaches, and restoration initiatives, alongside an in-depth review of enabling legal, economic, and technological tools. Drawing on interdisciplinary insights and real-world case studies, the chapter emphasizes adaptive, inclusive, and resilient models for conserving the Earth's biological richness in the 21st century.

Forests and biodiversity are essential for ecological stability and human well-being, but they confront unprecedented threats in developing countries. This chapter investigates the multifaceted interactions among environmental decline, socio-economic stresses, inadequate governance, and global market forces that drive biodiversity loss and deforestation. It underscores the difficulties of reconciling

valuable insights regarding local biodiversity, ecosystem services, and sustainable agricultural practices that have evolved over generations. By bridging indigenous knowledge with modern agroforestry techniques, this approach not only promotes biodiversity but also strengthens community resilience against climate change and environmental degradation. This study emphasizes the importance of participatory methods in the design and implementation of agroforestry systems, ensuring that the voices of indigenous peoples are central to conservation efforts.

This study sought to explore the role of traditional leaders in forest conservation in Ghana as part of broader global sustainability efforts. Drawing on oral traditions from Edwenase in the Amansie Central District, media reports of the Dormaa Anti-Galamsey operation, and relevant scholarly works, the paper demonstrates the ambivalent role of chiefs as both custodians and exploiters of biodiversity. While the Kobiri River and sacred forests of Edwenase highlight the enduring significance of taboos, rituals, and ancestral lands, the unchecked spread of galamsey reveals the erosion of cultural authority and the monetization of communal resources. By contrast, the Dormaa case illustrates how chiefs can mobilize customary legitimacy and collaborate with state institutions to enforce environmental laws, with strong public support. The study concludes that sustainable forest governance requires bridging traditional authority with statutory structures to protect biodiversity, preserve cultural heritage, and strengthen Ghana's environmental policy framework.

This chapter explores the transformative role of geospatial technologies in forest and biodiversity management. By leveraging tools such as GIS, remote sensing, spatial modeling, and UAVs, conservation efforts have become more data-driven, responsive, and precise. The integration of participatory mapping and indigenous knowledge enhances data quality and supports ethical, inclusive governance models. Geospatial platforms enable real-time monitoring, adaptive management, and early warning systems for environmental threats. While these advances offer immense promise, challenges such as data fragmentation, capacity gaps, and ethical concerns remain. The chapter concludes with recommendations for scaling innovations through inclusive policies, open data, and capacity building to ensure equitable and effective conservation outcomes.

This chapter explored how remote sensing and drone technologies transformed forest monitoring and management practices between 2015 to 2025. The objective was to assess their role in evaluating forest health, biodiversity, and carbon storage. A key research gap addressed was the limited understanding of how integrated platforms, combining satellites, drones, LiDAR, and vegetation indices, could offer more precise and scalable forest data. Using a conceptual framework and the Systematic Literature Review (SLR) technique for literature review, the study analyzed tools such as Landsat, Sentinel, MODIS, NDVI, EVI, and LiDAR to examine their applications in tracking deforestation, estimating biomass, and detecting illegal logging. The findings showed that merging satellite imagery with drone and ground-based sensors significantly improved the accuracy and frequency of forest assessments. Integration with GIS and machine learning further enhanced decision-making capabilities. Multi-source remote sensing enhances forest conservation, climate monitoring, policy development.

This chapter examines the crucial role of ecological modeling in biodiversity conservation and its associated health implications. It discusses how various modelling approaches, including species distribution models, agent-based simulations, and systems dynamics, are employed to forecast ecological responses to anthropogenic pressures such as habitat fragmentation, climate change, and pollution. The chapter emphasizes the connection between biodiversity loss and public health, highlighting the increased risks of zoonotic diseases, reduced ecosystem services, and deterioration in mental health. Integrating ecological and epidemiological data enables the identification of conservation priorities, assessment of intervention effectiveness, and support for informed decision-making for sustainable ecosystem management. The chapter advocates for interdisciplinary frameworks that combine

Preface

Biodiversity is not easy to understand and is also difficult to explain. This book explores the feeling of biodiversity, mentioning the present situation of biodiversity with the problems, remedial measures and future direction for conservation initiatives. Biodiversity, forest, plants, animals, including insects, birds, microorganisms and people are closely linked, especially when humans have a huge demand for forest and biodiversity. Biodiversity itself is very caring and provides all the support to people, but only people are very selfish and causing damage to the environment unnecessarily. Destruction of habitats, killing animals and plants are common issues everywhere. People are making food and houses without thinking about the threats to biodiversity and the environment.

The vicious cycle of modernization and tertiary demand is dominating our society, making people toxic in their minds, culture, practices, behaviors and consumption patterns. Conservation of forest and biodiversity and practices are new challenges for producers and consumers that make a critical challenge to ecosystems as recognized as a global priority. The creation and dissemination of biodiversity knowledge is important for students, educators, practitioners and researchers. Textbooks provide functional resources for gaining new knowledge with updated information. These are essential teaching supports within any educational framework. Active learning, united with liberal education, not only simplifies degree accomplishment but also prepares individuals with up-to-date, world-class knowledge. The global population is confronted with the associated risks of biodiversity, climate change, deforestation and the environment. It needs diverse biodiversity and ecosystems to reduce pollution of air, water, and soil. This means that there is an urgent need to safeguard biodiversity while managing forest, rivers, agriculture and food sustainably. It is aligned with the huge demand of a growing population, especially to fulfill the crying need for food and other necessities.

Innovative research and dynamic research on biodiversity protection and conservation are important in the world. This book builds the capacity of the global community to face biodiversity conservation-related challenges in the real world.

The book contains the following objectives: i) to explore the objectives and strategies of biodiversity conservation; ii) to connect the cultural, social, environmental, and economic dimensions of biodiversity; iii) to identify the problems and challenges faced by the community, along with the available options for sustainable conservation of forest and biodiversity; and iv) to guide the students, researchers, and practitioners to ensure collective efforts for innovations.

The book included eighteen chapters in three distinctive sections with very coherent diverse knowledge and practices. First section explores problem and challenges of biodiversity protection and forest conservation. Second section explains about the biodiversity conservation strategies and finally the approaches of forest and biodiversity monitoring and management in section three.

The objective is to present inclusive knowledge for students, serving as a textbook for both undergraduate and graduate programs in Ecosystems, Biodiversity Conservation, Climate Change, Ecological Agriculture, Regenerative Agriculture, Environmental Science, Environment and Development Studies, and Sustainability across various universities.

CHAPTER OVERVIEW

It is important to know the biodiversity and forest conservation perspectives, while the ignorance may create complexity to understand everything. **Chapter one** provides clear idea on the cultural values and role of communities protecting biodiversity. Different aspects of forest conservation and biodiversity protection have been mentioned in this chapter. This chapter makes the readers happy to understand all sorts of conservation issues.

Chapter two is the essence of forest as the harbor of biodiversity. No more enough forest in the world especially in the developing countries like Asia. The stolen forest has no space to breathe inside the forest. The chapter presents the present scenario of the forest and forest biodiversity in Asia and developing countries. There is no last forest to contribute in forest biodiversity. Anthropogenic activities and dilemma accelerate the situation clumsier. It indicates the future disaster where no more single tree in the forest to welcome the birds or animals to stay there or can provide food.

In **chapter three**, it mentions the threats of biodiversity are everywhere. The highlights of the chapter is more close to the causes and effects of deforestation that leading towards climate change issues and it is not so far away from the consequences of climate change impacts. There are many challenges we need to face every day to meet our demand from forest and surroundings for our survival. Our demand is becoming limit less and disrupting the socio-cultural norms and core ethics.

Chapter four exemplifies the role of environmental drivers and media protecting forest and biodiversity in China is important. The chapter investigates how the environmental concerns of youth and their engagement with media affect their intentions and behaviors regarding the sharing of forest protection information on social media. Environmental concern and social media engagement positively influence the youth's intention to share environmental protection information on social media, which subsequently encourages actual sharing behavior. Environmental concern and media engagement not only increase the intention to share but also facilitate the actual sharing of forest protection information.

While **chapter five** presents the vive of wood wide web in forest communication. In conservation mechanism, the plants communicate with others for their greater awareness and survival in the community even during hardship. Parental defense is so strong safeguarding the offspring from the environmental threats. The concealed realm beneath our feet conceals remarkable secrets. The subterranean networks that link trees and fungi, referred to as the "Wood Wide Web," illustrate that forests are more than assemblages of trees. It presents vibrant communities that engage in communication, collaborate, and provide mutual support. These natural interconnections remind wonder us to reconsider the definitions of intelligence and life. These networks are not merely fascinating—they are essential for life on Earth. They play a crucial role in safeguarding biodiversity, enhancing climate resilience, and facilitating the recovery of forests following disturbances. In their absence, ecosystems would face significant challenges, and the equilibrium of nature would be disrupted. To safeguard this living web, collaboration among biology, ecology, technology, philosophy, and indigenous knowledge is imperative. The forest has been conveying messages through its roots and fungi for an extended period. Now, as scientific understanding advances, we are only beginning to learn how to heed these messages.

Diverse challenges and problems of forest biodiversity conservation have been identified in developing countries (**chapter six**) that may help the readers to understand the context and situation. Chapter seven establishes the persistent inadequacy of conservation policies in developing nations is primarily a systemic failure of governance and political economy rather than an ecological issue. The existing framework, marked by institutional fragmentation, the political subjugation of the environmental sector, and a centralized management approach, consistently jeopardizes long-term sustainability by prioritizing short-term economic benefits over ecological integrity. It is evident that to facilitate this transition and achieve authentic policy effectiveness, a comprehensive three-pronged roadmap is imperative. Local communities and Indigenous governance frameworks is crucial to conserve Traditional Ecological Knowledge (TEK) necessary for effective, locally-tailored rule-making.

Chapter seven looks at the conflicting issues between human and animals with special focus on elephant in India. Both humans and elephants face mortality, resulting in the disruption of countless livelihoods. Conflict represents a persistent and systemic aspect of their coexistence. Poaching, collision, cruelty and killing of animals are threats to the elephant. Kindness, humanity and co-existence may ensure rich biodiversity in any given place. This chapter provides glimpse on the role of elephant in large seeds dispersal, safeguarding the boarders. Elephants are not just sizable, captivating mammals; they are, from an ecological perspective, essential ecosystem engineers and keystone species. This classification underscores their significant and frequently irreplaceable influence on the structure, composition, and resilience of the environments they occupy in both Asia and Africa. Tackling this complex issue requires a strategic shift from reactive and fragmented approaches to more holistic, landscape-scale, and justice-focused solutions.

Biodiversity conservation strategies are important. **Chapter eight** discusses the way of sustainable biodiversity conservation process. Preserving biodiversity represents one of the most pressing and intricate challenges to us. It needs a systemic approach, a long-term dedication, and profound collaboration across various stakeholders, disciplines, sectors, and cultures. This chapter has examined diverse strategies for biodiversity conservation, which include protected areas (PAs), restoration efforts, policy frameworks, and innovative technologies. It emphasizes that no single solution can address the problems. Diverse strategies are required considering the context, ethical, ecological, and social factors are crucial. The future of biodiversity conservation hinges on the integration of scientific knowledge with values, local insights with global initiatives, and ecological comprehension with political determination.

Chapter nine addresses the problems related to forest and biodiversity conservation in developing nations adopting sustainable solutions. Conundrum of biodiversity conservation is multifacets and country specific with attainable local solution. The crisis of conservation impacting forests and biodiversity in developing nations is deeply intertwined with socio-economic, political, and ecological elements. Forests and biodiversity are vital ecosystems that facilitate climate regulation, water cycles, cultural heritage, and livelihoods. The interconnected issues ranging from deforestation, habitat fragmentation, illegal trade, climate change, to the marginalization of indigenous populations underscore the complexity of achieving effective conservation in low- and middle-income countries. Conservation challenges in the Global South are multidimensional. Environmental degradation is a result of ecological mismanagement; it is closely associated with factors such as poverty, insecurity in land tenure, and population growth. The protection of biodiversity must be integrated into broader frameworks of sustainable development, human rights, and social equity. It is essential to recognize and empower stakeholders through legal reforms, participatory planning, and decentralized resource management. Addressing the conservation dilemma in developing nations requires global solidarity, with wealthier countries supporting ecological stewardship in biodiversity-rich areas constrained by economic limitations.

In **chapter ten**, forest microbimes have been discussed in multidisciplinary way that emphasizes the hidden pillars of biodiversity mentioning roles and threats of conservation strategies. The conservation of forest biodiversity has significantly neglected the important role of microbial communities (forest microbiome), which are essential for maintaining ecosystem health. Most of them are bacteria, fungi, and other microorganisms is responsible for key processes such as nutrient cycling, soil formation, and plant growth, all of which are important for the overall functioning of ecosystems. All threats are from human activities, including deforestation, land use change, and climate change, resulting in a decline in microbial diversity that has far-reaching implications for ecosystem stability. A shift is required to incorporate microbial conservation into forest management and restoration efforts, as overlooking this aspect undermines the effectiveness of existing strategies. Forest microbiome is essential for meeting global biodiversity and climate objectives. Microbial communities residing within forest ecosystems are not simply inhabitants; rather, they serve as crucial architects of forest structure, function, and resilience. Forest microbiome is an important component of biodiversity, as it plays a key role in recyçling essential nutrients, establishing soil fertility, directly enhancing plant growth, and providing defense against pathogens.

Chapter eleven discusses the vertical approaches to enrich biodiversity. Both longitudinal and horizontal coverage is a good option for dense populated areas. Accumulation and valuing of indigenous knowledge may strengthen the biodiversity conservation initiatives through enhancing community resilience. The exploration of indigenous agroforestry systems as a viable alternative for implementing sustainability initiatives is framed by several pressing issues, including climate change, deforestation, biodiversity loss, rising inequality, and worsening poverty. Historically, these indigenous systems have sustainably supplied food, timber, shelter, fuel, and contributed to biodiversity conservation, among other services. However, they are currently facing significant stress. On the other hand, the implementation of technical measures aimed at mitigating climate change impacts and enhancing carbon sequestration may inadvertently exacerbate social challenges. Climate-smart agroforestry has been introduced to leverage traditional agroforestry knowledge as a foundation for the application of modern agroforestry systems in Eastern Africa. In contrast, traditional perennial crop-based agroforestry systems, along with complementary mixed rainfed and wetland agriculture, have been examined across a diverse array of agro-ecological and agricultural intensification gradients in Mesoamerica. In both scenarios, it is posited that agroforestry offers numerous opportunities to reconcile biodiversity conservation with the needs of local communities adapting to a changing environment, provided that the broader socio-economic context is taken into consideration. Biodiversity plays a vital role in the functionality of ecosystems and the delivery of services. Nevertheless, agricultural intensification and climate change pose new threats to biodiversity, and their widespread occurrence is likely to rapidly undermine the adaptive capacity of existing agro ecosystems and the ecosystem services they provide. This situation underscores the necessity for research focused on transcending resilience approaches that can effectively integrate biodiversity into agricultural and climate adaptation strategies. The evolutionary trajectory of existing agricultural biodiversity within a specific region, in relation to its geo-physical context, serves as a foundation for examining the enduring 'agro ecological memory' present in that area.

In chapter twelve, community knowledge, local and traditional knowledge is essential for conserving forest and protecting biodiversity in Ghana has been documented. Function of traditional leaders in forest conservation efforts in Ghana, contributing to wider global sustainability initiatives. Sacred forests of Edwenase underscore the lasting importance of taboos, rituals, and ancestral lands; the rampant proliferation of galamsey indicates a decline in cultural authority and the commercialization of communal resources. The chapter concludes that effective forest governance necessitates the integration of traditional authority with legal frameworks to safeguard biodiversity, maintain cultural heritage, and enhance Ghana's environmental policy landscape.

Chapter thirteen describes the geospatial technologies currently in use have been integrated into the management of forests and biodiversity, fundamentally altering the perception, control, and management of ecosystems. It encompasses GIS, remote sensing, spatial modeling, and UAVs, play a crucial role in conservation planning and implementation. They facilitate the observation and interpretation of complex ecological patterns across various spatial and temporal scales, enabling near-real-time monitoring of deforestation and advanced forecasting of climate change impacts on community composition. This discussion highlights not only the individual strengths of each tool but also the transformative potential that arises from the interaction between these technologies, participatory processes, and community-based strategies. The incorporation of local and indigenous knowledge through participatory mapping enhances the quality of spatial data and promotes conservation efforts. Community-based forest management, supported by spatial platforms and mobile applications, has achieved remarkable success in various regions, underscoring the importance of inclusive governance and local agency in achieving ecological outcomes. Geospatial tools have revolutionized monitoring, evaluation, and adaptive management. Satellite imagery and ground observations related to conservation over time can be utilized to reassess the effectiveness of conservation plans, providing feedback for continuous improvement. Once remote-sensing-based early warning systems are established, threats such as forest fires, illegal logging, and biodiversity loss can be minimized.

In the context of globalization, the digital monitoring system of remote forest and biodiversity conservation is important (**chapter fourteen**). The incorporation of remote sensing technologies, including satellites and drones, into terrestrial tools has revolutionized the monitoring of forest health and forestry operations. The immediate availability of high-resolution, detailed information facilitates the application of remote sensing technologies in a way that offers unmatched capabilities for monitoring forest ecosystems, assessing forest cover loss, evaluating biodiversity, and measuring carbon stock, all of which are crucial for achieving sustainable forest management and addressing climate change. The integration of satellite, UAV, and ground data has opened new avenues for overcoming forest-related challenges and enhancing the understanding of forest dynamics, especially in remote access regions. Findings are helpful for policymakers, researchers, and forest managers. A key takeaway is that remote sensing technologies can facilitate the creation of more effective conservation strategies, thereby enhancing decision-making processes related to forest management and supporting climate change mitigation policies.

Ecological modeling has discussed in **chapter fifteen**. Such modeling contains a significant ability to characterize and forecast the behavior of ecological systems, as well as the nature and disposition of the ecological benefits these systems provide to humanity. It is important to deploy applied biomathematicians and statisticians but also for naturalists, ecologists, and others who have more expertise and experience on a multi-faceted understanding of the relationship between taxon biology and ecology. In alignment with the importance of this book, applied macro ecology is characterized by the careful construction of ecological models that illustrate and manage the interactions between taxon carriers and the natural resources and processes.

It believes that this book has a contribution to understanding and advancement of research in biodiversity conservation, practices and managing the forest and biodiversity. Welcome suggestions from readers to improve the content and structure of this work.

Md. Shafiqul Islam
International University of Business Agriculture and Technology, Dhaka, Bangladesh

Acknowledgment

The successful completion of this book can be achieved from the extensive assistance and support received from various individuals. I'm thankful to my eldest brother Enginner Abu Hanif Miah for his endless inspiration and encouragement in writing this book.I wish to express my profound gratitude to Assistant Professor Dr. Mohammed Shawkat Hossain (Editorail Advisory Board Member), Dafodil International University and Assistant Professor Dr. Sohag Mia (Editorial Advisory Board Member) Institute of Forestry and Environmental Sciences, Chittagong University for their rigorous review and constructive feedbacks to evaluate each chapter. They were ready to extend their intellect, and unwavering support throughout this process. I also extend my appreciation to all contributors for their excellent writings and contribution to this book and make it possible to reach the readers and users of this book. My heartfelt thanks go to Professor Dr. Mohammed Aatur Rahman for his encouragement during the writing process. Numerous colleagues from the International University of Business Agriculture and Technology (IUBAT), have provided essential assistance, without which this book would not have been achievable. My special thanks go to Shamnur Rahman who helped in organizing the table of contents. Additionally, I am thankful to Gerlinde Staffler for her ongoing inspiration, as well as to the community members in the study areas who generously shared their knowledge, making this research possible. I am deeply indebted to them for their unwavering support. I express my gratitude to the officials of NGOs and community people for their valuable time and for supplying useful data and information that contributed significantly to this work. It is my pleasure to let my greatest gratitude to IGI Global publisher for offering me to write and participate in completion of the book.

Editorial Advisory Board

Introduction

Biodiversity is the more pressing issues in the world. Several studies mentioned that the world is segmented into 5 Global Ecological Domains and 20 Ecological Zones based on the components of ecosystems (FAO,2001; Simons, 2005; Mac Dicken, 2015).Bangladesh is under two ecological zones such as Tropical rain forest-GEZ (33%) and Tropical moist deciduous forest -GEZ(67%) of the Tropical Domains of Global Ecological Domains. National Forest and Tree Resource Assessment- NFA (2007) reported that the central north-eastern and south-eastern forests are under Tropical rain forest-GEZ. Our world already has lost 517 million hectares (Mha) of forest cover completing the century, near to 13 percent of world forest cover in 2000. It was observed that forest cover has been decreasing and it was recorded that 13.4 million hectares (Mha) forest cover loss in 2001 and 29.6 million hectares (Mha) in 2024 (World Resources Institute, 2025).

Biodiversity and ecosystem have great role in agricultural production, rebuilding ecosystem, re-shaping the society and supplying pollinating agents for more production. Many factors are exacerbating the concern problems including overpopulation, knowledge gaps, lack of appropriate policy of biodiversity, trending few quality research and research transformation. It needs many efforts to protect the biodiversity, conserve local culture and community by building capacity and awareness for the conservation of forest. It is essential to perceive the conservation objectives of forest biodiversity, people and culture by putting emphasis on forest conservation. Many problems are facing in the developing countries related to biodiversity conservation. It's not only confined in the developing countries, is also challenging for the developed countries as well. We are bridging the gaps with advance technologies, hybrid species and so-called research results transformation among us but the reality is far away. Threats concerning biodiversity is not affecting only conservation initiatives, is also affecting forest due to climate change. Most of the vulnerable socio-economic groups are at risk due to rapid deforestation, habitat loss and climate change vulnerability. The huge task of biodiversity conservation is not very easy and it needs help from community, NGOs, concerned department, research organizations, universities, private organizations, financial institutions, the state and other stakeholders. Cultural beliefs and values, community trust and practices are important for biodiversity conservation. Old aged cultural practices are sustaining to protect biodiversity and it is continuing over decades.

PERSPECTIVE OF FOREST AND BIODIVERSITY CONSERVATION

Biodiversity is a feel word and nobody can make it visible to others as evidence. One needs to realize or feel to perceive biodiversity. Biodiversity is the diversity of life or variation of life or species. It has great importance to the people and environment. As the harbour, forest is the great place to conserve biodiversity. People and cultural aspect influence the conservation of biodiversity. Religious beliefs and cultural values are also important to succeed the conservation initiatives in the world. Few communities are conserving biodiversity for their own interest using the indigenous knowledge for their survival. Many developing countries such as Bangladesh, India and Indonesia have high biodiversity. The conservation of Biodiversity is connected with people, culture and communities and their efforts in forest management. It is essential to know the cultural context of people, community and objective of forests management. Cultural perspectives and religious beliefs may help in biodiversity protection. It is the time to raise community voices with their cultural practices to protect the biodiversity for their own needs.

Problems and Threats of Forest and Biodiversity Conservation

We are living in the spikes of problem and threats especially regarding biodiversity protection and forest conservation. There are many problems including lack of institutional support, policy, infrastructure, coordination, capacity, data, technology, information, monitoring, regulatory complexity, innovation, governance, stakeholder engagement, community participation and political commitments that are downing the conservation strength. Fragmented cooperation, inconsistency in policy and bureaucracy, lack of inter institutional support, lack of specified conservation objectives and fringe benefits, inadequate funds, low policy support and ineffective implementation strategies are restricting protection initiatives of forest and biodiversity. Problems, challenges and threats are very contextual and vary from location to location and country to country based on conservation goals and objectives.

Technological Advancement for Forest and Biodiversity Conservation

Advance technologies have many applications in protecting biodiversity and forest conservation. Identification, implementation, monitoring, evaluation and predicting biodiversity and forest status necessitates the use of advance technologies is unavoidable. It has diverse use in forest ecosystems monitoring, restoring, and protecting. Remote sensing gives large-scale, multi-temporal data for calculating land cover, identifying deforestation rate, and assessing forest health (Blumenfeld, 2020). Landsat and Sentinel both are useful to make continuous observations using longitudinal resolutions of 10 to 30 meters to analyzing vegetation patterns. Additionally, GIS collect biophysical, socio-economic, and administrative data to support evidence-based decision-making in conservation planning. Advanced tools such as LiDAR can produce 3D structural forest data, and UAVs or drones for localized restoration efforts (Oehmcke et al., 2021; Reiersen et al., 2022; Morin et al., 2023; Nguyen & Saha, 2024). Communities can co-produce spatial data through participatory mapping using mobile GIS platforms, where satellite access and accuracy is limited. All technologies are effective if we can integrate with diverse data. Sentinel-1, ALOS PALSAR penetrates cloud cover to creating model for mangrove and wetland (Shit et al., 2020; A. Kumar et al., 2022).

PRESENT STATUS OF BIODIVERSITY

Stolen Forest and Biodiversity

The stolen forest indicates three activities including the activities of deforestation, illegal cutting of forest and forest encroachment. According to Food and Agriculture Organization-FAO (2020), human destroy 7.3 million hectares forest land in a year. Deforestation reduces biodiversity and weakens the environmental stability. Forest encroachment is leading land use change as a driving force within the forest. Illegal cutting or logging reduce the biodiversity by causing disturbance to wood wide web and forest micro biomes and make the forest vulnerable to exist.

Longitudinal Aspects of Biodiversity Conservation and Forest Management

Spatial or longitudinal thinking can connect geospatial technologies to evaluate the status of biodiversity and help in planning and implementation of present and future course of action efficiently. It's like an instrument to tackle biodiversity loss and forest degradation. Such advance thinking and utilization of geospatial technologies can help in decision making process by providing information based on evidence. World Resources Institute -WRI (2025) reported that geospatial technologies and longitudinal thinking can help in connecting the problems, creativity and future course of action for biodiversity conservation and forest management.

Climate Change and Deforestation

Many climate stressors increased drought frequency, coupled wildfire regimes, and outbreaks the pest attack. Cumulative droughts have affected tree growth, decreased survivality rates, and inclined landscapes to mega fires (Bowman, 2016). Warm temperature increases the growth seasons of many insects and pathogens for forest pests that destroy forest stands. Moreover, heavy rainfall accelerates soil erosion and landslides in the mountain areas, disturbing root stability and restricting ecosystem services including water cycling and sediment storage. Biodiversity loss amalgams such pressure from abiotic sources. Myers (2000) reported that tropical forests possess fifty percent or more terrestrial species, which show narrow habitat and low dispersal volumes. Population fragmentation creates "extinction debts," and interrupts mutuality interactions including pollination and seed dispersal (Wearn, 2012). In addition to ecological disturbance and deforestation due to climate change, it has deep socio-cultural distractions.

FRAMEWORK FOR BIODIVERSITY CONSERVATION AND FOREST MANAGEMENT

Hidden Pillars of Biodiversity: Forest Biomes

Forest microbiomes represent very rich biodiversity especially for functional microorganisms. It consists of diverse communities including bacteria, virus, fungi, archaea and protists. It has great influence on soil and plant roots (rhizosphere), bacterial communities (taxonomic diversity), and fungal association (symbiotic for plant nutrition). Forest micro biomes have important role in forest niche by influencing phyllosphere (Bacteria, fungi), litter (bacteria, fungi), dead wood (fungi), soil (bacteria, fungi), and rhizosphere (bacteria, fungi) to perform primary ecological function. It is essential to protect biodiversity enriching ecological health and social identity.

Biodiversity Stewardship

We have the stewardship of the planet but we have no right to destroy our planet. Human populations have built mutual relationships with forest ecosystems over eras (Kar, Choudhary, and Singh, 2022). Such mutual relationships are linked with culture, cultural heritage, as they use local wisdom and indigenous practices and knowledge that help conservation and sustainable management of biodiversity (Silvestro et al., 2022). All practices and knowledge conceded from generation to generation focusing the ecosystems, informing practices that helps to make a balance between needs of human and sustainability of ecosystems (Sun et al., 2022).

People, Communities, and Cultural Heritage

Role of local communities and their traditional ecological knowledge in protecting forest biodiversity are useful to ensure sustainability in resource management (Kolluri et al., 2022). Indigenous and local communities have intimate knowledge about the natural environment, continuing sustainable practices from generation to generation over the decades (Ukoba and Jen, 2022). The concepts of cultural landscapes deal with cultural heritage and biodiversity connecting the people, culture and heritage.

Wood Wide Web: A New Techniques of Forest Communication

Wood wide web is an interconnected system of communication or it may call nerve system of the forest. Dr. Sunzanne Simrad introduced the phrase 'Wood Wide Web'. She is a forest ecologist and studied the communication system of plants and she showed that plant can send messages, share resources to others and warn the plants from the attack of insects or natural calamities like drought. In reality, it means that a world of cooperation and connection. It gives us the indication that how much biodiversity is important! Old one is mother tree making aware and taking care about the young seedlings to survive during hardship even. Trees built their network system to provide support to other by maintaining high biodiversity.

Monitoring of Biodiversity Using Technology (Remote Sensing, Drone, AI)

Technological advancement is important to monitor biodiversity by ensuring forest management. Most of the natural forests are located in the remote areas and nature has the pivotal design to protect the natural resources in the most inaccessible locations. Advanced and new efficient technologies are required to monitor biodiversity and forest management. These new technologies such as Ground Based

Sensors (LiDAR-Light Detection and Ranging), Satellite based system (Landsat, Sentinel, MODIS and NDVI), and Drones (equipped with cameras, LiDAR and multispectral sensors) have been examined in-depth analysis for forest canopy, tree health and biodiversity. Remote sensing devices can collect remote data by satellite and airborne systems including drones or aircraft. The use of advance technologies is getting popularity in highly remote areas.

CONSERVATION STRATEGIES

Geospatial Approaches to Biodiversity Conservation

Geospatial strategies are important to conserve biodiversity. Many countries including Nepal, Peru, Philippines, Tanzania, Mexico, Pakistan and Indonesia are using Geospatial technologies for planning, implementing of the activities related to biodiversity protection and forest management. Above mentioned countries adopted community-based forest management using several spatial tools such as Remote Sensing, Drones, GIS, GPS and many more.

Good Governance and Biodiversity Conservation

Good governance in biodiversity conservation is the essential element for forest management. Formal institutions, authority, law, policy and process are unavoidable practices to protect biodiversity. Responsibility, transparency, accountability are essentials in forest governance structures. Cultural values, indigenous knowledge and rights, access to the information and resources, structured administration is important to protect biodiversity and forest conservation. Governments use regulatory framework for longitudinal planning, conservation zoning as the key tools to reduce ecological disturbance. Governance should blend with the policy formation to implement inter-agency cooperation, and community engagement.

Indigenous Wisdom and Agro Forestry to Conserve Biodiversity

Agro forestry is one of the traditional approaches to go ahead with the conservation of biodiversity. The system relies on indigenous knowledge system (IKS) by integrating diverse landscape of cultural and religious values, ethno botanical, and biophysical, economic, socio- political and cultural knowledge. In this system, the community use agro forestry system to conserve biodiversity integrating indigenous knowledge in the diverse fields such as use of traditional knowledge, medicinal plant use, agricultural practices, land use practices, waste management practices, spiritual beliefs, extreme environment event management, resources sharing practices, agriculture and fisheries management.

Ecological Modeling for Forest Biodiversity Conservation

Ecological modeling is the mode of mathematical expression of natural phenomena; represent mathematical equations, computer simulations, or direct models by using media. Most models used in ecology using numerous media and software in order to simulate numerical value and behavior of ecosystems under environmental drivers. Outputs from these models can define species distribution maps, graphical and analytical visualizations model as generated from model simulations. Ecological modeling has an impact on basic and applied ecological research. Ecological modeling can provide a link to ecological theory and ecosystem changes observed by measuring and observation of species-area relationships, dominance-diversity patterns, and species associations.

Forest Co-Management

Forest co-management includes the shared responsibility of forest management among the stakeholders (local communities, non-government organizations and government agencies) to bring benefits for all through engaging NGOs and private sectors in many ways. Co-management initiatives may bring expertise, resources, and global experience to forest management for the forest-dependent communities helping them in decision-making processes. Forest co-management is an effective approach through the active participation of stakeholders. Many countries are managing many projects including forest biodiversity conservation. Co-management is the distribution of tasks and responsibility among diverse stakeholders. Few studies reported that forest co-management is reducing tree cover loss, dependency and removal of other forest resources from the forest (Islam *et al.,* 2019; Chowdhury *et al.,* 2022).

Sense of responsibility is important in biodiversity conservation. Forest based small scale enterprises (FBSSEs) include the processing of non-timber forest products (NTFPs), cultivation of medicinal plant, eco-tourism, honey collection, fishing, handicrafts from bamboo and cane, and sustainable timber harvesting. Such enterprises are labor-intensive and the community can use local knowledge, skills and family labors for many other economic activities by conserving forest and biodiversity to meet their own needs. FBSSEs are the potential economic pillar to make the community capable to diversify livelihood options, reduce dependency on resource extraction, processing and marketing. The effort needs capacity building training, financial support, market access, and institutional partnerships to scale up and contribute to local economy.

Single responsibilities and initiatives are not enough, even it is not easy to perform and sometimes it doesn't bring the success in protecting biodiversity and forest conservation. It needs partnership and collaboration from different stakeholders. Forest provides livelihood options to millions of people especially the indigenous people, forest dwellers and local communities in developing countries as well as in the world. Partnership is important to protect biodiversity and conservation of forest. Partnership could be developed in various forms such as collaborations between local communities and government agencies to national, multinational and international initiatives involving non-governmental organizations (NGOs), private organizations, international institutions, and indigenous people(Banday et al., 2021). Considering the context, partnerships are important in forest biodiversity conservation and management. It needs sustainable practices, resource sharing and distribution, and stop deforestation. Mutual trust, shared responsibilities, stakeholder engagement, and effective communication are important in building partnership.

Md. Shafiqul Islam

International University of Business Agriculture and Technology, Bangladesh

REFERENCES

Banday, M., Islam, M. A., Pala, N. A., Rashid, M., Ahmad, P. I., Rather, M. M., & Raja, R. (2021). Livelihood security and forest resource extraction by forest fringe communities in Indian Himalayan Region. In *Diversity and dynamics in forest ecosystems* (pp. 163–194). Apple Academic Press. DOI: 10.1201/9781003145318-8

Blumenfeld, J. (2020). *Wildfires Can't Hide from Earth Observing Satellites*. NASA EOSDIS.

Bowman, D. M., Balch, J., Artaxo, P., Bond, W. J., Cochrane, M. A., D'antonio, C. M., DeFries, R., Johnston, F. H., Keeley, J. E., Krawchuk, M. A., Kull, C. A., Mack, M., Moritz, M. A., Pyne, S., Roos, C. I., Scott, A. C., Sodhi, N. S., & Swetnam, T. W. (2011). The human dimension of fire regimes on Earth. *Journal of Biogeography*, *38*(12), 2223–2236. DOI: 10.1111/j.1365-2699.2011.02595.x PMID: 22279247

Chowdhury, F. I., Islam, K., Faroque, M. A., Islam, K. N., Rahman, M. F., Arif, M. T. A., Nath, T. K., & Jashimuddin, M. (2022). Assessing the impacts of co-management on protected area landscape under socio-imagery lens: Evidence from Bangladesh. *Journal of Sustainable Forestry*, *41*(7), 553–572. DOI: 10.1080/10549811.2020.1747497

FAO. (2001). *Global Forest Resources Assessment 2000 – Main Report* [= FAO Forestry Paper 140]. FAO. DOI: 10.1126/science.1219013

Islam, K. N., Rahman, M. M., Jashimuddin, M., Hossain, M. M., Islam, K., & Al Faroque, M. (2019). Analyzing multi-temporal satellite imagery and stakeholders' perceptions to have an insight into how forest co-management is changing the protected area landscapes in Bangladesh. *Forest Policy and Economics*, *101*, 70–80. DOI: 10.1016/j.forpol.2019.01.011

Kar, A. K., Choudhary, S. K., & Singh, V. K. (2022). How can artificial intelligence impact sustainability: A systematic literature review. *Journal of Cleaner Production*, *376*, 134120. DOI: 10.1016/j.jclepro.2022.134120

Kumar, A., Ekka, P., Patra, S., Kumar, G., Kishore, B. S. P. C., Kumar, R., & Saikia, P. (2022). Geospatial perspectives of sustainable forest management to enhance ecosystem services and livelihood security. In *Advances in Remote Sensing for Forest Monitoring* (pp. 10–42). Wiley. DOI: 10.1002/9781119788157.ch2

MacDicken, K. G. (2015). Global forest resources assessment 2015: What, why and how? *Forest Ecology and Management*, *352*, 3–8. DOI: 10.1016/j.foreco.2015.02.006

Morin, D., Planells, M., Mermoz, S., & Mouret, F. (2023). Estimation of forest height and biomass from open-access multi-sensor satellite imagery and GEDI Lidar data: high-resolution maps of metropolitan France. In *arXiv [stat.ML]*. arXiv. https://arxiv.org/abs/2310.14662

Myers, N., Mittermeier, R. A., Mittermeier, C. G., Da Fonseca, G. A. B., & Kent, J. (2000). Biodiversity hotspots for conservation priorities. *Nature*, *403*(6772), 853–858. DOI: 10.1038/35002501 PMID: 10706275

Nguyen, A., & Saha, S. (2024). Machine learning and multi-source remote sensing in forest carbon stock estimation: A review. In *arXiv [cs.LG]*. arXiv. https://arxiv.org/abs/2411.17624

Oehmcke, S., Li, L., Trepekli, K., Revenga, J., Nord-Larsen, T., Gieseke, F., & Igel, C. (2021). Deep learning based 3D point cloud regression for estimating forest biomass. In *arXiv [cs.CV]*. arXiv. https://arxiv.org/abs/2112.11335

Reiersen, G., Dao, D., Lütjens, B., Klemmer, K., Amara, K., Steinegger, A., Zhang, C., & Zhu, X. (2022). ReforesTree: A dataset for estimating tropical forest carbon stock with deep learning and aerial imagery. In *arXiv [cs.CV]*. arXiv. https://arxiv.org/abs/2201.11192

Shit, P. K., Pourghasemi, H. R., Das, P., & Bhunia, G. S. (Eds.). (2020). *Spatial modeling in forest resources management: Rural livelihood and sustainable development* (1st ed.). Springer Nature. DOI: 10.1007/978-3-030-56542-8

Silvestro, D., Goria, S., Sterner, T., & Antonelli, A. (2022). Improving biodiversity protection through artificial intelligence. *Nature Sustainability*, *5*(5), 415–424. DOI: 10.1038/s41893-022-00851-6 PMID: 35614933

SIMONS, H. (2005). Global Ecological Zoning for the FAO Global Forest Resources Assessment 2000 Entwicklung einer globalen ökologischen Zonierung für die weltweite Bewertung der Waldressourcen im Jahr 2000 seitens der FAO. Anwendung und Auswertung der Karte der natürlichen Vegetation Europas Application and Analysis of the Map of the Natural Vegetation of Europe, 55.

Sun, Z., Sandoval, L., Crystal-Ornelas, R., Mousavi, S. M., Wang, J., Lin, C., Cristea, N., Tong, D., Carande, W. H., Ma, X., Rao, Y., Bednar, J. A., Tan, A., Wang, J., Purushotham, S., Gill, T. E., Chastang, J., Howard, D., Holt, B., & John, A. (2022). A review of earth artificial intelligence. *Computers & Geosciences*, *159*, 105034. DOI: 10.1016/j.cageo.2022.105034

Wearn, O. R., Reuman, D. C., & Ewers, R. M. (2012). Extinction debt and windows of conservation opportunity in the brazilian amazon. *Science*, *337*(6091), 228–232. DOI: 10.1126/science.1219013 PMID: 22798612

World Resources Institute. (2025). *Global Forest Watch*. https://www.globalforestwatch.org

Chapter 1

Perspectives of Forest Conservation, Culture, Communities, and Biodiversity Protection

Md. Shafiqul Islam
https://orcid.org/0000-0002-5547-046X
International University of Business Agriculture and Technology, Bangladesh

ABSTRACT

Biodiversity is the count of both species (plants and animals) that exist in a given area over a period of time. Biodiversity is invisible and just like a "feel word", no one can see it or touch it or hold it, only feel it by the heart. Biodiversity is also counted both economic and non-economic benefits and losses. Again it covers both direct benefits and indirect benefits to the environment and human. The developing countries, especially the tropical countries including Bangladesh, India and Indonesia was rich in biodiversity in the world. Biodiversity conservation is directly linked with culture and communities, forest types and forest management. It is important in understanding the cultural context of forests and community engagement. Community involvement and their cultural efforts may ensure protection of biodiversity. All these efforts are the key drivers to make success of conservation of biological resources of the countries. It is important to raise the voices of the forest communities with their cultural practices. Therefore, the need of review the biodiversity and forest conservation strategies with the communities' cultural participation that have effectively contribute in protecting biodiversity. The study used qualitative data with document analysis and literature review method. Results showed that local people participation and their cultural practices helpful in protection of biodiversity and forest conservation. The ethnic communities are losing their occupation and migrating

DOI: 10.4018/979-8-3373-3992-4.ch001

to the urban areas for their wellbeing or livelihood activities. They are sourcing new employment opportunities due to loss of biodiversity and lost forest. Many threats and reasons were identified by the respondents as the reasons of biodiversity loss. Even though the lost forest could not able to accommodate their existing culture which are useful to protect biodiversity and forest conservation. Some myths, beliefs and taboos were prevailed in the society that may hindrance the conservation efforts. The results concluded that effective biodiversity policies must reflect the cultural practices of the forest communities since they change makers of conservation. It is important to intricate local communities and their cultural practices in protecting biodiversity and forest conservation. It gives us important message to take care about the people and take care of the earth to maximize the benefits of biodiversity. Effective and applicable biodiversity policy and forest conservation policy can save the last forest and the lost people of the communities.

1. INTRODUCTION

Forests are the natural home and more economically potential production system than others. It protects biodiversity; restore ecosystem and harbor of wildlife, including microorganisms, avian, insects, reptiles and wild plants. Four types of forests are found in Bangladesh likely: i) Mixed to evergreen hill forest; ii) Sal deciduous forest; (c) Natural mangrove forest; and iv) manmade mangrove plantation. Additionally, other three types of forest contribute to the biodiversity such as v) Fresh water swamp forest; vi) Homestead forests; and vii) Village forest. Forests are utilizing lands for environmental, economic, social, and political factors (Corticeiro et al., 2025). Increasing the value of forests, it needs policies to protect biodiversity, culture and communities by conserving forest in participatory ways. Numerous Indigenous groups in Bangladesh (including both flat and hilly areas) encounter serious ecological issues that endanger their livelihoods, cultural traditions, and general health. One of the major issues is the land degradation and destruction of natural habitats caused by deforestation, agricultural land conversion, unsustainable resource extraction, and unsustainably enforced tourism (Islam & O'Donell, 2020; Grafton & Melancon, 2020; Tuhiwaj Smith, 2012; Majumder et al., 2018). These actions greatly harm traditional sustainable land-based practices and their ecosystems, reduce biodiversity, and undermine essential ecosystem services for the survival of Indigenous communities, including clean water, fertile soil, and natural flood protection. Community participation and orest stakeholders' engagement are important to ensure a realistic strategy for the forest conservation in developing countries. About 70% of all plants and animals on Earth live in tropical forests, which are also vital for the provision of essential ecosystem goods and services

(Gardner et al. 2009; Laurance, 2007). According to Byron and Arnold (1999), almost 500 million people reside in tropical forests and rely on them in some way for their livelihood. Deforestation rates are high in the tropical region, leading to a rapid loss of biodiversity and wild habitats, despite the important role that tropical forests play in people's lives and the environment (Geist and Lambin, 2002). One of the primary causes of greenhouse gas emissions (GHG), tropical deforestation contributes significantly to global warming and accounts for over 20% of all anthropogenic GHG emissions (Baccini et al., 2012).

Almost 49 ethnic groups inhabit in Bangladesh with their different culture and customs. Many ethinc people or forest communities are surrounded by forest. Ten different ethnic groups reside in Bangladesh's coastline region, despite the Bangalee being the country's largest ethnic group (Rashid & Hasan, 2020). According to Kamal, et al., (2001), the coastal districts are home to the Chakma, Khyang, Marma, Munda, Murang, Rakhain, Tanchangya, Tripura, Mahato, and Pundra Khatrio ethnic groups. There are currently 45 different types of wildlife in the nearby woodlands, which the Mro people hunt for food and other everyday use. It was discovered that eight species including three birds and five mammals were in danger of going extinct in the area as a result of overharvesting (Chowdhury et al., 2014).

2. BACKGROUND

The application and suspected effectiveness of co-management in Bangladeshi forest management have evaluated critically in a study (Baten, 2024). Co-management has become increasingly popular around the world. It is effective for its collaborative approach and community participation to solve the conflicts concerning natural resources, but it is also criticized for its design faults, vulnerability to elite. Sustainability and realism of its success, especially with different factors and dynamics and effective community participation in decision-making processes, remain despite two decades of implementation and reports of successful results in managing Bangladesh's protected forest areas. The investigation explores various facets, exposing situations in which achievements may appear delicate while ignoring the fundamental problems of power imbalances among stakeholders. It emphasizes how important it is to build social capital in co-management and livelihood frameworks in order to promote sustainability and institutionalization. Effective communication is required to intricate interactions of institutional, social, and natural difficulties to the use of adaptive management techniques. Ways to measures engaged for the natural resource management are based on market centric theoretical framework. A study reported that the biodiversity threatened primarily due to lack of market and pessimistic externality (Sandmo, 2015; Perrings et al., 1992). It makes sense that by calculating the economic value of the resources and developing exchange rules related to biological resource protection, valuation techniques can offer valuable insights to support policy initiatives (Costanza & Folke, 1997; Pearce & Pearce, 2001; Bräuer, 2003; Kumar, 2005; Barbier, 2007; McAfee & Shapiro, 2010; Hahn et al., 2015). Since the implementation of a formal property rights regime may effectively manage natural resources, while the lack of property rights leads to resource degradation, institutional economists have augmented this understanding (Gibson et al., 2000; Bergseng & Vatn, 2009; Ituarte-Lima et al., 2014). On the other hand, some political economy assessments argue that the existence of overlapping property rights regimes leads to conflicting management and degradation of resources. It draws attention to the hierarchical relationships that exist in society and illuminates the political aspects of the resources management regime. It makes the case that, in the context of a non-cooperative solution, institutional arrangements (property rights) are susceptible to certain political and economic aspects resulting from accumulation by various agents. Additionally, it highlights the functions of official political institutions and the narratives around the shifts in environmental services (Robbins, 2012). The bio-environmental relationship in the context of power distribution to production activities and its connection to ecological analysis can be described via the lens of such literature (Greenberg & Park, 1994). It provides attention on the degradation of natural resources and loss of biodiversity is not only about the market but also about unequal power distribution by the stakeholders over forest management cultures

2.1 Justification of the Study

Bangladesh forest covers three vegetation types in three different ecosystems and locations such as, hill forests; Sal forests and mangrove forests. Bangladesh was categorically very rich in biodiversity and agro-biodiversity including plants and animals due to geographical settings. In addition to 3,000 crop types, it is expected to have around 8,000 rice varieties. In addition, the Bangladesh Agricultural Research Council has documented almost 3000 types of pulses, 781 types of oilseeds, 3516 vegetables, 156 types of spices, and 89 types of fruits (Baten et al., 2018). Different types of agrochemicals and industrial wastes (effluents) lead depletion ecosystems and biodiversity. However, constructing PAs alone does not produce the intended conservation outcome in many tropical regions, rendering the PAs system ineffectual (Geldmann et al., 2013; Clark et al., 2013). The lack of alternative income-generating options for those who have historically relied on forests for their livelihoods and the exclusion of locals from PA governance are the main causes of this (Mukul et al., 2016, 2012a). Furthermore, many PAs are susceptible, especially in the developing tropics, due to changes in land use surrounding them, agricultural development, illicit logging, fuel wood, and fodder gathering (Mondal & Nagendra, 2011; Karanth & DeFries, 2010; DeFries et al., 2007; Ervin, 2003). From both the protected and unprotected areas, forests are still being degraded due to deforestation, firewood collection, and encroaching on boundaries. Community encroachment into forest areas, at least partially, has happened to a significant degree in many forested areas in Bangladesh. The amount of fragmentation, deforestation, and forest degradation is large, more than 65% of the forestlands are estimated to be under this (Gain 2002). Researcher and civil society activities, including NGOs and indigenous people, have managed to raise national and international attention to these issues (Gain, 2002).

2.2 Objective

The main objective of this study is to assess the vulnerability of forests, biodiversity and communities. It gives the focus on the following specific or interim objectives:

- Know the present status of biodiversity and forest communities in the study areas
- Explore the reasons of biodiversity losses and threats to community people
- Identify the linkages and network among forest conservation, culture, communities and biodiversity
- Find out the challenges of forest and biodiversity conservation as faced by the community.
- Suggest potential solutions and recommendation to protect biodiversity, culture and communities through forest conservation.

2.3 Literature Review

The sixth major extinction event on Earth is being brought on by human activity, which is accelerating the loss of biological variety worldwide at rates 100–1,000 times faster than they were before humans. The local collection of food and fuel, never create huge impacts on forest or biodiversity (Lenzen et al., 2012).Growing resource demands have resulted in the destruction of forest landscapes due to hunting, wood removal, pollution, fragmentation, and other human influences, as well as the removal of natural forest for agriculture (Foley et al., 2005; Song et al., 2018). Since many sensitive species depend on intact forest landscapes and "primary" forests (Gibson et al., 2011; Betts et al., 2017), these pressures are having an effect on forest biodiversity (Newbold et al., 2013, 2016; Phillips et al., 2017). It should be mentioned that the map also emphasizes the biodiversity significance of the significant amount of deforestation that has taken place in temperate regions over the past 18 years, even though areas of particularly significant forest biodiversity loss occurred in the tropics during this period (a reflection of the higher species richness and density of endemic species within the tropics). Due primarily to extensive logging and fires, the forest cover in Scandinavia, Russia, Canada, and the USA has significantly decreased (Curtis et al., 2018). The protection of biodiversity and the reduction of climate change both depend heavily on the world's woods (Hill et al., 2019). Several studies mentioned about the loss of biodiversity and their underlying causes. The researchers mentioned about direct and indirect causes of biodiversity loss. Direct causes such as hunting, trade, consumption, use of wild animals, human animal conflict, noise from running vehicles, irresponsible tourism, habitat degradation, illegal cutting, fuel collection. Agricultural activities, clearing of undergrowth, gazing animals, and land erosion are major threats to the biodiversity and forest conservation (Table 1). On the other hand, indirect causes including poor capacity, poor coordination, lack of policy implication, poorly managed forest, corruption, climate change, community migration poses risk to the forest and biodiversity conservation.

Table 1. Causes of biodiversity losses

SL #	Causes of Biodiversity losses	References
1	**Direct causes**	
a	Hunting and poaching	Foster-Turley et al., 2016
b	Local and international trade	Hasan, 2015
c	Local consumption	Hasan, 2015
d	Use of living wild animals	Foster-Turley et al., 2016; Hasan, 2015
e	Human wildlife conflict	Rawshan et al, 2014: Feroz et al., 2011
f	Moving vehicle surrounding	Carr and Fahrig, 2001
g	Tourism disturbance	Foster-Turley et al., 2016
h	Degradation of habitat	Dey, 2018
i	Illegal extraction of resources	Foster-Turley et al., 2016
j	Fuel wood extraction	Foster-Turley et al., 2016
k	Expansion of agriculture in the forest areas	Foster-Turley et al., 2016; Dey, 2018
l	Undergrowth modification	Dey, 2018
m	Livestock grazing	Foster-Turley et al., 2016; Khan, 2018
n	Erosion of land	Feeroz, 2013
2	**Indirect causes**	
a	Poor capacity	Foster-Turley et al., 2016
b	Lack of coordination	Foster-Turley et al., 2016

continued on following page

Table 1. Continued

SL #	Causes of Biodiversity losses	References
c	Policy gap	Foster-Turley et al., 2016; Karanth, 2002
d	Lacking from law enforcement	Foster-Turley et al., 2016
e	Poorly managed forest	Mukul, 2014
f	Corruption	Hasan, 2015
g	Climate change	Payo et al, 2016; Mukul et al., 2019
h	Migration	Feeroz, 2013;

Source: compiled by the author

These literatures tell you the consequences of deforestation alerts that may be addressed to protect biodiversity and conserve forest by adopting strategic planning almost in developing countries. Nevertheless, current products only depict tree cover and fail to capture the significant spatial heterogeneity in forest ecological value. These scenarios make us alert for incorporating biodiversity and forest conservation issues in local practices and policy implication.

3. METHODOLOGY

This study was based on literature review and qualitative primary data. Primary data were collected using focus group discussion, Key Informant's interview and observation. Secondary data were collected using published journal articles, books, periodicals, internal official documents and social media. Primary data were collected discussing with key personnel including, Divisional Forest Officer (DFO), Bangladesh Forest Department, NGO Representative from Aranyak Foundation, 3 local community people (2 male and 1female). Five Focus Group Discussions were carried out with 3 female groups and two male groups. This study was based on literature review and qualitative primary data. Primary data were collected using focus group discussion, Key Informant's interview and observation. Secondary data were collected using published journal articles, books, periodicals, internal official documents and social media. Primary data were collected discussing with key personnel including, Divisional Forest Officer (DFO), Bangladesh Forest Department, NGO Representative from Aranyak Foundation, 3 local community people (2 male and 1female). Five Focus Group Discussions were carried out with 3 female groups and two male groups. Secondary data on forest conservation, protection of biodiversity and the role of communities and their culture in forest conservation focusing developing countries, the relevant peer reviewed journal articles were carried out to get present status and conservation initiatives in the respective fields. In case of qualitative data the following questions were administered to get more information from the respondents.

Question 1. How and when do you think the culture of forest communities plays a role in preserving biodiversity and forest conservation?

The answer of this question elicits the importance of cultures and communities and their contribution for forest conservation and protection of biodiversity.

Question 2. What are the context and contextual issues of forest and biodiversity conservation in Bangladesh and as well as in the developing countries?

The answer of this question represents the present status of biodiversity and forest conservation initiatives in the study context.

Question 3. What are the ways that anthropogenic causes impact forest conservation, biodiversity, and forest communities, and what are the solutions?

This question brings the answers of biodiversity and forest conservation challenges to provide the potential solutions and practical evidence to protect biodiversity and forest communities from the negative consequences.

Question 4. What roles do various stakeholders play in creating connections and enhancing forest communities' ability to preserve culture, biodiversity, and forest conservation?

The answer of this question explores the strategies to develop network-linkages and build partnership among the stakeholders to perform their active role in protecting biodiversity, communities and conserving forest.

Question 5. What are the potential solutions in protecting communities, cultures, and biodiversity and forest conservation?

Answer of the question provides practical solutions and strategic directions for future biodiversity and forest for community and its importance to everyone.

4. RESULTS

4.1 Present Status of Forest, Biodiversity and Communities

A prolonged history of forest cultural legacy, complete with unique knowledge, beliefs, rituals, and cultural practices, has also been passed down to the communities that rely on different forest products that include firewood, barks, wood, fruits, leaves, fishes, and honey . The social landscape of these forest areas, with its inhabitants and resources, is encapsulated in the intricate web of interactions between natural and cultural interfaces. Furthermore, the social and cultural fabric is entwined with elements such as climate change, inadequate governance, and over-exploitation of natural resources, resource seizure, and weak institutions. The ancient customs of the communities that depend on marine resources are being threatened by growing polarization and disparities in economic distribution and resource access (Rashid & Hasan, 2020). It was mentioned by the participants from FGDs and KIIs that biodiversity and ecosystem are going down with the time and modernization. Mostly developing countries are under pressure due to high population pressure and anthropogenic activities. If we consider the climatic zone, it showed that tropical countries were rich in biodiversity. Table 2 showed that there are five climatic zones on this earth including tropical zone; equatorial zone; temperate zone; tundra zone; and polar climatic zone. The Tropical and Equatorial climatic zone have high biodiversity still now. As for examples Bangladesh and India from tropic showed high biodiversity. Indonesia and Brazil from Equatorial are also showed high biodiversity from very beginning. The temperate zone including Europe and USA has moderate biodiversity. On the other hand, Siberia and Norway from Tundra demonstrated low biodiversity and Antarctica from polar region showed very low biodiversity. Considering all climatic zone, especially the developing countries are losing biodiversity in alarming rate.

Table 2. Status of Biodiversity based on climatic zone

SL#	Climatic Zone	Status of Biodiversity	Country for examples	Numbers of season
1	Tropical		Bangladesh, India	6 Seasons
2	Equatorial		Indonesia, Brazil	4 Seasons
3	Temperate		Europe, USA	4 Seasons
4	Tundra		Siberia, Norway	3 Seasons
5	Polar	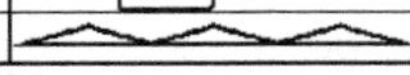	Antarctica	2 Seasons

Source: Author, 2024

High – 
Moderate – 
Low – 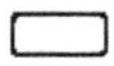
Very low = 

Hill forest situated in the Chittagong Hill Tracts and occupied the southeast region. It covers an area 6, 80,000 hectares land in the northeast and southeast in Bangladesh. The major plantspecies of the forests are *Dipterocarpusturbinatus*; *D. pilosus*; *Swintonia floribunda*; *Hopea odorata*; *Syzygiumgrande*; *Salmalia insignis*; *Lophopetalum fimbriatum*; and *Duabungasonnerationides*. Hill forest is conserving animal species including wild mammals (Asian elephants, Asiatic black bear, hoolock gibbon, sambar deer, wild dog, leopard), and other threatened species in the world.

Sal forest roughly covers an area of 1,20,000 hectares (0.81% of total land mass of the Bangladesh). Sal (Shorea robusta) is the main species of this forest followed by *Butea monosperma*, *Careya arborea*, *Terminalia belerica*, *Terminalia chebula*, *Dilleniapentagyna*, *Aphanamixispolystachya*, *Streblus asper*, and *Phyllanthus emblica*

The Sundarban is the largest chank of single natural mangrove forest in Bangladesh with an area of 601,700 hectares are natural mangroves. The main species is Sundri (*Heritiera fomes) followed by Excoecariaagallocha*, *Sonneratiaapelata*, *Avicennia officinalis*, *Avicennia alba*, *Hibiscus tiliaceus*, *Phoenix paludosa*, and *Acrostichumaureum.* The sundarban rich in biodiversity, it contains 334 species of plants and 269 species of wild vertebrates. The fauna including the Bengal tiger, Ganges and Irrawaddy dolphins and saltwater crocodiles, a number of threatened bird species, and at least 176 species of fish were found in the Sundarbans.The manmade or plantation Sundarban is with an area of 200,000 hectares of land.

Fresh water swamp forest situated in the law land of Bangladesh and it is flood or water tolerant forest. The forest occupied by major plant species including *Barringtonia acutangula*, *Millettia pinnata*, *Saccharum spontaneum*, *Phragmites karka*, *Acanthus ilicifolius*, *Alpinia allugas*, and *Schumannianthusdichotoma.* It is the home of many wild including wading birds, reptiles, and amphibians.

Homestead forests are situated every corners of the country and it has huge contribution to the biodiversity as well as towards the nation. It covers by the many species including *Mangifera indica*, *Artocarpus heterophyllus*, *Syzygiumcumini*, *Bambusa* spp. *Areca catechu*, and *Musa* sp. Major animal species including jungle cat, fishing cat, golden jackal, Bengal fox, mongoose, and civets were found in the homestead forest.

Village forest is managed by the indigenous people and it is extended at near to the homestead forest. We can found this type of forest in southeastern hilly areas. Many wild animals are found in this forest.

4.2 Causes of Biodiversity Loss

Two reasons are lying behind the loss of biodiversity such as nature itself changing the pattern of species distribution for natural causes. On the other hand, anthropogenic activities that caused biodiversity loss and primarily responsible of human activities. Respondents mentioned couple of reasons for biodiversity loss. Diverse factors are responsible for biodiversity loss and deforestation. Forest communities especially the minority people are living with trees inside the forest. They are like a cohabitant with each other and these communities are taking care of trees, animals, and forest for their sustenance. As for example in Bangladesh the *Khasi* people are cultivating betel leaf with trees without harming forest and biodiversity. Loss of cultural values of communities; illegal cutting of trees; migration of forest community people; deforestation & disturbance; practicing unsustainable agriculture everywhere; human settlement and building industries in the forest areas; land fragmentation of the previous ownership; uncontrolled tourism in the forest and protected areas; pollution; changes in hydrological system; inviting exotic and invasive species; political instability of many countries; lack of political wish; illegal hunting and killing of wild animals; climate change issues and corruptions are leading towards biodiversity loss (Table 3).

Table 3. causes of biodiversity loss

SL #	Causes of Biodiversity losses	Anthropogenic	Natural
1	Loss of cultural values of the forest communities		
2	Elicit cutting of trees and encroachment of forest		
3	Migration of forest communities toward urban and peri-urban areas		
4	Deforestation and disturbance		
5	Chemical agricultural practices reducing agro biodiversity		
6	Changes of land use pattern through housing and industrialization		
7	Land fragmentation and habitat loss		

continued on following page

Table 3. Continued

SL #	Causes of Biodiversity losses	Anthropogenic	Natural
8	Uncontrolled tourism		
9	Pollution		
10	Monoculture in the plantation		
11	By changing hydrological system		
12	Introduction of exotic species		
13	Political instability or lack of political will		
14	Illegal hunting and killing of wild animals		
15	Climate change issues		
16	Corruptions		

Source, author, 2024

4.3 Interface of Culture, Communities, Forest and Biodiversity Conservation

Forest offers a variety of ecosystem services, including provisioning services (honey, a non timber forest products), cultural services (tourism, and educational research), regulatory services (flood control, climate regulation, and protection from disasters like cyclones and storm surges), and supporting services (nutrient cycling, and biodiversity). Rashid and Hasan (2020) reported that *Rakhains* surround their homesteads and housing clusters with several tree species (para). According to the findings, "trees around the dwellings" provide the benefits that include i)trees that encircle homes slow down the wind that strikes them during cyclones; ii)they stop water from flowing during tidal surges; iii)if someone can cling to the trees, they may be able to survive when people fall into tidal surges during cyclones; and iv) coconut trees assist in providing drinkable water during the post-cyclone water crisis. In 2007, during the "Sidr," the actual application of such trees around homes and villages was proved when "Sundarbans" (mangrove forests) lessened the cyclone's strength and scale. In reality, communities and their cultures both are inextricably linked. A study reported that the diversity of species is declining less hurriedly in the land of indigenous people than others (Swiderska, 2020). The study found that forest communities are showing their cultural values and knowledge to save wild animals and protecting biodiversity. It means that the 370 million to 500 million indigenous people have a significant role in protecting biodiversity (Swiderska, 2020). The issues of communities, cultures or indigenous knowledge are not limited only in the developing countries, extending from southern China to the Americas. We can categorize the world into three distinct segments including human and domesticated; the wild; and the sacred and ancestral. Communities, cultures and local knowledge are diminished through modernization, commercialization, insecure right to the land and resources, migration and lack of cultural education.

4.4 Role of Indigenous Knowledge, Communities and their Cultural Values Protecting Biodiversity and Forest Conservation

It was said by the community people "We are the forest people" we never do any activities to destroy our forest. It is our forest; we have the right that inherent from one to other over many decades. This is an important reason why they culturally guard the forests. This can be found in the presence of several oral regulations which until now are still believed, for example, it is forbidden to throw waste inside forest, and it is not permissible for residents to have open defecation using forests (Table 4). This type of multiplier continues to survive in their daily lives. This is part of their strategy to protect the forest indirectly because it is so important that the river is maintained so that the forest as a water source naturally takes place. It was mentioned by the participants of the focus group that it never allows them to clear the forest for their necessities, no encroachment and deforestation even. Some myths like culture exists among them, the menstrual women are not allowed inside forest to collect wild food from the forest. The never evacuate or dump waste inside the forest and they respect their culture to follow all regulation exist among the communities. Biodiversity and forests are important for community people to sustain. The *Khasi* community people are cultivating betel leaf and it is coming to them from their ancestors. Any unnecessary killing of wild animals and cutting of trees are not accepted by the community. Culturally forest dwellers rely on forest for their traditional wild grown food (animals, insect, plants and fishes). Many community people engage themselves through alternative livelihoods. It may be handicrafts, wild orchid collection, making hat and broom using forest raw materials. They are producing different binding materials sun grass and other climbers. It is proven that forest communities are protecting biodiversity and conserving forest for their own existence and sustainability of cultures.

Table 4. Role of communities and their culture protecting biodiversity and forest

SL#	Role of indigenous knowledge and culture	Explanation
a	It is not permissible to clear the forest	Forest strongly protect communities from storm &cyclones
b	Women with menstruation period never enter inside the forest for any activities	It will not bring wellbeing of forest and biodiversity
c	It is not allowed to throw or keep household waste inside the forest	If it is done then nature or forest will be angry with them
d	Illegal cutting of trees are not allowed	They believe that illegal cutting of trees will bring punishment for them
e	Killing of wild animals within the forest is prohibited	If do anyone then who will be the natural guard of the forest? Wild animals are guard of the forest to protect it
f	Cutting of Forest trees are prohibited	*Khasi* people cultivate betel leaf with trees and lead their livelihood by protecting biodiversity and forest
g	Access to wild food from plants and animals	The stolen forest makes the communities vulnerable and leads towards food insecurity. Communities are always conserving forest and biodiversity to secure their regular bread from forests.
h	It is our home said by the forest communities	Forests are the home of ethnic people, and they never wish to make it fragile to live and they have strong mindset to protect biodiversity and conservation of forest with their common interest
i	Options for livelihood diversification	The community are protecting biodiversity and forest from encroachment to get raw materials supply from the forest for their alternative livelihood.
J	Water and recycling of water	The stream and lakes inside the forest is the major sources of water for the communities. It is only possible to ensure recycling water by protecting forest.
k	Sources of natural medicine	Community people need several plants and animals for their primary healthcare (healing system) and they are always conserving all the species (biodiversity) culturally.
l	Prediction on agricultural and other cultural activities by looking some flowers, plants and animals	Forest communities have the tradition to start planting, harvesting by looking flowers, plants, animals or sound of the animals, insects and birds. In this respect, the community is protecting biodiversity and forests.
l	Ecological balance	Community people believe that only forest can made balance between abiotic and biotic components if it is protected.

Source, author 2024

4.5 Problem and Challenges

Challenges of biodiversity conservation are many folds, it's intricate with community, culture and forest conservation issues altogether.

- Unemployment of forest communities and the surrounding of forest people
- Conflict settings between Forest Department and community people
- Political power utilized to encroached the forest
- Over extraction of resources

- Poor management
- Lack of research focus on the concerned issue
- Lack of awareness and motivation towards biodiversity conservation

Additionally many challenges were mentioned by the respondents based on their experience and observation. Some of these are relevant to biodiversity and forest conservation issues and many issues are related to communities and their cultures. Few of them are technical and policy related issues. Degradation of forest, wetland, and protected areas is one of them. Management of forest is still very traditional; there are no sound management strategies with long term vision. Inclusion of sustainable Agro- forestry Technology is faraway. Over population and rapid urbanization is creating tremendous pressure on biodiversity and forest to fulfill their needs. The wholesale demand of trees is reducing tree cover in a country. It was mentioned by the respondents that over extraction of resources especially medicinal plant species, non timber forest products (NTFPs) such as Bamboo, canes, mashroom, orchieds, sungrass, honey and other species. In the mangrove forest or coastal forest salinity Intrusion due to shrimp cultivation and low Freshwater stream in the Ganges is disturbing the plant species of the mangrove forest.As for example the destruction has been taken place in Chakaria and Teknaf mangrove forest by clearing the forest and conversion agricultural land intoshrimp farming nearby forest. Subsequently, the frequency of natural disasters (flood, flash flood, storms, cyclones, tidal surges, drought are increasing that leads towards threats to the biodiversity. Indiscriminate use of different types chemicals (fertilizers, insecticides, pesticides, herbicides, weedicides, and growth hormones), pollutions affecting biodiversity. Unplanned land development and land use changing destroy the forest and biodiversity. Low lying areas are made up raise land to build houses. Construction of road and development activities across the forest is enhancing biodiversity loss. Dry season flow in the stream, lakes and rivers are affecting forest cover and biodiversity. Different types of chemicals and wastes from households are polluting water through runoff. Wild animals hunting, trapping, and disturbance of animals is one of the threats to the forest conservation. Auction and management of open fisheries exacerbate the biodiversity. Inadequate efforts for conservation are a huge pressure for the stolen forest. There is lack of conservation, and propagation of rare, threatened and endagangerd species. Introduction of hybrid or exotic species is diminishing the base of biodiversity. The competition of the exotic and locally improved and hybrid varieties are making critical situation for biodiversity. Illegal and encroachment of the fallow forest lands are converting for agricultural uses. There is lack of trained and skilled manpower to manage the forest considering biodiversity. The initiatives is also lacking from appropriate software to monitor the status biodiversity and forest using information technology. Mono culture plantation is the treat for biodiversity. The local furniture industry's demand is met profitably by monoculture trees. Additionally, they extract more water from the earth. As a result, other nearby trees is deprived of adequate moisture. As a result, other smaller plants like grasses, herbs, and trees are unable to thrive in the region. As a result, the Khasi think that other companies and the forest Department should prohibit monoculture plantations in the woods.

4.6 Human Dependency on Forest

The ethnic (Khasi) people have a spiritual bond with wild birds and animals. The community is familiar with their cuisine, seasonal habits, fruits, and way of life. The forest is full of vines and trees. There are a lot of shaded trees. They offer birds, lizards, and insects protection. According to (Schlingmann et al., 2021), Lawachara park is home to 246 birds, 20 species of mammals, 6 species of reptiles, 4 types of amphibians, and 167 plants. The Park has a flagship attraction thanks to around 18 families of Hollock gibbons; other important species are the Capped Langur, Slow Loris, Pig-tailed Macaque, Orange-bellied Himalayan Squirrel, and Barking Deer. There are 6 tea gardens and 30 villages surrounding the park, two of which are Khashia Punji located within the park, all of which rely on the park's natural resources (Schlingmann et al., 2021).Other species that rely on it will either lose their homes or perish if a shade tree is felled. Every species contributes significantly to the maintenance of the surrounding ecosystem. For instance, woodpeckers don't nest in every tree. They drill holes with great care and are extremely picky. At that time, the hornbill was a prevalent species in Lawachara, and it told locals when the seasons changed. In the past, the Khasi people understood seasonal calendars, weather predictions, and alerts about natural disasters based on the movement of animals and their availability. Due to changes in environmental sustainability and the destruction of forests and wildlife, we can no longer do this.

4.7 Solutions to Protecting Biodiversity, Culture, Communities and Forest

Biodiversity can be managed in a sustainable way only if the management plans take proper account of the socio-economic context. This may include a requirement to develop jobs based on the sustainable exploitation of biodiversity, and ensuring that biodiversity conservation is carried out in a socially acceptable and cost-effective manner, in such a way that, wherever possible, it benefits the local population(Rahman et al., 2015).This research examines the culture and communities life, with an emphasis on their reliance on forest products. Their livelihoods are greatly affected, however, by the complex interplay of changes in environmental sustainability, illegal human behavior, and developmental programs. The modification of landscapes, especially through hill destruction, has noticeable impacts on local climatic patterns, negatively affecting the area's ecosystem (Boyce & Neale, 2006).

It was mentioned by the respondents to give a good number of solutions to protect biodiversity, culture and people through conservation of forest.

Ensuring management of forests for production: The solutions are only categorized as the nature-based solutions to protect biodiversity benefits rather than commercialization or trade off.

Restoring the forests: Restoration plays a key role to conserve forest within the nature by engaging all stakeholders, especially the communities their cultures. .

Agroforestry: It offers a sustainable approach through increasing food production, fuel and wood production regardless ofreduction ofexisting land and aincrease population, by maximizing the use of vertical space.

Sharing the benefits or incentives: It is essentialsto protect our last forest and biodiversity by sharing benefits from forest revenues or providing direct cash incentives to protect the lost forest and biodiversity for the community

Homegardens: Homegardens or homesteads are widespread and play a vital role in diversity and food security at every home of Bangladesh. A study reported that the application of mixed cropping pattern and year round production system can provide diverse food to the community with essential nutrients. It ensures the application of indigenous knowledge, increase production access them to more income and employment opportunities (Ferdous et al., 2016).

Wild food: All types of forest provide wild resources or food round the year. Many people engage themselves for collection, marketing and consumption as a daily diet. There are no seasonality problems; the community people can collect seasonal wild resources at their convenience. Baudron et al., (2019) reported that forest, community and cultures have co-existence with the harmony.

The escalating tourism industry and continuous modern development are making contemporary deforestation and forest degradation worse, which is harming the Khasi Indigenous community in Bangladesh (Majumder et al., 2018).

Conflicts Settings: It is common when something become a resource within Remote Island. A study mentioned that conflict occurs between different beneficiaries after growing of resources in a mudflat. A case study discussed the issues of plantation in uninhabited mudflat island, when it was growing as a resource, a conflict setting was observed and the Forest Department treated it as a National Park, to resolve the conflicts (Iftekhar and Takama, 2008).

Community lead restoration and protection of the biodiversity on which the communities rely on forest such as reforesting swamp forests, digging up dried-up wetlands, and restocking fish took place alongside these capacity building initiatives, with community guards shielding the restored forests from illegal use. Mechanisms for sharing the benefits of natural resources were created, along with co-management strategies. Community management can foster social cohesion and create chances for cultural, educational, and interactive engagement.

5. DISCUSSION

According to the Convention of Biological Diversity -CBD (1992), it is a commitment to keep its statutory, administrative, and legal mechanism to conserve, sustainable use and fair and unbiassed sharing of biological resources at everywhere(Arora, 2022). The study showed that the size of the forest is becoming smaller day by day and forest are lacking from the rich biodiversity. Total forest area in Bangladesh is 2,600,000 hectares, covering 17% of the total landmass. The Bangladesh Forest Department (BFD) owns and manages roughly 61.52% of the 1,600,000 hectares of forestlands, the deputy commissioner (executive chief of district) is in charge of 26.80% of the unclassified forest (697,000 hectares), and the community controls 10.38% of the land (270,000 hectares) as private woodland and community forests(Foster-Turley et al., 2016; Dey 2018). Bangladesh has one of the lowest per capita forestlands in the world (Mukul & Quazi, 2009), while having a rich biodiversity. This is mostly because of the country's dense urban population. Additionally, the nation has had 2600 hectares of deforestation, which is among the highest rates in South Asia. A study mentioned that we are aware that other anthropogenic pressures, including hunting, exploitation, and climate change, are also significant, but they will only be taken into account in our research to the degree that land use or population density act as stand-ins (Hill et al., 2019). Even while biodiversity is greatly impacted by climate change, over the brief time period that our research focuses on, it is impossible to separate the effects of a changing climate from those of land use change and human populations. In Asia, significant sums of money are being spent to develop tree plantations on degraded terrain. These programs which frequently have political motivations, aim to improve the economy and the environment. Planting in the vacant forest area or enrich planting using non native species following monocultures, create risk to indigenous biodiversity. Both plants and animals connected with grasslands, scrub and other immediate seral vegetation will repeatedly be missing (Sayer et al., 2004). These cultures need to be protected as the priority to conserve biodiversity indigenous cultures. Since the terms 'environmental sustainability' and 'resilience' are seen as colonial, many community members choose to describe their land-based practices, land rights, and land-based spirituality as their forms of environmental sustainability or resilience. Numerous community members urged us as researchers to prioritize community land rights in relation to their 'environmental sustainability' or 'resilience'. Numerous community members expressed concern that many researchers and policymakers seek to impose outside viewpoints on the community's routine land-based activities in the name of 'environmental sustainability' or 'resilience'. Accordingly, we centered our conclusions in this article on the needs of the community, as per their recommendations (Datta et al., 2024). The preservation of forests is considered essential for the survival of Indigenous cultures, especially those that are closely related to forest life (Freitas et al., 2020). To lessen deforestation, Grainer additionally contends that sustainable forest management, directed by government-regulated logging methods and extensive scientific research, is promoted (Freitas et al., 2020). Urban biodiversity especially in Dhaka megacity is very important for human settlement. A study mentioned that anthropogenic causes including rapid urbanization, wetland transformation, pollution, huge construction work, mass vehicle travels through using the parks and gardens as a short passageway have a damaging effect on urban biodiversity and sustainable city (Jaman et al., 2021). Miss-waste management has the detrimental impacts on urban biodiversity by reducing the number of species (Alom & Khan, 2014).

6. CONCLUSION AND RECOMMENDATIONS

6.1 Conclusion

Even when the forest department took legal action by declaring protected areas, habitat conversions have not been sufficiently controlled, and most of the lands inside South Asia's forest protected areas have been impacted by human activity in some way (Clark et al., 2013).Overall, we discovered that Bangladesh's forest protected areas are currently insufficient in terms of both number and coverage to save the nation's rapidly declining biodiversity. Additionally, the many forest ecosystems that require conservation are not well represented in the forest protected areas. The Bangladesh Forest Department's lack of capability and infrastructure, as well as the local population's limited participation and assistance (mostly through project interventions), are further impeding the long-term viability and prosperity of the nation's forest protected area systems. It was mentioned by the respondents that forest cover reduced remarkably in Bangladesh over last two decades. According to a study, the impact of clearing forested habitat is indicated by the ecological significance of the tree-cover loss between 2000 and 2018 (Hill et al., 2019). Bangladesh's forests and biodiversity act as a buffer against climate change and offer valuable natural items to the local populace as well as for sale in marketplaces. Additionally, forests protect soil from excessive erosion and preserve the integrity of freshwater supplies and watersheds (Uddin et al., 2013). Domestic and international tourism to the nation's natural areas has been promoted for its ability to support ecosystems' sustainable growth and to offer incentives for protecting these areas rather than turning them into other uses. Biodiversity can be managed in a sustainable way only if the management plans have proper consideration of the socio-economic context. This may include a requirement to develop jobs based on the sustainable exploitation of biodiversity, and ensuring that biodiversity conservation is carried out in a socially acceptable and cost-effective manner, in such a way that, wherever possible, it benefits go the local communities (Rahman, 2015). However, modern development, reforestation, deforestation, and environmental sustainability change impact their way of life. The Khasi are essential to environmental and ecological protection. In their environmental management, they utilize Indigenous knowledge. They are aware that cultivating betel leaves in the woods allows them to coexist with wild animals, trees, plants, and the natural world. As a result, Khasi forests do not embrace modernization, such as reforestation (mono-culture), tourism, and manmade infrastructure development. The Khasi wish to maintain the forest and nature unspoiled free from the influence of outsiders. Therefore, no development initiative should be designed or carried out without first consulting the local population. Decision-making on forest concerns, especially at the local and national levels, should include the Khasi community. In order to educate the public about the role of Indigenous peoples in environmental conservation, the Bangladeshi report and educational curriculum should include Indigenous voices. This study promotes the integration of cultural viewpoints into forest conservation policies and actions by amplifying the voices of the ethnic community, their culture, and knowledge and practices. It highlights the need for a more inclusive and culturally aware strategy to environmental sustainability, one that recognizes and honors the resilience of community peoples, such as the Khasis in Bangladesh. Application of Artificial Intelligence (AI) can bring positive benefits in many ways from planning, implementing, monitoring and evaluation of biodiversity and forest. A study has provided emphasis on the use of updated technologies, tools and even AI for enhancing biodiversity conservation initiatives (Ullah et al., 2025).

6.2 Recommendations

The general recommendations were made in the following points:

- Working to preserve indigenous cultures is necessary in order to attain the rich diversity of nature.
- Indigenous peoples' rights to land, natural resources, and traditional knowledge must be acknowledged and protected by governments in accordance with the law.
- At all levels of biodiversity conservation initiatives, Indigenous peoples must be completely and successfully engaged.

- It is also essential for strengthening assistance for disadvantaged and marginalized indigenous peoples in order to meet the UN Sustainable Development Goals.
- All of the areas that are contributing to their decline must include both indigenous cultures and biodiversity.

Appropriate and restructured Biodiversity Rules and Regulations should be enforced and implanted to ensure conservation initiatives. Law and Act on current Biodiversity conservation process should be amended considering all drivers. Traditional knowledge may provide indications of uses of elements of biodiversity that could lead to sustainable exploitation and low-impact uses of natural and semi-natural areas. The Ministry of Environment and Forestry should have a separate Wildlife Cell or Department. Anthropogenic activities frequently affect biodiversity conservation. Over harvesting, illegal hunting of wild animals and extraction of wild plants should be restricted to protect biodiversity. At least some Major Forest Area, Wetlands and portion of River should be marked as Sanctuary for Bio-diversity Conservation under the Direct Supervision of Government. Unsustainable development and unsustainable consumption lead to biodiversity loss or ecosystem damage viii. Water Pollution by any means should be Minimized and Existing Rules and Regulations on Water Pollution should be Effectively Enforced. Effluent Treatment Plant has to set up to the Outlet o Use of Pesticide in the Agricultural Fields should be reduced and Biological Control of Pest should be introduced. The Height of Chimney of the Brickfield must by beyond 120ft with a Scrubber in its Outlet. Brickfield must be Established Maintaining a Reasonable Distance and Position. Clearing and burning of forest of hills for Jhum (shifting) cultivation should be stopped. To create mass awareness on biodiversity and forest conservation, mass media including newspapers, social links, Television, Leaflets should be used among the stakeholders and society. Private and NGOs can lead the role to disseminate knowledge and make the community aware on biodiversity and forest conservation. More research may bring the changes in wider scale. Theissue related to biodiversity and forest conservation should be paid proper attention by the public and policy-makers. Additionally, the study made specific recommendation addressing policy, community and future research.

6.2.1 Policy Level Recommendation

According to a study total 102 policies (58 Government of Bangladesh funded, and 44 donors funded) and 848 strategic tasks were recognized connecting to forest biodiversity issues in Bangladesh throughout the period of 1992–2013 (Rahman et al., 2016). Although there are many restrictions, Bangladesh has a number of laws and regulations that protect the country's current forest and wildlife species. We hope that these laws will be properly implemented in order to support the country's distinctive biodiversity and varied wildlife. The Forest Act of 1927; the Wildlife (Conservation and Security) Act of 2012; the Environment Conservation Act of 1995; the Environment Protection Act of 2000; the Climate Change Trust Act of 2010; the Biological Diversity Act of 2012; the Forest Policy of 1994; the Environment Policy of 1992; the Wetland Policy of 1998; and the Coastal Zone Policy of 2005 are some of the significant laws and policies pertaining to the conservation of Bangladesh's forests and wildlife. The nation must put out great effort to put these laws into effect and rid the political system of corruption. The power dynamics of public bureaucracies, which are the main players in the country's forest biodiversity policy, are therefore worth examining. Additionally, this is a requirement for more meaningful research on the influence of the key players in Bangladesh especially on numerous land-use disputes. Therefore, the study needs to examine how policies pertaining to biodiversity in forests affect the power capacities of the primary relevant bureaucracies in Bangladesh. This covers both pertinent domestic bureaucracy and those of foreign donors. The power capabilities of the various rival bureaucracies in this field will be revealed by such an analysis, which is essential for finding possible partners for any successful biodiversity strategy (Rahman et al., 2016). In anticipation of recent years, the conservation and management of biodiversity and forest resources have given less attention in comparison to other ecosystems. In the era of the blue economy and connected SDGs targets, the government has emphasized on the exploitation of coastal forest and biodiversity (Islam & Shamsuddoha, 2018; Ortolano et al., 2017). In the linking of forest biodiversity, the Bangladesh government has paid special attention on protection of biodiversity through forest conservation.

6.2.2 Community Level Recommendation

It is important to motivate the community people to stay inside or nearby forest for the protection of biodiversity. Incentive option should be made available to conserve forest by playing their active role considering their cultural values. Train them to build their capacity on the technical aspect of biodiversity, forest conservation and importance. Learning from the communities should be documented and it may conserve for the future application.

6.2.3 Recommendation for Future Research

Specific research is important for the protection of biodiversity, community, culture and conservation of forest. The research is not only for the academic or researcher, the research for all stakeholders including the forest communities to disseminate the knowledge among them for future application and practices at the field level. Future research is needed to explain how the forest community people are addressing the problem of forest conservation in relation to environmental protection, identity, and ancestral land. As a result, it is essential to acknowledge Indigenous land rights for Bangladeshi Indigenous and minority peoples in order to promote forest and biodiversity conservation.

REFERENCES

Alom, M. M., & Khan, M. Z. H. (2014). Environmental and social impact due to urban drainage problems in Dhaka City, Bangladesh. [IJEAT]. *International Journal of Engineering and Advanced Technology, 1*(6), 2249–8958.

Arora, T. (2022). Legal framework for conservation of biodiversity in India. In *Biodiversity in India: Status, Issues and Challenges* (pp. 589–610). Springer Nature Singapore. DOI: 10.1007/978-981-16-9777-7_23

Baccini, A. G. S. J., Goetz, S. J., Walker, W. S., Laporte, N. T., Sun, M., Sulla-Menashe, D., Hackler, J., Beck, P. S. A., Dubayah, R., Friedl, M. A., Samanta, S., & Houghton, R. (2012). Estimated carbon dioxide emissions from tropical deforestation improved by carbon-density maps. *Nature Climate Change, 2*(3), 182–185. DOI: 10.1038/nclimate1354

Barbier, S. (2007). *Influence de la diversité, de la composition et de l'abondance des essences forestières sur la diversité floristique des forêts tempérées* (Doctoral dissertation, Université d'Orléans).

Baten, M. A. (2024). Revisiting the success of co-management in forest protected areas of Bangladesh: Untangling the interplay of power, participation, and community empowerment. *Environmental Policy and Governance, 34*(6), 649–662. DOI: 10.1002/eet.2114

Baudron, F., Tomscha, S. A., Powell, B., Groot, J. C. J., Gergel, S. E., & Sunderland, T. (2019). Testing the Various Pathways Linking Forest Cover to Dietary Diversity in Tropical Landscapes. *Frontiers in Sustainable Food Systems, 3*, 97. DOI: 10.3389/fsufs.2019.00097

Bergseng, E., & Vatn, A. (2009). Why protection of biodiversity creates conflict–Some evidence from the Nordic countries. *Journal of Forest Economics, 15*(3), 147–165. DOI: 10.1016/j.jfe.2008.04.002

Betts, M. G., Wolf, C., Ripple, W. J., Phalan, B., Millers, K. A., Duarte, A., Butchart, S. H. M., & Levi, T. (2017). Global forest loss disproportionately erodes biodiversity in intact landscapes. *Nature, 547*(7664), 441–444. DOI: 10.1038/nature23285

Boyce, C., & Neale, P. (2006). *Conducting in-depth interviews: A guide for designing and conducting in-depth interviews for evaluation input* (Vol. 2). Watertown, MA: Pathfinder international.

Bräuer, I. (2003). Money as an indicator: To make use of economic evaluation for biodiversity conservation. *Agriculture, Ecosystems & Environment*, *98*(1-3), 483–491. DOI: 10.1016/S0167-8809(03)00107-5

Carr, L. W., & Fahrig, L. (2001). Effect of road traffic on two amphibian species of differing vagility. *Conservation Biology*, *15*(4), 1071–1078. DOI: 10.1046/j.1523-1739.2001.0150041071.x

Chowdhury, M. S. H., Izumiyama, S., Nazia, N., Muhammed, N., & Koike, M. (2014). Dietetic use of wild animals and traditional cultural beliefs in the Mro community of Bangladesh: An insight into biodiversity conservation. *Biodiversity (Nepean)*, *15*(1), 23–38. DOI: 10.1080/14888386.2014.893201

Clark, N. E., Boakes, E. H., McGowan, P. J. K., Mace, G. M., & Fuller, R. A. (2013). Protected areas in South Asia have not prevented habitat loss: A study using historical models of land-use change. *PLoS One*, *8*(5), e65298. DOI: 10.1371/journal.pone.0065298

Corticeiro, S., Vieira, H., Almeida, M., Laranjeiro, D., Lillebø, A., & Oliveira, B. R. (2025). Assessing Forestry Needs and Challenges in Portugal: Insights from the Sector Interested Parties. *Forests*, *16*(3), 501. DOI: 10.3390/f16030501

Costanza, R., & Folke, C. (1997). Valuing ecosystem services with efficiency, fairness and sustainability as goals. *Nature's services: Societal dependence on natural ecosystems*, 49-70.

Curtis, P. G., Slay, C. M., Harris, N. L., Tyukavina, A., & Hansen, M. C. (2018). Classifying drivers of global forest loss. *Science*, *361*(6407), 1108–1111. DOI: 10.1126/science.aau3445

Datta, R., Singha, R., & Hurlbert, M. (2024). Indigenous land-based perspectives on environmental sustainability: Learning from the Khasis indigenous Community in Bangladesh. *Sustainability (Basel)*, *16*(9), 3678. DOI: 10.3390/su16093678

DeFries, R., Hansen, A., Turner, B. L., Reid, R., & Liu, J. (2007). Land use change around protected areas: Management to balance human needs and ecological function. *Ecological Applications*, *17*(4), 1031–1038. DOI: 10.1890/05-1111

Dey, T. K. (2018). Introduction to the wildlife of Bangladesh and management techniques. *Lalmatia, Dhaka, Bangladesh: Nature. Conservation & Society*, •••, 1148.

Ervin, J. (2003). Rapid assessment of protected area management effectiveness in four countries. *Bioscience*, *53*(9), 833–841. DOI: 10.1641/0006-3568(2003)053[0833:RAOPAM]2.0.CO;2

Feeroz, M. M. (2013). Biodiversity of Protected Areas of Bangladesh, Vol. III: Teknaf Wildlife Sanctuary. *BioTrack. Arannayak Foundation. Dhaka, Bangladesh.*

Feroz, M. M., Hasan, M. K., & Khan, M. M. H. (2011). Biodiversity of Protected Areas of Bangladesh, Vol. I, Rema-Kalenga Wildlife Sanctuary. *BioTrack, Arannayk Foundation, Dhaka, 214pp.*

Ferdous, Z., Datta, A., Anal, A. K., Anwar, M., & Khan, A. S. M. M. R. (2016). Development of home Garden Model for Year Round Production and Consumption for Improving Resource-Poor Household Food Security in Bangladesh. *NJAS Wageningen Journal of Life Sciences*, *78*(1), 103–110. DOI: 10.1016/j.njas.2016.05.006

Foley, J. A., Defries, R., Asner, G. P., Barford, C., Bonan, G., Carpenter, S. R., Chapin, F. S., Coe, M. T., Daily, G. C., Gibbs, H. K., Helkowski, J. H., Holloway, T., Howard, E. A., Kucharik, C. J., Monfreda, C., Patz, J. A., Prentice, I. C., Ramankutty, N., & Snyder, P. K. (2005). Global consequences of land use. *Science*, *309*(5734), 570–574. DOI: 10.1126/science.1111772

Foster-Turley, P., Das, R., Hasan, M. K., & Hossain, P. R. (2016). Bangladesh tropical forests and biodiversity assessment. *United States Foreign Assistance Act. Section*, *118*, 119.

Freitas, C. T., Lopes, P. F., Campos-Silva, J. V., Noble, M. M., Dyball, R., & Peres, C. A. (2020). Co-management of culturally important species: A tool to promote biodiversity conservation and human well-being. *People and Nature*, *2*(1), 61–81. DOI: 10.1002/pan3.10064

Gain, P. (2002). *Bangladesh environment: facing the 21st Century. Society of Environment and Human Development*. SEHD.

Gardner, T. A., Barlow, J., Chazdon, R., Ewers, R. M., Harvey, C. A., Peres, C. A., & Sodhi, N. S. (2009). Prospects for tropical forest biodiversity in a human-modified world. *Ecology Letters*, *12*(6), 561–582. DOI: 10.1111/j.1461-0248.2009.01294.x

Geldmann, J., Barnes, M., Coad, L., Craigie, I. D., Hockings, M., & Burgess, N. D. (2013). Effectiveness of terrestrial protected areas in reducing habitat loss and population declines. *Biological Conservation*, *161*, 230–238. DOI: 10.1016/j.biocon.2013.02.018

Gibson, C. C., McKean, M. A., & Ostrom, E. (Eds.). (2000). *People and forests: Communities, institutions, and governance*. Mit Press. DOI: 10.7551/mitpress/5286.001.0001

Gibson, L., Lee, T. M., Koh, L. P., Brook, B. W., Gardner, T. A., Barlow, J., Peres, C. A., Bradshaw, C. J. A., Laurance, W. F., Lovejoy, T. E., & Sodhi, N. S. (2011). Primary forests are irreplaceable for sustaining tropical biodiversity. *Nature, 478*(7369), 378–381. DOI: 10.1038/nature10425

Grafton, E., & Melançon, J. (2020). The dynamics of decolonization and Indigenization in an era of academic "reconciliation". *Decolonizing and indigenizing education in Canada*, 135-154.

Hasan, M. K. (2015). Mammal Management strategy with emphasis on Human-Wildlife conflicts in Bangladesh. *Dhaka, Bangladesh: Bangladesh Forest Department*, 112.

Hill, S. L., Arnell, A., Maney, C., Butchart, S. H., Hilton-Taylor, C., Ciciarelli, C., Davis, C., Dinerstein, E., Purvis, A., & Burgess, N. D. (2019). Measuring forest biodiversity status and changes globally. *Frontiers in Forests and Global Change, 2*, 70. DOI: 10.3389/ffgc.2019.00070

Iftekhar, M. S., & Takama, T. (2008). Perceptions of Biodiversity, Environmental Services, and Conservation of Planted Mangroves: A Case Study on NijhumDwip Island, Bangladesh. *Wetlands Ecology and Management, 16*(2), 119–137. DOI: 10.1007/s11273-007-9060-8

Islam, M. M., & Shamsuddoha, M. D. (2018). Coastal and marine conservation strategy for Bangladesh in the context of achieving blue growth and sustainable development goals (SDGs). *Environmental Science & Policy, 87*, 45–54. DOI: 10.1016/j.envsci.2018.05.014

Islam, M. M., & Shamsuddoha, M. D. (2018). Coastal and marine conservation strategy for Bangladesh in the context of achieving blue growth and sustainable development goals (SDGs). *Environmental Science & Policy, 87*, 45–54. DOI: 10.1016/j.envsci.2018.05.014

Islam, M. S., & O'Donnell, E. (2020). Legal rights for the Turag: Rivers as living entities in Bangladesh. *Asia Pacific Journal of Environmental Law, 23*(2), 160–177. DOI: 10.4337/apjel.2020.02.03

Jaman, M. F., Sarker, A. R., Alam, M., Rahman, M., Rabbe, F., Rana, A. S., & Hossain, S. (2021). Species diversity, distribution and habitat utilization of urban wildlife in a megacity of Bangladesh. *Biodiversity Journal, 12*(3), 635–653. DOI: 10.31396/Biodiv.Jour.2021.12.3.635.653

Kamal, M., Quazi, A. R., & Akhter, S. (2001). *Resource Use by Indigenous Communities in the Coastal Zone*. PDO-ICZMP.

Karanth, K. K., & DeFries, R. (2010). Conservation and management in human-dominated landscapes: Case studies from India. *Biological Conservation, 143*(12), 2865–2869. DOI: 10.1016/j.biocon.2010.05.002

Karanth, U. K. (2002). Mitigating human-wildlife conflicts in southern Asia. *Making parks work: strategies for preserving tropical nature*, 250-264.

Khan, M. M. H. (2018). *Photographic guide to the wildlife of Bangladesh*. Arannayk Foundation.

Kumar, A. (2005). *Biodiversity & Conservation*. APH Publishing.

Laurance, W. F. (2007). Forest destruction in tropical Asia. *Current Science*, •••, 1544–1550.

Lenzen, M., Moran, D., Kanemoto, K., Foran, B., Lobefaro, L., & Geschke, A. (2012). International trade drives biodiversity threats in developing nations. *Nature, 486*(7401), 109–112. DOI: 10.1038/nature11145

Majumder, N. M., Rahman, M. M., & Monjur-Ul-Haider, M. (2018). *Modernity and Ethnic Dynamics: A Study on Khasi Community in northeast Bangladesh. Man and Culture*. Springer.

McAfee, K., & Shapiro, E. N. (2010). Payments for ecosystem services in Mexico: Nature, neoliberalism, social movements, and the state. *Annals of the Association of American Geographers, 100*(3), 579–599. DOI: 10.1080/00045601003794833

Mondal, P., & Nagendra, H. (2011). Trends of forest dynamics in tiger landscapes across Asia. *Environmental Management, 48*(4), 781–794. DOI: 10.1007/s00267-011-9720-6

Mukul, S. A. (2014). Biodiversity conservation and ecosystem functions of traditional agroforestry systems: case study from three tribal communities in and around Lawachara National Park. In Chowdhury, M. S. H. (Ed.), *Forest conservation in protected areas of Bangladesh: policy and community development perspectives* (pp. 171–179). Springer. DOI: 10.1007/978-3-319-08147-2_8

Mukul, S. A., Alamgir, M., Sohel, M. S. I., Pert, P. L., Herbohn, J., Turton, S. M., Khan, M. S. I., Munim, S. A., Reza, A. H. M. A., & Laurance, W. F. (2019). Combined effects of climate change and sea-level rise project dramatic habitat loss of the globally endangered Bengal tiger in the Bangladesh Sundarbans. *The Science of the Total Environment, 663*, 830–840. DOI: 10.1016/j.scitotenv.2019.01.383

Mukul, S. A., & Quazi, S. A. (2009). Communities in conservation: protected area management and enhanced conservation in Bangladesh.

Mukul, S. A., Rashid, A. M., & Khan, N. A. (2016). Forest protected area systems and biodiversity conservation in Bangladesh. *Preprints*.

Mukul, S. A., Rashid, A. Z. M. M., Quazi, S. A., Uddin, M. B., & Fox, J. (2012a). Local peoples' response to co-management in protected areas: A case study from Satchari National Park, Bangladesh. *Forests, Trees and Livelihoods*, *21*(1), 16–29. DOI: 10.1080/14728028.2012.669132

Mukul, S. A., Rashid, A. Z. M. M., Uddin, M. B., & Khan, N. A. (2016). Role of non-timber forest products in sustaining forest-based livelihoods and rural households' resilience capacity in and around protected area: A Bangladesh study. *Journal of Environmental Planning and Management*, *59*(4), 628–664. DOI: 10.1080/09640568.2015.1035774

Newbold, T., Hudson, L. N., Arnell, A. P., Contu, S., De Palma, A., Ferrier, S., Hill, S. L. L., Hoskins, A. J., Lysenko, I., Phillips, H. R. P., Burton, V. J., Chng, C. W. T., Emerson, S., Gao, D., Pask-Hale, G., Hutton, J., Jung, M., Sanchez-Ortiz, K., Simmons, B. I., & Purvis, A. (2016). Has land use pushed terrestrial biodiversity beyond the planetary boundary? A global assessment. *Science*, *353*(6296), 288–291. DOI: 10.1126/science.aaf2201

Newbold, T., Scharlemann, J. P. W., Butchart, S. H. M., Sekercioglu, Ç. H., Alkemade, R., Booth, H., & Purves, D. W. (2013). Ecological traits affect the response of tropical forest bird species to land-use intensity. *Proceedings. Biological Sciences*, *280*(1750), 20122131. DOI: 10.1098/rspb.2012.2131

Ortolano, L., Sánchez-Triana, E., & Ferdausi, S. A. (2017). Strategy for adapting to climate change and conserving biodiversity in the Bangladesh Sundarbans. *Climate and Development*, *9*(4), 325–336. DOI: 10.1080/17565529.2016.1167660

Payo, A., Mukhopadhyay, A., Hazra, S., Ghosh, T., Ghosh, S., Brown, S., Nicholls, R. J., Bricheno, L., Wolf, J., Kay, S., Lázár, A. N., & Haque, A. (2016). Projected changes in area of the Sundarban mangrove forest in Bangladesh due to SLR by 2100. *Climatic Change*, *139*(2), 279–291. DOI: 10.1007/s10584-016-1769-z

Pearce, D. W., & Pearce, C. G. (2001). The value of forest ecosystems.

Phillips, H. R. P., Newbold, T., & Purvis, A. (2017). Land-use effects on local biodiversity in tropical forests vary between continents. *Biodiversity and Conservation*, *26*(9), 2251–2270. DOI: 10.1007/s10531-017-1356-2

Rahman, M. M., Kabir, M. E., Jahir Uddin Akon, A. S. M., & Ando, K. (2015). High Carbon Stocks in Roadside Plantations Under Participatory Management in Bangladesh. *Global Ecology and Conservation, 3*, 412–423. DOI: 10.1016/j.gecco.2015.01.011

Rahman, M. R. (2015). Causes of biodiversity depletion in Bangladesh and their consequences on ecosystem services. *American journal of environmental protection, 4*(5), 214-236.

Rahman, M. S., Sarker, P. K., & Giessen, L. (2016). Power players in biodiversity policy: Insights from international and domestic forest biodiversity initiatives in Bangladesh from 1992 to 2013. *Land Use Policy, 59*, 386–401. DOI: 10.1016/j.landusepol.2016.09.013

Rashid, S., & Hasan, M. T. (2020). Cultural Heritage at the Nexus of Sea, Forest, and People: Insights into the Lifeworld of Coastal People in Bangladesh.

Rawshan, K., Feeroz, M. M., Giordano, A. J., & Hasan, M. K. (2014). *Use of the media to characterize recent trends in human-wildlife conflict in Bangladesh: 1990-2010. The Festschrift on the 50th Anniversary of the IUCN Red List of Threatened Species. Compilation of Papers and Abstracts*. IUCN.

Sandmo, A. (2015). The early history of environmental economics. *Review of Environmental Economics and Policy, 9*(1), 43–63. DOI: 10.1093/reep/reu018

Sayer, J., Chokkalingam, U., & Poulsen, J. (2004). The restoration of forest biodiversity and ecological values. *Forest Ecology and Management, 201*(1), 3–11. DOI: 10.1016/j.foreco.2004.06.008

Schlingmann, A., Graham, S., Benyei, P., Corbera, E., Sanesteban, I. M., Marelle, A., & Reyes-García, V. (2021). Global patterns of adaptation to climate change by Indigenous Peoples and local communities. A systematic review. *Current Opinion in Environmental Sustainability, 51*, 55–64. DOI: 10.1016/j.cosust.2021.03.002

Song, X.-P., Hansen, M. C., Stehman, S. V., Potapov, P. V., Tyukavina, A., Vermote, E. F., & Townshend, J. R. (2018). Global land change from 1982 to 2016. *Nature, 560*(7720), 639–643. DOI: 10.1038/s41586-018-0411-9

Swiderska, K. (2020). Protecting indigenous cultures is crucial for saving the world's biodiversity. *The Conversation, 14*(02).

Tuhiwai Smith, L. (2012). Decolonizing methodologies: Research and indigenous peoples.

Uddin, M. S., Van Steveninck, E. D. R., Stuip, M., & Shah, M. A. R. (2013). Economic valuation of provisioning and cultural services of a protected mangrove ecosystem: A case study on Sundarbans Reserve Forest, Bangladesh. *Ecosystem Services*, *5*, 88–93. DOI: 10.1016/j.ecoser.2013.07.002

Ullah, F., Saqib, S., & Xiong, Y. C. (2025). Integrating artificial intelligence in biodiversity conservation: Bridging classical and modern approaches. *Biodiversity and Conservation*, *34*(1), 45–65. DOI: 10.1007/s10531-024-02977-9

Chapter 2
Stolen Forest:
Breathing Inside a Forest

Ferdous Ahmed
https://orcid.org/0000-0001-8737-8142
International University of Business Agriculture and Technology, Bangladesh

Md. Sazid Hasan
International University of Business Agriculture and Technology, Bangladesh

ABSTRACT

Forests stand alone for maintaining all vital functions of biodiversity of wildlife and climate regulation as well as carbon sequestration and water filtration. This chapter investigates the effects of climate change on forest degradation specifically intended for ecosystems and communities as well as indigenous populations whose forests cost both culturally and economically. However, the aims of this chapter papers are: (a) To analyze the extent and drivers of forest loss; (b) examine the ecological consequences of forest degradation on biodiversity; (c) assessing the socio-economic and cultural impacts of deforestation on local communities; (d) to propose sustainable strategies and policy recommendations. In deforestation alone (excluding regrowth), tropical primary forests are still declining at rate of several million hectares/year. However, in 2023, primary tropical forest lost 3.7 million hectares. Simultaneously, in tropical regions especially (Amazon, Congo, Southeast Asia), in some cases 50-90% of logging may be illegal or driven by weak governance.

1. INTRODUCTION

Ecological experts classify forests as vital ecosystems because they protect 75% of terrestrial biodiversity which is essential for preserving global biodiversity (Achieng et al. 2023). The Earth's climate depends on forests because they absorb carbon and

DOI: 10.4018/979-8-3373-3992-4.ch002

manage the natural water cycle while sustaining ecological stability (Adom, et al. 2024). Through their influence on water management and flood reduction as well as nutrient cycling and air and water purification functions forests improve basic human survival on Earth (Asabere, 2020). In addition to their protective role trees function as natural barriers that shield the soil from erosion while strengthening the ecosystem to resist environmental changes that happen during climate change thus ensuring long-term stability (Hallaj et al., 2024). Forests demonstrate critical value for all aspects of life by fulfilling numerous essential functions which unite their ecological worth with their importance to human needs (Werdel et al., 2024). Human activities have caused unprecedented destruction of forests through three primary actions combined into the "Stolen Forest" category that includes deforestation alongside illegal logging and encroachment (Ward et al., 2024). Annually humans destroy trees covering 7.3 million hectares of forested territory which compromises the biological structure of these ecosystems (FAO, 2020). Deforestation produces extensive environmental damage because it degrades ecosystem functions while causing severe biodiversity reduction then weakens environmental stability through its emission of 20 percent of Earth's greenhouse effect (Chiriacò et al., 2024). The elimination of forests by humans weakens their carbon storage ability while eliminating important materials needed for survival including nourishment and drugs and basic resources (Bonan, 2008).

In Bangladesh, the Sundarbans which is the largest mangrove forest covering approximately 10,000 km2. It serves as a critical natural barrier, providing storm surge protection, nutrient trapping and biodiversity support (Rahman et al., 2025). Over recent years, this ecosystem has undergone measurable degradation driven by sea-level rise, coastal erosion and human encroachment, for instance roughly 210 km2 of land has been lost since 1964 as the seas have encroached on the delta (CIWEM, 2025). Remote sensing studies covering 2017-2023 indicate increasing fragmentation of mangrove cover in the Bangladeshi portion of the Sundarbans, signaling declining structural integrity of this vital ecosystem (Roy et al., 2025). These changes not only effect the species habitat but also the livelihoods of millions of people who rely on forest resources in the delta region.

Climate change exists in a mutual cycle with forest degradation which produces complex effects between the two entities. Forests absorb and store a net 7.6 billion metric tonnes of carbon dioxide (CO2) annually (Wear & Wibbenmeyer, 2024), which is about twice as much carbon as they emit, helping to mitigate climate change by removing CO2 from the atmosphere (Mitchard et al., 2023). Forest degradation requires immediate attention because the feedback process shows its significant role in advancing climate change and destroying worldwide ecological systems. Much like the "Stolen Forest" framework shows the way indigenous and local communities are pushed aside to the detriment of their traditional ecological knowledge (TEK) which has protected forest ecosystems for thousands of years (Tran et al., 2024). Weakening the traditional forest management system which Indigenous peoples developed to conserve biodiversity while ensuring ecosystem longevity (UNDP, 2024). Traditional knowledge and cultural practices of local communities come under severe risk due to external forces such as land grabs and commercial exploitation which disregards their stewardship and results in forest destruction (Mongabay, 2024). Regions facing environmental risks contain these communities who experience severe effects from modified precipitation behavior along with diminishing agriculture territory through environmental changes (Nakashima & Roué, 2002).

The world recognizes the importance of integrating Indigenous Knowledge Systems (IKS) into forest management. The IPCC reports within international climate agreements promote the incorporation of TEK into environmental policies because it enhances restoration and conservation programs (IPCC, 2023). Sustainable forest management now has a holistic framework that results from uniting TEK with modern scientific methods which has generated worldwide achievements in restoration and biodiversity conservation efforts (Urton, 2024). Through their alignment TEK and modern approaches create adaptive climate resilience strategies that simultaneously maintain cultural traditions and ecological contributions of local communities to build durable forest ecosystems. The ecological damage from forest degradation transforms into social adversity that indigenous people together with local communities must endure. Local communities which sustain ties with forests for their dwelling places as well as cultural practices along with economic activities lead climate adaptation by using their acquired knowledge to create suitable strategies (Bhatt et al., 2024). The implementation of nature-based solutions needs to focus on group involvement to restore and protect forests because indigenous and local communities need to be the center of decision-making (IPCC, 2023). The research investigates how traditional ecological knowledge should integrate with scientific methods to develop sustainable forest management plans that promote social and environmental sustainability.

2. PROBLEM STATEMENT

Forests are often described as the lungs of the Earth, providing vital oxygen, regulating the climate, and sustaining biodiversity. Yet, these lungs are being stolen, fragmented by deforestation, illegal logging, and unsustainable land use. The paradox of "breathing inside a forest" highlights the tension between the life-giving role of forests and the suffocating impact of human exploitation. While people enter forests seeking oxygen, peace, and connection with nature, excessive extraction and encroachment threaten to strip forests of their ability to breathe for themselves and the planet.

Globally, forests are disappearing at alarming rates, taking with them the critical ecosystem services that maintain ecological balance, carbon sequestration, and biodiversity (FAO, 2022; IPCC, 2023). Between 2015 and 2020, approximately 10 million hectares of forest were lost annually, with devastating consequences for biodiversity, carbon storage, and local livelihoods (FAO 2020). This stolen forest phenomenon represents not only an environmental crisis but also a socio-cultural one, as communities, wildlife, and future generations lose access to the life-supporting functions of these ecosystems.

Communities that rely on forests for clean air, water, and cultural identity increasingly find themselves in environments where the essence of breathing is threatened. The central challenge lies not only in halting deforestation, but also in ensuring equitable access, sustainable management, and ecological restoration (Tran et al., 2024). Ultimately, the core problem is the imbalance between ecological preservation and anthropogenic pressure, an imbalance that undermines the ability of forests to sustain life. Addressing this requires an integrated understanding of socio-ecological systems and the development of frameworks that protect forests as living, interdependent systems for future generations.

3. OBJECTIVES

- To assess the extent and drivers of forest loss, including deforestation, illegal logging and land-use change at both global and regional scales, with special attention to tropical and mangrove ecosystems.
- To examine the ecological consequences of forest degradation, focusing on biodiversity decline, carbon storage loss and the disruption of climate-regulating processes.
- To evaluate the socioeconomic and cultural impacts of forest loss on Indigenous and local communities, particularly concerning livelihoods, equity and traditional ecological knowledge.
- To propose integrated strategies and policy recommendations for forest conservation, restoration, and sustainable management that align with global climate commitments and Sustainable Development Goals (SDGs).

4. LITERATURE REVIEW

This literature review synthesizes recent scholarship (2019–2025) on forest loss, its ecological and social consequences, and responses relevant to the "stolen forest breathing inside a forest" framing. It highlights major drivers of forest loss, key ecosystem and human impacts, promising restoration and governance responses, and gaps where the proposed study can contribute. FAO's (2024a) flagship report provides the latest global overview of forest cover, trends in deforestation and forest degradation, and emphasizes innovations (technological, financial, institutional) needed to reverse loss. It contextualizes how forest ecosystems are being "stolen" through diverse pressures, and points to opportunities for evidence-based policy and institutional change. However, Pan et al. (2024) synthesized three decades of ground-based and remote-sensing data to estimate global forest carbon fluxes, including contributions from intact and regrowth forests. It shows that while forests remain net carbon sinks (~3.56 ± 0.37 Pg C yr^{-1}), shifts in contributions from intact vs. regrowth forests raise concerns about long-term resilience amid degradation pressures. It underscores the idea that forest "breathing" capacity is at risk under escalating extraction. In addition, Adom et al. (2024) reviewed the interactions between deforestation, illegal logging, and climate dynamics, highlighting how forest loss exacerbates climate vulnerability, biodiversity decline, and socio-ecological disruption. It draws attention to the cascading risks when the forest's regulation services are compromised. Tonouéwa et al. (2024) focused on the methods for tracing timber supply chains to detect illegal logging activity. It frames illegal extraction as a hidden driver of forest loss — a "stealth" theft of forest function and value. Improvements in traceability are argued as essential to restoring the forest's integrity. On the other hand, Damnyag et al. (2024) studied the empirical quantifies stumpage, carbon, and fiscal revenue losses due to illegal logging in Ghana. It illustrates a concrete example of how forest extraction erodes both ecological and economic capital, reinforcing the "stolen forest" metaphor. Another FAO's (2022) study sets out three strategic pathways halting deforestation, restoring degraded lands, and sustainable use as an integrated approach to forest recovery. It emphasizes socio-ecological balance and institutional innovation, offering a blueprint aligned with your study's aims.

This earlier edition of FAO-UNEP (2020) explored the interrelations among forest ecosystems, biodiversity, and human well-being, including policy innovations and case studies on forest conservation and sustainable use. It forms a useful baseline for understanding how forest functions (including "breathing") have already been under stress. FAO. (2024b) edited, policy-oriented version of the full 2024 report, this summary is useful for drawing out key messages and recommended interventions in a more accessible style, emphasizing forest governance and innovation. Moreover, Mitchard et al. (2023) addressed the methodological flaws in evaluating carbon-offset forest projects, highlighting the danger of misestimating impact on forest carbon and overpromising climate benefits. It serves as a methodological caution relevant to assessing "breathing capacity" in restoration and mitigation process. However, Achieng et al. (2023) focused on the African tropical ecosystems, this review discusses challenges in biodiversity monitoring, data gaps, governance constraints, and the urgency to link biodiversity loss to ecosystem service decline. It supports your framing by reinforcing that "breathing" is tied to diversity and ecosystem integrity. Nevertheless, John et al. (2025) compiled a global database of reforestation efforts and assesses reliability of spatial data in monitoring projects. The integrity of reforestation data is vital when evaluating how well forests can regain their "breathing" capacity after loss.

Table 1. Reasons for stolen forest by Region

Country/Region	Main Reasons for Forest Loss ("Stolen Forest")	% or Scale	Reference
Brazil (Amazon)	Agricultural expansion (cattle ranching, soy), illegal logging, fires, weak enforcement	Brazil lost ~1.2 million ha of primary forest in 2022 alone; ~60% of global tropical primary forest loss occurs in Amazon	World Resources Institute (WRI, 2025); FAO (2022)
Indonesia	Oil palm expansion, timber plantations, illegal logging, peatland drainage	Indonesia reduced loss recently but still loses ~200,000 ha annually; historically one of top 3 deforestation hotspots	FAO (2022); Forest Declaration Assessment (2023)
Congo Basin (DRC, Cameroon, Gabon, etc.)	Fuelwood/charcoal extraction, shifting cultivation, weak governance, illegal logging	DRC lost ~500,000 ha in 2022; second only to Brazil in tropical forest loss	Global Forest Watch (2025); Achieng et al. (2023)
West Africa (Ghana, Côte d'Ivoire)	Cocoa expansion, illegal logging, artisanal mining	Ghana lost ~60% of original forest cover in last century; illegal logging accounts for ~70% of timber trade in Ghana	Damnyag et al. (2024); FAO (2020)
Southeast Asia (Myanmar, Cambodia, Laos)	Timber extraction, conflict-driven deforestation, commercial plantations	Cambodia had world's highest rate of deforestation between 2001–2015; Myanmar continues to lose >100,000 ha annually	Adom et al. (2024); Tonouéwa et al. (2024)
Bangladesh	Encroachment for settlements, agriculture, fuelwood dependency, industrial expansion	Forest cover ~11% (down from 20% in 1980s); Sundarbans face pressure from shrimp farming & illegal felling	FAO (2020); Local forestry reports
Latin America (outside Brazil – Peru, Bolivia, Colombia)	Coca cultivation, cattle ranching, infrastructure projects	Peru lost ~150,000 ha of forest in 2022, mainly in Amazonian regions	WRI (2025); Forest Declaration Assessment (2023)

4.1 Illicit Felling and the "Stolen Forest"

Illicit felling (illegal logging, timber laundering, clandestine clearing) encompasses unauthorized cutting, falsified permits, illegal land conversion, and the laundering of illegally cut timber into formal markets (Global Witness, 2024). Recent global syntheses highlight that forest crimes are often multi-faceted, organized, and tightly linked to corruption and other transnational crimes which makes illicit felling distinct from regulated harvests and especially harmful to local rights and ecosystem integrity (John et al, 2025). Large-scale monitoring platforms and forensic investigations document that illicit felling remains a major component of global forest loss. Remote-sensing and alert systems reported record tropical primary forest losses driven by fires, expansion and clandestine clearing in 2023–2024, with hotspots in the Amazon, Congo Basin, and parts of Southeast Asia and Latin America (World Resources Institute, 2025). Investigative reporting and supply-chain studies continue to reveal illegally sourced timber entering international markets, undermining governance improvements (Dingkuhn et al., 2025). Key drivers include commodity expansion (pasture, soy, oil palm, cocoa), high global demand for luxury hardwoods, weak governance, corruption, and links with other illicit economies (Shrestha, Chaudhary, & Shrestha,2025). UNODC and INTERPOL analyses emphasize that the convergence of organized criminal networks and governance gaps allows illicit felling to scale and persist, particularly where enforcement capacity and land-tenure clarity are low. Illicit felling often targets high-value and large trees (high-grading), causing disproportionate impacts on forest structure, carbon stocks, and biodiversity compared with equivalent area loss (Cabernard, et al. 2024). Such selective, clandestine extraction accelerates functional decline (reduced carbon sequestration, altered microclimate) and degrades the forest's capacity to "breathe" for humans and climate alike (Debie, E. et al. (2025). These functional effects are emphasized in global assessments and carbon-project critiques that warn about over- or mis-attribution of carbon benefits when illegality is not accounted for.

Economic analyses and investigative reports document large fiscal and stumpage losses where illegal timber bypasses royalties and taxes (Nellemann & INTERPOL Environmental Crime Programme, 2012). Numerous recent investigations show illegal or mis-documented timber from high-deforestation regions reaching US and European markets, often through laundering chains that exploit certification loopholes, weak checks, or falsified paperwork. However, this market leakage creates incentives for further clandestine extraction (Tran et al., 2024). Illicit felling commonly removes customary access, undermines local livelihoods, and erodes trust in institutions. Criminology and human-rights literature therefore frame forest crime not only as environmental harm but as dispossession and slow violence a literal "theft" of community resources, memory, and future subsistence. This justice dimension is central to the "stolen forest" concept and to calls for reparative governance. Interdisciplinary psychology and environmental humanities work shows that ecological grief and solastalgia (distress at environmental change at home) are common reactions to ecosystem loss (Damnyag et al., 2024). When loss is illicit i.e., caused by criminal actors or corrupt decisions the emotional response is often compounded by anger, betrayal, and moral outrage. Recent reviews recommend integrating psychometric and narrative measures of grief and moral emotion into socio-ecological studies of forest crime (Khosravi Mashizi & Escobedo, 2025). Therefore, for mitigation process advances in remote sensing, near-real-time deforestation alerts (GLAD/GFW), forensic wood identification, and traceability (RFID, chemical fingerprinting, blockchain pilots) improve detection and supply-chain transparency. However, literature and policy analyses note important limits: technological adoption is uneven, traceability at origin remains challenging, and laundering tactics adapt meaning technical systems must be paired with governance and legal reforms.

4.2 Illegal Logging and the "Stolen Forest"

Illegal logging refers to the unauthorized harvesting, processing, and trade of timber, often involving violations of national laws and international agreements (Mongabay, 2023). It encompasses activities such as logging without permits, logging in protected areas, and falsifying documents to launder timber into legal markets. However, this illicit trade undermines conservation efforts and contributes to significant environmental degradation (Franco-Moraes et al., 2023). Illegal logging remains a pervasive issue across tropical and temperate forests. In regions like the Amazon Basin, the Congo Basin, and Southeast Asia, illegal logging is often linked to organized crime, corruption, and weak governance. For instance, in the Democratic Republic of Congo, over 90% of logging activities are reported to be illegal, exacerbating deforestation and biodiversity loss (Tonouéwa et al., 2024). Moreover, Illegal logging timber extraction and associated trade that violates national laws or international agreements remains a leading driver of forest loss and degradation worldwide (Chiriacò, et al., 2024). Beyond measurable losses of biomass and biodiversity, illegal logging produces profound social injustices: displaced communities, captured governance, and dispossessed cultural landscapes that many scholars and activists describe as "stolen forests." (Bélisle, et al. 2025). This essay synthesizes empirical evidence on the scale and drivers of illegal logging, explains the mechanisms through which criminal timber economies operate, outlines ecological and social consequences, and evaluates policy responses and governance reforms that seek to protect forests and the rights of forest-dependent peoples.

Forests are multifunctional: they sequester carbon, host biodiversity, underpin livelihoods, regulate water cycles, and sustain cultural identities (FAO, 2020). Yet, illegal logging ranging from small-scale illicit felling to large-scale organized extraction erodes these functions while transferring the benefits of timber to illegal actors and costs to local people and states. The metaphor "stolen forest" captures both the physical removal of trees and the dispossession of rights, access, and stewardship exercised by indigenous and local communities (Baral et al., 2025). Understanding illegal logging requires attention to supply- and demand-side drivers, the institutional context that permits illegality, and the variety of ecological and social harms it causes. Measuring illegal logging is difficult because it is, by definition, hidden in part of the formal economy and in informal markets (World Resources Institute, 2025). However, remote-sensing analyses have documented widespread forest-cover loss, while ground-based investigations and trade-traceability analyses reveal that a substantial share of timber in both domestic and international markets originates from unlawful sources (Takahashi & Soma, 2025).Spatially, illegal logging hotspots are concentrated in tropical regions (Amazonia, Central Africa, Southeast Asia) and in countries with weak governance or high demand for high-value timber (Global Forest Watch, 2025).

4.3 Forest Encroachment

Forest encroachment remains a critical issue in both tropical and temperate regions, driven by a complex interplay of socio-economic pressures, institutional weaknesses, and environmental changes. Recent studies (2023–2025) provide nuanced insights into these dynamics, particularly in South and Southeast Asia (Mongabay, 2023). Encroachment is often driven by poverty, land scarcity, and the need for agricultural land. A study in Sudurpaschim Province, Nepal, identified key drivers of forest encroachment between 1990 and 2020 (Poudel et al., 2025). The research utilized machine learning classifiers to analyze field data, revealing that socio-economic pressures significantly contribute to forest loss in the region. In Bangladesh's Chittagong Hill Tracts (CHT), forest degradation is exacerbated by complex institutional arrangements and land disputes with local communities (Chittagong Hill Tracts Study, 2023). These issues hinder effective forest and landscape restoration efforts, despite the government's commitment to restore 0.75 million hectares of forests (Ahammad et al., 2023). Climate change impacts, such as altered rainfall patterns and increased frequency of extreme weather events, exacerbate forest degradation. In the Sundarbans, Bangladesh's coastal mangrove forest, biodiversity faces extreme threats due to legal and institutional gaps, highlighting the need for stronger policy implementation (Islam, 2025). Encroachment into forested areas often leads to habitat destruction, threatening biodiversity. For instance, the encroachment of orchards on the Hyrcanian forest edge has enhanced bird communities initially; however, continued land use can lead to a decrease in biodiversity.

Forest loss contributes significantly to carbon emissions. In 2023 alone, Bangladesh lost 20.2 kilo-hectares of natural forest, releasing 11.6 million tonnes of CO_2 emissions (Al-Bawwat et al, 2023). Co-management approaches, integrating local communities in forest governance, have shown positive impacts on forest cover. In Bangladesh's Sal Forest, community-based forest management led to increased forest cover and improved livelihoods for forest-dependent communities (Tran et al., 2024). Advancements in remote sensing technologies have enhanced the ability to monitor forest encroachment. Studies employing machine learning and multi-source remote sensing data have improved the accuracy of forest carbon stock estimation, aiding in better forest management and conservation planning. However, addressing forest encroachment requires a multifaceted approach, including strengthening institutional frameworks, promoting community-based forest management, and leveraging technological advancements for monitoring and enforcement. Continued research and policy innovation are essential to mitigate the impacts of forest encroachment and promote sustainable forest management practices.

4.4 Forest Land Use Conversion

Forest land-use conversion (FLUC), the conversion of forest to agriculture, pasture, plantations, mining sites, settlements, or infrastructure remains a dominant pathway of global forest loss with major consequences for biodiversity, carbon stocks, hydrology and livelihoods. Recent scholarship has shifted from describing area loss to unpacking multi-scalar drivers (supply chains, policy, tenure), spatial heterogeneity (different "hotspots"), and the cascading social-ecological impacts of conversion (Silva-Junior et al., 2023). Global and regional analyses show that patterns of conversion vary: tropical South America (Amazon, Cerrado), Southeast Asia (Indonesia, Malaysia), and parts of Central Africa remain primary hotspots, while temperate regions experience more localized, often agricultural or urban conversions. Recent reports document record-high losses in some savanna/forest mosaics (e.g., Brazil's Cerrado) and renewed pressure in Amazonian frontiers, driven by commodity expansion, fires and land-grabbing. Conversion reduces local species richness and functional diversity, fragments habitat, and lowers ecosystem carbon stocks literature syntheses estimate substantial carbon losses following conversion and long recovery times if left to regenerate (Ahammad et al., 2023) Biodiversity-oriented metrics that integrate production-chain impacts show that land-use change remains a leading cause of recent biodiversity decline. Remote sensing (high-resolution satellite time-series), land-cover change models, and footprint approaches that link consumption to production have improved attribution of forest conversion

(Cabernard et al., 2024). New studies combine satellite detection with on-the-ground socio-economic datasets and trade-linked models to identify drivers and actors behind conversion. However, detection limits (smallholder clearing, understory degradation) and attribution uncertainty remain challenges. Research increasingly foregrounds the social dimensions of conversion: impacts on Indigenous and local communities, contested tenure, livelihood trade-offs, and unequal distribution of benefits and harms (Silva-Junior et al. (2023). Studies highlight how policy or corporate zero-deforestation pledges may produce unintended outcomes (leakage, shifting frontiers) if not paired with effective local governance and tenure security.

Contemporary literature evaluates a range of interventions: protected areas and Indigenous territories (shown to reduce conversion in many contexts), jurisdictional sourcing initiatives, commodity-specific moratoria and corporate commitments, payments for ecosystem services and restoration programs. While some policies have reduced deforestation regionally, evidence shows mixed effectiveness unless interventions address underlying drivers (market demand, tenure insecurity, corruption). Global assessments and reports call for integrated supply-chain governance, stronger law enforcement, support for sustainable livelihoods, and financing for restoration. Very recent analyses (2023–2025) document concerning rebounds in some areas (e.g., parts of the Amazon and Cerrado) and note that wildfire-driven losses can rival commodity-driven conversion in certain years linking climate extremes to land-use outcomes. There is growing attention to how finance (banking, investments) and international trade policies accelerate or could mitigate FLUC.

Table 2. Recent Literatures on Forest Land Use Conversion (FLC)

Author(s)	Year	Method	Region	Key Finding	Gap
Cabernard et al.	2024	Supply-chain footprint modeling, biodiversity assessment	Global	Linked biodiversity impacts to global consumption-driven land-use change	Uncertainty in footprint models; limited local validation
IPCC AR6 WG3 (Ch.7)	2022	Assessment report, synthesis of global literature	Global	Synthesized evidence on AFOLU, mitigation, and land-use change drivers	Limited sub-national detail; rapidly evolving trends not fully captured
Hallaj et al.	2024	Remote sensing, ecological modeling	Global/tropical focus	Quantified forest land-use change impacts on biodiversity and ecosystem services	Needs more integration of social drivers with ecological models
Fragoso-Medina	2024	Field surveys, landscape ecology	Mexico (tropical forests)	Conversion to agriculture reduces tree density and species richness	Long-term recovery dynamics understudied

continued on following page

Table 2. Continued

Author(s)	Year	Method	Region	Key Finding	Gap
Chaves	2024	Remote sensing, hotspot analysis	Brazil (AMACRO frontier)	Identified AMACRO as a new Amazonian deforestation hotspot	Future monitoring and governance effectiveness uncertain
Global Forest Review (WRI)	2024	Satellite data analysis	Global	Reported global forest loss trends and primary drivers (agriculture, fire, commodity crops)	Attribution to smallholders vs. industrial actors still limited
Sylvester et al.	2024	Food system modeling, trade analysis	Global	Connected global food system expansion to deforestation and land-use change	Limited regional specificity; policy implications need local testing
Debie et al.	2025	Carbon stock measurement, synthesis	Africa (biosphere reserves)	Showed major carbon stock loss due to agricultural expansion	Uncertainty in carbon recovery pathways after conversion
Chiriacò et al.	2024	GHG inventory, scenario modeling	Europe/global	Quantified emissions linked to deforestation and land-use change	Spatial heterogeneity in emissions poorly constrained
Global Witness	2024	Policy report, investigative research	Brazil (Cerrado)	Exposed Cerrado as deforestation hotspot linked to soy and beef	Independent monitoring of corporate commitments needed
WRI Insight	2025	Remote sensing & trade data integration	Global	Highlighted main commodities driving forest loss in 2025	Unclear how policy measures interact with trade drivers
Forest Declaration Assessment	2024	Policy evaluation, global progress assessment	Global	Reviewed progress toward 2030 deforestation commitments	Showed commitments insufficient; monitoring accountability lacking
USDA Southern Research Station	2025	Land-use data analysis	Southeastern USA	Documented forest conversion to pasture and cropland in the US South	More socio-economic driver analysis required
Reuters	2024	News report synthesizing research & expert commentary	Global (tropics)	Summarized recent tropical deforestation drivers and trends	Lacks methodological depth; relies on secondary sources

4.5 Traditional Ecological Knowledge in Forest Management

Traditional Ecological Knowledge (TEK) the place-based, intergenerational ecological knowledge held by Indigenous Peoples and local communities is increasingly recognized as essential for effective, equitable, and resilient forest management. Recent literature (2022–2025) emphasizes TEK's contributions to biodiversity conservation, fire and pest management, landscape restoration, ecosystem-service stewardship, and monitoring, while also interrogating power imbalances, epistemic tensions, and legal/ institutional barriers to meaningful co-management. Contemporary reviews frame TEK not as static folklore but as dynamic knowledge systems that combine observational learning, ritual practice, resource governance institutions (taboos, sacred sites), and adaptive management techniques. Scholars stress that TEK is both normative (rules about use) and technical (specific practices for burning, selective harvesting, species husbandry), and that it must be treated as a plural knowledge source complementary to, not merely ancillary to, Western science. This reframing underpins recent calls for epistemic pluralism in policy and research. Several empirical and synthesis studies document measurable ecological benefits where TEK-guided practices are applied. Indigenous territories and community-managed forests frequently show lower deforestation, fewer large fires, and higher biodiversity retention compared with neighboring lands governed under conventional state rules a pattern robustly documented in the Amazon and other regions. TEK-informed practices such as fine-scale patch burning, seasonal harvesting rules, and sacred-site protection create heterogeneity in fuel loads, conserve key habitat patches, and sustain culturally important species.

A growing body of work highlights TEK's value in wildfire-prone systems. Locally practiced low-intensity burning regimes and fuel break customs have reduced catastrophic fire risk in Mediterranean, boreal and tropical savanna-forest mosaics where such practices remain active or have been revived through joint programs. Knowledge-transfer projects and service-learning initiatives have shown promise in hybrid training that pairs TEK holders with formal fire managers to reintroduce controlled burning in a socially acceptable way. Recent studies explore spatially explicit ways to align TEK with restoration planning and ecosystem-service mapping. Integrative approaches combine community preferences and place-based knowledge with remote-sensing and ecosystem-quality metrics to prioritizes restoration sites that maximize ecological function and cultural values simultaneously. This joint approach both improves ecological outcomes and strengthens local legitimacy for restoration projects. Multiple analyses emphasize that TEK's effectiveness depends heavily on secure land tenure, legal recognition of customary rights, and meaningful participation in decision-making. Where Indigenous and community forest rights are recognized and enforced, TEK-based governance institutions from sacred groves to customary bans often deliver conservation co-benefits. Conversely, policy initiatives that instrumental TEK without addressing tenure insecurity, benefit-sharing, or epistemic respect risk tokenism and can exacerbate marginalization. The literature converges on practical prescriptions: recognize and legally secure Indigenous and community forest rights; support community-led monitoring and management;

co-produce management plans that respect cultural practices; finance TEK-led restoration and fire-management programs; and reform institutional decision-making to enable equitable co-management. Several case studies show these steps improve both ecological resilience and social wellbeing, but success depends on redistributing decision-making power rather than mere participatory consultation.

5. METHODOLOGY

This study adopts a qualitative-phenomenological approach to explore both the ecological and human dimensions of "stolen forests"—forests impacted by illegal logging, encroachment, and unsanctioned extraction. The methodology combines environmental assessment, ethnographic observation, and spatial analysis to capture the ecological, social, and psychological experiences of these forest spaces. This methodology ensures a holistic, multi-disciplinary approach, integrating ecological science, social research, and phenomenological observation to understand the complex phenomenon of stolen forests. This highlights the following:

- **Qualitative focus:** Understanding human perceptions, experiences, and socio-cultural narratives related to stolen forests.
- **Phenomenological focus:** Investigating the sensory and experiential aspects of "breathing inside a forest" affected by illicit activities.

Expected Outcomes

- ***Ecological Insights:*** Spatial patterns of forest degradation and biodiversity loss.
- ***Socio-Cultural Insights:*** Community perceptions, dependence, and resistance related to stolen forests.
- ***Experiential Insights:*** Sensory narratives illustrating the intangible aspects of forest loss, contributing to environmental awareness and policy advocacy.

However, this chapter aims to initiate future work in Bangladesh based on the following methodology to measure the real scenario for stolen forest.

Figure 1. Schematic method for future extended research in Bangladesh

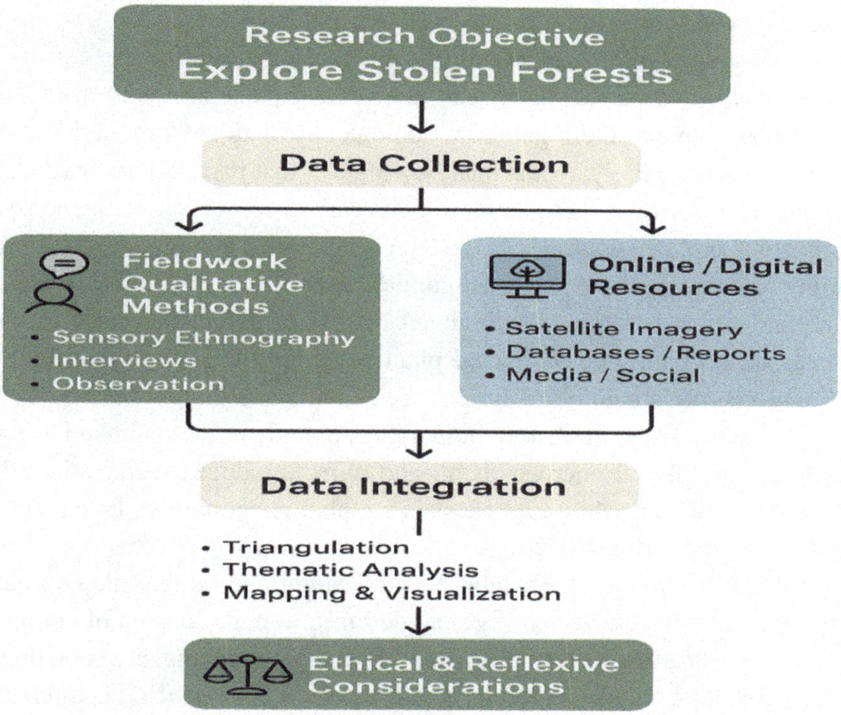

6. DISCUSSION

6.1 Extent and Drivers of Forest Loss in the Context of Global and Local Ecosystems

Forests are among the most vital ecosystems on Earth, covering approximately 31 percent of the planet's land area and providing critical ecosystem services including biodiversity conservation, carbon storage, hydrological regulation, and cultural sustenance (Hallaj et al., 2024). However, they are under intense pressure from human activity. Forest loss comprising deforestation, degradation, illegal logging, and land-use change has accelerated over the past decades, with profound consequences for global climate stability and local livelihoods. This essay examines the extent of forest loss globally and locally, and analyzes the key drivers underpinning this transformation. Satellite-based monitoring systems have revealed that the tropics alone lost over 11 million hectares of tree cover in 2021, with primary tropical forests accounting for a significant proportion of this loss. Over the past 30 years, the Amazon Basin, the Congo Basin, and Southeast Asia have emerged as global hotspots of deforestation. While some temperate regions (e.g., Europe and parts of North America) have seen net forest expansion due to afforestation and forest management, these gains are ecologically distinct from the loss of old-growth forests. Alarmingly, recent analyses suggest that parts of the Amazon have transitioned from being a carbon sink to a net carbon source, signaling ecosystem destabilization at planetary scale. At the local level, forest loss manifests differently depending on ecological and socio-economic contexts. In Brazil and Bolivia, cattle ranching and soy cultivation are dominant drivers of deforestation. In Indonesia and Malaysia, oil palm and pulpwood plantations are the leading culprits, while in the Congo Basin, smallholder agriculture and artisanal logging drive extensive forest degradation. In South Asia's Sundarbans, mangrove loss is linked to shrimp aquaculture, agricultural encroachment, and climate-induced salinization. These examples illustrate that while forest loss is a global phenomenon, its patterns and impacts are locally differentiated.

Agricultural conversion both subsistence farming and large-scale commercial agriculture is the primary driver of global deforestation. Expansion of commodity crops such as soy, palm oil, and maize, along with cattle ranching, has been linked to large-scale forest clearance, particularly in the tropics. This demand is often driven by international markets, embedding local deforestation in global trade networks. Logging, both legal and illegal, remains a significant contributor to forest degradation and loss. Selective logging reduces biomass density and leaves forests more vulnerable to fire and disease, while large-scale illegal timber trade fuels corruption and undermines governance. It is estimated that illegal logging accounts for 15–30 percent of the global timber trade, with higher proportions in tropical countries. Beyond immediate tree removal, logging roads open access to remote forest frontiers, enabling further agricultural conversion and settlement. Infrastructure projects including roads, mining, and hydropower dams—contribute to forest loss by directly clearing land and indirectly facilitating further deforestation. Mining in the Amazon and Congo Basin, for instance, has created deforestation hotspots, often accompanied by pollution and social displacement. Urban expansion also converts peri-urban forests into settlements, particularly in rapidly growing regions of Asia and Africa. Forest fires, both natural and human-induced, are increasingly important drivers. Intentional burning to clear land often escapes control, leading to large-scale forest loss. Climate change intensifies this feedback loop by increasing drought frequency and severity, making degraded forests more fire-prone. The Amazon's susceptibility to fire exemplifies this interaction between human and climatic drivers. Underlying all proximate drivers are systemic governance challenges. Weak enforcement, insecure land tenure, and corruption facilitate illegal logging and unregulated land-use change. Where local communities lack rights or participation in forest governance, external actors often appropriate land, exacerbating both ecological degradation and social injustice.

Figure 2. Drivers of Tree Cover Loss by Regions (2001-2024)

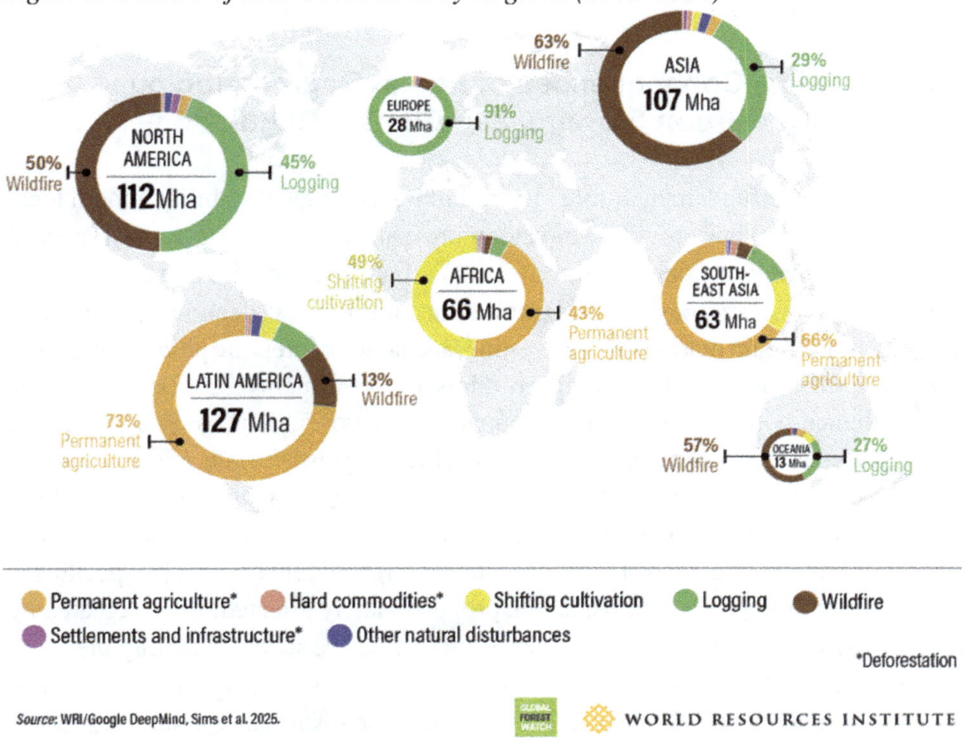

Source: (Sims, 2025)

The dynamics of forest loss illustrate the interdependence between global processes and local realities. Commodity markets and international demand for beef, soy, and palm oil link deforestation in the Amazon and Southeast Asia to consumer behavior in Europe, North America, and China. Similarly, global climate change exacerbates local vulnerabilities, such as increased fire risk or salinization of mangrove ecosystems. Conversely, local governance innovations, such as community forestry in Nepal or Indigenous land rights in Brazil, demonstrate that local actors can play pivotal roles in resisting global deforestation pressures. Forest loss is a multifaceted challenge operating across scales. Globally, it threatens biodiversity, carbon storage, and climate stability, while locally, it undermines livelihoods, cultural practices, and ecological resilience. The drivers—agriculture, logging, land-use change, fire, and governance failures—are deeply interconnected, linking distant markets to immediate local realities. Effective responses must therefore integrate global policy frameworks (e.g., zero-deforestation supply chains, climate agreements) with local empowerment strategies (tenure security, community forest management, Indigenous rights). Only by addressing both the structural and proximate causes can societies mitigate the extent of forest loss and safeguard forests as ecological and cultural commons.

6.2 Ecological Consequences of Forest Degradation on Biodiversity, Carbon Storage, and Climate Regulation

Forests play a fundamental role in maintaining ecological stability at both local and global scales. They provide habitat for more than 80 percent of terrestrial species, store approximately 45 percent of terrestrial carbon, and regulate hydrological and climatic systems. Yet, beyond outright deforestation, forest degradation selective logging, fires, fragmentation, overharvesting, and land-use pressure poses a subtler but equally significant threat. Degradation diminishes forest quality, ecosystem services, and resilience without necessarily eliminating tree cover. This essay assesses the ecological consequences of forest degradation, focusing on three interrelated domains: biodiversity, carbon storage, and climate regulation. Degradation reduces habitat quality by opening canopies, fragmenting landscapes, and altering microclimates. Species reliant on intact forest interiors, such as large mammals, canopy specialists, and old-growth-dependent birds, are disproportionately affected. Edges created by logging roads and clearing increase exposure to invasive species, hunting pressure, and predation. Over time, species richness declines and ecological communities become homogenized, dominated by generalist or disturbance-tolerant species. Biodiversity loss extends beyond species counts to the breakdown of ecological processes. The disappearance of pollinators, seed dispersers, and apex predators undermines regeneration cycles and trophic structures. For example, reduced dispersal of large-seeded trees following the decline of primates and birds in degraded tropical forests leads to shifts in species composition toward less carbon-dense species. This has cascading impacts on both ecosystem resilience and carbon storage. Degraded forests are less able to respond to climate shocks, disease outbreaks, or pest infestations. Lower species and genetic diversity reduce adaptive capacity, making ecosystems more vulnerable to collapse. In contrast, intact forests provide greater ecological insurance against environmental variability.

Selective logging and fuelwood collection reduce aboveground biomass densities. Even moderate degradation can result in 20–50 percent reductions in carbon stocks compared to intact forests. Unlike deforestation, which releases carbon immediately, degradation often creates chronic emissions over decades, steadily weakening the global carbon sink. Disturbance of soils through logging, compaction, and drainage accelerates soil carbon loss. In peatland forests of Indonesia and the Congo Basin, degradation and fire release massive quantities of greenhouse gases, turning these ecosystems into persistent carbon sources. Repeated disturbance lowers the capacity of forests to recover carbon stocks. Secondary forests may regrow, but their composition is often less diverse and less carbon-dense, reducing long-term sequestration potential. This undermines global efforts to meet climate mitigation targets through forest-based strategies such as REDD+ (Reducing Emissions from Deforestation and Forest Degradation). Forests regulate local and regional climates through evapotranspiration, cloud formation, and rainfall recycling. Degradation reduces canopy cover and transpiration, altering precipitation patterns. In the Amazon, cumulative forest degradation has already diminished rainfall, raising the risk of drought and savanna transitions. Degraded forests, with drier understories and fragmented landscapes, are more prone to fires. Fire events not only release large carbon emissions but also further degrade habitats, creating a feedback loop of ecosystem instability. Similarly, fragmented forests are more vulnerable to storm damage, pests, and disease outbreaks.

Figure 3. Adapting forest plantations to climate change: strategies, challenges, and opportunities in Malaysia

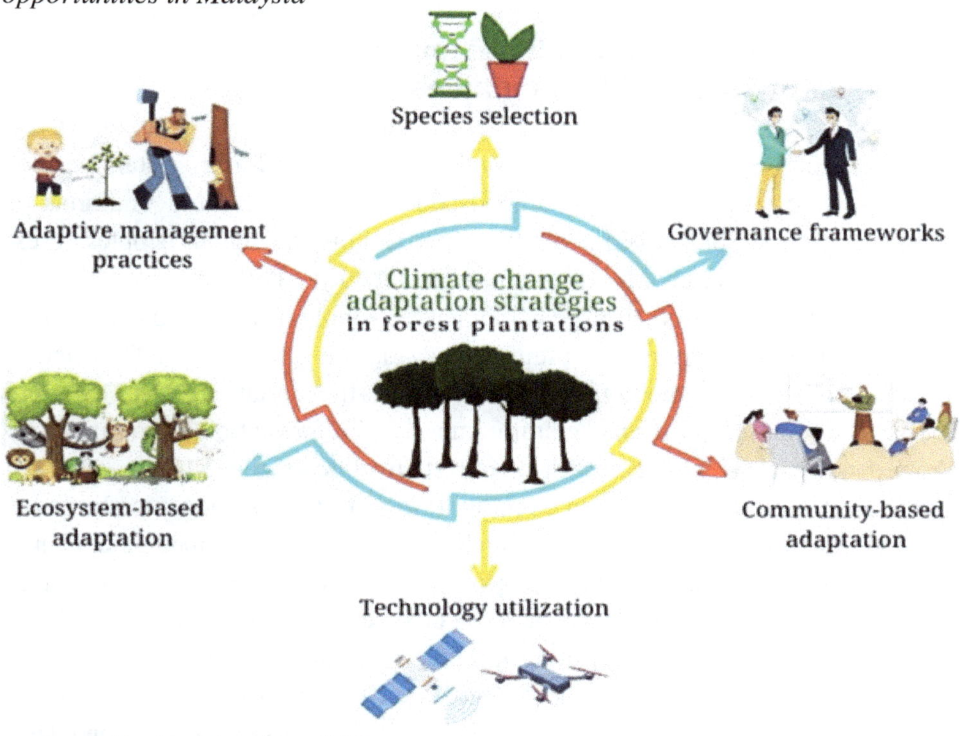

Source: Nurul Asyiqin and Ong (2025)

The weakening of tropical forest carbon sinks undermines global climate stability. Recent evidence suggests that parts of the Amazon now emit more carbon than they absorb due to degradation and fires, signaling a potential tipping point. If unchecked, such transitions could shift forests from climate regulators to amplifiers of global warming. The consequences of forest degradation are deeply interconnected. Biodiversity supports carbon storage: diverse forests with varied species compositions are more productive and resilient, maintaining higher biomass. Conversely, carbon loss accelerates climate change, which feeds back into biodiversity stress through rising temperatures, altered rainfall, and more frequent disturbances. Climate regulation, in turn, is a function of both carbon storage and ecological integrity. The degradation of any one domain thus undermines the others, producing cascading risks at multiple scales. Forest degradation is an insidious driver of ecological decline, reducing biodiversity, diminishing carbon storage, and destabilizing climate regulation. While deforestation has long been recognized as a critical environmental issue, degradation poses equally severe risks, often hidden behind the illusion of continued tree cover. Protecting intact forests, strengthening community forest management, and reducing pressures from illegal logging and unsustainable land use are essential to maintaining the ecological functions upon which both local livelihoods and global climate stability depend. In the face of escalating environmental crises, safeguarding forest quality is as urgent as preserving forest area.

6.3 Socio-Economic and Cultural Impacts of Deforestation on Local Communities

Forests constitute one of the most vital ecosystems on Earth, sustaining biodiversity while supporting the livelihoods, identities, and cultural traditions of millions of people. For local communities—particularly Indigenous peoples, forest dwellers, and rural households—forests provide not only material resources but also intangible cultural values that shape worldviews, cosmologies, and social cohesion. Yet, deforestation driven by agricultural expansion, illegal logging, infrastructure projects, and global commodity demand undermines these relationships. This essay assesses the socio-economic and cultural impacts of deforestation on communities who depend on forest resources, with attention to livelihood vulnerability, economic marginalization, social disruption, and cultural erosion. Deforestation directly diminishes the availability of forest products such as timber, fuelwood, medicinal plants, fruits, and bush meat. In rural and Indigenous societies, these resources constitute critical components of subsistence and cash incomes. Loss of forest cover reduces access to food security buffers, pushing households into precarious dependence on external markets and wage labor. In regions such as the Amazon and Southeast Asia, forest clearance for soy, palm oil, or cattle ranching displaces traditional land uses, creating economic marginalization for smallholders and pastoralists. The economic benefits of deforestation often accrue to external actors—logging companies, agribusinesses, or state elites—while costs are borne disproportionately by local communities. This exacerbates inequality and entrenches cycles of poverty. Where land tenure is insecure, communities may lose traditional territories without compensation. Furthermore, the influx of migrants into deforested frontiers often generates land conflicts, resource scarcity, and labor exploitation.

Forest clearance can lead to physical displacement of communities, either through land grabs, mining projects, or conservation enclosures. Forced migration severs communities from ancestral landscapes and social safety networks, resulting in economic vulnerability. Young people frequently migrate to urban centers in search of employment, eroding intergenerational transmission of ecological knowledge and cultural traditions. Forests are not merely physical resources; they are living archives of Indigenous knowledge and cosmologies. Traditional ecological knowledge concerning plant uses, forest management, and spiritual practices—depends on continuous interaction with intact ecosystems. Deforestation reduces opportunities for knowledge practice and transmission, eroding cultural resilience. For many forest peoples, forests are sacred spaces that embody ancestral spirits, origin stories, and rituals. Sacred groves, burial grounds, and ceremonial landscapes are often the first to be degraded by commercial exploitation. The destruction of these sites weakens collective identity, diminishing spiritual relationships with land and disrupting cultural continuity. Cultural ties to forests are often contested when external interests dominate land use. Communities may experience cultural dislocation when denied participation in decision-making processes about forest management. This fuels tensions between local groups and external actors, as well as intra-community divisions when individuals are co-opted into commercial exploitation. The erosion of cultural cohesion undermines community governance and adaptive capacity.

Figure 4. Deforestation Impacts

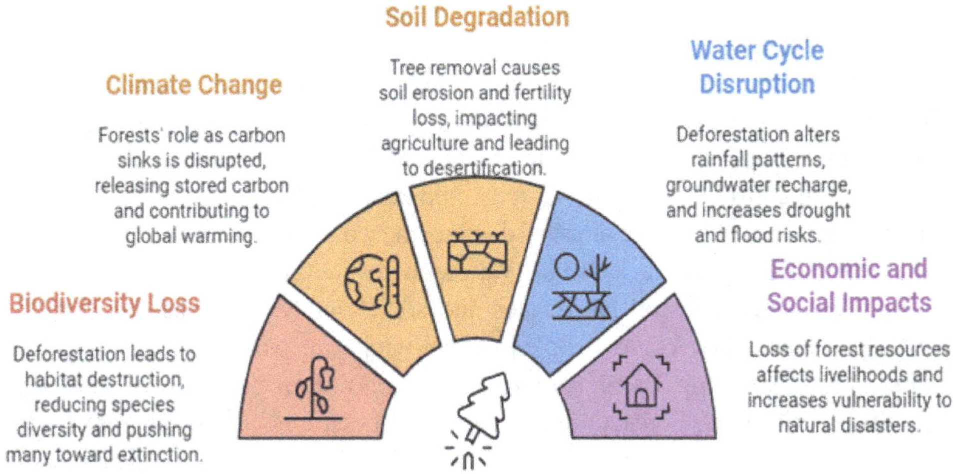

Source: Tracextech & Tracextech, 2024

The socio-economic and cultural dimensions of deforestation are deeply interwoven. Economic dispossession weakens the material basis for cultural practices, while cultural erosion undermines community solidarity necessary for economic resilience. For example, when communities lose access to medicinal plants, they not only face health insecurity but also the decline of ritual and spiritual practices tied to healing traditions. Similarly, livelihood shocks may compel households to engage in unsustainable resource extraction, perpetuating a cycle of forest degradation and cultural loss. Deforestation is more than an ecological crisis; it is a profound socio-economic and cultural disruption for communities who depend on forests for their survival and identity. While forests provide timber and carbon storage for the global economy, their destruction erodes local livelihoods, deepens inequalities, and fragments cultural traditions. Any strategy to address deforestation must therefore transcend conservation and carbon metrics to include the protection of community rights, tenure security, and cultural survival. Integrating local voices, strengthening community forest management, and respecting Indigenous knowledge systems are essential to safeguarding not only the world's forests but also the societies and cultures that give them life and meaning.

6.4 Sustainable Strategies and Policy Recommendations for Protecting, Restoring, and Managing Forests as Living Systems for Future

Forests are critical living systems that sustain biodiversity, regulate climate, and provide livelihoods for millions of people worldwide. They act as ecological infrastructure supplying clean water, storing carbon, and offering cultural and spiritual value. Yet, they face unprecedented threats from deforestation, illegal logging, climate change, and unsustainable land-use practices. Protecting forests is therefore not only a matter of conservation but of safeguarding planetary health and intergenerational equity. This essay explores sustainable strategies and policy recommendations to protect, restore, and manage forests as dynamic living systems for future generations. Strong governance is essential to forest protection. National laws against illegal logging, mining, and encroachment need effective enforcement mechanisms, supported by monitoring technologies such as satellite imagery and drones. International agreements such as the UN REDD+ program and the Convention on Biological Diversity (CBD) offer global frameworks that countries must domesticate and enforce. Protected areas, when effectively managed, remain a cornerstone of forest conservation. However, "fortress conservation" alone is insufficient. Recognizing Indigenous peoples' and local communities' land rights has proven to yield better ecological outcomes while ensuring social justice. Community-managed forests in Nepal and Latin America illustrate how co-management can balance ecological integrity with local livelihoods.

Figure 5. Sustainable Development Goals (SDGs) Integration for Protecting, Restoring, and Managing Forest

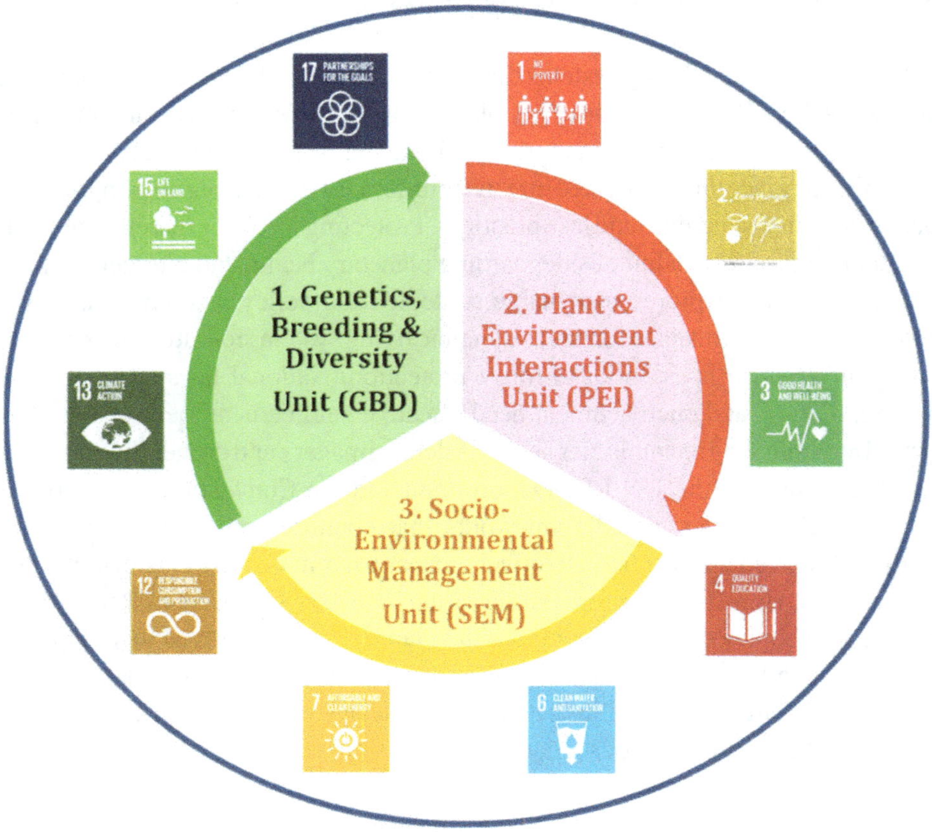

Source: Koutika et al. (2022)

Deforestation is often driven by global demand for commodities like palm oil, soy, timber, and beef. Enforcing sustainable certification schemes (e.g., Forest Stewardship Council, RSPO) and ensuring supply chain traceability can reduce market incentives for deforestation. Restoration must go beyond tree planting. Ecological restoration emphasizes the recovery of native species, soil fertility, and hydrological cycles. Programs such as the Bonn Challenge aim to restore 350 million hectares of degraded land by 2030. However, monoculture plantations often undermine biodiversity and carbon sequestration potential; diverse native forests should be prioritized. Forest restoration can be embedded in climate adaptation and mitigation policies. Reforestation, afforestation, and assisted natural regeneration help sequester carbon while reducing disaster risks such as floods, landslides, and droughts. Incorporating forests into green infrastructure planning enhances urban resilience as well. Restoration requires long-term investment. Green bonds, carbon markets, and public–private partnerships provide avenues for financing. For instance, carbon offset schemes can channel funds into forest recovery while incentivizing emissions reduction globally. Forests must be managed as complex adaptive systems rather than resource stockpiles. Adaptive management integrates scientific knowledge with traditional ecological knowledge (TEK), allowing for flexible responses to environmental changes. Involving local communities in decision-making fosters stewardship and cultural continuity. Forests should not be treated as isolated ecosystems but integrated into national development agendas. Agroforestry systems, sustainable timber harvesting, and eco-tourism illustrate how forests can contribute to economic growth without compromising ecological integrity. Forest management strategies must align climate mitigation goals with biodiversity conservation. Intact and restored forests are more resilient to climate stressors, offering dual benefits for both carbon storage and ecosystem stability.

7. CONCLUSION

The notion of the *stolen forest* captures not only the physical appropriation of land through deforestation, illegal logging, and encroachment, but also the dispossession of ecological functions, cultural identities, and intergenerational rights bound to forest ecosystems. To "breathe inside a forest" is to recognize forests as living systems that sustain climate regulation, biodiversity, and human survival, while simultaneously acknowledging the suffocation imposed by human exploitation. This paradox underscores that the crisis of stolen forests is not solely environmental but deeply socio-political and ethical. As forests are degraded, communities lose more than resources—they lose spiritual connections, traditional ecological knowledge, and cultural continuity. Ecologically, the theft manifests in collapsing biodiversity, diminished carbon sinks, and destabilized climate systems. Socio-economically, it fuels inequality, disempowerment, and ecological injustice, disproportionately affecting Indigenous peoples and marginalized communities who depend most on these ecosystems. Therefore, the urgency of addressing stolen forests lies in reframing them as spaces of relational interdependence rather than commodities. Protecting and restoring them requires integrating scientific monitoring, participatory governance, and cultural recognition into policy and practice. To breathe inside a forest is to breathe with it to honor its agency, rhythms, and capacity to regenerate. Ultimately, securing the future of forests is inseparable from securing the possibility of future human life, culture, and planetary balance. Forests are living systems dynamic, complex, and interdependent with human well-being. Protecting, restoring, and managing them requires holistic strategies that balance ecological integrity with social and economic needs. Sustainable policies must not only prevent further destruction but also revive degraded ecosystems and empower communities. The challenge of forest governance is therefore not merely ecological but deeply ethical, requiring societies to honor obligations to future generations. By treating forests as more than resources as life-support systems we can secure their capacity to breathe and sustain the planet for centuries to come.

8. POLICY RECOMMENDATIONS

In response to the increase of deforestation, illegal logging and forest degradation. This chapter is proposing policy recommendation that aligns with Sustainable Development Goals (SDG). The goal is to promote holistic framework where forest conservation must be a part of a nation's development, community empowerment and global climate resilience.

Strengthen Legal Frameworks and Enforcement: Develop and enforce forest protection laws with strict penalties for illegal logging, land encroachment and timber laundering.

Recognize and Secure Ingenious Community Forest Rights: Legally Recognize Indigenous and ensure land tenure security to prevent land grabbing.

Promote Sustainable and Transparent Supply Chain: Implement Deforestation free-certification, blockchain-based traceability and corporate accountability mechanism to eliminate illegal logging.

Invest in Ecological Restoration and Nature-Based Solutions: Prioritize native ecosystem restoration, support community-led reforestation and link verified recovery efforts to carbon credit markets.

Mobilize Green and Climate Finance: Establish blended National Forest Restoration Funds, expand access to green bonds and incentivize agroforestry through fiscal support.

Empower Local Communities through Education and Technology: Train local stakeholders in forest monitoring, promote eco-tourism enterprises and deploy digital tools for participatory forest protection.

Foster Knowledge Co-Production and Global Cooperation: Strengthen collaborating among scientist, policymakers and indigenous groups through shared research hubs and regional forest resilience partnerships.

REFERENCES

Achieng, A. O., Arhonditsis, G. B., Mandrak, N. E., Febria, C. M., Opaa, B., Coffey, T. J., Masese, F. O., Irvine, K., Ajode, Z. M., Barasa, J. E., Obiero, K., & Kaunda-Arara, B. (2023). Monitoring biodiversity loss in rapidly changing Afrotropical ecosystems: An emerging imperative for governance and research (arXiv preprint arXiv:2303.13984). https://doi.org//arXiv.2303.13984.DOI: 10.48550

Adom, R. K., Reid, M., Afuye, G. A., & Simatele, M. D. (2024). Assessing the Implications of Deforestation and Climate Change on Rural Livelihood in Ghana: A Multidimensional Analysis and Solution-Based Approach. *Environmental Management, 74*(6), 1124–1144. DOI: 10.1007/s00267-024-02053-6

Ahammad, R., Hossain, M. K., Sobhan, I., Hasan, R., Biswas, S. R., & Mukul, S. A. (2023). Social-ecological and institutional factors affecting forest and landscape restoration in the Chittagong Hill Tracts of Bangladesh. *Land Use Policy, 125*, 106478. DOI: 10.1016/j.landusepol.2022.106478

Al-Bawwat, A. K., Jurado, F., Gomaa, M. R., & Cano, A. (2023). Availability and the possibility of employing wastes and biomass materials energy in Jordan. *Sustainability (Basel), 15*(7), 5879. DOI: 10.3390/su15075879

Asabere, S. B. (2020). *Urbanisation, Land Use and Soil Resource: Spatio-Temporal Analyses of Trends and Environmental Effects in Two Metropolitan Regions of Ghana (West Africa)* (Doctoral dissertation, Dissertation, Göttingen, Georg-August Universität, 2020).

Baral, S., Subedi, S., Ford, R., Rai, R. K., Khanal Chhetri, B. B., Ojha, S., Tiwari, K. R., Kafle, G., Puri, L., & Rai, D. (2025). Resilience pathways through community-based forest management for navigating the triple planetary crisis. *Trees, Forests and People*, *22*, 100972. DOI: 10.1016/j.tfp.2025.100972

Bélisle, A. C., Mantyka-Pringle, C., Boulanger, Y., Croteau, B., Wapachee, A., Drapeau, L.-J., Desrochers, M., & Asselin, H. (2025). Indigenous knowledge, forest landscape modeling, and the cumulative effects of environmental changes. *Ecological Applications*, *35*(2), e70053. DOI: 10.1002/eap.70053

Bhatt, H., Jugran, H. P., & Pandey, R. (2024). Cultural ecosystem services nexus with socio-cultural attributes and traditional ecological knowledge for managing community forests of Indian Western Himalaya. *Ecological Indicators, 166*, 112379. DOI: 10.1016/j.ecolind.2024.112379

Bonan, G. B. (2008). Forests and climate change: forcings, feedbacks, and the climate benefits of forests. *science, 320*(5882), 1444-1449.

Cabernard, L., Pfister, S., & Hellweg, S. (2024). Biodiversity impacts of recent land-use change driven by increases in agri-food imports. *Nature Sustainability*, *7*(11), 1512–1524. DOI: 10.1038/s41893-024-01433-4

Chiriacò, M. V., Galli, N., Santini, M., & Rulli, M. C. (2024). Deforestation and greenhouse gas emissions could arise when replacing palm oil with other vegetable oils. *The Science of the Total Environment*, *914*, 169486. DOI: 10.1016/j.scitotenv.2023.169486

CIWEM. (2025). The Sundarbans: Battling climate change in the world's largest mangrove forest. Chartered Institution of Water and Environmental Management. https://www.ciwem.org/news/the-sundarbans-battling-climate-change-in-worlds-largest-mangrove-forest

Damnyag, L., Bampoh, A. A., Obiri, B. D., & Oduro, K. A. (2024). Assessment of revenue loss from illegal logging in Ghana's informal timber sector. *International Forestry Review*, *26*(1), 16–28. DOI: 10.1505/146554824838457907

Dingkuhn, E. L., O'Sullivan, L., Grady, C. A., de Klerk, E., & Schulte, R. P. O. (2025). Land-use governance: The interplay of social, market, and policy drivers – A global systematic review. *Earth System Governance*, *25*, 100275. DOI: 10.1016/j.esg.2025.100275

FAO. (2020). Global Forest Resources Assessment 2020: Main report. Rome: Food and Agriculture Organization of the United Nations. https://www.fao.org/interactive/forest-resources-assessment/2020/en/

FAO. (2022). The State of the World's Forests 2022: Forest pathways for green recovery and building inclusive, resilient and sustainable economies. Rome: Food and Agriculture Organization. https://openknowledge.fao.org/items/4c8bd12f-d6b8-4755-a82f-1284c41bf012

FAO. (2024a). The State of the World's Forests 2024: Forest-sector innovations towards a more sustainable future. Rome: Food and Agriculture Organization of the United Nations. https://openknowledge.fao.org/items/ec487897-97b5-43ec-bc2e-5ddfc76c8e85

FAO. (2024b). *Summary for policymakers: The state of the world's forests 2024*. FAO., DOI: 10.4060/cd1211en

Forest Declaration Assessment. (2023). Forest Declaration Assessment 2024: Forests under fire – Global assessment of commitments to halt deforestation by 2030. https://forestdeclaration.org/resources/forest-declaration-assessment-2024

Fragoso-Medina, M. del C., Navarrete-Segueda, A., Ceccon, E., & Martínez-Ramos, M. (2024). Effects of the forests–agriculture conversion on the availability and diversity of forest products in a neotropical rainforest region. *Trees, Forests and People*, *15*, 100481. DOI: 10.1016/j.tfp.2023.100481

Global Forest Watch. (2025). Indigenous and community forests indicators: Global Forest Review 2024–2025. World Resources Institute.https://gfr.wri.org/social-governance-issues-indicators/indigenous-community-forests

Global Witness. (2024). The Cerrado crisis: Brazil's deforestation frontline. https://www.globalwitness.org/en/campaigns/forests/the-cerrado-crisis-brazils-deforestation-frontline

Hallaj, Z., Bijani, M., Karamidehkordi, E., Yousefpour, R., & Yousefzadeh, H. (2024). Forest land use change effects on biodiversity, ecosystem services and human well-being: A systematic analysis. *Environmental and Sustainability Indicators*, *23*, 100445. DOI: 10.1016/j.indic.2024.100445

Intergovernmental Panel on Climate Change (IPCC). (2023). Chapter 7: Agriculture, forestry and other land uses (AFOLU). In Climate change 2022: Mitigation of climate change. Contribution of Working Group III to the Sixth Assessment Report of the Intergovernmental Panel on Climate Change (pp. 747–860). Cambridge University Press. https://doi.org/DOI: 10.1017/9781009157926.009

Islam, M. Z. (2025). A contemporary analysis of policy implementation for coastal mangrove forest management in the Bangladesh Sundarbans. *Marine Policy*, *171*, 106460. DOI: 10.1016/j.marpol.2024.106460

John, A., Allotey, S., Koebe, T., Tyukavina, A., & Weber, I. (2025). A global dataset of location data integrity-assessed reforestation efforts. https://doi.org//arXiv.2508.11349.DOI: 10.48550

Khosravi Mashizi, A., & Escobedo, F. (2025). The role of traditional ecological knowledge and ecosystem quality in managing ecosystem services. *Scientific Reports*, *15*(1), 31510. DOI: 10.1038/s41598-025-17611-z

Koutika, L. S., Matondo, R., Mabiala-Ngoma, A., Tchichelle, V. S., Toto, M., Madzoumbou, J. C., & Nzila, J. D. D. (2022). Sustaining forest plantations for the United Nations' 2030 Agenda for sustainable development. *Sustainability (Basel)*, *14*(21), 14624. DOI: 10.3390/su142114624

Mitchard, E. T. A., Carstairs, H., Cosenza, R., Saatchi, S. S., Funk, J., Nieto Quintano, P., Brade, T., McNicol, I. M., Meir, P., Collins, M. B., & Nowak, E. (2023). Serious errors impair an assessment of forest carbon projects: A rebuttal of West et al. (2023). arXiv Preprint arXiv:2312.06793. https://doi.org//arXiv.2312.06793 ,DOI: 10.48550

Mongabay. (2023). Shrinking civil space and persistent logging: 2023 in review in Southeast Asia. https://news.mongabay.com/2023/12/shrinking-civil-space-and-persistent-logging-2023-in-review-in-southeast-asia/

Mongabay. (2024). The year in tropical rainforests: 2024. https://news.mongabay.com/2024/12/the-year-in-tropical-rainforests-2024/

Nellemann, C., & INTERPOL Environmental Crime Programme. (2012). Green carbon, black trade: Illegal logging, tax fraud and laundering in the world's tropical forests (Rapid Response Assessment). United Nations Environment Programme / GRID-Arendal. https://wedocs.unep.org/handle/20.500.11822/8030

Nurul Asyiqin, A. B., & Ong, K. H. (2025). Adapting forest plantations to climate change: Strategies, challenges, and opportunities in Malaysia. *New Forests*, *56*(3), 35. DOI: 10.1007/s11056-025-10105-w

Pan, Y., Birdsey, R. A., Phillips, O. L., Houghton, R. A., Fang, J., Kauppi, P. E., Keith, H., Kurz, W. A., Ito, A., Lewis, S. L., Nabuurs, G.-J., Shvidenko, A., Hashimoto, S., Lerink, B., Schepaschenko, D., Castanho, A., & Murdiyarso, D. (2024). The enduring world forest carbon sink. *Nature*, *631*(8021), 563–569. DOI: 10.1038/s41586-024-07602-x

Poudel, T. R., Aryal, P. C., Khan, M. T., Roberts, N. J., Poudel, M., & Shrestha, D. P. (2025). Forest structure, diversity, and regeneration in a community-managed forest of Nepal: A model for carbon sequestration and sustainable management. *Plant-Environment Interactions*, *6*(2), e70044. DOI: 10.1002/pei3.70044

Rahman, K. S., Dana, N. H., Rahman, M. M., Mondal, H. S., Chen, L., & Islam, M. N. (2025). Degradation of mangrove forests in the Sundarbans: An assessment based on perspectives of mangrove resource collectors using the DPSIR framework. *Trees, Forests and People*, *19*, 100769. DOI: 10.1016/j.tfp.2024.100769

Reuters. (2025). Amazon fires drive unprecedented global forest loss in 2024, report says. https://www.reuters.com/sustainability/cop/amazon-fires-drive-unprecedented-global-forest-loss-2024-report-says-2025-05-21

Roy, S. K., Mojumder, P., Chowdhury, M. A. A., & Hasan, M. M. (2025). Evaluating mangrove forest dynamics and fragmentation in Sundarbans, Bangladesh using high-resolution Sentinel-2 satellite images. *Global Ecology and Conservation, 58*, e03493. DOI: 10.1016/j.gecco.2025.e03493

Silva-Junior, C. H. L., Silva, F. B., Arisi, B. M., Mataveli, G., Pessôa, A. C. M., Carvalho, N. S., Reis, J. B. C., Silva Júnior, A. R., Motta, N. A. C. S., Moreira e Silva, P. V., Ribeiro, F. D., Siqueira-Gay, J., Alencar, A., Saatchi, S., Aragão, L. E. O. C., Anderson, L. O., & Melo, M. (2023). Brazilian Amazon indigenous territories under deforestation pressure. *Scientific Reports, 13*(1), 5851. Advance online publication. DOI: 10.1038/s41598-023-32746-7

Sims, M. (2025, June 12). *New data shows what's driving forest loss around the world.* World Resources Institute. https://www.wri.org/insights/forest-loss-drivers-data-trends

Sylvester, J. M., Gutiérrez-Zapata, D. M., Pérez-Marulanda, L., Vanegas-Cubillos, M., Bruun, T. B., Mertz, O., & Castro-Núñez, A. (2024). Analysis of food system drivers of deforestation highlights foreign direct investments and urbanization as threats to tropical forests. *Scientific Reports, 14*(1), 15179. Advance online publication. DOI: 10.1038/s41598-024-65397-3

Takahashi, T., & Soma, T. (2025). Local ecological knowledge, reforestation, and vegetation recovery: A remote sensing-based assessment in Gandaki Province, Nepal. *Mountain Research and Development, 45*(1), 00015. Advance online publication. DOI: 10.1659/mrd.2024.00015

Tonouéwa, J. F. M., Biaou, S. S. H., Assèdé, E. S. P., Agossou, H., & Balagueman, R. O. (2024). Timber traceability: Determining effective methods to combat illegal logging in Africa: A review. *Trees, Forests and People, 18*, 100709. DOI: 10.1016/j.tfp.2024.100709

Tracextech, & Tracextech. (2024, April 30). ASEAN deforestation regulations and sustainable supply chains in Southeast Asia. Blockchain for Food Safety, Traceability and Supplychain Transparency. https://tracextech.com/asean-deforestation-regulations-in-southeast-asia/

Tran, T. M. A., Gagnon, V. S., & Schelly, C. (2024, October 8). A review of traditional ecological knowledge in resilient livelihoods and forest ecosystems: Lessons for restoration sciences and practices. Michigan Technological University. https://doi.org/.DOI: 10.1080/14728028.2024.2408725

Urton, G. (2024). *Andean cosmos*. Ediciones El Lector.

USDA Southern Research Station. (2025). Recent land-use change and forest conversion to pasture in the Southeastern U.S. https://www.fs.usda.gov/pnw/pubs/gtr802/Vol1/pnw_gtr802vol1_alig.pdf

Ward, M., Ashman, K., Lindenmayer, D. B., Legge, S., Kindler, G., Cadman, T., Fletcher, R., Whiterod, N., Lintermans, M., Zylstra, P., Stewart, R., Thomas, H., Blanch, S., & Watson, J. E. (2024). Shifting baselines clarify the impact of contemporary logging on forest-dependent threatened species. *Conservation Science and Practice*, *6*(9), e13185. DOI: 10.1111/csp2.13185

Wear, D. N., & Wibbenmeyer, M. (2024). Land-use change, no-net-loss policies, and effects on carbon dioxide removals. *Carbon Balance and Management*, *19*(1), 40. DOI: 10.1186/s13021-024-00287-9

Werdel, T. J., Matarrita-Cascante, D., & Lucero, J. E. (2024). State of Traditional Ecological Knowledge in the wildlife management profession. *The Journal of Wildlife Management*, *88*(5), e22579. DOI: 10.1002/jwmg.22579

World Resources Institute. (2025). Fires drove record-breaking tropical forest loss in 2024. Global Forest Review. https://gfr.wri.org/latest-analysis-deforestation-trends

World Resources Institute. (2025). New data shows what's driving forest loss around the world (supply chain & commodity links). https://www.wri.org/insights/forest-loss-drivers-data-trends

Chapter 3
Threats and Challenges of Deforestation, Climate Change, and Socio-Cultural Disruption

Niamjit Das
https://orcid.org/0000-0001-7619-8306
Center for Natural Resource Studies, Dhaka, Bangladesh & Department of Forestry and Environmental Science, School of Agriculture and Mineral Sciences, Shahjalal University of Science and Technology, Sylhet, Bangladesh

ABSTRACT

Forest ecosystems are facing an unprecedented convergence of threats from deforestation, climate change, and socio-cultural disruption. This chapter investigates key drivers of forest loss—agriculture, logging, and infrastructure—and quantifies their impacts on fragmentation and biodiversity decline. It analyzes how climate change amplifies these pressures via altered climates, intensified wildfires, and pest outbreaks, disrupting species composition and ecosystem function. Socio-cultural impacts are explored, including displacement of Indigenous communities, erosion of traditional land practices, and shifting values undermining local stewardship. By integrating ecological and social dimensions, the chapter highlights the complex interplay of these threats and advocates holistic conservation strategies. Adaptive management, participatory governance, cross-scale collaboration, and innovative financing are emphasized as pathways for enhancing resilience and addressing global forest challenges.

DOI: 10.4018/979-8-3373-3992-4.ch003

INTRODUCTION

Forests stand at the crossroads of planetary health and human well-being. Covering nearly one-third of Earth's terrestrial surface and storing more than 650 gigatons of carbon, they regulate climate, sustain biodiversity, and underpin the livelihoods of hundreds of millions of people (Sarre, 2020). Yet despite this critical role, global forest loss has continued at an alarming pace: between 1990 and 2020, more than 420 million hectares of forest disappeared, largely due to agricultural expansion, logging, infrastructure development, and urban sprawl (Hansen et al., 2013; Curtis et al., 2018). This sustained deforestation undermines carbon sequestration, erodes ecosystem integrity, and reduces resilience to external shocks. Understanding these multifaceted threats requires an integrated perspective that spans biophysical processes, socio-economic drivers, governance systems, and cultural dimensions (Das, 2025a).

Figure 1. Threats and challenges for deforestation, climate change, and socio-cultural disruption.

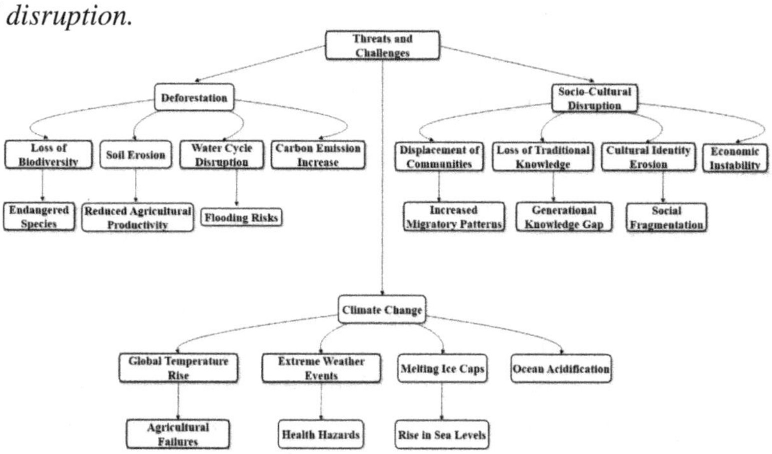

Human transformation of forest landscapes has accelerated in the past half-century. Industrial-scale agriculture now accounts for nearly 40 percent of tropical deforestation, with commodity chains linking remote ecosystems to global markets (Curtis et al., 2018). Smallholder clearing for subsistence farming and fuelwood adds further complexity, producing mosaics of degraded patches and remnant fragments (Kaimowitz & Angelsen, 1998). Expanding road networks and peri-urban growth intensify edge effects—altered microclimates, invasive species, and habitat fragmentation—that degrade interior forest ecosystems. In selective-logging zones, unsustainable extraction may remove as much as 80 percent of aboveground biomass, compromising regeneration capacity and exposing forests to further conversion pressures (Kuck et al., 2021).

Overlaying this deforestation crisis is the accelerating pace of climate change. Rising greenhouse gas concentrations have already pushed atmospheric CO_2 to record highs, altering energy balances and intensifying weather extremes. Forests, long viewed as carbon sinks, face a shifting baseline where sink strength may decline as temperatures climb and precipitation patterns shift (Bonan, 2008). Climate feedback loops—such as peatland desiccation, permafrost thaw, and fire-carbon cycles—threaten to shift forests from net carbon sinks to net carbon sources, amplifying warming trends (IPCC, 2022).

Climate-driven stressors manifest in heightened drought frequency, intensified wildfire regimes, and novel pest outbreaks. In Mediterranean-type biomes and boreal regions alike, multiyear droughts have diminished tree growth, increased mortality rates, and predisposed landscapes to megafires (Bowman et al., 2011). Warming extends the active seasons for insects and pathogens, facilitating range expansions of forest pests that decimate native stands. Extreme precipitation events, meanwhile, trigger soil erosion and landslides in steep terrain, undermining water regulation and sediment retention services.

Biodiversity loss compounds these abiotic pressures. Tropical forests harbor more than half of terrestrial species, many with narrow habitat affinities and limited dispersal capacities (Myers et al., 2000). Fragmentation isolates populations, creates so-called "extinction debts," and disrupts mutualistic interactions such as pollination and seed dispersal (Wearn et al., 2012). The loss of keystone taxa—from large-bodied frugivores to canopy-specialist epiphytes—can trigger cascading effects that reshape community composition for decades. As functional diversity wanes, forests lose their capacity to buffer against environmental fluctuations, jeopardizing both ecological stability and human dependence.

Beyond ecological ramifications, deforestation and climate change inflict profound socio-cultural disruptions. Indigenous peoples and local communities have stewarded forests for millennia, weaving intricate knowledge systems that sustain resource use, governance, and spiritual practice (Ostrom, 2015). Yet land-use policies often marginalize customary tenure, eroding governance structures that once balanced use and conservation. Displacement from ancestral territories severs cultural ties, undermines food sovereignty, and exacerbates social inequalities. The resulting loss of biocultural diversity diminishes adaptive capacity and dissolves the social fabric critical for collective action. Gendered impacts are particularly acute, as women often bear disproportionate burdens in fuelwood collection, food security, and community governance, while youth—emerging as critical actors in forest stewardship—remain under-recognized in policy frameworks.

Land tenure conflicts further intensify displacement pressures. In regions where legal frameworks remain ambiguous, competing claims by agribusiness, mining interests, and conservation projects breed tension (Nasi et al., 2008). Communities frequently lack the resources to defend their rights in formal courts, resulting in forced evictions or coerced agreements that favor external actors. These dynamics fuel cycles of poverty, illicit land grabbing, and violent confrontation, impeding sustainable management efforts and breeding mistrust between stakeholders.

Governance responses to these intertwined threats are often fragmented and reactive. National policies may prioritize short-term economic gains over long-term ecosystem health, while international agreements struggle with enforcement gaps and perverse incentives (Lambin & Meyfroidt, 2011). Participatory frameworks that engage local knowledge, gender-inclusive policies, and equitable benefit-sharing have shown promise; however, scaling such models remains challenging. Strengthening institutions to integrate ecological science, social justice, and transparent accountability is imperative for forging resilient pathways.

Resilience emerges at the nexus of socio-ecological interactions. Landscapes managed under community governance frequently exhibit lower deforestation rates, higher biodiversity indices, and more robust carbon stocks than strictly protected or privately owned counterparts (Balvanera et al., 2019). Adaptive co-management, which blends traditional ecological knowledge with scientific monitoring, fosters learning cycles that can anticipate and mitigate emerging threats. Such approaches underscore the principle that forest conservation cannot be decoupled from local livelihoods, cultural integrity, or intergenerational equity.

This chapter examines the complex web of deforestation, climate change, and socio-cultural disruption through a socio-ecological lens. Figure 1 provides a visual synthesis of the chapter's core themes, illustrating how ecological, climatic, and cultural disruptions are interdependent rather than isolated. The chapter traces historical trajectories of forest loss, dissects key climatic feedbacks, and illuminates the lived experiences of communities on the front lines. By synthesizing insights from ecology, climate science, governance, and political ecology (Das, 2025a), it identifies leverage points for transformative, equity-centered action. Subsequent sections will explore mitigation strategies, adaptation frameworks, and governance innovations that hold promise for sustaining forests and the people who depend upon them.

DEFORESTATION: SCOPE, DRIVERS, AND ECOLOGICAL IMPACTS

Scope of Deforestation

Deforestation refers to the permanent conversion of forested land to non-forest uses, distinguishing it from temporary disturbances such as selective logging or natural fires. The crisis is intensifying: in 2024 alone, 6.7 million hectares of tropical primary rainforest were lost—the highest in at least two decades—largely due to massive fires exacerbated by climate change and El Niño. This loss released 3.1 gigatonnes of CO_2, equivalent to India's annual fossil fuel emissions (WRI, 2025). The permanence of such conversion, coupled with accelerating climatic feedbacks, underscores the urgency of integrated strategies that address both ecological processes and human dimensions.

Proximate Drivers of Deforestation

The immediate causes of deforestation are diverse but interconnected. They include agricultural expansion, logging, infrastructure development, and urbanization, as summarized in Table 1.

Table 1. Drivers of Deforestation.

Driver Type	Specific Driver	Scale/Region	Mechanism	Reference
Proximate	Agricultural expansion	Amazon, Congo, Southeast Asia	Clearing for cattle, oil palm, smallholder crops	Curtis et al., 2018
Proximate	Logging (legal & illegal)	Global tropics	Selective and clear-cut extraction opening forest fronts	Gibbs et al., 2002
Proximate	Infrastructure development	Amazon, Borneo	Road building and hydroelectric dams fragment habitats	Gibbs et al., 2002
Underlying	Commodity demand & prices	Global	Market driven frontier expansion for export commodities	Lambin & Meyfroidt, 2011
Underlying	Weak governance/ enforcement	Various	Corruption, unclear land rights facilitating illegal clearing	Colchester & Boscolo, 2006
Underlying	Population growth & migration	Africa, Southeast Asia	Land pressure for settlement, fuelwood, shifting agriculture	Turner et al., 2003

Underlying Drivers and Multiplier Effects

Beneath proximate causes lie structural and systemic drivers. Global commodity demand links local deforestation to distant markets, while weak governance and insecure land tenure create conditions for land grabs and illegal exploitation. Population growth and migration intensify pressures on forest frontiers, while climate change acts as a force multiplier, exacerbating droughts, fires, and pest outbreaks that accelerate forest loss. These drivers interact synergistically, producing multiplier effects: for example, agricultural expansion combined with weak governance can trigger violent land conflicts, while urbanization coupled with climate extremes increases vulnerability to floods, heat stress, and food insecurity.

Historical Trends and Global Patterns

Human societies have altered forest cover since the dawn of agriculture, but the pace and scale of loss have accelerated dramatically in the past half-century. Between 1990 and 2020, tropical regions experienced the most rapid forest loss in recorded history (Geist & Lambin, 2002). Figure 2 provides a clear snapshot of deforestation patterns, helping readers visualize the scale and trajectory of forest loss. By illustrating trends across regions, the figure reinforces the chapter's central argument: deforestation is accelerating and deeply intertwined with climate and cultural disruption. Global deforestation exhibits pronounced spatial and temporal variability. Early clearing concentrated along rivers and floodplains, but mechanized logging, road building, and global commodity markets have since pushed forest frontiers deeper into once-remote landscapes.

- ***Southeast Asia***: Peatland forests are highly vulnerable; drainage canals, illicit logging, and recurrent fires have released vast carbon stocks, compounding regional air pollution crises (Hosonuma et al., 2012).
- ***Amazon Basin***: Cattle ranching and mechanized soy cultivation now drive as much as 40% of gross deforestation, converting primary forests into pastures or monocultures (Bernhard et al., 2024).
- ***Africa***: Loss is more diffuse, with smallholders clearing woodlands for subsistence farming and fuelwood. In the Democratic Republic of Congo, charcoal demand has become a major driver, alongside industrial logging concessions (Fearnside, 2005).

Policy responses have shown mixed results. Brazil reduced Amazon deforestation by over 70% between 2004 and 2012 through law enforcement, satellite monitoring, and conservation incentives (DeFries et al., 2010), but losses rebounded after 2015 due to policy reversals. Indonesia's 2011 moratorium on primary forest conversion slowed some clearing, yet weak enforcement and peatland drainage continue to undermine progress (Meyfroidt et al., 2010). By contrast, several temperate nations have achieved net forest gains through afforestation, secondary growth, and sustainable forestry certification (Seydewitz et al., 2023).

Agricultural Expansion and Frontier Logging

Agricultural expansion remains the largest proximate driver of tropical deforestation. Commercial operations account for nearly 40% of gross forest loss, with cattle ranching dominant in Latin America, oil palm in Southeast Asia, and mechanized grain production in West Africa (Bernhard et al., 2024). The conversion process often follows a predictable trajectory: initial clearing with chainsaws or bulldozers, burning of residual biomass, and establishment of pastures or monocultures. Fire, often used as a low-cost tool, can escape control under dry conditions, igniting peat deposits or adjacent forest patches and producing haze episodes that affect human health at regional scales (Hosonuma et al., 2012).

Smallholder agriculture also contributes significantly to forest fragmentation. While individual clearings are modest, their cumulative footprint can rival that of commercial ventures. On steep slopes, intensified cropping leads to soil erosion and downstream sedimentation. Yet polyculture systems often retain higher biodiversity and carbon than monocultures, offering potential pathways for integrating smallholder livelihoods with conservation objectives (Kaimowitz & Angelsen, 1998).

Frontier logging frequently precedes agricultural incursions. Selective extraction of high-value timber species opens forest interiors to new road networks, lowering transport costs for both logs and subsequent farm products. In the Brazilian Amazon, selective logging can remove a substantial share of aboveground biomass, leaving canopy gaps that alter microclimates and soil moisture regimes (Fearnside, 2005). These openings facilitate invasive grasses and pioneer species, raising fire susceptibility and limiting regeneration of late-successional trees.

Figure 2. Variation of deforestation patterns in different regions. Top-left: Kerala, India; Top-right: Amazon Rainforest, Brazil; Bottom-left: Sumatra, Indonesia; Bottom-right: Mount Kilimanjaro, Tanzania. Source: Global Forest Watch, 2024.

Illegal logging exacerbates these impacts by bypassing sustainable harvest regulations and reforestation requirements. In Southeast Asia, estimates suggest that 50–90% of logging operations are illicit, exporting contraband wood products to international markets (Geist & Lambin, 2002). This widespread "timber leakage" undermines governance, erodes state revenues, and deprives local communities of customary rights. Strengthening chain-of-custody systems, improving remote sensing monitoring, and imposing trade restrictions on illegally harvested timber have shown promise, though enforcement remains a persistent challenge (Sarre, 2020).

Ecological Impacts: Carbon and Climate Feedbacks

The ecological consequences of deforestation are profound. Forests act as major carbon sinks, storing over 650 gigatons of carbon globally. Their removal releases vast amounts of greenhouse gases, undermining global climate goals. Recent studies show that parts of the Amazon Basin are shifting from carbon sink to source, releasing more CO_2 than they absorb due to fire, drought, and land-use pressures (Lovejoy & Nobre, 2019). This shift risks crossing tipping points where forests lose their stabilizing role in the climate system, amplifying global warming.

Ecological Impacts: Biodiversity and Ecosystem Function

Tropical forests harbor more than half of the world's terrestrial species despite covering only about 12 percent of land area (Seydewitz et al., 2023). Their loss drives biodiversity decline and disrupts intricate ecological networks. Habitat fragmentation isolates populations, disrupts pollination and seed dispersal, and creates "extinction debts" where species decline long after disturbance. The loss of keystone species—such as large frugivores in the Amazon or elephants in African forests—triggers cascading effects that reshape ecosystems.

Within these forests, many organisms occupy highly specialized niches—pollinators tied to flowering cycles, frugivores dispersing large-seeded trees, and epiphytes dependent on old-growth canopy structure (Myers et al., 2000). When forests are cleared or fragmented, these mutualistic relationships unravel, leading to population declines, local extinctions, and erosion of ecosystem services such as pollination and nutrient cycling. Edge effects extend hundreds of meters into forest interiors, favoring generalist and invasive species over specialists. Small fragments (<100 ha) often lose up to 75 percent of their large mammal species within a decade, with cascading effects on seed dispersal and plant recruitment (Berenguer et al., 2024). Biodiverse forests are also more resilient to climate stress. As Das and Sarker (2015) demonstrate, species richness enhances productivity and ecological balance, while Das (2024) highlights that structurally and compositionally diverse ecosystems buffer against climatic fluctuations and anthropogenic pressures. Quantitative modeling of diameter and height distributions provides further insights into stand structure and regeneration dynamics, which are critical for assessing long-term sustainability (Das et al., 2018).

Soil, Water, and Ecosystem Services

Deforestation undermines critical ecosystem services. The removal of deep-rooted trees disrupts water uptake and transpiration, altering evapotranspiration rates and potentially shifting local rainfall patterns. Exposed soils are prone to nutrient leaching, compaction, and erosion, diminishing agricultural productivity and increasing sediment loads in downstream rivers. Retaining riparian buffer zones can mitigate sedimentation and maintain water quality, but such practices are often overlooked in rapid conversion processes (DeFries et al., 2005).

Carbon dynamics also shift dramatically. Primary forests store 150–200 megagrams of carbon per hectare in aboveground biomass; when cleared, much of this carbon is combusted or decomposes, releasing CO_2 and other greenhouse gases. Secondary regrowth on abandoned pastures can recapture carbon stocks, but recovery is slow and incomplete: after 20 years, regrowth typically restores only 40–60 percent of original carbon stocks (Rudel et al., 2005). This underscores the irreplaceable role of old-growth forests in long-term climate mitigation.

Socio-Cultural and Health Dimensions

The ecological impacts of deforestation are inseparable from socio-cultural disruptions. Indigenous peoples and local communities, who have stewarded forests for millennia, face displacement, loss of livelihoods, and erosion of cultural identity. Gendered impacts are particularly acute: women often bear disproportionate burdens in fuelwood collection and food security, yet remain underrepresented in governance. Youth migration to urban centers disrupts intergenerational knowledge transfer, weakening adaptive capacity. Deforestation also heightens health risks. Forest fragmentation and wildlife trade increase the likelihood of zoonotic disease spillovers, as highlighted by the COVID-19 pandemic (UNEP, 2023). These socio-cultural and health dimensions reveal that deforestation is not only an ecological crisis but also a profound human challenge.

The scope of deforestation is vast and multifaceted, driven by agricultural, industrial, and socio-economic forces. Historical trends reveal shifting epicenters of forest loss, from riverine clearings in early agricultural societies to modern commodity frontiers in the Amazon, Congo Basin, and Southeast Asia. Commercial agriculture and frontier logging act in concert to strip forest cover, while smallholder activities exacerbate fragmentation at the local scale. The ecological consequences—carbon emissions, biodiversity erosion, disruption of mutualistic networks, soil degradation, and altered hydrological cycles—reverberate through ecosystems and human societies alike. A nuanced understanding of these processes is essential for crafting policies and interventions that halt forest loss, safeguard ecosystem services, and ensure that solutions are grounded in equity and justice.

CLIMATE CHANGE: FEEDBACK, EXTREMES, AND VULNERABILITY

Forests, Carbon Cycling, and Feedbacks

Forests are central to the global carbon cycle, acting as both sinks and potential sources of greenhouse gases. Old-growth tropical forests store vast amounts of carbon above and below ground, buffering the climate system. Primary forests typically hold 150–200 Mg C ha^{-1} aboveground and 50–100 Mg C ha^{-1} belowground (Pan et al., 2011), while forest soils can store up to three times more carbon than aboveground biomass (Jobbágy & Jackson, 2000). Through photosynthesis, forests absorb atmospheric CO_2 and store it as biomass, with tropical forests displaying the world's highest Net Primary Production (14 Mg C ha^{-1} yr^{-1}). Secondary regrowth on abandoned land also recovers carbon, though recovery rates vary widely.

Human disturbances such as deforestation, selective logging, and fragmentation weaken these sinks. Reduced canopy cover lowers NPP, exposed soils accelerate microbial decomposition, and fragmentation introduces edge effects that raise mortality. Collectively, these changes risk transforming forests from net sinks into net sources of carbon (Houghton, 2005). Table 2 offers a structured lens into the feedback mechanisms and consequences of climate change, such as rising temperatures, altered precipitation patterns, and increased frequency of extreme weather events.

Table 2. Climate-Change Feedbacks and Impacts.

Feedback Mechanism	Process	Consequence	Example / Evidence
Temperature Rise	Higher global mean temperatures increase vapor pressure deficits, forcing stomatal closure and reducing photosynthesis.	Reduced growth, carbon starvation, and greater vulnerability to pests and fire.	Global mean surface temperature +1.1 °C above pre industrial; projected +1.5 °C by mid-century (Anderegg et al., 2015).
Drought Stress	Prolonged water deficits impair photosynthesis, deplete carbohydrate reserves, and trigger hydraulic failure.	Tree mortality, reduced carbon uptake, and ecosystem collapse.	2012–2016 "hot drought" in western North America killed >100 million conifers, releasing ~360 Tg C (Williams et al., 2013).
Fire–Carbon Feedback	Rising heat and vegetation stress prolong fire seasons and increase fuel loads.	More frequent, intense fires release CO_2, damage soils, and hinder regeneration.	Amazon 2024: 6.64 M ha degraded, 643 Mt CO_2 released (Beuchle et al., 2025); Australia's 2019–20 "Black Summer" burned >10 M ha, releasing 715 Tg CO_2.

continued on following page

Table 2. Continued

Feedback Mechanism	Process	Consequence	Example / Evidence
Pest–Pathogen Outbreaks	Warmer winters extend insect and pathogen ranges.	Large scale forest die offs, reduced carbon storage, altered species composition.	Bark beetle outbreaks in boreal/ temperate forests (Bentz et al., 2010).
Extreme Rainfall & Flooding	Intense precipitation events destabilize soils and slopes.	Landslides, erosion, and loss of forest cover.	South Asia 2023: extreme rainfall triggered landslides and erosion (UNEP, 2023).
Tipping Points	Crossing ecological thresholds (e.g., Amazon dieback, boreal conversion).	Irreversible biome shifts, massive carbon release, loss of resilience.	Amazon Basin shifting from sink to source (Lovejoy & Nobre, 2019; Beuchle et al., 2025).

Reinforcing feedbacks intensify this vulnerability. Drought mortality suppresses photosynthesis and triggers die-offs (McDowell et al., 2008). Fire–carbon cycles are increasingly evident: in 2024, the Amazon experienced its worst fire season in two decades, with 6.64 million hectares degraded and 643 million tons of CO_2 released (Beuchle et al., 2025). Pest–pathogen outbreaks, fueled by warmer winters, are devastating boreal and temperate forests (Bentz et al., 2010). These feedbacks create self-reinforcing cycles that accelerate ecosystem decline. Crossing ecological tipping points—such as Amazon dieback or boreal forest conversion to grassland—would unlock vast carbon stores and disrupt global climate regimes (Lovejoy & Nobre, 2019).

Climate Extremes and Forest Vulnerability

The year 2024 marked a turning point, with extreme weather reaching unprecedented levels. Record-breaking heatwaves, droughts, wildfires, and floods displaced millions and caused thousands of deaths (Ripple et al., 2024). In the Amazon, the 2023/24 drought surpassed previous records, intensifying fire vulnerability (World Weather Attribution, 2024). In South Asia, extreme rainfall triggered landslides and soil erosion (UNEP, 2023). In boreal regions, megafires and pest outbreaks are transforming forests into grass–shrub mosaics (Kurz et al., 2008).

- ***Temperature rises and drought:*** Global mean surface temperatures have already risen by ~1.1 °C above pre-industrial levels, with projections reaching 1.5 °C by mid-century. Elevated temperatures increase vapor pressure deficits, forcing stomatal closure that conserves water but limits photosynthesis (Anderegg et al., 2015). Prolonged droughts impair growth, deplete carbohydrate reserves, and elevate risks of hydraulic failure. In western North

America, the 2012–2016 "hot drought" killed over 100 million conifers, releasing 360 Tg C (Williams et al., 2013).

- ***Fire regimes:*** Rising temperatures and vegetation stress prolong fire seasons and increase fuel loads. Globally, burned area rose by 35 percent between 2001 and 2020 (Andela et al., 2019). Fires are more intense, consuming more biomass and damaging soils to greater depths. Fire-sensitive ecosystems such as the Amazon face risks of entering self-reinforcing fire–carbon cycles (Celis et al., 2023). Australia's "Black Summer" fires of 2019–2020 burned over 10 million hectares, killed an estimated 3 billion animals, and released 715 Tg CO_2 (Ford et al., 2018; West et al., 2020).

Biodiversity, Species Shifts, and Community Reassembly

Climate extremes disrupt biodiversity and ecosystem functioning. Species are shifting poleward and upslope, with montane species in the Himalayas and Andes moving 20–40 meters per decade (Lenoir et al., 2010). Temperate species such as beech (Fagus spp.) have expanded northward by up to 10 km per decade in Europe (Bertrand et al., 2011). Yet many species cannot adapt swiftly, creating a "climatic debt" (Pearson & Dawson, 2003).

Phenological mismatches—such as earlier flowering and delayed pollinator activity—reduce reproductive success (Kudo & Cooper, 2019). The loss of keystone species, from large frugivores in the Amazon to elephants in African forests, triggers cascading effects that reshape ecosystems (Berenguer et al., 2024). As functional diversity declines, forests lose their capacity to buffer against environmental fluctuations, undermining both ecological stability and human dependence.

Community reassembly is also underway. Earlier leaf-out exposes foliage to late frosts, while mismatched fruiting and disperser activity disrupt regeneration. Native specialists are increasingly displaced by generalists and invasive exotics, reshaping entire ecosystems.

Socio-Cultural Vulnerabilities

The impacts of climate change on forests are inseparable from socio-cultural vulnerabilities. Indigenous peoples and local communities, who depend on forests for livelihoods and cultural identity, face heightened risks from displacement, food insecurity, and governance failures (Purwestri & Lusiana, 2025). Gendered impacts are particularly acute: women often bear disproportionate burdens in fuelwood collection and food provisioning during droughts, yet remain underrepresented in decision-making (Agarwal, 2010). Youth migration to urban centers erodes intergenerational knowledge transfer, weakening adaptive capacity. Climate stress and forest fragmentation also increase the risk of zoonotic spillovers, as highlighted by the COVID-19 pandemic (UNEP, 2023).

Assisted Migration and Evolutionary Responses

To counter range contractions, some conservationists advocate assisted migration—relocating species to habitats projected to become suitable under future climates. While this may prevent extinctions, it carries risks, including the introduction of pests, pathogens, or disruptive traits, and potential genetic swamping of local populations. Natural evolutionary processes, such as phenotypic plasticity, provide some buffering capacity, but for long-lived tree species the pace of climate change may exceed their adaptive potential (Aitken et al., 2008).

Vulnerability Assessment and Adaptive Capacity

Assessing forest vulnerability requires understanding the interaction between exposure, sensitivity, and adaptive capacity (Turner et al., 2003). Exposure refers to the magnitude of climate change at a given location, sensitivity reflects how forests respond, and adaptive capacity denotes the ability of ecosystems and stakeholders to adjust. Climate distribution models identify vulnerability hotspots: in the Amazon, warming beyond 3 °C could push 60% of lowland forests outside their current climatic niches (Malhi et al., 2009), threatening long-term ecosystem integrity.

Management Strategies for Resilience

Adaptive management promotes both biological and structural diversity. Mixed-species plantings enhance functional redundancy, while retaining legacy trees and snags supports habitat stability and nutrient cycling. Ecological corridors facilitate migration and gene flow, while assisted natural regeneration—such as planting drought-tolerant genotypes and improving soils—helps restore degraded areas. Traditional knowledge also plays a critical role. Indigenous fire practices, such as mosaic burning, reduce fuel loads, while locally adapted seed sourcing preserves genetic diversity (Clark et al., 2015). Integrating scientific and traditional approaches strengthens resilience by combining innovation with time-tested practices.

Policy, Finance, and Justice Dimensions

Adaptation is most effective when supported by robust policy frameworks and financial mechanisms. Payments for Ecosystem Services (PES) incentivize conservation, while climate-smart forestry guidelines recommend adjusting harvest schedules to future conditions. Parametric insurance tools provide rapid compensation for climate-triggered losses. Global mechanisms such as the Paris Agreement and REDD+ channel funding toward climate-resilient forestry, though their success depends on transparent governance and equitable benefit-sharing (Griscom et al., 2017).

Climate extremes disproportionately affect the Global South, where adaptive capacity is often weakest. In 2024, small island states and developing countries experienced the highest number of dangerous heat days, threatening health and livelihoods (Ripple et al., 2024). These inequities underscore the importance of climate justice: while tropical communities bear the brunt of climate impacts, they contribute least to global emissions. Addressing these disparities requires embedding equity and rights-based approaches into climate and forest governance (IPCC, 2022).

Climate change is no longer a distant threat but a present force reshaping forests and societies. Reinforcing feedbacks, extreme events, and socio-ecological vulnerabilities interact in complex ways, creating cascading risks that undermine both ecosystems and human well-being. Recognizing these dynamics is essential for designing integrated strategies that combine mitigation, adaptation, and resilience, while ensuring justice for the communities most affected.

SOCIO-CULTURAL DISRUPTION: COMMUNITIES, LIVELIHOODS, AND GOVERNANCE

Forest ecosystems are more than ecological assemblies—they are deeply tied to the homes, spirituality, and economic vitality of countless communities. As deforestation and climate change accelerate, these connections unravel, causing profound disruptions to local well-being, cultural identity, and governance equity. This section explores three interwoven dimensions of socio-cultural disruption: the erosion of Indigenous and local ecological knowledge (TEK), rising land tenure conflicts and displacement, and the evolution of legal and participatory frameworks that attempt to balance conservation with community needs. Table 3 provides a structured overview of how environmental degradation—particularly deforestation and climate change—translates into tangible disruptions for communities.

Table 3. Socio-Cultural Disruptions and Community Impacts.

Disruption Dimension	Underlying Cause	Community Impact	Example Case	Reference
Loss of TEK	Land alienation, language loss	Erosion of adaptive practices and food security	Dayak swidden knowledge displaced by plantations	Nettle & Romaine, 2000
Tenure insecurity	Customary vs. formal law clashes	Forced displacement, livelihood loss	Rubber concessions overlapping Khmer villages	Colchester & Boscolo, 2006
Cultural dislocation	Resettlement, migration	Social fragmentation, mental health stress	Bali dam–induced relocations	Damnyag & Saastamoinen, 2012
Livelihood transformation	Market integration, youth migration	Dependence on wage labor, urban poverty	Sahel pastoralists moving to cities	de et al., 2014

Forests as Cultural Landscapes

For Indigenous Peoples and Local Communities (IPLCs), forests are cultural landscapes as much as ecological systems. They provide food, medicine, shelter, and spiritual meaning, while serving as sites of cultural continuity where TEK is transmitted across generations (Rahman & Alam, 2016; UNDP, 2024). Yet deforestation and climate change are eroding these cultural fabrics, producing cascading disruptions that extend far beyond ecological loss (Kazi, 2024).

Livelihood Disruptions and Inequality

Deforestation undermines the livelihoods of rural and forest-dependent populations. In Sub-Saharan Africa, charcoal production and smallholder agriculture remain critical income sources, but forest degradation reduces long-term productivity and deepens poverty traps. In Southeast Asia, the expansion of oil palm plantations has displaced smallholders, forcing many into precarious wage labor. These processes exacerbate inequality, as benefits accrue to agribusiness and elites while costs are borne by marginalized groups. Women are disproportionately affected, as they often shoulder the burden of fuelwood collection, water fetching, and food provisioning, yet remain underrepresented in decision-making (Agarwal, 2010).

Indigenous and Local Knowledge Under Threat

TEK embodies generations of ecological understanding, guiding adaptive management, resource partitioning, and biodiversity conservation. Practices such as rotational farming and sacred grove taboos among the Kayapó in Brazil or the Dayak in Borneo help preserve ecological balance (Stott, 1999). Customary tenure systems align harvesting with regeneration, as seen in India and the Philippines, while ethnobotanical studies reveal that IPLCs safeguard numerous culturally valuable species (Balick & Cox, 2021).

However, TEK is increasingly threatened by land alienation, language loss, and modernization pressures. Expansion of commercial plantations in Southeast Asia has disrupted Dayak traditions (Colchester & Boscolo, 2006), while the erosion of Indigenous languages accelerates the loss of nuanced ecological vocabulary. Formal education systems that marginalize native languages further weaken intergenerational transmission. The decline of TEK carries severe consequences: communities lose food security as traditional agroforestry systems are abandoned, spiritual well-being suffers as displacement severs cultural ties, and adaptive capacity to climate variability is diminished.

Land Tenure Conflicts and Displacement

Secure land tenure is the foundation of sustainable forest stewardship, yet national frameworks often clash with customary systems. Customary tenure, rooted in collective use and oral agreements, frequently lacks legal recognition, leaving IPLCs vulnerable to state-sanctioned concessions.

Case studies from the Mekong Region, Amazon Basin, and Africa's Miombo Woodlands illustrate this dynamic. In Cambodia and Laos, land concessions for rubber and sugarcane have sparked violent evictions (Schmink & Wood, 1992). In Brazil, inadequate territorial demarcation has exposed Indigenous groups to illegal logging. Post-colonial nationalization in southern Africa undermined local authority, fueling unsustainable charcoal production and enforcement backlashes (Damnyag et al., 2012).

Displacement triggers economic marginalization, forcing families into precarious labor markets and reducing access to traditional livelihoods. It also leads to cultural dislocation, where rituals and practices tethered to ancestral land are disrupted—as exemplified by resettlements linked to infrastructure projects in Bali (Rice & Ward, 1996). Political exclusion often follows, with displaced communities lacking the resources to pursue justice or restoration, fueling cycles of unrest and criminalization (Peluso & Vandergeest, 2011).

Emerging Legal Frameworks

In response to these challenges, global frameworks have sought to safeguard customary rights. The UN Declaration on the Rights of Indigenous Peoples (UNDRIP) affirms the rights of IPLCs to participate in resource governance and mandates Free, Prior, and Informed Consent (FPIC) (Chhatre & Agrawal, 2009). The

FAO's Voluntary Guidelines on the Responsible Governance of Tenure (VGGT) encourage documentation of traditional rights and equitable conflict resolution (FAO, 2012). Additionally, REDD+ safeguards under the Warsaw Framework aim to embed tenure security and Indigenous rights into climate mitigation projects, though implementation remains uneven.

Governance Failures and Fragmentation

Despite these frameworks, governance responses are often fragmented, privileging short-term economic gains over long-term resilience. Corruption, elite capture, and weak enforcement undermine conservation policies, while incoherence across forestry, agriculture, and mining exacerbates conflicts. Infrastructure corridors in Africa and South Asia, promoted as development pathways, frequently accelerate forest loss and displacement. These governance failures deepen mistrust between communities and state institutions, fueling cycles of resistance and conflict (Barr et al. 2014).

Policy, Enforcement, and Participatory Approaches

Recognizing the shortcomings of top-down conservation strategies, policymakers and practitioners are increasingly embracing participatory governance models that integrate community rights with sustainable forest management. This reflects a growing acknowledgment that local stakeholders are active stewards, not passive beneficiaries.

Community-Based Forest Management (CBFM)

CBFM transfers authority from central agencies to local user groups through co-management frameworks. These arrangements empower communities to decide forest use and share benefits.

- Nepal's Community Forestry Program has enabled over 22,000 user groups to manage 1.6 million hectares since the 1990s, reducing deforestation, improving water yield, and enhancing livelihoods through equitable revenue sharing (Marshall & Stokes, 2014).
- Mexico's Ejidos, communal lands established during agrarian reforms, have evolved into platforms for payment for ecosystem services (PES), particularly in Oaxaca's cloud forests.

- Ghana's Modified Taungya System blends agroforestry with reforestation by allowing cocoa farmers to cultivate degraded reserves, sharing timber revenues while receiving training in sustainable practices (Abada et al., 2016).

Participatory Monitoring and Enforcement

Empowering communities to monitor forests and enforce conservation rules enhances accountability and reduces enforcement costs.

- In the Peruvian Amazon, the Monitoring of the Andean Amazon Project (MAAP) uses satellite imagery and mobile alerts to detect illegal activity, enabling rapid community response.
- In Guatemala's Maya Biosphere Reserve, over 200 trained community rangers conduct patrols and deliver environmental education, contributing to a 40% reduction in illegal extraction since 2010.
- In India's Joint Forest Management (JFM) program, forest user committees oversee benefit distribution and flag corruption, reinforcing trust in state authorities (Chhatre & Agrawal, 2009).

Policy Instruments and Incentive Schemes

A diverse range of policy tools align environmental goals with community interests:

- Payments for Ecosystem Services (PES): Costa Rica's national program is a global benchmark, credited with a 52% increase in forest cover from 1987–2013.
- Certification schemes: The Forest Stewardship Council (FSC) and FairWild provide market advantages for sustainably sourced products, requiring adherence to rigorous social and ecological criteria (Mattos et al., 2010).
- Climate finance mechanisms: REDD+ has mobilized resources such as the $50 million directed to Shinyanga, Tanzania, supporting village land-use planning, agroforestry, and income diversification linked to carbon credits.

Despite their promise, participatory approaches face persistent challenges:

- Power asymmetries can lead to elite capture, marginalizing vulnerable groups, including women and lower castes (Agarwal, 2010).
- Policy incoherence arises from overlapping mandates across forestry, mining, and agriculture, creating confusion that impairs enforcement (Sikor & Lund, 2009).

- Financial sustainability is often lacking, as many initiatives depend on short-term project funding without long-term budgeting or self-sustaining enterprises (Ostrom, 2015).

Emerging Dimensions: Urbanization, Youth, and Health

Urbanization is an emerging driver of socio-cultural disruption. Peri-urban forests in India and Africa are being cleared for housing and infrastructure, displacing marginalized groups without adequate compensation. Youth migration to cities disrupts intergenerational knowledge transfer, while forest fragmentation and wildlife trade increase the risk of zoonotic disease spillovers, as highlighted by COVID-19 (UNEP, 2023). These dynamics illustrate that deforestation is not only an environmental crisis but also a public health and social justice issue.

Governance Innovations and Community Resistance

Despite these challenges, promising innovations are emerging. In Peru's Amazon, participatory monitoring combining satellite alerts with community patrols has reduced illegal logging by 40% since 2015. Social movements and community-led resistance are increasingly recognized as critical forces in forest protection, demonstrating that resilience emerges when governance integrates rights, equity, and local knowledge (Hareesh & Shailaja, 2025).

Socio-cultural disruption lies at the heart of the deforestation crisis. It manifests in livelihood insecurity, displacement, cultural erosion, governance failures, and health risks. Yet it also reveals pathways for resilience through rights-based governance, gender-inclusive policies, and community-led innovation. Addressing these disruptions requires moving beyond ecological framings to embrace forests as socio-ecological systems, where cultural integrity and ecological sustainability are inseparable.

INTERCONNECTED DRIVERS AND MULTIPLIER EFFECTS

Complexity of Deforestation Drivers

Deforestation is rarely the outcome of a single factor. Instead, it emerges from a web of interconnected drivers that reinforce one another across ecological, economic, and socio-political domains. Proximate drivers such as agricultural expansion, logging, infrastructure development, and urbanization are shaped and amplified by underlying forces including global commodity demand, insecure land tenure, weak governance, and demographic change. These interactions create multiplier effects, where the combined impact of multiple drivers exceeds the sum of their parts, producing cascading risks for both ecosystems and societies (Bernhard et al., 2024).

Agriculture and Global Commodity Chains

Agricultural expansion remains the dominant proximate driver of deforestation, but its impacts are magnified by global commodity chains. Soy and beef production in Brazil, palm oil in Indonesia, and cocoa in West Africa are not only local land-use issues but also tied to distant consumer markets in Europe, China, and North America. These chains stimulate infrastructure expansion (roads, ports, processing facilities), encourage land speculation, and displace smallholders and Indigenous peoples. Conservation in one region can also lead to leakage, where pressures shift elsewhere—as seen when Indonesia's palm oil moratorium redirected expansion to Papua New Guinea and parts of Africa (Meyfroidt et al., 2010).

Urbanization and Peri-Urban Growth

Urbanization is an increasingly important but underexplored driver. Expanding cities consume peri-urban forests for housing, roads, and industrial zones. In India, peri-urban expansion has become a leading cause of forest conversion, with cascading effects on water security and air quality (MoEFCC, 2023). Multiplier effects emerge from the synergy between urban demand and rural supply chains: growing urban populations increase demand for charcoal, timber, and agricultural products, intensifying rural deforestation. At the same time, urban sprawl reduces ecological buffers, heightening vulnerability to floods, heatwaves, and zoonotic spillovers (UNEP, 2023).

Governance, Tenure, and Illicit Economies

Weak governance and insecure land tenure amplify the impacts of other drivers. Where property rights are unclear, forests become vulnerable to land grabs, illegal logging, and extractive concessions. These dynamics intersect with illicit economies—wildlife trafficking, mining, and narcotics cultivation—that thrive in governance vacuums (Peluso & Vandergeest, 2011). Multiplier effects arise when illegal logging networks finance armed groups, exacerbating cycles of violence and displacement, or when carbon finance mechanisms prioritize global emission targets over local rights, creating new conflicts (Purwestri & Lusiana, 2025).

Interactions between Land-Use Change and Climate Forcing

The conversion of forests to agricultural or pastoral land alters surface energy balances, hydrological cycles, and atmospheric composition, feeding back onto local and regional climates (Bonan, 2008). Forest removal reduces evapotranspiration, decreasing atmospheric moisture and rainfall downwind, which can extend dry seasons and exacerbate drought conditions. In the Amazon Basin, large-scale deforestation has been linked to a 20% decline in dry-season precipitation in adjacent regions, increasing fire risk and mortality among remnant trees (Devaraju et al., 2015).

These coupled processes can trigger runaway feedback loops. Reduced canopy cover elevates ground temperatures, accelerating soil organic matter decomposition and CO_2 emissions (Davidson, 2009). In palm oil landscapes of Southeast Asia, drought-induced fires on drained peatlands release 1.5–2.5 gigatons of CO_2 annually, rivaling emissions from entire industrial sectors (Gaveau et al., 2014). Such events demonstrate how land-use decisions and climate anomalies interact to unleash large-scale carbon release events, while declines in plantation quality under climate stress further disrupt local economies and cultural ties.

Feedback Loops Intensifying Socio-Cultural Disruption

Social systems can absorb shocks, but intensifying ecological stressors overwhelm cultural coping mechanisms (Adger, 2003). In the African Sahel, deforestation of parklands for charcoal and agriculture heightens temperature extremes, reduces forage for pastoralists, and prompts migration to urban centers. These migrations erode kinship networks and TEK transmission, weakening social capital that once mediated resource sharing and conflict resolution. As customary institutions degrade, land-use monitoring and enforcement falter, opening forests to illegal logging and poaching (Crooks & Sanjayan, 2006).

In Amazonian Indigenous territories, the interplay of drought-induced fire outbreaks and cattle ranching has disrupted ceremonial hunting cycles, precipitating food shortages and ritual crises (Celis et al., 2023). When food-sharing norms collapse under scarcity, social cohesion fractures, increasing the likelihood of violent disputes over resources (De Sherbinin et al., 2012). Escalating conflict discourages investment in conservation and hinders sustainable development, locking communities into cycles of ecological degradation and social fragmentation.

Cross-Scale Interactions and Tipping Points

Local disturbances interacting with global drivers can precipitate abrupt regime shifts at ecosystem and landscape scales (Kurz et al., 2008). In boreal forests, bark beetle outbreaks—facilitated by warmer winters—combine with logging to exceed critical fuel load thresholds, triggering megafires that convert forests to grass–shrub mosaics (Flannigan et al., 2009). The resulting shift reduces regional carbon uptake and alters albedo, feeding back to regional warming and further fire risk.

In the Congo Basin, fragmentation beyond 30% core area disrupts animal-mediated seed dispersal, compromising regeneration at landscape scale (Laurance et al., 2002). Models reveal early warning signals—such as rising variance in tree cover anomalies or slowing recovery rates—that can herald tipping points (Dakos et al., 2008). Incorporating such indicators into monitoring platforms enables pre-emptive management, such as targeted fuel breaks or accelerated restoration in high-risk zones (Carpenter et al., 2011).

Integrated System Modeling and Scenario Analysis

To navigate the complexity of intersecting threats, researchers employ integrated assessment models (IAMs) and spatially explicit land-system models that couple socioeconomic drivers with biophysical processes. The Dinamica EGO platform, for example, has simulated how alternative agricultural policies combined with RCP8.5 climate projections influence deforestation trajectories and carbon emissions in the Brazilian Amazon (Soares-Filho et al., 2006). Scenarios suggest that reducing pasture expansion by 30% and restoring 10 million hectares could avoid 0.7 Gt CO_2 yr^{-1} by 2050 while safeguarding community livelihoods.

Participatory scenario workshops enrich model outputs by integrating IPLC priorities and risk perceptions. In Viet Nam's Central Highlands, workshops with ethnic minority coffee growers informed land-allocation models, leading to co-developed agroforestry landscapes that balance carbon sequestration with cash crop production (Ferreira De Athayde et al., 2006). This approach revealed synergies—such as shade trees that conserve soil, improve coffee yields, and enhance carbon stocks—while also highlighting trade-offs, including reduced short-term income under strict canopy retention.

Implications for Policy and Management

Recognizing synergies and compounding effects demands policies that transcend siloed interventions. Multilevel governance frameworks can align agricultural, forestry, and climate objectives by:

- **Integrating** national REDD+ strategies with agricultural extension services to promote climate-smart practices and forest-friendly value chains (Springate-Baginski & Wollenberg, 2010).

- **Coordinating** land-use zoning across ministries to ensure protected areas, community forests, and concessions collectively maintain connectivity thresholds.
- **Embedding** social safeguards—such as FPIC and gender-inclusive benefit sharing—into climate finance mechanisms to protect vulnerable groups and preserve TEK reservoirs (Colchester & Lohmann, 1993).

Financial instruments should leverage carbon markets to scale ecosystem restoration. Blended finance models—combining concessional loans, grants, and private capital—can support landscape investment platforms that channel resources to local cooperatives, enabling restoration of millions of hectares and creation of green jobs by 2030. Performance-based payments tied to multiple indicators (carbon, biodiversity, livelihoods) incentivize co-benefit delivery and reduce perverse incentives that prioritize single metrics (Ostrom, 2015).

The intertwined nature of deforestation, climate change, and socio-cultural disruption underscores the need for integrated responses that address multiple drivers and feedbacks. By mapping the synergies and compounding effects across ecological and social systems, this chapter can identify leverage points—such as maintaining landscape connectivity, safeguarding customary governance, and deploying early-warning systems—that interrupt vicious cycles. Future chapters will explore practical frameworks for implementing these cross-sectoral strategies, emphasizing the co-creation of knowledge and equitable governance as foundations for resilient forest landscapes.

MITIGATION, ADAPTATION, AND PATHWAYS TO RESILIENCE

Forests occupy a central role in mitigating climate change, adapting to its myriad impacts, and fostering human and ecological resilience. Effective pathways intertwine strategies that reduce greenhouse gas emissions, strengthen forest and community capacity to withstand disturbances, and integrate social and ecological objectives across landscapes. This section examines leading mitigation approaches—ranging from policy instruments to land-use management—followed by adaptation measures for ecosystems and societies. I then explore integrated landscape approaches, innovative financing and governance mechanisms, and adaptive learning frameworks that together chart resilient pathways.

Forests as Climate Solutions

Forests are both frontline victims of climate change and indispensable allies in addressing it. They absorb nearly one third of anthropogenic CO_2 emissions annually, yet deforestation and degradation release around 21% of global greenhouse gases (UNFCCC 2024). Effective strategies must therefore combine mitigation (reducing emissions and enhancing sinks), adaptation (building resilience to climate stressors), and transformative pathways that integrate ecological, social, and governance dimensions.

Mitigation through Protection and Restoration

Protecting intact forests remains the most cost-effective mitigation strategy. The Paris Agreement's REDD+ framework provides a global mechanism for reducing emissions from deforestation and degradation, while enhancing conservation and sustainable management (UNFCCC 2024). Yet critiques highlight risks of greenwashing, inequitable benefit sharing, and weak safeguards. Future REDD+ must embed rights-based approaches and ensure Free, Prior, and Informed Consent (FPIC) for Indigenous and local communities.

Forest restoration also offers significant mitigation potential. Global estimates suggest that forest based Natural Climate Solutions (NCS) could deliver up to 37% of cost-effective CO_2 mitigation needed by 2030 (Kübler and Günter 2024). Initiatives such as the Bonn Challenge and AFR100 aim to restore millions of hectares, but restoration must prioritize native species and ecological integrity rather than monoculture plantations.

Climate-Smart Forestry and Innovation

Emerging approaches such as Climate Smart Forestry (CSF) integrate ecological forestry, carbon forestry, and smart technologies. By using AI, LiDAR, and digital twin models, CSF enables real time monitoring, early warning systems, and adaptive management (Wang et al. 2025). These innovations enhance carbon sequestration, improve resilience to climate stressors, and support transparent carbon accounting. However, equitable access to these technologies remains a challenge in the Global South, where capacity building and finance are urgently needed.

Mitigation Strategies: Curbing Emissions and Enhancing Carbon Stocks

REDD+ and Supply□Chain Interventions

Reducing Emissions from Deforestation and Forest Degradation (REDD+) has emerged as the premier framework for financially rewarding countries and communities that curb forest-sector emissions and enhance carbon stocks through conservation, sustainable management, and restoration. Under the Warsaw Framework, REDD+ countries develop national strategies, establish forest reference emission levels, and build robust Monitoring, Reporting, and Verification (MRV) systems to quantify outcomes. Jurisdictional approaches—linking subnational programs with national frameworks—leverage economies of scale, strengthen governance, and align policy instruments for land-use zoning, law enforcement, and incentive distribution (Štreimikienė, 2014). Table 4 serves as a strategic pivot in the chapter—shifting the focus from threats to solutions and resilience-building.

Table 4. Mitigation, Adaptation & Pathways to Resilience.

Strategy Type	Strategy/Tool	Threat Addressed	Mechanism/Approach	Outcome & Reference
Mitigation	REDD+ & zero-deforestation pledges	Deforestation-driven emissions	Results-based payments, supply-chain traceability	Reduced forest loss; Springate-Baginski & Wollenberg, 2010
Mitigation	Landscape restoration (Bonn Challenge)	Degraded lands, carbon gap	Passive regeneration, active planting	~1.5–3 Pg C sequestered by 2050; Silver et al., 2000
Adaptation	Ecosystem-based adaptation (mangroves, peatlands)	Coastal flooding, fire risk	Habitat restoration, buffer enhancement	Increased storm protection; Silver et al., 2000
Adaptation	Community forestry & agroforestry	Drought, land degradation	Local governance, mixed-species systems	Improved yields & governance; Marshall & Stokes, 2014
Integrated resilience	Landscape approaches & multi-stakeholder coalitions	Combined ecological & socio-cultural threats	Spatial zoning, benefit-sharing, adaptive management	Balanced land uses & livelihoods; Sayer et al., 2013

Private-sector "zero-deforestation" commitments complement REDD+ by targeting commodity supply chains. Major buyers of palm oil, soy, cattle, and pulp and paper trace raw materials to origin and impose no-conversion policies on suppliers (Štreimikienė, 2014). Certification schemes—such as the Forest Stewardship Council (FSC), Roundtable on Sustainable Palm Oil (RSPO), and ProTerra soy standard—verify compliance with environmental and social criteria, offering price premiums and market access to participating producers. Though challenges persist in ensuring transparency, smallholder inclusion, and avoidance of "leakage," combining public REDD+ finance with corporate sourcing policies forms a powerful bulwark against frontier-driven clearing.

Landscape Restoration and Reforestation

Global restoration pledges under the Bonn Challenge and the New York Declaration on Forests aim to restore 350 million hectares of degraded lands by 2030 (Silver et al., 2000). Restoration modalities span the spectrum from passive natural regeneration to active enrichment planting and assisted natural regeneration (ANR). Passive approaches—protecting regrowth from fire, grazing, and illegal extraction—often yield high biodiversity gains at low cost, especially near forest remnants (Mendes et al., 2021). Active interventions—such as mixed-species plantations of native trees, enrichment of species important for timber or non-timber products, and ANR via invasive species control—accelerate structural complexity and carbon accumulation (Crouzeilles et al., 2017).

Meta-analyses estimate that restoration of 350 million hectares could sequester 1.5–3.0 petagrams of carbon by mid-century, offsetting up to 10 percent of the emissions gap under current Nationally Determined Contributions (NDCs) (Marvin et al., 2023). Regionally tailored programs—China's Grain-for-Green on the Loess Plateau, Brazil's Atlantic Forest Pact, and Ethiopia's Regreening the Highlands—demonstrate that restoration co-benefits include improved soil fertility, water regulation, and rural livelihoods (Shi et al., 2021).

Climate-Smart Agriculture and Agroforestry

Agricultural expansion remains a primary driver of forest loss. Climate-Smart Agriculture (CSA) and agroforestry enhance on-farm productivity while building soil carbon stocks and reducing pressure on forest frontiers (FAO, 2016). Conservation tillage, cover cropping, and integrated crop–livestock systems raise yields by 20–30 percent and sequester 0.5–2.0 Mg C ha^{-1} yr^{-1} in soils (Lal, 2004). Silvopastoral systems—integrating fodder trees with livestock grazing—store up to 40 Mg C ha^{-1} aboveground while improving animal productivity (Montagnini & Nair, 2004). Farmer-Managed Natural Regeneration (FMNR) in the Sahel rehabilitates parklands by protecting stumps and root suckers, driving a tenfold increase in tree density at negligible cost and capturing ~0.35 Mg C ha^{-1} yr^{-1} (Korir et al., 2025).

CSA and agroforestry reduce the "deforestation footprint" per unit of agricultural output and diversify income streams—bolstering farmer resilience to market and climate shocks. Scaling these practices depends on extension services attuned to local conditions, farmer cooperatives for input and output aggregation, and financial incentives—such as carbon credits for on-farm sequestration (Grantham, 2012).

Adaptation Measures: Strengthening Ecosystem and Community Resilience

Ecosystem-Based Adaptation (EbA)

Ecosystem-based Adaptation (EbA) leverages biodiversity and ecosystem services to moderate climate impacts, enhance livelihoods, and conserve species and habitats (Tiwari et al., 2022). Mangrove restoration along tropical coastlines attenuates storm surges by up to 66%, reduces wave energy, and supports fisheries that sustain coastal economies (McLaughlin et al., 2002). In Vietnam's Mekong Delta, re-establishing 20,000 ha of mangroves increased fish yields by 50%, sequestered ~6.2 Mg C ha^{-1} yr^{-1} in peat soils, and safeguarded livelihoods against cyclonic winds.

In high-altitude Andean systems, paramo grassland protection maintains water regulation services, sustaining year-round streamflow for agriculture and urban users. Restoration of drained peatlands in Indonesia's Central Kalimantan prevents fires, subsidence, and carbon loss, while preserving water quality for downstream communities (Gaveau et al., 2014). These cases underscore that intact and restored ecosystem provide cost-effective "green infrastructure," often outperforming engineered alternatives.

Community-Driven Forestry and Adaptive Co-Management

Secure land tenure and participatory governance strengthen adaptive capacity. Community forestry programs—such as Mexico's ejidos—empower local groups to manage forests, set harvesting rules, and reinvest revenues in social services. These arrangements typically yield 20–30% increases in forest cover, improved fuelwood and fodder availability, and stronger social cohesion (McDougall, 2008).

Adaptive co-management melds traditional ecological knowledge with scientific monitoring. In Australia's Kakadu National Park, Aboriginal fire stewardship—early dry-season mosaic burns—reduces late-season megafires, lowers greenhouse gas emissions, and sustains biodiversity (Murphy et al., 2015). Collaborative monitoring platforms integrate satellite data with ranger observations, enabling rapid response to illegal incursions, fire outbreaks, and pest infestations.

Disturbance regimes are increasingly recognized as critical drivers of forest dynamics, shaping both biodiversity outcomes and regeneration capacity. In South Asian Sal (Shorea robusta) forests, for example, disturbance gradients have been shown to restructure stand composition, alter regeneration pathways, and reduce long-term resilience (Das, 2025b). These findings highlight that forest degradation is not a uniform process but one mediated by the intensity and frequency of disturbance, with cascading implications for ecosystem services and community livelihoods. Integrating such insights into management frameworks underscores the need for site-specific strategies that account for both ecological thresholds and socio-cultural contexts.

Community-Based Adaptation and Equity

Community-based forest management (CBFM) has demonstrated success in enhancing adaptive capacity. Gender-inclusive governance strengthens resilience, as women's knowledge of seed systems, medicinal plants, and food security is vital during crises (Agarwal, 2010). Youth engagement is equally important: in Latin America and Africa, youth-led forest stewardship initiatives are emerging as critical forces for innovation and advocacy.

Urban and Peri-Urban Adaptation

Urban and peri-urban forests are increasingly recognized as adaptation assets. They reduce heat island effects, regulate stormwater, and provide cultural and recreational services. Integrating green infrastructure into city planning—such as Delhi's urban forest corridors or Nairobi's Karura Forest—demonstrates how adaptation can extend beyond rural landscapes (MoEFCC, 2023). Embedding forest resilience into urban climate strategies is particularly urgent in rapidly urbanizing regions of Asia and Africa.

Livelihood Diversification and Social Safety Nets

Diversifying income and strengthening social protection are vital for communities facing climate risks. Ecotourism, sustainable non-timber forest product (NTFP) enterprises, and value-added agroforestry goods—such as shade-grown coffee, honey, and essential oils—can raise household incomes by 40–60% while maintaining forest cover above 70% (Scherr & McNeely, 2008). Mobile platforms and cooperatives facilitate market access, price information, and logistics, reducing transaction costs and empowering small producers (Giel, 2008).

Non-timber forest products remain a cornerstone of livelihood security in many forest-dependent regions. In northeastern Bangladesh, Das (2014) documented that rural households exhibit high levels of dependency on NTFPs, with seasonal fluctuations shaping both subsistence use and market sales. Such reliance underscores the dual role of NTFPs as subsistence buffers and sources of cash income, highlighting their importance in poverty alleviation and resilience building.

Crop insurance and public works programs—when designed to support conservation—provide safety nets that cushion climate shocks without encouraging land-use change. In Ethiopia's Productive Safety Net Programme (PSNP), food and cash transfers are linked to watershed rehabilitation and tree planting, generating livelihood benefits alongside ecosystem restoration.

Agroforestry has also emerged as a promising adaptation strategy that balances food production with ecological sustainability. In Bangladesh, experiments with Jatropha-based agri-silviculture systems demonstrated that soybean (*Glycine max*) yields increased significantly when integrated with tree cover compared to sole cropping systems (Das, 2015). These findings highlight the potential of diversified land-use practices to enhance farmer incomes, restore soil fertility, and reduce pressure on natural forests. By embedding crop production within tree-based systems, communities can simultaneously address livelihood needs and strengthen ecological resilience.

Integrated Landscape Approaches

Landscape approaches coordinate land uses—production, conservation, restoration, and settlement—to optimize synergies and manage trade-offs across scales (Sayer et al., 2013). Key elements include:

- *Spatial Planning:* Zoning maps allocate areas for protected forests, sustainable agriculture, restoration corridors, and settlements based on biophysical and socio-economic criteria (Adelaja et al., 1989).
- *Multi-Stakeholder Platforms:* Convening governments, Indigenous organizations, private sector, NGOs, and researchers fosters shared vision, conflict resolution, and joint decision-making.
- *Common Metrics:* Monitoring frameworks track carbon, biodiversity, water, and livelihood indicators to ensure benefits and costs are equitably shared (Murphy & Heberling, 1994).

In Ghana's cocoa landscape, participatory mapping with farmers, industry, and conservationists designated 30% forest retention, 40% intensification of existing farmland through agroforestry, and 30% restoration zones. Over five years, deforestation dropped by 20%, cocoa yields rose 15%, and farmer incomes increased by 12% (Murphy & Heberling, 1994).

Innovative Financing and Governance

Blended Finance and Green Bonds

Mobilizing the estimated US $200 billion per year needed for land-sector transformation requires innovative finance. Blended finance pools concessional grants, soft loans, and private capital to de-risk investments in sustainable landscapes. The Green Climate Fund's pilot "Landscape Program in Brazil" earmarked US $150 million—25% concessional—for agroforestry, restoration, and technical assistance for smallholders. Green bonds—debt instruments earmarked for environmental projects—have financed large-scale reforestation. Chile's US $500 million green bond funds the planting of 150,000 ha of native species, backed by sovereign guarantees and performance-linked coupons.

Payments for Ecosystem Services (PES) and Insurance

PES schemes remunerate land stewards for carbon, water, biodiversity, and landscape beauty. Costa Rica's national PES program, financed through a fuel tax, pays landowners up to US $90 ha^{-1} yr^{-1} for forest protection and agroforestry, doubling forest cover between 1996 and 2016. Kenya's Upper Tana–Nairobi Water Fund pools US $10 million annually from water utilities and hydropower firms to pay farmers for riparian restoration, reducing sedimentation by 40% and securing US $500 million in water revenues (Saravanan et al., 2024).

Parametric insurance products—indexed to precipitation, temperature, or fire severity—offer swift payouts when triggers are met. In Mexico's Yucatán, a pilot fire-index insurance covered 20,000 ha of community forest, disbursing US $1 million after the 2019 fire season to finance rapid replanting.

Tenure Security and Digital Governance

Legal recognition of customary land rights is foundational. Liberia's pilot blockchain land registry validated claims for 50 community forests, reducing disputes by 35% and unlocking US $5 million in sustainable timber investments. In Peru, digital traceability platforms using QR codes and blockchain link cacao cooperatives to global buyers, embedding social and environmental covenants in smart contracts that automate premium payments (Saravanan et al., 2024).

Adaptive Management, Learning, and Early Warning

Resilience demands dynamic learning cycles of planning, action, monitoring, and reflection (Turner et al., 2003). Adaptive management frameworks—such as the Forest Resilience Alliance in Oregon—bring together federal agencies, tribes, and NGOs to co-design fire risk maps, pilot prescribed burns, and evaluate ecological and social outcomes annually (Foro Económico Mundial, 2021).

Early warning systems integrate satellite analytics, local sensors, and community reporting. Europe's EFFIS platform aggregates fire danger indices, while Spain's Catalonia region combines automated alerts with community brigades to reduce large fire incidents by 45% between 2000 and 2020. In the Amazon, the Monitoring of the Andean Amazon Project (MAAP) uses high-resolution imagery and local networks to detect deforestation incursions within 72 hours.

Scenario planning workshops engage stakeholders in stress-testing landscape plans under divergent futures—varying commodity prices, climate scenarios, and policy regimes. In Vietnam's Central Highlands, shifting from coffee monocultures to diverse agroforestry buffered incomes against a 25% yield decline under 2 °C warming while enhancing watershed protection (Brown et al., 2016).

Mitigation and adaptation in forest landscapes are mutually reinforcing. REDD+, corporate commitments, and restoration curb emissions and revive carbon stocks, while EbA, community forestry, and livelihood diversification strengthen capacity to absorb shocks. Integrated landscape approaches and innovative finance align incentives and mobilize resources. Tenure security, digital governance, and early warning systems safeguard rights and accelerate responses. Central to success are adaptive management and co-production of knowledge, ensuring that strategies evolve with shifting baselines. By weaving these into equity-centered frameworks, societies can sustain forest functions, safeguard cultural heritage, and chart resilient futures in an era of escalating change.

FUTURE RESEARCH DIRECTIONS

Forest ecosystems face a web of interacting threats that demand nuanced, forward-looking research. Addressing these challenges requires both deepening our understanding of complex processes and innovating methods that integrate ecological, climatic, and socio-cultural dimensions. The following priorities outline pathways toward more resilient and equitable forest futures.

1. Towards Integrated Socio-Ecological Research
    - *Quantify synergistic impacts:* Develop coupled socio-ecological models to assess how deforestation, climate change, and cultural disruption amplify one another's effects on ecosystem services.
    - *Long-term studies:* Establish multi-decadal research sites to monitor biodiversity, carbon stocks, and community practices over time.
    - *Socio-cultural dynamics:* Map social networks and cultural values to understand how norms shape land-use decisions, conservation adoption, and resilience.
2. High-Resolution Monitoring and Digital Innovation
    - *Remote sensing and AI:* Harness drones, satellite imagery, and machine learning to detect early-warning signals of degradation, illegal logging, and land-use change at fine scales.

- *Digital twin forests:* Expand the use of virtual forest models and soil biodiversity monitoring systems (e.g., MESOSCAN) to simulate ecosystem responses under different scenarios.
- *Citizen science:* Integrate community-based platforms to ground-truth remote sensing data, especially in Global South regions where data gaps remain acute.

3. Climate–Forest Feedbacks and Tipping Points
   - *Quantify thresholds and risks:* Assess the likelihood of Amazon dieback, boreal biome shifts, and peatland collapse under varying warming scenarios.
   - *Investigate compound extremes:* Examine how drought, fire, pests, and disease outbreaks interact to trigger cascading ecosystem failures.
   - *Expand monitoring networks:* Strengthen long-term ecological observatories in underrepresented regions such as Sub-Saharan Africa and South Asia.
4. Socio-Cultural Dimensions and Equity
   - *Gender and youth dynamics:* Examine how gendered divisions of labor shape vulnerability and adaptive capacity, and how Indigenous women's leadership strengthens governance.
   - *Traditional ecological knowledge (TEK):* Document, revitalize, and integrate TEK into governance and adaptation strategies.
   - *Intergenerational transfer:* Assess how youth migration and urbanization affect cultural continuity and forest stewardship.
5. Urbanization, Health, and Emerging Risks
   - *Urban–forest interfaces:* Evaluate how expanding cities reshape forest landscapes, alter hydrological cycles, and increase demand for timber and charcoal.
   - *Public health linkages:* Investigate how forest fragmentation and wildlife trade heighten zoonotic disease risks, integrating ecology, epidemiology, and urban planning.
6. Governance, Policy, and Finance
   - *Rights-based approaches:* Evaluate the effectiveness of Free, Prior, and Informed Consent (FPIC) and Indigenous land titling in strengthening conservation outcomes.
   - *Assess carbon finance:* Critically examine REDD+, PES, and green bonds, with attention to risks of greenwashing, elite capture, and long-term sustainability.
   - *Digital governance:* Explore blockchain traceability, participatory monitoring apps, and other innovations to enhance transparency and accountability.

7. Integrated Modeling and Scenario Analysis
    - *Develop hybrid models:* Combine process-based ecology, agent-based socio-economic dynamics, and machine learning.
    - *Participatory scenario planning:* Incorporate community perspectives into model design to ensure outputs are scientifically robust and socially relevant.
    - *Policy foresight:* Apply system dynamics to evaluate alternative land-tenure frameworks, governance reforms, and incentive schemes.
8. Pathways to Transformative Resilience
    - *Shift from risks to solutions:* Move beyond documenting threats to designing transformative interventions.
    - *Identify leverage points:* Target opportunities in supply chains, governance, and finance that can disrupt destructive feedbacks.
    - *Scale community-based adaptation:* Test models that integrate TEK, gender equity, and youth leadership.
    - *Narratives for mobilization:* Explore cultural framings that inspire collective action and embed justice and equity into resilience strategies.

Future research must transcend siloed approaches, embracing systems thinking that reflects the interdependence of ecological, climatic, and socio-cultural processes. By combining cutting-edge technology with traditional knowledge, strengthening equity and governance, and designing transformative solutions, research can chart pathways toward regenerative and resilient forest futures.

SOLUTIONS AND RECOMMENDATIONS

The intertwined crises of deforestation, climate change, and socio-cultural disruption demand solutions that are systemic, inclusive, and forward-looking. Past approaches often treated ecological, economic, and social challenges in isolation, but evidence shows that only integrated strategies can deliver durable outcomes. Solutions must therefore combine policy reform, adaptive management, community empowerment, innovative finance, and technological innovation, while embedding equity and justice at their core.

1. Policy and Institutional Reforms
    - *Integrated land-use zoning:* Clearly delineate conservation, agriculture, and development zones to reduce conflict and fragmentation.
    - *Secure land tenure:* Provide legal recognition and streamlined registration of Indigenous and community land rights to strengthen stewardship.

- *Mainstream forest targets:* Embed conservation goals into national climate commitments (e.g., NDCs) and development plans.
- *Cross-sectoral coordination:* Establish governance bodies that align forestry, agriculture, infrastructure, and social welfare policies, supported by anti-corruption safeguards and strong enforcement.

2. Adaptive Management and Ecosystem-Based Strategies
    - *Ecosystem-based management:* Prioritize biodiversity, carbon storage, and water regulation simultaneously.
    - *Iterative monitoring:* Integrate real-time ecological and social data into adaptive decision-making cycles.
    - *Risk reduction:* Strengthen fire management, pest control, and early-warning systems through landscape-scale coordination.
    - *Climate-smart restoration:* Promote mixed-species plantations, natural regeneration, and Indigenous-led practices such as mosaic burning to enhance resilience.
3. Community Engagement and Traditional Knowledge
    - *Co-design programs:* Ensure conservation initiatives are developed with Indigenous and local communities, guaranteeing equitable benefit-sharing.
    - *Revitalize TEK:* Document, validate, and scale practices such as sacred groves, rotational agroforestry, and customary dispute-resolution systems.
    - *Capacity building:* Provide participatory training on sustainable harvesting, restoration, and governance.
    - *Inclusive governance:* Integrate gender and youth perspectives, recognizing their unique roles in ecological knowledge and stewardship.
4. Financing Mechanisms and Economic Incentives
    - *Payments for ecosystem services (PES):* Reward communities for maintaining forest cover and carbon stocks.
    - *Green bonds and blended finance:* Mobilize private capital for reforestation, sustainable forestry, and community-based development.
    - *Carbon markets and REDD+:* Channel international finance into local conservation, embedding safeguards to prevent greenwashing and elite capture.
    - *Livelihood diversification:* Support microfinance, eco-tourism, and non-timber forest products to reduce dependence on deforestation.
5. Technological Innovations and Monitoring
    - *Remote sensing and AI:* Detect illegal logging, degradation, and land-use change in real-time.

- *Community-based apps:* Equip rangers with mobile tools to report incursions, fires, and biodiversity observations.
- *Open-access dashboards:* Synthesize ecological, climatic, and socioeconomic indicators for policymakers and practitioners.
- *Drone and GIS mapping:* Guide restoration, corridor planning, and habitat connectivity assessments.
- *Blockchain traceability:* Secure supply chains for forest-risk commodities to ensure compliance with deforestation-free commitments.

6. Cross-Sector Collaboration and Capacity Building
    - *Transdisciplinary networks:* Bridge ecology, climate science, social sciences, and economics to co-produce knowledge.
    - *Multi-stakeholder platforms:* Convene governments, NGOs, private sector actors, and communities to align incentives and resolve conflicts.
    - *Training programs:* Build skills in adaptive management, participatory governance, and conflict resolution for forest professionals and community leaders.
    - *Global exchanges:* Expand South–South and North–South collaborations to share best practices and innovations.
7. Embedding Justice and Equity
    - *Climate justice lens:* Prioritize equitable benefit-sharing, Indigenous rights, and gender-responsive governance frameworks.
    - *Legitimacy and stewardship:* Ensure inclusive governance structures that foster long-term stewardship and resilience.
    - *Ethical safeguards:* Prevent exploitation and ensure transparency in finance, policy, and conservation programs.

CONCLUSION

Deforestation and climate change are deeply interconnected crises with ecological, social, and cultural consequences. Forest loss drives carbon emissions, biodiversity decline, and ecosystem disruption, while displacing communities, eroding traditional knowledge, and widening inequalities. Climate change acts as a force multiplier, intensifying these risks through droughts, fires, and extreme events. Yet solutions exist. Rights-based governance, community forestry, ecosystem-based adaptation, and innovative finance can protect forests while sustaining livelihoods. Embedding justice, gender equity, and youth participation makes strategies more inclusive and durable. Ultimately, safeguarding forests requires integrated approaches that recognize them as socio-ecological systems. By aligning mitigation, adaptation, and resilience pathways, societies can protect forests not only as carbon sinks but also as cultural and ecological commons—essential foundations for a just and sustainable future.

REFERENCES

Abada Mbolo, M. M., Zekeng, J. C., Mala, W. A., Fobane, J. L., Djomo Chimi, C., Tangboulou Ngavounsia, , Nyako, C. M., Menyene, L. F. E., & Tamanjong, Y. V. (2016). The role of cocoa agroforestry systems in conserving forest tree diversity in the Central region of Cameroon. *Agroforestry Systems*, *90*(4), 577–590. DOI: 10.1007/s10457-016-9945-8

Adelaja, A., Derr, D., & Rose-Tank, K. (1989). Economic and equity implications of land-use zoning in suburban agriculture. *Journal of Agricultural Ethics*, *2*(2), 97–112. DOI: 10.1007/BF01826926

Adger, W. N. (2003). Social capital, collective action, and adaptation to climate change. *Economic Geography*, *79*(4), 387–404. DOI: 10.1111/j.1944-8287.2003.tb00220.x

Agarwal, B. (2010). *Gender and green governance: The political economy of women's presence within and beyond community forest*. Oxford University Press. DOI: 10.1093/acprof:oso/9780199569687.001.0001

Aitken, S. N., Yeaman, S., Holliday, J. A., Wang, T., & Curtis-McLane, S. (2008). Adaptation, migration or extirpation: Climate change outcomes for tree populations. *Evolutionary Applications*, *1*(1), 95–111. DOI: 10.1111/j.1752-4571.2007.00013.x

Andela, N., Morton, D. C., Giglio, L., Paugam, R., Chen, Y., Hantson, S., Van Der Werf, G. R., & Randerson, J. T. (2019). The Global Fire Atlas of individual fire size, duration, speed and direction. *Earth System Science Data*, *11*(2), 529–552. DOI: 10.5194/essd-11-529-2019

Anderegg, W. R. L., Hicke, J. A., Fisher, R. A., Allen, C. D., Aukema, J., Bentz, B., Hood, S., Lichstein, J. W., Macalady, A. K., McDowell, N., Pan, Y., Raffa, K., Sala, A., Shaw, J. D., Stephenson, N. L., Tague, C., & Zeppel, M. (2015). Tree mortality from drought, insects, and their interactions in a changing climate. *The New Phytologist*, *208*(3), 674–683. DOI: 10.1111/nph.13477

Balick, M. J., & Cox, P. A. (2021). *Plants, people, and culture: The science of ethnobotany* (2nd ed.). CRC Press, Taylor & Francis Group.

Balvanera, P., Pfaff, A., Viña, A., Frapolli, E. G., & Hussain, S. A. (2019). *IPBES global assessment report on biodiversity and ecosystem services*. IPBES Secretariat.

Barr, C., Barney, K., & Laird, S. A. (2014). Governance failures and the fragmentation of tropical forests. In Kettle, C. J., & Koh, L. P. (Eds.), *Global forest fragmentation* (pp. 132–157). CAB International. DOI: 10.1079/9781780642031.0132

Bentz, B. J., Régnière, J., Fettig, C. J., Hansen, E. M., Hayes, J. L., Hicke, J. A., Kelsey, R. G., Negrón, J. F., & Seybold, S. J. (2010). Climate change and bark beetles of the western United States and Canada: Direct and indirect effects. *Bioscience, 60*(8), 602–613. DOI: 10.1525/bio.2010.60.8.6

Berenguer, E., Armenteras, D., Lees, A. C., Fearnside, P. M., Alencar, A., Almeida, C., Aragão, L., Barlow, J., Bilbao, B., Brando, P., Bynoe, P., Finer, M., Flores, B. M., Jenkins, C. N., Silva, C.Jr, Smith, C., Souza, C., García-Vilacorta, R., & Nascimento, N. (2024). Drivers and ecological impacts of deforestation and forest degradation in the Amazon. *Acta Amazonica, 54*(spe1), e54es22342. https://doi.org/ DOI: 10.1590/1809-4392202203420

Bernhard, K. P., Shapiro, A. C., & Hunt, C. A. (2024). Drivers of tropical deforestation: A global review of methodological approaches and analytical scales. *Biodiversity and Conservation, 33*(1), 1–29. DOI: 10.1007/s10531-023-02747-z

Bertrand, R., Lenoir, J., Piedallu, C., Riofrío-Dillon, G., De Ruffray, P., Vidal, C., Pierrat, J.-C., & Gégout, J.-C. (2011). Changes in plant community composition lag behind climate warming in lowland forests. *Nature, 479*(7374), 517–520. DOI: 10.1038/nature10548

Beuchle, R., Bourgoin, C., Carreiras, J., Heinrich, V., & Achard, F. (2025). *Deforestation and forest degradation in the Amazon: Update for year 2023 and assessment of humid forest regrowth.* Publications Office of the European Union., DOI: 10.2760/7304684

Bonan, G. B. (2008). Forests and climate change: Forcings, feedbacks, and the climate benefits of forests. *Science, 320*(5882), 1444–1449. DOI: 10.1126/science.1155121

Bowman, D. M. J. S., Balch, J., Artaxo, P., Bond, W. J., Cochrane, M. A., D'Antonio, C. M., DeFries, R., Johnston, F. H., Keeley, J. E., Krawchuk, M. A., Kull, C. A., Mack, M., Moritz, M. A., Pyne, S., Roos, C. I., Scott, A. C., Sodhi, N. S., & Swetnam, T. W. (2011). The human dimension of fire regimes on Earth. *Journal of Biogeography, 38*(12), 2223–2236. DOI: 10.1111/j.1365-2699.2011.02595.x

Brown, I., Martin-Ortega, J., Waylen, K., & Blackstock, K. (2016). Participatory scenario planning for developing innovation in community adaptation responses: Three contrasting examples from Latin America. *Regional Environmental Change, 16*(6), 1685–1700. DOI: 10.1007/s10113-015-0898-7

Carpenter, S. R., Cole, J. J., Pace, M. L., Batt, R., Brock, W. A., Cline, T., Coloso, J., Hodgson, J. R., Kitchell, J. F., Seekell, D. A., Smith, L., & Weidel, B. (2011). Early warnings of regime shifts: A whole-ecosystem experiment. *Science, 332*(6033), 1079–1082. DOI: 10.1126/science.1203672

Celis, N., Casallas, A., Lopez-Barrera, E. A., Felician, M., De Marchi, M., & Pappalardo, S. E. (2023). Climate change, forest fires, and territorial dynamics in the Amazon rainforest: An integrated analysis for mitigation strategies. *ISPRS International Journal of Geo-Information, 12*(10), 436. DOI: 10.3390/ijgi12100436

Chhatre, A., & Agrawal, A. (2009). Trade-offs and synergies between carbon storage and livelihood benefits from forest commons. *Proceedings of the National Academy of Sciences of the United States of America, 106*(42), 17667–17670. DOI: 10.1073/pnas.0905308106

Clark, K. L., Skowronski, N., & Gallagher, M. (2015). Fire management and carbon sequestration in pine barren forests. *Journal of Sustainable Forestry, 34*(1-2), 125–146. DOI: 10.1080/10549811.2014.973607

Colchester, M., & Boscolo, M. (2006). *Justice in the forest: Rural livelihoods and forest law enforcement.* Center for International Forestry Research.

Colchester, M., & Lohmann, L. (Eds.). (1993). *The struggle for land and the fate of the forests.* World Rainforest Movement; Ecologist; Zed Books.

Crooks, K. R., & Sanjayan, M. A. (Eds.). (2006). *Connectivity conservation.* Cambridge University Press. DOI: 10.1017/CBO9780511754821

Crouzeilles, R., Ferreira, M. S., Chazdon, R. L., Lindenmayer, D. B., Sansevero, J. B. B., Monteiro, L., Iribarrem, A., Latawiec, A. E., & Strassburg, B. B. N. (2017). Ecological restoration success is higher for natural regeneration than for active restoration in tropical forests. *Science Advances, 3*(11), e1701345. DOI: 10.1126/sciadv.1701345

Curtis, P. G., Slay, C. M., Harris, N. L., Tyukavina, A., & Hansen, M. C. (2018). Classifying drivers of global forest loss. *Science, 361*(6407), 1108–1111. DOI: 10.1126/science.aau3445

Dakos, V., Scheffer, M., Van Nes, E. H., Brovkin, V., Petoukhov, V., & Held, H. (2008). Slowing down as an early warning signal for abrupt climate change. *Proceedings of the National Academy of Sciences of the United States of America, 105*(38), 14308–14312. DOI: 10.1073/pnas.0802430105

Damnyag, L., Saastamoinen, O., Appiah, M., & Pappinen, A. (2012). Role of tenure insecurity in deforestation in Ghana's high forest zone. *Forest Policy and Economics, 14*(1), 90–98. DOI: 10.1016/j.forpol.2011.08.006

Das, N. (2014). Assessment of dependency levels of the forest community people livelihoods through non timber forest products in the north eastern region of Bangladesh. *International Journal of Forest Usufructs Management, 15*(1), 61–69.

Das, N. (2015). Comparative growth analysis and yield performance of *Glycine max* under *Jatropha curcas* based agrisilviculture system of agroforestry in the northern part of Bangladesh. *Journal of Forestry*, *2*(2), 14–23. DOI: 10.18488/journal.101/2015.2.2/101.2.14.23

Das, N. (2024). Tree species diversity, composition and structure in the tropical moist deciduous forest of Kadigarh National Park, Mymensingh, Bangladesh. *Asian Journal of Forestry*, *8*(1), 41–49. DOI: 10.13057/asianjfor/r080104

Das, N. (2025a). *Biodiversity in forest ecosystems: Patterns, drivers, and conservation strategies*. Nova Science Publishers., DOI: 10.52305/AZJY2712

Das, N. (2025b). Disturbance gradients reshape biodiversity, stand structure, and regeneration dynamics in Sal (*Shorea robusta*) forests of Dharmapur National Park, Bangladesh. *Preprints*. https://doi.org/DOI: 10.20944/preprints202509.2279.v1

Das, N., Mahapatra, C. K., Biswas, S. K., Das, P., & Majumdar, A. (2018). Suitability of the normal, log-normal and weibull distributions for modeling diameter and height distributions of *Swietenia Mahagoni* plantations in Bangladesh. *Journal of Biology and Nature*, *8*(4), 146–155.

Das, N., & Sarker, S. K. (2015). Tree species diversity and productivity relationship in the central region of Bangladesh. *Journal of Forestry*, *2*(2), 24–33. DOI: 10.18488/journal.101/2015.2.2/101.2.24.33

Davidson, E. A. (2009). The contribution of manure and fertilizer nitrogen to atmospheric nitrous oxide since 1860. *Nature Geoscience*, *2*(9), 659–662. DOI: 10.1038/ngeo608

De Sherbinin, A., Levy, M., Adamo, S., MacManus, K., Yetman, G., Mara, V., Razafindrazay, L., Goodrich, B., Srebotnjak, T., Aichele, C., & Pistolesi, L. (2012). Migration and risk: Net migration in marginal ecosystems and hazardous areas. *Environmental Research Letters*, *7*(4), 045602. DOI: 10.1088/1748-9326/7/4/045602

DeFries, R., Hansen, A., Newton, A. C., & Hansen, M. C. (2005). Increasing isolation of protected areas in tropical forests over the past twenty years. *Ecological Applications*, *15*(1), 19–26. DOI: 10.1890/03-5258

DeFries, R. S., Rudel, T., Uriarte, M., & Hansen, M. (2010). Deforestation driven by urban population growth and agricultural trade in the twenty-first century. *Nature Geoscience*, *3*(3), 178–181. DOI: 10.1038/ngeo756

Devaraju, N., Bala, G., & Modak, A. (2015). Effects of large-scale deforestation on precipitation in the monsoon regions: Remote versus local effects. *Proceedings of the National Academy of Sciences of the United States of America, 112*(11), 3257–3262. DOI: 10.1073/pnas.1423439112

Fearnside, P. M. (2005). Deforestation in Brazilian Amazonia: History, rates, and consequences. *Conservation Biology, 19*(3), 680–688. DOI: 10.1111/j.1523-1739.2005.00697.x

Ferreira De Athayde, S., Mosimann Da Silva, G., Kaiabi, J., Kaiabi, M., Rocha De Souza, H., Ono, K., & Bruna, E. M. (2006). Participatory research and management of arumã by the Kaiabi people in the Brazilian Amazon. *Journal of Ethnobiology, 26*(1), 36–59. DOI: 10.2993/0278-0771(2006)26[36:PRAMOA]2.0.CO;2

Flannigan, M. D., Krawchuk, M. A., De Groot, W. J., Wotton, B. M., & Gowman, L. M. (2009). Implications of changing climate for global wildland fire. *International Journal of Wildland Fire, 18*(5), 483–495. DOI: 10.1071/WF08187

Food and Agriculture Organization of the United Nations (FAO). (2012). *Voluntary guidelines on the responsible governance of tenure of land, fisheries and forests in the context of national food security*. FAO.

Food and Agriculture Organization of the United Nations (FAO). (2016). *Climate change, agriculture and food security*. FAO.

Ford, B., Val Martin, M., Zelasky, S. E., Fischer, E. V., Anenberg, S. C., Heald, C. L., & Pierce, J. R. (2018). Future fire impacts on smoke concentrations, visibility, and health in the contiguous United States. *GeoHealth, 2*(8), 229–247. DOI: 10.1029/2018GH000144

Foro Económico Mundial. Marsh & McLennan, SK Group, & Zurich Insurance Group. (2021). *The global risks report 2021 insight report* (16th ed.). World Economic Forum.

Gaveau, D. L. A., Salim, M. A., Hergoualc'h, K., Locatelli, B., Sloan, S., Wooster, M., Marlier, M. E., Molidena, E., Yaen, H., DeFries, R., Verchot, L., Murdiyarso, D., Nasi, R., Holmgren, P., & Sheil, D. (2014). Major atmospheric emissions from peat fires in Southeast Asia during non-drought years: Evidence from the 2013 Sumatran fires. *Scientific Reports, 4*(1), 6112. DOI: 10.1038/srep06112

Geist, H. J., & Lambin, E. F. (2002). Proximate causes and underlying driving forces of tropical deforestation. *Bioscience, 52*(2), 143–150. DOI: 10.1641/0006-3568(2002)052[0143:PCAUDF]2.0.CO;2

Global Forest Watch. (2024). Deforestation data and imagery. World Resources Institute. https://www.globalforestwatch.org

Grantham, R. (2012). Incentivizing conservation and customer behavior through pricing signals. *Proceedings of the Water Environment Federation, 2012*(1), 108–113. DOI: 10.2175/193864712811694569

Griscom, B. W., Adams, J., Ellis, P. W., Houghton, R. A., Lomax, G., Miteva, D. A., Schlesinger, W. H., Shoch, D., Siikamäki, J. V., Smith, P., Woodbury, P., Zganjar, C., Blackman, A., Campari, J., Conant, R. T., Delgado, C., Elias, P., Gopalakrishna, T., Hamsik, M. R., & Fargione, J. (2017). Natural climate solutions. *Proceedings of the National Academy of Sciences of the United States of America, 114*(44), 11645–11650. DOI: 10.1073/pnas.1710465114

Hansen, M. C., Potapov, P. V., Moore, R., Hancher, M., Turubanova, S. A., Tyukavina, A., Thau, D., Stehman, S. V., Goetz, S. J., Loveland, T. R., Kommareddy, A., Egorov, A., Chini, L., Justice, C. O., & Townshend, J. R. G. (2013). High-resolution global maps of 21st-century forest cover change. *Science, 342*(6160), 850–853. DOI: 10.1126/science.1244693

Hareesh, G., & Shailaja, N. (2025). Deforestation in the 21st century: Social impacts and community consequences. *Contemporaneity of Language and Literature in the Robotized Millennium, 7*(2), 1–12. REST Publisher. https://restpublisher.com/wp-content/uploads/2025/04/34.-Deforestation-in-the-21st-Century-Social-Impacts-and-Community-Consequences.pdf

Hosonuma, N., Herold, M., De Sy, V., De Fries, R. S., Brockhaus, M., Verchot, L., Angelsen, A., & Romijn, E. (2012). An assessment of deforestation and forest degradation drivers in developing countries. *Environmental Research Letters, 7*(4), 044009. DOI: 10.1088/1748-9326/7/4/044009

Houghton, R. A. (2005). Aboveground forest biomass and the global carbon balance. *Global Change Biology, 11*(6), 945–958. DOI: 10.1111/j.1365-2486.2005.00955.x

Intergovernmental Panel on Climate Change (IPCC). (2022). *Climate change 2022: Impacts, adaptation and vulnerability. Contribution of Working Group II to the Sixth Assessment Report of the Intergovernmental Panel on Climate Change* (Pörtner, H.-O., Roberts, D. C., & Tignor, M., Eds.). Cambridge University Press., DOI: 10.1017/9781009325844

Jobbágy, E. G., & Jackson, R. B. (2000). The vertical distribution of soil organic carbon and its relation to climate and vegetation. *Ecological Applications, 10*(2), 423–436. DOI: 10.1890/1051-0761(2000)010[0423:TVDOSO]2.0.CO;2

Kaimowitz, D., Angelsen, A., & Center for International Forestry Research. (1998). *Economic models of tropical deforestation: A review*. CIFOR.

Kazi, M. R. (2024). Forests support Indigenous cultures: A vital relationship for sustainability. *Nourish the Planet*. https://nourishtheplanet.com/publication/forests-support-indigenous-cultures-a-vital-relationship-for-sustainability/

Korir, J., Munyao, C., Kigen, C., Too, P., & Chesire, M. (2025). Harnessing nature's resilience through farmer-managed natural regeneration for climate change adaptation in the central rift, Kenya. *International Journal of Environmental Sciences, 8*(Special Issue 1), 80–99. https://doi.org/DOI: 10.47604/ijes.3163

Kübler, D., & Günter, S. (2024). Forest restoration for climate change mitigation and adaptation. In P. Katila, C. J. P. Colfer, W. De Jong, G. Galloway, P. Pacheco, & G. Winkel (Eds.), *Restoring forests and trees for sustainable development: Policies, practices, impacts, and ways forward* (pp. 135–159). Oxford University Press. https://doi.org/DOI: 10.1093/9780197683958.003.0006

Kuck, T. N., Silva Filho, P. F. F., Sano, E. E., Bispo, P. D. C., Shiguemori, E. H., & Dalagnol, R. (2021). Change detection of selective logging in the Brazilian Amazon using X-band SAR data and pre-trained convolutional neural networks. *Remote Sensing (Basel), 13*(23), 4944. DOI: 10.3390/rs13234944

Kudo, G., & Cooper, E. J. (2019). When spring ephemerals fail to meet pollinators: Mechanism of phenological mismatch and its impact on plant reproduction. *Proceedings of the Royal Society B: Biological Sciences, 286*(1904), 20190573. https://doi.org/DOI: 10.1098/rspb.2019.0573

Kurz, W. A., Dymond, C. C., Stinson, G., Rampley, G. J., Neilson, E. T., Carroll, A. L., Ebata, T., & Safranyik, L. (2008). Mountain pine beetle and forest carbon feedback to climate change. *Nature, 452*(7190), 987–990. DOI: 10.1038/nature06777

Lal, R. (2004). Soil carbon sequestration to mitigate climate change. *Geoderma, 123*(1–2), 1–22. DOI: 10.1016/j.geoderma.2004.01.032

Lambin, E. F., & Meyfroidt, P. (2011). Global land use change, economic globalization, and the looming land scarcity. *Proceedings of the National Academy of Sciences of the United States of America, 108*(9), 3465–3472. DOI: 10.1073/pnas.1100480108

Laurance, W. F., Lovejoy, T. E., Vasconcelos, H. L., Bruna, E. M., Didham, R. K., Stouffer, P. C., Gascon, C., Bierregaard, R. O., Laurance, S. G., & Sampaio, E. (2002). Ecosystem decay of Amazonian forest fragments: A 22-year investigation. *Conservation Biology, 16*(3), 605–618. DOI: 10.1046/j.1523-1739.2002.01025.x

Lenoir, J., Gégout, J., Guisan, A., Vittoz, P., Wohlgemuth, T., Zimmermann, N. E., Dullinger, S., Pauli, H., Willner, W., & Svenning, J. (2010). Going against the flow: Potential mechanisms for unexpected downslope range shifts in a warming climate. *Ecography, 33*(2), 295–303. DOI: 10.1111/j.1600-0587.2010.06279.x

Lovejoy, T. E., & Nobre, C. (2019). Amazon tipping point: Last chance for action. *Science Advances, 5*(12), eaba2949. Advance online publication. DOI: 10.1126/sciadv.aba2949

Malhi, Y., Aragão, L. E. O. C., Galbraith, D., Huntingford, C., Fisher, R., Zelazowski, P., Sitch, S., McSweeney, C., & Meir, P. (2009). Exploring the likelihood and mechanism of a climate-change-induced dieback of the Amazon rainforest. *Proceedings of the National Academy of Sciences of the United States of America, 106*(49), 20610–20615. DOI: 10.1073/pnas.0804619106

Marshall, N. A., & Stokes, C. J. (2014). Influencing adaptation processes on the Australian rangelands for social and ecological resilience. *Ecology and Society, 19*(2), 14. Advance online publication. DOI: 10.5751/ES-06440-190214

Marvin, D. C., Sleeter, B. M., Cameron, D. R., Nelson, E., & Plantinga, A. J. (2023). Natural climate solutions provide robust carbon mitigation capacity under future climate change scenarios. *Scientific Reports, 13*(1), 19008. DOI: 10.1038/s41598-023-43118-6

Mattos, P. P. D., Salis, S. M., Braz, E. M., & Crispim, S. M. A. (2010). Sustainable management of natural forests in Pantanal region, Brazil. *Ciência Florestal, 20*(2), 321–333. DOI: 10.5902/198050981855

McDougall, C. (Ed.). (2008). *Forests of learning: Experiences from research on an adaptive collaborative approach to community forestry in Nepal; a synthesis of lessons from the Adaptive Collaborative Management Research Project in Nepal, 1999–2002 and 2004–2007*. Center for International Forestry Research.

McDowell, N., Pockman, W. T., Allen, C. D., Breshears, D. D., Cobb, N., Kolb, T., Plaut, J., Sperry, J., West, A., Williams, D. G., & Yepez, E. A. (2008). Mechanisms of plant survival and mortality during drought: Why do some plants survive while others succumb to drought? *The New Phytologist, 178*(4), 719–739. DOI: 10.1111/j.1469-8137.2008.02436.x

McLaughlin, S., McKenna, J., & Cooper, J. A. G. (2002). Socio-economic data in coastal vulnerability indices: Constraints and opportunities. *Journal of Coastal Research, 36*, 487–497. DOI: 10.2112/1551-5036-36.sp1.487

Mendes, L. M., César, R. G., Uezu, A., Beltrame, T. P., Rodriguez, L. C. E., Gomes, H. B., & Cullen, L. (2021). Large canopy and animal-dispersed species facilitate natural regeneration in tropical forest restoration. *Restoration Ecology*, *29*(4), e13406. DOI: 10.1111/rec.13406

Meyfroidt, P., Rudel, T. K., & Lambin, E. F. (2010). Forest transitions, trade, and the global displacement of land use. *Proceedings of the National Academy of Sciences of the United States of America*, *107*(49), 20917–20922. DOI: 10.1073/pnas.1014773107

Ministry of Environment. Forest and Climate Change (MoEFCC). (2023). *Annual report 2023–24*. Government of India. https://www.moef.gov.in/uploads/2023/05/Annual-Report-English-2023-24.pdf

Montagnini, F., & Nair, P. K. R. (2004). Carbon sequestration: An underexploited environmental benefit of agroforestry systems. *Agroforestry Systems*, *61–62*(1–3), 281–295. DOI: 10.1023/B:AGFO.0000029005.92691.79

Murphy, B., Edwards, A., & Meyer, C. (Mick), & Russell-Smith, J. (Eds.). (2015). *Carbon accounting and savanna fire management*. CSIRO Publishing. https://doi.org/DOI: 10.1071/9780643108523

Murphy, D. J., & Heberling, M. E. (1994). Purchasing strategies for environmental restoration projects. *International Journal of Physical Distribution & Logistics Management*, *24*(5), 45–52. DOI: 10.1108/09600039410064026

Myers, N., Mittermeier, R. A., Mittermeier, C. G., Da Fonseca, G. A. B., & Kent, J. (2000). Biodiversity hotspots for conservation priorities. *Nature*, *403*(6772), 853–858. DOI: 10.1038/35002501

Nasi, R., Brown, D., Wilkie, D., Bennett, E., Tutin, C., van Tol, G., & Christophersen, T. (2008). *Conservation and use of wildlife-based resources: The bushmeat crisis* (CBD Technical Series No. 33). Secretariat of the Convention on Biological Diversity; Center for International Forestry Research.

Nettle, D., & Romaine, S. (2000). *Vanishing voices: The extinction of the world's languages*. Oxford University Press. DOI: 10.1093/oso/9780195136241.001.0001

Ostrom, E. (2015). *Governing the commons: The evolution of institutions for collective action*. Cambridge University Press. DOI: 10.1017/CBO9781316423936

Pan, Y., Birdsey, R. A., Fang, J., Houghton, R., Kauppi, P. E., Kurz, W. A., Phillips, O. L., Shvidenko, A., Lewis, S. L., Canadell, J. G., Ciais, P., Jackson, R. B., Pacala, S. W., McGuire, A. D., Piao, S., Rautiainen, A., Sitch, S., & Hayes, D. (2011). A large and persistent carbon sink in the world's forests. *Science, 333*(6045), 988–993. DOI: 10.1126/science.1201609

Pearson, R. G., & Dawson, T. P. (2003). Predicting the impacts of climate change on the distribution of species: Are bioclimate envelope models useful? *Global Ecology and Biogeography, 12*(5), 361–371. DOI: 10.1046/j.1466-822X.2003.00042.x

Peluso, N. L., & Vandergeest, P. (2011). Political ecologies of war and forests: Counterinsurgencies and the making of national natures. *Annals of the Association of American Geographers, 101*(3), 587–608. DOI: 10.1080/00045608.2011.560064

Purwestri, R. C., & Lusiana, B. (2025). Editorial: Socioeconomic implications of forests and forestry in a changing climate and governance. *Frontiers in Forests and Global Change, 8*, 620331. DOI: 10.3389/ffgc.2025.1620331

Rahman, M. H., & Alam, K. (2016). Forest dependent Indigenous communities' perception and adaptation to climate change through local knowledge in the protected area—A Bangladesh case study. *Climate (Basel), 4*(1), 12. DOI: 10.3390/cli4010012

Rice, R. A., & Ward, J. R. Smithsonian Migratory Bird Center, & Natural Resources Defense Council. (1996). *Coffee, conservation, and commerce in the Western Hemisphere: How individuals and institutions can promote ecologically sound farming and forest management in northern Latin America*. Smithsonian Migratory Bird Center; Natural Resources Defense Council.

Ripple, W. J., Wolf, C., Gregg, J. W., Rockström, J., Mann, M. E., Oreskes, N., Lenton, T. M., Rahmstorf, S., Newsome, T. M., Xu, C., Svenning, J.-C., Pereira, C. C., Law, B. E., & Crowther, T. W. (2024). The 2024 state of the climate report: Perilous times on planet Earth. *Bioscience, 74*(12), 812–824. DOI: 10.1093/biosci/biae087

Rudel, T. K., Coomes, O. T., Moran, E., Achard, F., Angelsen, A., Xu, J., & Lambin, E. (2005). Forest transitions: Towards a global understanding of land use change. *Global Environmental Change, 15*(1), 23–31. DOI: 10.1016/j.gloenvcha.2004.11.001

Saravanan, T., Ramesh, G., Hussein, R. R., Bansal, S., Kumar, Y. L., & Krishnakumar, K. (2024). Blockchain-based smart contracts for sustainable supply chain practices. In *2024 International Conference on IoT, Communication and Automation Technology (ICICAT)* (pp. 1138–1143). IEEE. https://doi.org/DOI: 10.1109/ICICAT62666.2024.10923000

Sarre, A. (Ed.). with Food and Agriculture Organization of the United Nations. (2020). *Global forest resources assessment, 2020: Main report*. Food and Agriculture Organization of the United Nations.

Sayer, J., Sunderland, T., Ghazoul, J., Pfund, J.-L., Sheil, D., Meijaard, E., Venter, M., Boedhihartono, A. K., Day, M., Garcia, C., Van Oosten, C., & Buck, L. E. (2013). Ten principles for a landscape approach to reconciling agriculture, conservation, and other competing land uses. *Proceedings of the National Academy of Sciences of the United States of America, 110*(21), 8349–8356. DOI: 10.1073/pnas.1210595110

Scherr, S. J., & McNeely, J. A. (2008). Biodiversity conservation and agricultural sustainability: Towards a new paradigm of 'ecoagriculture' landscapes. *Philosophical Transactions of the Royal Society of London. Series B, Biological Sciences, 363*(1491), 477–494. DOI: 10.1098/rstb.2007.2165

Schmink, M., & Wood, C. H. (1992). *Contested frontiers in Amazonia*. Columbia University Press.

Seydewitz, T., Pradhan, P., Landholm, D. M., & Kropp, J. P. (2023). Deforestation drivers across the tropics and their impacts on carbon stocks and ecosystem services. *Anthropocene Science, 2*(1), 81–92. DOI: 10.1007/s44177-023-00051-7

Shi, P., Li, Z., Li, P., Zhang, Y., & Li, B. (2021). Trade-offs among ecosystem services after vegetation restoration in China's Loess Plateau. *Natural Resources Research, 30*(3), 2703–2713. DOI: 10.1007/s11053-021-09841-5

Sikor, T., & Lund, C. (2009). Access and property: A question of power and authority. *Development and Change, 40*(1), 1–22. DOI: 10.1111/j.1467-7660.2009.01503.x

Silver, W. L., Ostertag, R., & Lugo, A. E. (2000). The potential for carbon sequestration through reforestation of abandoned tropical agricultural and pasture lands. *Restoration Ecology, 8*(4), 394–407. DOI: 10.1046/j.1526-100x.2000.80054.x

Soares-Filho, B. S., Nepstad, D. C., Curran, L. M., Cerqueira, G. C., Garcia, R. A., Ramos, C. A., Voll, E., McDonald, A., Lefebvre, P., & Schlesinger, P. (2006). Modelling conservation in the Amazon basin. *Nature, 440*(7083), 520–523. DOI: 10.1038/nature04389

Springate-Baginski, O., & Wollenberg, E. (2010). *REDD, forest governance and rural livelihoods: The emerging agenda*. CIFOR.

Stott, P. A. (1999). *Tropical rain forest: A political ecology of hegemonic mythmaking*. IEA Environment Unit.

Štreimikienė, D. (2014). Residential energy consumption trends, main drivers and policies in Lithuania. *Renewable & Sustainable Energy Reviews, 35*, 285–293. DOI: 10.1016/j.rser.2014.04.012

Tiwari, A., Rodrigues, L. C., Lucy, F. E., & Gharbia, S. (2022). Building climate resilience in coastal city living labs using ecosystem-based adaptation: A systematic review. *Sustainability (Basel), 14*(17), 10863. DOI: 10.3390/su141710863

Turner, B. L.II, Kasperson, R. E., Matson, P. A., McCarthy, J. J., Corell, R. W., Christensen, L., Eckley, N., Kasperson, J. X., Luers, A., Martello, M. L., Polsky, C., Pulsipher, A., & Schiller, A. (2003). A framework for vulnerability analysis in sustainability science. *Proceedings of the National Academy of Sciences of the United States of America, 100*(14), 8074–8079. DOI: 10.1073/pnas.1231335100

United Nations Development Programme (UNDP). (2024). Indigenous knowledge is crucial in the fight against climate change – here's why. *UNDP Climate Promise*. https://climatepromise.undp.org/news-and-stories/indigenous-knowledge-crucial-fight-against-climate-change-heres-why

United Nations Environment Programme (UNEP). (2023). *Annual report 2023*. UNEP. https://www.unep.org/resources/annual-report-2023

United Nations Framework Convention on Climate Change (UNFCCC). (2024). *UN Climate Change Conference: COP29, Baku, Azerbaijan, 11–22 November 2024*. UNFCCC. https://unfccc.int/cop29

Wang, L., Rosenfeldt, M., Koutsaviti, A., Harizani, M., Zhao, Y., Leelahakorn, N., Frachon, A., Raadam, M. H., Miettinen, K., Pateraki, I., Ioannou, E., & Kampranis, S. C. (2025). Systematic biotechnological production of isoprenoid analogs with bespoke carbon skeletons. *Nature Communications, 16*(1), 2098. DOI: 10.1038/s41467-025-57494-2

Wearn, O. R., Reuman, D. C., & Ewers, R. M. (2012). Extinction debt and windows of conservation opportunity in the Brazilian Amazon. *Science, 337*(6091), 228–232. DOI: 10.1126/science.1219013

West, S., Visentin, D. C., Neil, A., Kornhaber, R., Ingham, V., & Cleary, M. (2020). Forging, protecting, and repairing community resilience informed by the 2019–2020 Australian bushfires. *Journal of Advanced Nursing, 76*(5), 1095–1097. DOI: 10.1111/jan.14306

Williams, A. P., Allen, C. D., Macalady, A. K., Griffin, D., Woodhouse, C. A., Meko, D. M., Swetnam, T. W., Rauscher, S. A., Seager, R., Grissino-Mayer, H. D., Dean, J. S., Cook, E. R., Gangodagamage, C., Cai, M., & McDowell, N. G. (2013). Temperature as a potent driver of regional forest drought stress and tree mortality. *Nature Climate Change*, *3*(3), 292–297. DOI: 10.1038/nclimate1693

World Resources Institute (WRI). (2025). *Climate adaptation investments yield massive returns: Evidence from 320 projects across 12 countries*. World Resources Institute. https://www.wri.org/news/release-wri-study-finds-climate-adaptation-investments-yield-massive-returns

World Weather Attribution (WWA). (2024). *When risks become reality: Extreme weather in 2024*. World Weather Attribution; Climate Central. https://www.worldweatherattribution.org/when-risks-become-reality-extreme-weather-in-2024/

KEY TERMS AND DEFINITIONS

Adaptive management: A structured, iterative process of decision-making in resource management that adjusts actions based on monitoring and new information.

Biodiversity: The variety of life at all levels—genetic, species, and ecosystem—whose richness underpins ecosystem function and resilience.

Climate change: Long-term alteration of temperature and precipitation patterns driven by greenhouse gas emissions, affecting ecosystem processes and species distributions.

Deforestation: The large-scale removal of trees and conversion of forested land to non-forest uses, leading to habitat loss and reduced carbon storage.

Ecosystem resilience: The capacity of a forest system to absorb disturbances, reorganize, and maintain essential functions in the face of changing environmental and social conditions.

Habitat fragmentation: The breaking up of continuous forest into smaller, isolated patches, which impairs species movement, reduces genetic diversity, and heightens edge effects.

Land tenure security: The recognition and enforcement of individuals' or communities' rights to use, control, and transfer land, which underlies sustainable stewardship.

Participatory governance: Inclusive decision-making frameworks that engage multiple stakeholders—government, local communities, NGOs—in designing and implementing forest policies.

Socio-cultural disruption: The breakdown or alteration of traditional social structures, practices, and values—often resulting from external pressures such as land dispossession.

Traditional knowledge: Indigenous and local communities' cumulative body of ecological insights, practices, and customs, passed down through generations to manage landscapes.

Chapter 4

The Influence of Environmental Concern and Media Engagement on Forest Protection Sharing Behaviour on Social Media Among Youth in Guangxi in China

Gan Liu
https://orcid.org/0009-0003-5272-6022
Universiti Sains Malaysia, Malaysia

Mohamad Saifudin Mohamad Saleh
https://orcid.org/0000-0002-1591-0002
School of Communication, Universiti Sains Malaysia, Malaysia

Yang Jiao
School of Communication, Universiti Sains Malaysia, Malaysia

Miao Huang
School of Animation and Digital Arts, Communication University of China, Nanjing, China

DOI: 10.4018/979-8-3373-3992-4.ch004

ABSTRACT

The main purpose of this study is to explore the influence of youth' environmental concern and media engagement on their intention and behavior of sharing forest protection information on social media. This study focus on the youth in Guangxi Province of China, and data will be collected through online questionnaires. In total, 476 valid questionnaires were collected. The collected data will be analyzed using the minimum structural equation model (PLS-SEM). The research results show that environmental concern and social media engagement have a positive impact on youth intention of sharing environmental protection information on social media, which in turn promotes actual sharing behavior. This study combines the environmental concern in environmental informatics with the media engagement, providing a novel interdisciplinary analytical framework. However, this study only focuses on cognitive variables and lacks emotional variables. Future researchers can consider conducting cross-regional comparative studies to enhance the broad applicability of this research.

1. INTRODUCTION

In recent decades, with the growth of population and income, the global demand and consumption of wood products have been increasing, and the volume of wood harvesting has also been on the rise (Mensah et al., 2025). Driven by human activities and natural disasters, the world's forest resources are disappearing at an alarming rate. The latest research shows that in 2024, a total of 67,000 square kilometers of forests were destroyed worldwide, an area equivalent to more than twice the size of Belgium (Josephson, 2024). Moreover, the rate of forest loss is accelerating in China. According to the statistical data of China from 1986 to 2020, approximately 83.3 million hectares of forests were damaged, accounting for 39.7% of China's forest area (Zebardast & Radaei, 2022). This level of destruction has never been seen in the past 20 years. Forests and grasslands, as important components of the Earth's ecosystem, play crucial roles in conserving soil and water, conserving water sources, regulating climate, and maintaining biodiversity (Lukin et al., 2021).

In recent years, with the improvement of people's environmental protection awareness and the development of forest tourism, the biological nature of forests has become increasingly rich. For example, Guangxi has cumulatively completed the cultivation of 1,055 hectares of mangroves and restored 3,147 hectares of existing mangroves since 2020 (Wang et al., 2024). By the end of 2024, the number of white-headed langur monkeys in and around nature reserves has increased from 598 in 2003 to over 1,400 (Krus, 2024). Due to its geographical location, Guangxi is rich in ecological resources (Tuo et al., 2021). Among the core indicators of China's forests, Guangxi ranks among the top in the country. Moreover, Guangxi has established an intelligent management system that integrates forest fire prevention monitoring, intelligent patrolling, unmanned aerial vehicle monitoring and inspection, etc (Zhang & Zhu, 2023). However, the overall forest is still shrinking on a large scale. From 2002 to 2024, Guangxi lost 3.51 kha of moist virgin forest, accounting for 0.12% of its total tree cover loss during the same period (Zhong et al., 2025). During this period, the total area of wet primary forests in Guangxi decreased by 6.0% (Chen et al., 2025).

Moreover, a large number of people, in pursuit of profits, wantonly destroy the forests, and the forests are still being reduced at a very fast rate (Kenrick et al., 2023). The significance of forests and the severity of their destruction have greatly increased people's environmental concern (Kumar et al., 2022). In previous studies, many researchers have found that environmental concern have a positive promoting effect on the intention and behavior of sharing environmental knowledge. For instance, in an empirical study on biodiversity, it was found that the environmental concern arising from biodiversity greatly enhanced people's intention and behavior to share information about biological conservation on social media (Wu et al., 2020). The youth have a relatively high level of knowledge and culture. Furthermore, precisely because Chinese young people have received a good education, they have a strong desire for knowledge. They show a very positive attitude towards forest protection and are willing to actively participate in protection actions related to forests and the environment. (Du et al., 2023). The promotion of forest protection behaviors through social media by youths is of great significance for the development of forest tourism and the protection of forests. Although Guangxi has achieved remarkable accomplishments in forest tourism and ecological protection in the past, its forest area is still experiencing significant reduction on a large scale.(Chen et al., 2025).

Previous research has demonstrated that young people's engagement with social media plays a significant role, particularly in relation to environmental protection. An empirical study by Moroccan scholars revealed that young people engage with environmental issues and news primarily through social media platforms (Hajri & Daife, 2024). Similarly, Andersson and Öhman (2017) explored the meaning-making characteristics of young people's conversations on environmental and sustainability issues within social media. Furthermore, the Ant Forest application developed by China's Alibaba has exerted significant influence by introducing gamification mechanisms. This encourages young people to advance ecological restoration in China's desertified regions through virtual interactions and stimulates their sharing of forest-related information (Zhang et al., 2022). Consequently, such social media-based interactive communication facilitates the formation of a public discourse environment for forest conservation within virtual spaces among young people, which subsequently translates into concrete environmental actions in the real world.

As a developing country, China has long recognized the significant role of social media in the field of environmental protection, including the sustainable development and protection of forests (Liu & Feng, 2021). Social media serves as the fastest way for people to obtain information about forests, and plays a crucial role in disseminating information related to forest disasters and forest protection. Media engagement refers to the interactive behaviors of individuals with media content through actions such as liking, sharing, commenting, watching, and participating in activities (Moran et al., 2020). Through various social media platforms, the public can obtain information, acquire knowledge of forest protection and improve environmental literacy at the same time. By posting content related to forest protection on social media, people not only achieve rapid dissemination and enhance their influence in a short period of time, but also significantly reduce the cost of public participation. This is of great benefit to the sustainable development of China's forests (Wu et al., 2022). Stimulated by various forest ecological problems, youth have the intention to use social media to share about forest protection and take forest information sharing behaviour on social media (Zhang et al., 2022).

However, existing research provides insufficient evidence regarding Chinese youth's forest protection behaviour on social media. In order to better understand the forest environmental protection behaviors of youth on social media, this study will mainly focus on Guangxi Province, China, and the main objective of this study is to investigate the influence of environmental concern and media participation on youth' intention and behavior to share forest protection information on social media. As an area with relatively well-developed ecological resources in southwestern China, Guangxi boasts a forest coverage rate of 49.7%, with a total timber volume of 860 million cubic metres and a forest area spanning 118,130 square kilometres (Guangxi Zhuang Autonomous Region Department of Ecology and Environment, 2024). Moreover, Guangxi possesses diverse ecosystem types, undertaking the dual responsibilities of economic development and ecological conservation. Investigating forest conservation behaviours among adolescents in Guangxi within social media environments not only aids in revealing the interactive relationship between local ecological contexts and digital communication, but also contributes to understanding the role of new media technologies in promoting public engagement in forest conservation and disseminating sustainable development concepts.

Moreover, this study drew on the Theory of Planned Behavior and the Protection Motivation Theory, taking into account both psychological factors and media factors. It constructed a brand-new framework that provides positive reference value for studying the forest protection behavior of youth, and also fills the gaps and deficiencies of previous research.

2. THEORETICAL BACKGROUND AND HYPOTHESIS DEVELOPMENT

2.1 Theoretical Foundation

This study is based on the Theory of Planned Behavior (TPB) and the Protection Motivation Theory (PMT), aiming to investigate the influence of environmental concern and media engagement on the intention of youth to share information about forest protection on social media, as well as their actual behavior of sharing such information.

The TPB is a classic socio-psychological model explaining the relationship between an individual's behavioural intention and actual behaviour. This theory posits that individual behaviour is primarily driven by behavioural intention, which is jointly influenced by three factors: attitude, subjective norm, and perceived behavioural control (Ajzen, 1991). This theory has been widely applied in understanding individuals' information-sharing behaviour (Chen & Yang, 2020). More specifically, TPB posits that when individuals hold positive attitudes towards a behaviour, perceive social support, and believe they possess the capability to perform it, their behavioural intention strengthens, thereby increasing the likelihood of actual action (Ajzen & Schmidt, 2020). Within the present study's context, TPB has been creatively extended to elucidate the mechanisms underpinning adolescents' intentions and behaviours regarding sharing forest conservation information on social media.

The PMT emphasises cognitive appraisal and protective motivation when individuals confront environmental threats. This theory posits that behavioural outcomes depend on both the perception of threat (threat appraisal) and the assessment of coping capabilities (coping appraisal) (Rogers, 1983). Moreover, The PMT elaborates on how the perception of environmental threats, such as deforestation, is triggered through emotional mechanisms and leads to corresponding behaviors (Kothe et al., 2019). Furthermore, according to the PMT theory, when youth have a strong perception of the damage and threats brought by forests, and they have a strong ability to participate in social media, and they believe that sharing information about forest protection on social media can significantly reduce forest destruction, then their desire to share information about forest protection on social media will be extremely strong (Zhang et al., 2025). Furthermore, the Theory of Planned Behavior also strengthens the exemplary role of sharing information about forest protection on social media in terms of attitudes and cognitive aspects, for young people. (Diepolder et al., 2024).

Environmental concern is regarded as a key variable in predicting information-sharing intentions and pro-environmental behaviour, denoting an individual's awareness and level of concern regarding environmental issues (Dunlap et al., 2000). Within this study, environmental concern is specifically conceptualised as the significant cognitive awareness and level of concern individuals hold towards the deterioration of external ecological environments (Shen & Wang, 2022). Ajzen (2011) emphasised that attitudes are determined by the totality of accessible behavioural beliefs linking actions to various outcomes and attributes, which are fundamentally cognitive and affective in nature. Consequently, within the Theory of Planned Behaviour (TPB), environmental concern may be regarded as an individual's overall evaluative stance towards environmental issues.

Academic literature offers multiple definitions of media engagement. One definition refers to consumers' participation in social media activation campaigns as the synchronous and holistic investment of cognitive, affective, and behavioural energy throughout the activity (Mirbagheri & Najmi, 2019). From a consumer perspective, social media engagement comprises three behavioural levels: consumption, contribution, and creation. These reflect the continuum of user involvement in activity-related content, progressing from passive information reception to active content interaction (Muntinga et al., 2011). Within this study, media engagement is defined as the manner in which youth users actively and bidirectionally disseminate information via social media (Yu & Ramaprasad, 2019).

Moreover, Rogers (1983) points out that an individual's threat appraisal and coping appraisal are jointly triggered by external environmental stimuli and internal psychological factors. Information from diverse sources may trigger cognitive processes underpinning protective motivation, ultimately leading individuals to develop protective intent. Consequently, this theoretical perspective aligns with the variables employed in the present study. Thus, environmental concern constitutes not merely an attitude but an affective cognitive response to environmental threats, capable of prompting individuals to adopt protective or communicative actions. Media engagement, meanwhile, reflects proactive coping and behavioural motivation arising from individuals' internal psychological resources.

By integrating PMT and TPB theories, this study not only validates the applicability of traditional behavioral models but also reveals new mechanisms for youth's environmental protection behaviors in the digital media era. Therefore, this study employs environmental concern and media engagement as independent variables, with the intention to share forest conservation information on social media serving as the mediating variable, and the actual behaviour of sharing such information as the dependent variable. This model aims to investigate the psychological transformation mechanism through which adolescents progress from environmental cognition to behavioural practice within the social media context. Based on this, the theoretical framework of this study is obtained and the framework can be seen in

Figure 1. The Theoretical Framework of Study

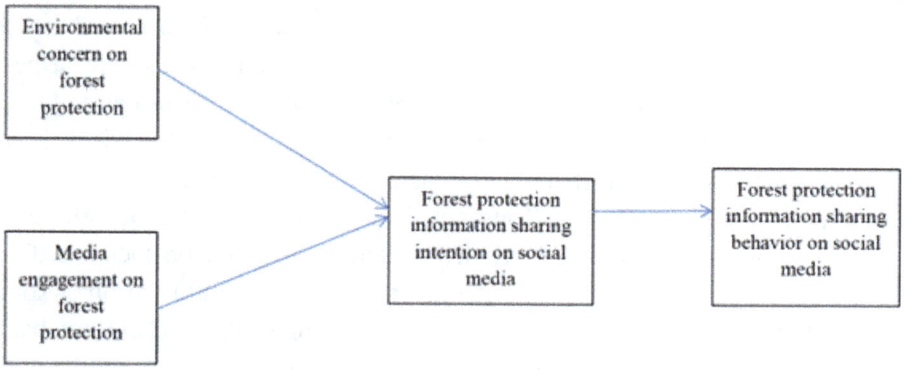

2.2 Hypotheses Development

Environmental concern mainly refer to an important perspective held by people regarding the phenomenon of the deterioration of the external ecological environment (Shen & Wang, 2022). They play an extremely crucial guiding role in people's further intentions and behaviors towards environmental protection (Zhang et al., 2025). For instance, in one study that examines the factors influencing youth intention to purchase green products, it was found that environmental concern have a positive promoting effect on the intention to purchase green products (Duong et al., 2022). Furthermore, another empirical study also has found that environmental concern can prompt youth actively utilize their own skills to promote environmental awareness (Shutaleva et al., 2021). Media engagement refers to the active involvement of individuals in media content, which includes their cognitive, emotional and behavioral interactions on digital platforms (Yu & Ramaprasad, 2019). In the field of environmental studies, media engagement typically refers to individuals using communication media to disseminate information related to the environment (Rodrigues et al., 2023). Active media engagement can encourage youth to actively utilize social media to disseminate ecological information (Saleh & Heinrichs, 2020). For instance, an empirical study focusing on teenagers found that those with good media engagement were more likely to actively use social media to spread information about environmental protection (Knupfer et al., 2023). Therefore, based on this, the hypotheses can be made:

H1: Environmental concern of youth about forest protection have a positive effect on the sharing of forest protection information intention on social media.

H2: The media engagement of youth on forest protection have a positive effect on the sharing of forest protection information intention on social media.

In the research on the relationship between behavior and intention, many researchers from various fields have conducted in-depth studies. It includes aspects such as ecological environment, flood dissemination, social media usage, and forest protection. For instance, in a study on environmental protection behaviors, it was found that the intention to protect the environment positively influences an individual's environmental protection actions (Kousar et al., 2022). Furthermore, according to a study that examines how users share environmental information on social media, it was found that the intention to share environmental protection information on social media positively influences users' sharing behavior (Chen, 2020). Therefore, based on this assumption, the following hypothesis was developed:

H3: The forest protection information sharing intention on social media have a positive effect on the forest protection information sharing behaviour.

3. METHOD

This research employ the survey method. The data was collected in the form of questionnaires, and the questionnaires were collected through the Wenjuanxing platform. The respondents of this study were mainly youth aged 18 to 35 from Guangxi. Guangxi is located in the southwestern part of China. It boasts extremely diverse biological resources and forest areas and serves as an important ecological barrier for China (Chen, 2020). However, its forests are also facing significant threats from destruction. Therefore, choosing Guangxi is not only extremely beneficial for understanding the role of youth in forest protection, but also provides a reference for other places. To ensure the validity of the sample, a questionnaire introduction was set before the start of the survey: Only those young people aged between 18 and 35 and who have lived in Guangxi for at least one year are eligible to participate. Tourists are not allowed to participate. Considering the vast territory of Guangxi in China, this study adopted an online questionnaire approach for data collection. The questionnaire was mainly distributed through the WeChat platform. The questionnaire collection period was from June 2nd, 2025 to June 22nd, 2025, lasting for 20 days.

In this study, the Likert five-point scale was employed to measure these items, with 1 representing "very dissatisfied" and 5 representing "very satisfied" (Alabi & Jelili,2023). Since all the samples were from China, the questionnaire was professionally edited and translated into both Chinese and English to ensure the validity of the questionnaire itself. The data will be collected using a purposing sampling method and analyzed using SmartPLS. All the scales used in the questionnaire were derived from previous studies. The specific scales are presented in Table 1.

Table 1. The items of this study

Construct	Item	Questionnaire Item	Source
Environment Concern	EC1	I worry about the effects of development on local forest ecosystems.	Maleknia et al., 2025
	EC2	I feel concerned about the loss of biodiversity in forest areas.	
	EC3	I feel a sense of urgency about addressing threats to forests.	
	EC4	I am troubled by the degradation of green spaces in my area.	
Media Engagement	ME1	I actively seek out news articles and reports on forest protection.	Maleknia et al., 2025
	ME2	I engage in discussions with friends and family about forest protection news and issues.	
	ME3	I stay updated on current forest protection events through various media channels.	
	ME4	I participate in online forums or groups related to forest protection advocacy.	
Forest information sharing intention	FISI1	I intend to share forest protection information on social media.	Shore et al., 2022
	FISI2	I predict that I would share forest protection information on social media.	
	FISI3	I plan to share forest protection information on social media.	
Forest information sharing behaviour	FISB1	I have posted forest protection information on my social media.	Chen, 2020
	FISB2	I have shared forest protection suggestions with others on social media.	
	FISB3	I have let others know about my forest protection behavior through social media.	

4. RESULTS

A total of 476 valid data were collected in this study, including 222 females and 254 males, accounting for 46.64% and 53.36% respectively. All the interviewed youth were between 18 and 35 years old. Among them, the group aged 21-25 was the largest, with 165 individuals (34.66%), while the group aged 31-35 was the smallest, with 81 individuals (17.02%). Furthermore, all the respondents had a bachelor's degree or above, indicating a relatively high level of knowledge. Among them, the largest number was those with a bachelor's degree, accounting for 62.6%, while the smallest number was those with a doctoral degree, accounting for 8.83%. Regarding the regions where the respondents come from, the majority (222, accounting for 46.64%) are from rural areas. Compared to urban areas, the forest resources in rural areas are also more abundant, which to some extent makes the sample more targeted. The specific demographic information can be found in Table 2.

Table 2. The Demographic Profile

Demographic Variables		Categories	Frequency	Percentage (%)
Gender		Female	222	46.64%
		Male	254	53.36%
Age		18-20	108	22.69%
		21-25	165	34.66%
		26-30	122	25.63%
		31-35	81	17.02%
Study level		Bachelor's or equivalent level	298	62.60%
		Master's or equivalent level	136	28.57%
		Doctoral or equivalent level	42	8.83%
Area		Urban areas	155	32.56%
		Rural area	222	46.64%
		City suburb	66	13.87%
		Other	33	6.93%

4.1 Measurement Model

This study mainly used the latest version 4.0 of SmartPLS to complete the data analysis through partial least-squares structural equation modelling (PLS-SEM). This study analyzed the data's Outer Loadings (OL), Cronbach's Alpha (CA), and Average Variance Extracted (AVE) to determine the proposed model and its reliability and validity. The specific indicators are shown in Table 3 and the measurement model with out loadings and CA can be seen in Figure 2. The outloading value of each data point is greater than 0.708, which conforms to the benchmark proposed by Hair et al. (2019). Therefore, it proves the rationality and effectiveness of each indicator. According to the suggestion of Bagozzi et al. (1991), when the factor load is greater than 0.5, the load of all items on the related construct is significant. Furthermore, the CR of all constructs is between 0.825 and 0.873, all of which are higher than the critical value of 0.7. This indicates that these constructs have internal consistency and reliability (Hale et al., 2014). For each structure, the AVE value exceeds the critical value of 0.7, indicating that these constructs have internal consistency and reliability (Hale et al., 2014).

Table 3. Reliability and Cross-loading

Constructs	Items	OL	CA	AVE
Environment Concern	EC1	0.820	0.825	0.657
	EC2	0.885		
	EC3	0.809		
	EC4	0.719		
Media Engagement	ME1	0.897	0.873	0.723
	ME2	0.840		
	ME3	0.839		
	ME4	0.822		
Forest information sharing intention	FISI1	0.863	0.835	0.746
	FISI2	0.858		
	FISI3	0.871		
Forest information sharing behaviour	FISB1	0.890	0.834	0.750
	FISB2	0.837		
	FISB3	0.870		

Figure 2. Measurement model with outer loadings and CA

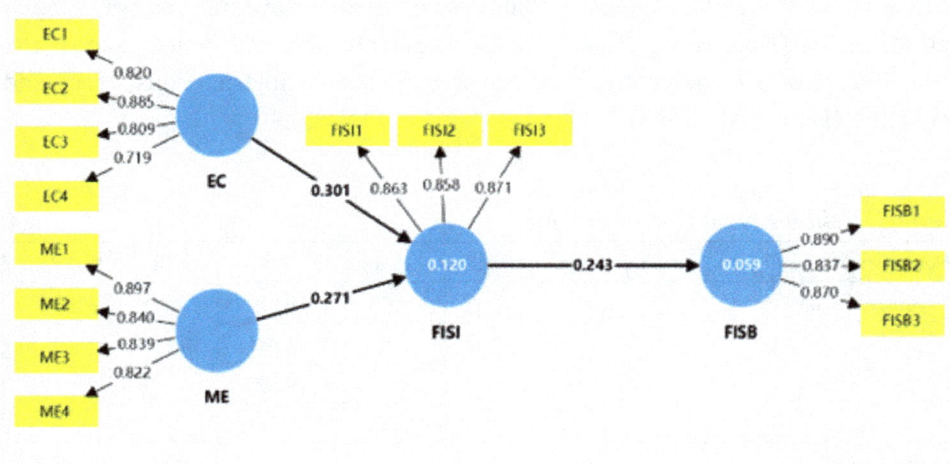

Discriminant validity refers to whether a concept can truly be differentiated from other concepts when based on empirical standards (Rönkkö & Cho, 2022). The Fornell-Larcker criterion and HTMT (heterotrait-monotrait ratio) are important ways to test the evaluation of discriminant standards. In this study, both of these two methods were used for evaluating the discriminant validity during the placement of the samples. Among them, Table 4 is the Heterotrait Monotrait (HTMT) Criterion, and Table 5 is the Fornell-Larcker Criterion. The AVE value of each construct was higher than the square multiple correlation coefficient of all other constructs, thereby confirming the discriminant validity.

Table 4. Discriminant validity (Heterotrait Monotrait (HTMT) Criterion)

	EC	FISB	FISI	ME
EC				
FISB	0.341			
FISI	0.337	0.268		
ME	0.419	0.306	0.289	

Table 5. Discriminant Validity (Fornell- Lacker Criterion)

	EC	FISB	FISI	ME
EC	0.810			
FISB	0.291	0.866		
FISI	0.301	0.243	0.864	
ME	0.367	0.272	0.271	0.850

4.2 Structural Model

Table 6 summarizes the results of all the hypothesis tests. All the hypotheses show positive support: H1 ($\beta = 0.233$, $t = 5.261$, $p = 0.322$, $p < 0.001$), H2 ($\beta = 0.185$, $t = 4.139$, $p = 0.001$), and H3 ($\beta = 0.243$, $t = 5.302$, $p < 0.0001$). Thus, all three proposed hypotheses (H1–H3) are found to be statistically significant and thus supported

Table 6. Path Coefficient of Hypothesis

Hypothesis Relationship	Type	Std. beta (β)	Standard deviation (STDEV)	T values	P values	Result
H1 EC -> FISI	Direct	0.233	0.044	5.261	0.000	Accept
H2 ME -> FISI	Direct	0.185	0.045	4.139	0.000	Accept
H3 FISI -> FISB	Direct	0.243	0.046	5.302	0.000	Accept

5. DISCUSSION

Empirical findings confirm that all hypotheses proposed in this study hold true. The results collectively validate the rationale for integrating TPB and PMT. Environmental concern, acting as both an attitude and media interaction, significantly enhances youth forest information sharing intention. Forest information sharing intention, in turn, substantially increases forest information sharing behaviour. Overall, the findings regarding environmental concern and media engagement as two distinct factors confirm the applicability of combining TPB and PMT to explain the formation of sharing intentions. Based on these overarching conclusions, the following discussion outlines theoretical contributions and practical implications.

The research findings show that environmental concern regarding the sharing of information about social media forest protection have a positive impact. The results demonstrate that the assessment of external threats and the youth's perception of their own efficacy in alleviating the destruction of forest destruction have, under the influence of these two factors, stimulated the intention of youth to share information about forest protection on social media. From TPB perspective, this study confirms environmental concern as an overall attitude towards adolescents' specific behaviour of sharing forest conservation information on social media. From PMT perspective, environmental concern is an adaptive response to perceived threats.

Consequently, heightened environmental concern can directly and effectively enhance adolescents' sharing intentions. Within the Chinese context, group responsibility and norms under collectivist culture reinforce the formation of emotions and values regarding environmental issues. That is to say, when confronted with various forest destruction incidents, they are more likely to develop environmental concern (Sokkar, 2024). This parallels the attitudinal formation process within the TPB framework, wherein overall attitudes persistently increase when individuals perceive both societal support and personal capability (Phang et al., 2023).

Concurrently, it is crucial to note that such environmental concern must remain within reasonable bounds. Excessive environmental anxiety may induce behavioural inhibition, thereby hindering their sharing of forest conservation information on social media (Chen et al., 2025).

Furthermore, in addition to being concerned about the degree of environmental concerns, it is also necessary to pay attention to verifying the authenticity of the forest protection information. If the information lacks scientific basis or merely exaggerates the facts of the forest crisis, people's willingness to share will also be reduced to a certain extent (Vasile & Iordăchescu, 2022). People, including the youth, tend to share the content that highlights the issues of forests and provides specific solutions. Apart from individual factors, social media also plays a significant role. The algorithmic recommendation mechanism of social media determines how many people can be exposed to the facts of forest destruction and threats (Ognibene et al., 2023). Although people have a strong awareness of environmental concerns, social media platforms do not encourage the dissemination of information on forest protection (Słupińska et al., 2022). As a result, the effectiveness of this dissemination will be greatly reduced, and it will also significantly lower people's willingness to share information about forest protection on social media.

In addition, this study mainly focuses on Guangxi, the southeastern region of China. This area has a relatively high forest coverage rate and is also one of the important ecological protection zones in China. Nevertheless, Guangxi is also facing ecological pressures such as deforestation, illegal hunting, and urban expansion (Yang et al., 2022). Against this background, the youth group has a relatively high level of knowledge and cultural literacy. Therefore, as a key force for future ecological civilization construction, their environmental attitudes and media behaviors are particularly important (Kasuma et al., 2018). The multi-ethnic cultures of Guangxi, such as the Zhuang and Yao ethnic groups, inherently possess an ecological ethical view of coexistence with nature (Wang et al., 2023). Therefore, youth in this cultural atmosphere may be more likely to form an ecological identity, which might also be one of the cultural roots that strongly influence the sharing intention of environmental concern (Du et al., 2023).

This research indicates that media engagement serves as a significant direct predictor of youth intention to share forest information. This directly validates the alignment between media engagement and the individual's internal proactive coping motivation within the KMT theoretical framework. Furthermore, reading environmental news, commenting and interacting under posts, liking and endorsing relevant content, and reposting or sharing material significantly enhances their willingness to disseminate forest information sharing intention.

As ecological topics gradually gained popularity on social platforms, youth not only obtained knowledge through these platforms but also achieved self-construction and identity dissemination through sharing and interaction This is closely related to the development of Chinese social media (Bailey et al., 2020). However, it is important to note that the media literacy of youth in different regions may vary, and thus it will have an impact on the youth final intention and behavior of sharing on social media (Wang et al., 2025). Moreover, Chinese youth have a high acceptance of information forms such as short videos, graphic notes, etc., on platforms like Xiaohongshu, Weibo, and Douyin (Li, 2025). These media forms themselves have strong dissemination capabilities and community interactivity, creating conditions for the sharing of environmental issues.

However, when youth use social media to participate in discussions about forest protection information, they may also be affected by algorithmic recommendations, making it difficult for them to access updated information. This is mainly because social media platforms recommend information based on each person's different preferences (Liao et al., 2021). Over time, it becomes difficult for them to access updated information, and thus highly homogeneous content may limit the dissemination of forest information when youth share it, preventing more people from accessing it, thereby weakening the influence of the forest information itself (Potnis et al., 2025).

Meanwhile, this study also verified the significant predictive effect of behavioral intentions on actual behaviors. That is, the more youth tend to want to share information about forest protection, the more they actually share it on social media. This result aligns with the Theory of Planned Behavior, which emphasizes that intention is the most direct predictor of behavior (Conner, 2020). Sharing behavior does not occur randomly but is based on clear cognition and motivation (Cai & Shi, 2022). This also reminds us that merely having attitudes or media involvement is not sufficient; it is necessary to prompt individuals to form clear behavioral intentions in order to promote the occurrence of actual actions (Liu et al., 2024). Therefore, in future forest protection publicity strategies, we should focus on stimulating individual motivation, clarifying behavioral goals, and providing easily operable media tools and action paths.

Moreover, unlike traditional communication channels, environmental protection communication on social media has the following characteristics: User-driven, their sharing behavior not only includes forwarding but also involves content creation and expression of personal opinions (Çakıcı et al., 2025). Additionally, social media has a strong algorithm recommendation mechanism (Saleh et al., 2021). Under the influence of this algorithm promotion mechanism, social media content tends to more frequently push content that users are interested in and reduce content that they are not interested in. This mechanism allows environmental protection topics to spread more readily among the groups that are concerned about such topics, thereby forming a "echo chamber effect" and further enhancing their influence (Moe et al., 2023). The positive impact of media participation on the willingness to share, as discovered in this study, is precisely based on this interaction logic of society, algorithms, and self-expression. During the the youth sharing the forest protection information process, young people are not only disseminators of information, but also participants in the topic and creators of their identities.

6. CONCLUSION

This study mainly focuses on the youth in Guangxi, China. Its aim is to explore the influence of environmental concern and media engagement on their intention and behavior of sharing information about forest protection on social media. The study reveals the psychological activities involved in youth' dissemination of forest protection information on social media. Environmental concern and media engagement not only enhance the intention to share, but also this intention sharing the forest protection information will drive the actual behavior to occur. This finding provides empirical support for the environmental communication in the social media era, and offers practical guidance for environmental education, the dissemination of information about forest protection on social media platforms, and the formulation of policies related to environmental issues. This study also constructed a new theoretical framework, which provides a reference for explaining the ecological communication behaviors of young people, and offers valuable references for future research on environmental communication and young people's media behaviors. In the future, cultural, emotional, and social structure dimensions should be further introduced to establish a more comprehensive model of young people's environmental communication behavior.

Although this study has achieved relatively clear empirical results, it also has significant limitations. For instance, the data of this research questionnaire comes from individuals. When filling out the questionnaire, they may be influenced by social expectations. Therefore, future research can combine actual social media behavior data such as sharing records and content analysis to enhance the objectivity and persuasiveness of the data. In fact, this study treated media engagement as a single variable without differentiating the characteristics of various social platforms. Since the communication mechanisms of platforms such as Weibo, Xiaohongshu, and Douyin have significant differences, future comparative studies on multiple platforms can be conducted to deeply explore how the platform ecosystem affects environmental protection behaviors. Finally, as the sample of this study is concentrated in Guangxi, which has a specific cultural and policy background, the applicability of the research conclusions in other provinces or other countries/regions needs to be further verified. In the future, cross-regional or cross-cultural comparative studies can be considered to enhance the broad applicability and theoretical universality of the research.

REFERENCES

Ajzen, I. (1991). The theory of planned behaviour'. *Organizational Behavior and Human Decision Processes, 50*(2), 179–211. DOI: 10.1016/0749-5978(91)90020-T

Ajzen, I. (2011). *The theory of planned behaviour: Reactions and reflections* (Vol. 26). Taylor & Francis.

Ajzen, I., & Schmidt, P. (2020). Changing Behavior Using the Theory of Planned Behavior. In M. S. Hagger, L. D. Cameron, K. Hamilton, N. Hankonen, & T. Lintunen (Eds.), The Handbook of Behavior Change (pp. 17-31). Cambridge University Press. DOI: DOI: 10.1017/9781108677318.002

Alabi, A. T., & Jelili, M. O. (2023). Clarifying likert scale misconceptions for improved application in urban studies. *Quality & Quantity, 57*(2), 1337–1350. DOI: 10.1007/s11135-022-01415-8

Andersson, E., & Öhman, J. (2017). Young people's conversations about environmental and sustainability issues in social media. *Environmental Education Research, 23*(4), 465–485. DOI: 10.1080/13504622.2016.1149551

Bagozzi, R. P., Yi, Y., & Phillips, L. W. (1991). Assessing construct validity in organizational research. *Administrative Science Quarterly, 36*(3), 421–458. DOI: 10.2307/2393203

Bailey, E. R., Matz, S. C., Youyou, W., & Iyengar, S. S. (2020). Authentic self-expression on social media is associated with greater subjective well-being. *Nature Communications, 11*(1), 4889. DOI: 10.1038/s41467-020-18539-w PMID: 33024115

Cai, Y., & Shi, W. (2022). The influence of the community climate on users' knowledge-sharing intention: The social cognitive theory perspective. *Behaviour & Information Technology, 41*(2), 307–323. DOI: 10.1080/0144929X.2020.1808704

Çakıcı, Z., Meriç, E., & Uysal, R. (2025). Environmental Communication Strategies on TikTok: A Study on the United Nations Environment Programme (UNEP). *Akdeniz Üniversitesi İletişim Fakültesi Dergisi*, (48), 95–120. DOI: 10.31123/akil.1625534

Chen, Y. (2020). An investigation of the influencing factors of Chinese WeChat users' environmental information-sharing behavior based on an integrated model of UGT, NAM, and TPB. *Sustainability (Basel), 12*(7), 2710. DOI: 10.3390/su12072710

Chen, Y., Xie, Y., Li, M., Xie, M., & Mo, W. (2025). The effects of various degrees of meteorological drought on forest vegetation in ecologically Fragile Karst Areas of Guangxi, China. *Environmental Technology*, *46*(9), 1333–1347. DOI: 10.1080/09593330.2024.2386864 PMID: 39128838

Conner, M. (2020). Theory of planned behavior. Handbook of sport psychology, 1-18. DOI: 10.1002/9781119568124

Diepolder, C. S., Weitzel, H., & Huwer, J. (2024). Exploring the impact of sustainable entrepreneurial role models on students' opportunity recognition for sustainable development in sustainable entrepreneurship education. *Sustainability (Basel)*, *16*(4), 1484. DOI: 10.3390/su16041484

Du, M., Chai, C. S., Di, W., & Wang, X. (2023). What affects adolescents' willingness to maintain climate change action participation: An extended theory of planned behavior to explore the evidence from China. *Journal of Cleaner Production*, *422*, 138589. DOI: 10.1016/j.jclepro.2023.138589

Dunlap, R. E., Van Liere, K. D., Mertig, A. G., & Jones, R. E. (2000). New Trends in Measuring Environmental Attitudes: Measuring Endorsement of the New Ecological Paradigm: A Revised NEP Scale. *The Journal of Social Issues*, *56*(3), 425–442. 10.1111/0022-4537.00176. DOI: 10.1111/0022-4537.00176

Duong, C. D., Doan, X. H., Vu, D. M., Ha, N. T., & Dam, K. V. (2022). The role of perceived environmental responsibility and environmental concern on shaping green purchase intention. *Vision (Basel)*. Advance online publication. DOI: 10.1177/09722629221092117

Guangxi Zhuang Autonomous Region Department of Ecology and Environment. (2024). Guangxi Ecological and Environmental Status Bulletin. Guangxi Zhuang Autonomous Region Department of Ecology and Environment. http://sthjt.gxzf.gov.cn/zfxxgk/zfxxgkgl/fdzdgknr/zdlyxxgk/hjzkgb/P020250613584678791273.pdf

Hair, J., Risher, J., Sarstedt, M., & Ringle, C. (2019). When to use and how to report the results of PLS-SEM. *European Business Review*, *31*(1), 2–24. Advance online publication. DOI: 10.1108/EBR-11-2018-0203

Hajri, O., & Daife, Y. (2024). *The role of social media in engaging young people in environmental issues. E3S Web of Conferences, Josephson, P. R. (2024). Hero Projects: The Russian Empire and Big Technology from Lenin to Putin*. Oxford University Press.

Kasuma, S. A. A., Saleh, M. S. M., Akhiar, A., & Ismail, Y. M. B. (2018). Malaysian university students' preferences of social media and LMS in academia. [IJVPLE]. *International Journal of Virtual and Personal Learning Environments*, *8*(1), 51–67. DOI: 10.4018/IJVPLE.2018010104

Kenrick, J., Rowley, T., & Kitelo, P. (2023). 'We are our land'—Ogiek of Mount Elgon, Kenya: Securing community tenure as the key enabling condition for sustaining community lands. *Oryx*, *57*(3), 298–312. DOI: 10.1017/S003060532300008X

Knupfer, H., Neureiter, A., & Matthes, J. (2023). From social media diet to public riot? Engagement with "greenfluencers" and young social media users' environmental activism. *Computers in Human Behavior*, *139*, 107527. DOI: 10.1016/j.chb.2022.107527

Kothe, E. J., Ling, M., North, M., Klas, A., Mullan, B. A., & Novoradovskaya, L. (2019). Protection motivation theory and pro-environmental behaviour: A systematic mapping review. *Australian Journal of Psychology*, *71*(4), 411–432. DOI: 10.1111/ajpy.12271

Kousar, S., Afzal, M., Ahmed, F., & Bojnec, Š. (2022). Environmental awareness and air quality: The mediating role of environmental protective behaviors. *Sustainability (Basel)*, *14*(6), 3138. DOI: 10.3390/su14063138

Kumar, R., Kumar, A., & Saikia, P. (2022). Deforestation and forests degradation impacts on the environment. In *Environmental degradation: Challenges and strategies for mitigation* (pp. 19–46). Springer International Publishing., DOI: 10.1007/978-3-030-95542-7_2

Li, F. (2025). Effects of consumers' engagement in pro-environment activities on social media on green consumption behaviour. *Asia Pacific Journal of Marketing and Logistics*, *37*(2), 403–421. DOI: 10.1108/APJML-09-2023-0862

Liao, S. H., Widowati, R., & Hsieh, Y. C. (2021). Investigating online social media users' behaviors for social commerce recommendations. *Technology in Society*, *66*, 101655. DOI: 10.1016/j.techsoc.2021.101655

Liu, M., & Feng, Q. (2021). Does social capital promote garbage classification? Evidence from China. *SAGE Open*, *11*(3), 21582440211040781. DOI: 10.1177/21582440211040781

Liu, M., Qu, Y., Wang, J., Liao, Y., Zheng, G., Guo, Y., & Liu, L. (2024). A 30-m annual grassland dataset from 1991 to 2020 for Inner Mongolia, China. *Scientific Data*, *11*(1), 1143. DOI: 10.1038/s41597-024-03990-x PMID: 39420016

Mensah, P., Pimenta, A. S., de Melo, R. R., Amponsah, J., Tuo, G., Chakurah, I., Ampadu, S. D., Buckman, I., Nikoi, M., Minkah, E., Miranda, N. O., & de Medeiros, P. L. (2025). The Global Supply Chain of Wood Products: A Literature Review. *Forests*, *16*(7), 1036. DOI: 10.3390/f16071036

Mirbagheri, S., & Najmi, M. (2019). Consumers' engagement with social media activation campaigns: Construct conceptualization and scale development. *Psychology and Marketing*, *36*(4), 376–394. 10.1002/mar.21185. DOI: 10.1002/mar.21185

Moe, H., Lindtner, S. S., & Ytre-Arne, B. (2023). Polarisation and echo chambers? Making sense of the climate issue with social media in everyday life.

Moran, G., Muzellec, L., & Johnson, D. (2020). Message content features and social media engagement: Evidence from the media industry. *Journal of Product and Brand Management*, *29*(5), 533–545. DOI: 10.1108/JPBM-09-2018-2014

Ognibene, D., Wilkens, R., Taibi, D., Hernández-Leo, D., Kruschwitz, U., Donabauer, G., Theophilou, E., Lomonaco, F., Bursic, S., Lobo, R. A., Sánchez-Reina, J. R., Scifo, L., Schwarze, V., Börsting, J., Hoppe, U., Aprin, F., Malzahn, N., & Eimler, S. (2023). Challenging social media threats using collective well-being-aware recommendation algorithms and an educational virtual companion. *Frontiers in Artificial Intelligence*, *5*, 654930. DOI: 10.3389/frai.2022.654930 PMID: 36699613

Phang, G., Ilham, Z., & Ilham, Z. (2023). Theory of planned behavior to understand pro-environmental behavior among Universiti Malaya students. *AIMS Environmental Science*, *10*(5), 691–708. DOI: 10.3934/environsci.2023038

Potnis, D., Tahamtan, I., & McDonald, L. (2025). Negative consequences of information gatekee** through algorithmic technologies: An Annual Review of Information Science and Technology (ARIST) paper. *Journal of the Association for Information Science and Technology*, *76*(1), 262–288. DOI: 10.1002/asi.24955

Rodrigues, J., Costa e Silva, E., & Pereira, D. I. (2023). How can geoscience communication foster public engagement with geoconservation? *Geoheritage*, *15*(1), 32. DOI: 10.1007/s12371-023-00800-5

Rogers, R. W. (1983). Cognitive and physiological processes in fear appeals and attitude change: A revised theory of protection motivation. In Cacioppo, J., & Petty, R. (Eds.), *Social psychophysiology: A sourcebook* (pp. 153–176). Guilford Press.

Rönkkö, M., & Cho, E. (2022). An updated guideline for assessing discriminant validity. *Organizational Research Methods*, *25*(1), 6–14. DOI: 10.1177/1094428120968614

Saleh, M. S. M., & Heinrichs, H. (2020). Media representation of environmental issues in Malaysia: Newspapers and environment non-government organisation newsletters. *Pacific Journalism Review*, *26*(2), 118–131.

Saleh, M. S. M., Kassim, N. M., Tukur, N. A., Mukhiar, S. N. S., & Balaraman, R. A. (2021). Sustainable universities as brand marketing for universities: A case of Universiti Sains Malaysia. *Media Watch*, *12*(1), 127–148. DOI: 10.15655/mw_2021_v12i1_205463

Shen, M., & Wang, J. (2022). The impact of pro-environmental awareness components on green consumption behavior: The moderation effect of consumer perceived cost, policy incentives, and face culture. *Frontiers in Psychology*, *13*, 580823. DOI: 10.3389/fpsyg.2022.580823 PMID: 35795420

Shutaleva, A., Martyushev, N., Nikonova, Z., Savchenko, I., Abramova, S., Lubimova, V., & Novgorodtseva, A. (2021). Environmental Behavior of Youth and Sustainable Development. Sustainability 2022, 14, 250.

Słupińska, K., Wieruszewski, M., Szczypa, P., Kożuch, A., & Adamowicz, K. (2022). Social media as support channels in communication with society on sustainable forest management. *Forests*, *13*(10), 1696. DOI: 10.3390/f13101696

Tuo, F., Gao, B., & Dong, Z. (2021). Technological breakthrough for the afforestation of populus euphratica in the mu us desert in China. *Sustainability (Basel)*, *13*(24), 13900. DOI: 10.3390/su132413900

Vasile, M., & Iordăchescu, G. (2022). Forest crisis narratives: Illegal logging, datafication and the conservation frontier in the Romanian Carpathian Mountains. *Political Geography*, *96*, 102600. DOI: 10.1016/j.polgeo.2022.102600

Wang, J., Balakrishnan, B., Wan, X., Yu, Q., & Ye, Q. (2025). The revival of spiritual practices: Factors influencing the "seeking deities and offering prayers" behavior of China's Generation Z on social media in an atheistic context. *Frontiers in Psychology*, *15*, 1485265. DOI: 10.3389/fpsyg.2024.1485265 PMID: 39949975

Wang, K., Jia, M., Zhang, X., Zhao, C., Zhang, R., & Wang, Z. (2024). Evaluating ecosystem service value changes in mangrove forests in Guangxi, China, from 2016 to 2020. *Remote Sensing (Basel)*, *16*(3), 494. DOI: 10.3390/rs16030494

Wu, W., Wang, W., & Zhang, M. (2022). Does internet public participation slow down environmental pollution? *Environmental Science & Policy*, *137*, 22–31. DOI: 10.1016/j.envsci.2022.08.006

Wu, Y., Xie, L., Yuan, Z., Jiang, S., Liu, W., & Sheng, H. (2020). Investigating public biodiversity conservation awareness based on the propagation of wildlife-related incidents on the Sina Weibo social media platform. *Environmental Research Letters, 15*(9), 094082. DOI: 10.1088/1748-9326/ab9ed1

Yang, Y., Chen, J., Lan, Y., Zhou, G., You, H., Han, X., Wang, Y., & Shi, X. (2022). Landscape pattern and ecological risk assessment in Guangxi based on land use change. *International Journal of Environmental Research and Public Health, 19*(3), 1595. DOI: 10.3390/ijerph19031595 PMID: 35162617

Yu, Y., & Ramaprasad, J. (2019). Engagement on Digital Platforms: A Theoretical Perspective. In ICIS (Vol. 1, No. 1, pp. 201-214).

Zebardast, L., & Radaei, M. (2022). The influence of global crises on reshaping pro-environmental behavior, case study: The COVID-19 pandemic. *The Science of the Total Environment, 811*, 151436. DOI: 10.1016/j.scitotenv.2021.151436 PMID: 34742989

Zhang, B., Hu, X., & Gu, M. (2022). Promote pro-environmental behaviour through social media: An empirical study based on Ant Forest. *Environmental Science & Policy, 137*, 216–227. DOI: 10.1016/j.envsci.2022.08.020

Zhang, B., Hu, X., & Gu, M. (2022). Promote pro-environmental behaviour through social media: An empirical study based on Ant Forest. *Environmental Science & Policy, 137*, 216–227. 10.1016/j.envsci.2022.08.020. DOI: 10.1016/j.envsci.2022.08.020

Zhang, Q., Wen, R., Hendra, L. B., & Ding, Z., & LC, R. (2025, April). Can AI Prompt Humans? Multimodal Agents Prompt Players? Game Actions and Show Consequences to Raise Sustainability Awareness. In *Proceedings of the 2025 CHI Conference on Human Factors in Computing Systems* (pp. 1-29). DOI: 10.1145/3706598.3713661

Zhang, Z., & Zhu, L. (2023). A review on unmanned aerial vehicle remote sensing: Platforms, sensors, data processing methods, and applications. drones, 7(6), 398. DOI: 10.3390/drones7060398

Zhong, X., Huang, X., Zhu, C., Wang, Y., Chapman, C. A., Garber, P. A., Chen, Y., & Fan, P. (2025). Science-based suggestions to save the world's rarest primate species Nomascus hainanus. *Science Advances, 11*(15), eadv4828. Advance online publication. DOI: 10.1126/sciadv.adv4828 PMID: 40215308

KEY TERMS AND DEFINITIONS

Algorithmic Recommendations: Algorithm recommendation is a commonly used computer algorithm in today's social media. It analyzes users' preferences to predict the content they would like and then pushes it to them. Algorithmic recommendation, on the one hand, is beneficial for users to discover the information they are interested in. However, on the other hand, it is prone to cause information ossification, preventing users from discovering new information.

Environmetal Concern: Environmental concern is a comprehensive term that describes and reflects the degree of worry that individuals, groups and society have towards the ecological environment.

Forest Information Sharing Behaviour: Forest information sharing behavior refers to the act of an individual sharing the knowledge they possess about forests through certain means. This kind of behavior not only enables more people to learn and acquire information about forests, but also helps to encourage more people to join the ranks of forest conservation.

Forest Information Sharing Intention: The intention to share forest information refers to the individual's desire to share the knowledge they possess about forests through certain channels, so that more people can be informed and understand. The willingness to share forest information is the foundation and prerequisite for driving such behavior.

Forest Protection: Forest protection is an important means and action for maintaining the health of forest ecosystems. Through forest protection, not only can ecological stability be maintained, but it also has a positive impact on sustainable development.

Media Engagement: Media participation refers to the way users utilize social media to disseminate information. Unlike traditional reading and browsing of information, it emphasizes an active and two-way form of participation.

Youth: The entire youth population. The classification of the youth varies from country to country. In China, those aged 14 to 28 are classified as “youth”, and the youth group represents the youthful vitality of a country.

Chapter 5
The Mystery of the Wood Wide Web in Forest Communication

Arun Agrawal
https://orcid.org/0000-0001-7233-6660
Institute of Technology and Management, Gwalior, India

Aruna Bajpai
https://orcid.org/0000-0002-9716-8680
Institute of Technology and Management, Gwalior, India

Jagveer Singh
Institute of Technology and Management, Gwalior, India

Anju Lata Gajpal
Kushabhau Thakre Patrakarita Avam Jansanchar Vishwavidyalaya, India

Manasi D. Karajgar
Dr. D.Y. Patil B-School, Pune, India

Tejaswi Kiran Hude
Shree Ramchandra College of Engineering, India

ABSTRACT

The "Wood Wide Web" is an underground network created by mycorrhizal fungi that connects trees and plants. This hidden system enables the sharing of nutrients, information, and warnings, showing that forests function as a unified, cooperative community rather than isolated trees. Trees use this network to help each other, adapt to environmental changes, and promote overall forest health. Although much remains unknown, this chapter explores the biological, ecological, and philosophical aspects of forest communication, discusses current research and tools, and highlights its role in supporting biodiversity and climate resilience. It invites a deeper appreciation of nature's intelligence and interconnectedness.

DOI: 10.4018/979-8-3373-3992-4.ch005

INTRODUCTION

Forests have traditionally been viewed as assemblages of individual trees competing for limited resources such as light, water, and nutrients (West, 2024). However, this competitive perspective only partially explains forest dynamics. Recent ecological research reveals that forests function as complex, interconnected systems where trees exchange information, share resources, and influence one another's growth and survival (Govaert et al., 2024). Beneath the forest floor lies an extensive and highly organized network of symbiotic relationships that mediate these interactions. This interconnection primarily occurs through mycorrhizal fungi, which establish mutualistic associations with plant roots (Priyadarshani et al., 2025). These fungi form extensive mycelial networks that link the roots of multiple trees, creating an underground infrastructure capable of transferring water, nutrients, and signaling compounds (Fioroni et al., 2025). This phenomenon, widely referred to as the *Wood Wide Web*, was first conceptualized by Dr. Suzanne Simard (Dervash et al., 2024). Simard's experimental work demonstrated that resource exchange among trees—particularly carbon and nitrogen—can occur through fungal pathways, suggesting that forest ecosystems may operate through interconnected processes rather than isolated competition. The concept of the Wood Wide Web challenges traditional competition-based ecological models by proposing a network-oriented understanding of forest ecosystems (Robinson et al., 2024). Mature trees, often termed *hub* or *source* trees, have been observed facilitating the establishment and resilience of nearby seedlings, particularly those sharing genetic similarities. Such mechanisms of resource allocation may enhance forest stability and regeneration (Bas et al., 2024). While these findings indicate cooperative dynamics, alternative interpretations suggest that these exchanges could arise from self-optimizing behavior rather than conscious cooperation, highlighting the ongoing debate between altruistic and self-regulatory ecological models. Understanding these below-ground interactions has direct implications for forest ecology, management, and conservation. In the context of accelerating climate change, deforestation, and habitat degradation, recognizing the role of mycorrhizal networks can inform sustainable forestry practices. Protecting these networks contributes to enhanced carbon sequestration, soil fertility, and water regulation—ecosystem services essential for long-term environmental resilience.

This chapter examines the scientific and conceptual development of the Wood Wide Web hypothesis and its broader ecological implications. It begins with a historical overview of plant communication theories, tracing how perceptions have evolved with advances in molecular ecology and symbiosis research. The following sections describe the mechanisms underlying mycorrhizal interactions, supported by empirical evidence from key studies, including Simard's pioneering experiments and subsequent replication attempts by independent researchers.

The later sections assess the ecological relevance of inter-plant signaling, focusing on biodiversity maintenance, regeneration processes, and ecosystem recovery. Critical perspectives are also included to evaluate whether observed cooperative behavior represents intentional signaling or passive biochemical diffusion. The chapter concludes with emerging research directions involving omics technologies, soil metagenomics, and long-term ecological monitoring aimed at resolving these debates.

Recent technological advances such as artificial intelligence, bioacoustics, and soil sensor networks are providing new methodologies to analyze communication dynamics within forest ecosystems. The final section connects these findings to policy frameworks such as the United Nations Sustainable Development Goals (SDGs 13 and 15), the Paris Agreement, and REDD+ initiatives, emphasizing how scientific understanding of below-ground networks can guide sustainable forest management and restoration strategies.

Ultimately, this chapter underscores that forests represent complex ecological networks shaped by continuous biological exchange. By analyzing the Wood Wide Web through empirical evidence and critical debate, it reframes forests as dynamic systems of interaction and adaptation, grounded in ecological interdependence rather than anthropomorphic cooperation.

HISTORICAL EVOLUTION OF FOREST COMMUNICATION THEORIES

For many years, scientists believed that plants were passive living things that simply competed for sunlight, water, and nutrients (Chai et al., 2024). Early ecological studies were based on this idea, and plants were seen as individual organisms reacting to their surroundings with little interaction. The idea that plants could communicate or show complex behaviour was difficult for many researchers to accept. Early claims about plant intelligence were often dismissed as emotional rather than scientific (Kingsland & Taiz, 2025). Even though Jagadish Chandra Bose showed in the early 1900s that plants respond to external stimuli, most scientists did not accept his work at the time because they felt the evidence was not strong enough (Kingsland & Taiz, 2025). For decades, plant science focused mainly on photosynthesis, growth, and reproduction.

This thinking changed in the late twentieth century. Research showed that soil fungi form close partnerships with plant roots (Bhupenchandra et al., 2024). These fungi build large underground networks that connect many plants and allow the movement of water, nutrients, and chemical messages. This suggested that plants interact far more than earlier believed. A major turning point came from the work of Dr. Suzanne Simard. Using carbon isotope tracing, she showed that trees can share carbon through fungal networks, especially from older trees to younger ones (Bas et al., 2024). She also found that trees attacked by pests send warning signals to nearby trees. These findings led to the term *Wood Wide Web*, comparing these fungal networks to an information-sharing system (Merckx et al., 2024).

These discoveries changed how scientists view forests. Research now explores how plants recognize relatives, share nutrients, and respond to stress signals. Together, these findings show that forests function as connected systems rather than isolated competitors.

UNDERSTANDING MYCORRHIZAL NETWORKS

Definition of mycorrhizae: mutualistic association between fungi and plant roots

Mycorrhizae are symbiotic associations between the roots of most land plants and certain soil fungi as shown in figure 1. The term "mycorrhiza" comes from Greek, with *mykes* meaning fungus and *rhiza* meaning root (Goswami et al., 2025). This reflects the close physical and functional connection between fungal hyphae and plant roots. In this mutual relationship, both partners benefit. The fungus receives carbohydrates produced by the plant through photosynthesis. In return, the plant gains improved access to water and essential nutrients such as phosphorus, nitrogen, and other minerals that the fungus extracts from the soil.

Figure 1. Mycorrhizal Networks

Fungal hyphae extend far beyond the plant's root zone, significantly increasing the surface area available for nutrient and water absorption. As a result, plants associated with mycorrhizal fungi often show greater resilience to drought, enhanced resistance to pathogens, and improved tolerance to harmful soil substances. The fungi benefit by obtaining organic carbon compounds from the plant, which sustain their growth and reproduction.

There are two principal types of mycorrhizae. **Ectomycorrhizae** form a sheath around the root surface and penetrate between root cells without entering them, while **arbuscular mycorrhizae** penetrate root cells and form specialized structures for nutrient exchange (Chen et al., 2024). These associations have existed for more than 400 million years and are fundamental to the functioning of many terrestrial ecosystems.

Beyond nutrient exchange, mycorrhizal networks also play a role in plant-to-plant interactions. They form extensive underground networks that connect multiple plants, enabling the transfer of resources and chemical signals. These networks are a key component of the ecological phenomenon known as the Wood Wide Web, providing the structural and functional basis for forest communication and cooperation.

Types of mycorrhizal fungi

Mycorrhizal fungi are mainly divided into two types based on how they connect with plant roots: ectomycorrhizal (ECM) and arbuscular mycorrhizal (AM) fungi. These two forms of partnership differ in how they share nutrients and support communication within forest ecosystems.

Ectomycorrhizal Fungi

Ectomycorrhizal fungi form close partnerships with many woody plants by growing around root tips and creating a hyphal mantle on the surface (Mrak et al., 2024). Their hyphae move between root cells without entering them, forming the Hartig net, which supports nutrient exchange. These fungi are effective at releasing nitrogen and phosphorus from complex organic matter and help decompose leaf litter. Species such as *Amanita*, *Boletus*, and *Russula* are common examples. They also create large underground networks that connect multiple trees, supporting nutrient transfer across forests and contributing to the Wood Wide Web, although some mechanisms still require further study.

Arbuscular Mycorrhizal Fungi

Arbuscular mycorrhizal fungi enter plant root cell walls and form branched arbuscules in the root cortex, increasing the surface area for nutrient exchange (Cargill et al., 2025). Belonging to Glomeromycota, they partner with nearly 80% of land plants, including grasses, herbs, and many tropical trees. They are highly effective at absorbing phosphorus from mineral soils, using fine hyphae that reach nutrients beyond the plant's root zone. Although they cannot break down organic matter, they excel at gathering inorganic nutrients. Like ectomycorrhizal fungi, they create underground networks that transfer nutrients and signals, though the exact mechanisms and ecological significance still require further study.

Mechanisms of nutrient exchange and information flow

The mycorrhizal network operates through sophisticated mechanisms that facilitate both nutrient exchange and information flow between plants, creating an intricate underground communication system that rivals the complexity of the internet (Xiaofan & Limpens, 2025).

Nutrient Exchange Mechanisms

The mycorrhizal partnership is a two-way exchange in which plants supply fungi with carbon compounds, mainly sugars produced through photosynthesis, because fungi cannot create these compounds on their own (Zhang et al., 2025). In return, fungi provide plants with essential nutrients such as phosphorus, nitrogen, and trace minerals. Their extensive hyphal networks and specialized enzymes allow them to access these nutrients more efficiently. This exchange occurs through active transport, where specific proteins move nutrients from fungal hyphae into plant root cells, while carbon flows from plant to fungus. The balance shifts with plant needs and environmental conditions. Although well documented, many findings come from controlled experiments. Ongoing debates question whether nutrient transfers reflect active cooperation or passive movement, highlighting the need for more field-based research.

Information Flow Networks

Mycorrhizal networks do more than exchange nutrients; they also transmit chemical signals that help plants respond to environmental stress. When a plant faces drought, pests, or disease, it releases chemical compounds into the fungal network, which travel through hyphae via cytoplasmic streaming and reach other plants within hours or days (Xiaofan & Limpens, 2025). These signals can trigger defence responses, guide root growth, or influence flowering and fruiting. Some studies also report electrical signalling across these networks. Although well documented, the mechanisms and ecological importance of this signalling remain debated. Researchers continue to examine whether these signals reflect active communication or passive biochemical movement.

Role in seedling establishment, immunity, and drought resistance

Mycorrhizal fungi are important partners that help plants survive in key ways. They support the growth of new seedlings, boost the plant's ability to fight off diseases, and help plants stay healthy during dry conditions.

Seedling Establishment

Young seedlings struggle to survive in dense forests, but mycorrhizal fungi significantly improve their chances by expanding the effective reach of their root systems (Toca et al., 2022). Through extensive hyphal networks, fungi increase access to water and nutrients far beyond what roots alone can reach, sometimes extending resource access up to a thousand times the root surface area. These networks link seedlings to the wider underground system, allowing them to receive circulating carbon and nutrients. In some forests, older "mother trees" transfer carbon, nitrogen, and phosphorus to young plants, supporting early growth when seedlings have limited photosynthetic ability. However, the level of support varies by species and environment, and questions remain about whether these exchanges reflect active cooperation or passive nutrient movement.

Enhanced Immunity

Mycorrhizal partnerships play an important role in protecting plants from diseases and soil-borne pathogens. These fungi produce chemical compounds that suppress harmful microbes and shield plant roots (Priyadarshani et al., 2025). Their presence in the soil also occupies space and resources, reducing opportunities for pathogens to spread. Mycorrhizal fungi further support plant immunity by triggering systemic acquired resistance, a process that prepares plants to respond more quickly to future stress or infection. Through shared fungal networks, stressed plants can send signals to neighboring plants, enabling them to activate defensive responses in advance. Although experimental studies support these effects, the exact mechanisms and consistency of immune enhancement vary with plant species, fungal partners, and environmental conditions. Ongoing debates question whether such signals represent active communication or passive by-products, highlighting the need for more field-based research.

Drought Resistance

Mycorrhizal networks help plants survive dry conditions by using fine fungal hyphae that reach water in tiny soil pores inaccessible to roots (Branco et al., 2022). These networks expand a plant's ability to obtain water and reduce dependence on moisture near the root zone. Mycorrhizal fungi also support water balance through osmotic adjustment and improved water use efficiency. However, drought resistance varies by fungal species, plant type, soil conditions, and environment. Some studies show strong benefits, while others report limited effects. More long-term field research is needed to determine whether these benefits result from active fungal processes or passive hydraulic movement.

How the network adapts and evolves over time

Mycorrhizal networks are highly adaptable. They can change their structure, function, and partnerships over time in response to environmental stress, seasonal changes, and long-term shifts in the ecosystem.

Dynamic Structural Adaptation

The structure of mycorrhizal networks is always changing, depending on how much nutrients are available and what plants need. When the soil is rich in nutrients, fungal threads grow quickly and form dense links between plants. But when resources are limited, the network removes less useful connections and strengthens the main routes where activity is highest. This helps save energy and works in a way similar to how the brain removes unused connections during development.

Seasonal changes also shape the network. In spring, as trees start photosynthesis again, carbon flows increase and fungal networks grow rapidly. In autumn, as plants slow down, the network becomes more focused. Fungi shift their energy toward making spores and preparing for winter. These regular changes help the network survive through different seasons and environmental challenges.

Partnership Evolution and Selection

Mycorrhizal networks are not fixed; they regularly evaluate and change their partnerships with plants. Fungi prefer to share nutrients with plants that consistently give them carbon in return. Plants that offer little or unreliable carbon may have their network links reduced over time (Figueiredo et al., 2021). This creates a natural system where cooperation is rewarded, and selfish behavior is discouraged.

Over longer periods, fungi and plants influence each other's evolution. Plants become better at sending and receiving chemical signals, while fungi improve their ability to pull nutrients from the soil and deliver them efficiently. This mutual development makes the entire network stronger, more efficient, and better able to survive changing conditions.

Adaptive Responses to Environmental Change

Mycorrhizal networks respond quickly to changes in their environment. After events like forest fires, droughts, or disease outbreaks, the remaining parts of the network adjust themselves to help the ecosystem recover. Fast-growing fungi form early connections that support new plant growth. Over time, these temporary links are replaced by stronger and more complex partnerships as the forest regrows.

Climate change is also influencing how these networks behave. Fungi are adapting by connecting with more types of plants and by becoming better at handling stress. Some fungal species are moving to new areas along with their plant partners. As temperatures and rainfall patterns shift, fungi follow the changing forests. This helps maintain the network even as the landscape evolves.

Examples from boreal, temperate, and tropical forests

Mycorrhizal networks take different forms in forests around the world. Each type of forest—whether tropical, temperate, or boreal—develops unique features based on its climate, soil, and the kinds of trees that grow there. These differences show how flexible and adaptable the partnerships between fungi and plants are, allowing them to thrive in a wide range of environments.

Boreal Forest Networks

The boreal forests of Canada, Alaska, and Siberia have some of the largest ectomycorrhizal networks in the world. These forests, filled with conifer trees like spruce, fir, and pine, rely on strong underground fungal connections to survive in poor soils and short growing seasons. In British Columbia, scientists have found that a single mother tree can support more than 200 smaller plants through these fungal links. Fungi such as *Suillus* and *Lactarius* play a key role by breaking down tough organic matter in acidic soils. These networks are highly cold-resistant and stay active even under snow, helping move carbon during the short times when trees can photosynthesize.

Temperate Forest Complexity

Temperate forests have the most varied mycorrhizal relationships, often combining both ectomycorrhizal and arbuscular fungi in one ecosystem. Forests with European beech and oak show how these connections can become complex and layered over time. In the Pacific Northwest, Douglas fir forests form vast underground networks where a single tree can connect to hundreds of others, even across different species. Fungi like *Rhizopogon* and *Tricholoma* help transfer resources and information between conifers and broadleaf trees. The region's mild, rainy climate allows these fungal networks to stay active all year long.

Tropical Forest Dynamics

Tropical rainforests have the most diverse mycorrhizal networks, linking thousands of plant species mainly through arbuscular mycorrhizal fungi. In the Amazon, just one handful of soil can hold dozens of fungal species from the Glomeromycota group. These networks work in fast-paced environments where nutrients cycle quickly and competition is intense. In Costa Rica's cloud forests, fungi connect plants from the ground to the treetops, including epiphytes, forming complex, 3D networks. Because of the high plant and fungal diversity, these systems are very resilient and can keep functioning even after storms, landslides, or human disruptions.

TREE-TO-TREE COMMUNICATION: METHODS AND MECHANISMS

The realization that trees engage in complex forms of communication has transformed our understanding of forests. Beneath the soil, fungal networks link trees and allow them to pass chemical signals to one another. Through these underground connections, they can share nutrients, sense potential dangers, and adjust their responses as a collective. This hidden system reveals how deeply interconnected forest life is and how thoughtfully these ecosystems function.

Biochemical Signaling via the Fungal Network

Trees mainly communicate through chemical signals passed along mycorrhizal fungal networks that connect their roots. These underground systems act like living cables, sending complex messages from one tree to another using fungal threads that reach beyond the roots. The fungi help carry and even translate these signals so different tree species can understand each other.

Trees release special chemicals like VOCs, phenolic compounds, and other molecules into the network (Rasheed et al., 2023). These messages move through the fungi by a fast internal flow called cytoplasmic streaming. Fungi can also adjust or boost the signals as they travel.

Every chemical blend carries its own message. Compounds such as terpenes can signal danger, while others convey information about nutrients or growth conditions. Together, these signals form a rich chemical vocabulary that allows trees and other plants to exchange important updates throughout the forest.

Transmission of Stress Signals

Forest trees have smart early warning systems that quickly alert nearby plants to dangers. When a tree is stressed by pests, disease, or harsh conditions, it sends alarm chemicals through the mycorrhizal network. These messages can travel fast—reaching other trees within hours (Pollastrini & Bussotti, 2023).

One clear example is pest attacks. When insects start eating a tree's leaves, the tree makes defensive chemicals and also sends a warning underground. Nearby trees pick up the signal and prepare by making their own toxins, boosting resin production, or changing their leaf chemistry so pests won't want to eat them.

Drought signals work in a similar way. Trees that are low on water send out stress messages. Other trees respond by using less water, growing roots deeper, or readying themselves for dry conditions. This teamwork helps the whole forest survive drought.

Trees also share warnings about infections. If a tree gets a fungal or bacterial disease, it releases both antimicrobial compounds and warning signals. Healthy trees then strengthen their defenses, helping stop the disease from spreading through the forest.

Nutrient Allocation Based on Kin Recognition

Trees have the surprising ability to recognize and support their own offspring and close relatives through underground fungal networks. This kin recognition system relies on unique chemical signals that help trees tell the difference between their own seedlings, unrelated trees, and even other species. Mature "mother trees" can detect their young by identifying specific chemical markers and will often give them more resources as a result.

This recognition process involves complex chemical sensing. Trees analyze molecular signals from the roots of nearby plants using the shared fungal network. These signals reveal genetic similarities, helping trees decide where to send nutrients. Studies have shown that mother trees can send up to 40% more carbon and nutrients to their own offspring compared to unrelated plants.

The balance between cooperation and competition plays a key role in how trees share resources. Trees tend to support those in their family group more generously. However, when growing near unrelated individuals, they may compete more directly for water, sunlight, and nutrients. Despite this, the fungal network often encourages cooperation during tough times—such as droughts or pest outbreaks—because survival of the whole community becomes more important than individual gain.

What makes this even more fascinating is that the kin recognition system isn't just short-term. Trees can maintain special connections with their own young over several years. These long-lasting bonds allow them to continue offering support across different growing seasons. In many cases, trees show patterns that look like they are supporting their close relatives, often directing helpful chemical signals or resources toward them. This behaviour suggests a form of kin-oriented cooperation within forest communities.

This natural preference, often called biological nepotism, helps closely related trees survive and keeps their family lines strong in the forest. It also shows that many quiet signals and exchanges happen underground, slowly shaping how the forest grows and stays balanced.

Carbon and Nitrogen Transfer

The mycorrhizal network helps trees share important nutrients like carbon and nitrogen. This underground system works like a forest-wide resource exchange. Trees support one another depending on their needs and the availability of resources. Carbon usually flows from trees that get a lot of sunlight to smaller trees growing in the shade. These include young seedlings and trees under the forest canopy. During the growing season, a large tree can pass several kilograms of carbon to nearby trees through fungal threads.

Nitrogen sharing works differently but follows a similar idea. Trees that grow in areas with more nitrogen or those that can fix nitrogen better pass some of it to other trees that have less. Fungi help move these nutrients. They gather nitrogen and carry it through their thread-like structures to trees that need it. This helps the forest maintain a balanced supply of nutrients.

This exchange goes both ways. Trees do not always give or always receive. Their roles change based on the season or their health. A tree under stress from drought or pests can receive extra nutrients from healthier neighbors. Later, when it becomes strong again, it can help other trees in return. This system helps the forest stay strong and healthy.

The process shows how trees work together. They do not grow alone. Instead, they form a connected group. The fungal network not only moves nutrients. It also creates a system of shared support. This helps the forest grow, recover, and stay alive during hard times. By helping each other through these hidden networks, trees increase the strength and stability of the whole forest.

Seasonal and Species-Specific Patterns of Communication

Tree communication changes with the seasons. These changes follow the growth cycles of forests and local weather conditions. In spring, trees become active again after winter. They start sharing more signals to coordinate their growth and manage how they use resources. Communication increases during this time.

Summer is the busiest season for tree communication. Trees send strong signals related to pest control, resource distribution, and continued growth. This helps them manage threats and keep the forest balanced.

In autumn, tree communication focuses on getting ready for winter. Trees share signals about when to shed leaves, move nutrients from leaves to roots, and prepare for cold weather. These signals help them survive the winter months (Simard, 2021).

Winter communication slows down. It does not stop completely. Evergreen trees, which stay active in cold weather, continue to share signals. Their activity is lower, but they still pass information through the underground network.

Different types of trees communicate in different ways. Each species uses its own method to survive and grow. Coniferous trees, like pines and firs, send signals all year. Deciduous trees, which lose their leaves, send more signals during the growing season and less in winter.

Some trees, called pioneer species, grow quickly and send more signals to take in resources fast. Climax species, which grow slowly and live longer, share resources in more careful and cooperative ways.

Trees also talk to other species. While trees mostly share information with their own kind, they can send signals to different species as well. In forests with many types of trees, communication systems adjust. These forests create shared ways to pass information, which helps the entire forest work together. This type of mixed-species communication makes forests stronger and more adaptable.

EXPERIMENTAL EVIDENCE AND LANDMARK STUDIES

The idea of forest communication through mycorrhizal networks developed over many years. Researchers conducted careful experiments to explore how trees interact. These studies questioned the old belief that trees live alone and do not help each other. The results showed clear proof that trees are connected underground. They share nutrients and information through fungal networks. This discovery precipitated a paradigm shift in the scientific understanding of forest dynamics. It also gave new insights into how plants behave and support one another.

Suzanne Simard's Carbon Isotope Experiments

Suzanne Simard's early research in the forests of British Columbia changed how we understand trees. She was the first to show that trees can share resources using mycorrhizal networks. In the 1990s, she ran a series of key experiments using carbon-14 and carbon-13 isotopes. These isotopes helped trace how carbon moved between different tree species in mixed forests (Moguel, 2022).

She placed radioactive carbon into paper birch trees and stable carbon isotopes into Douglas fir trees. This helped her observe carbon moving both ways between the two types of trees. Before this, many scientists believed trees only competed with each other. Her research proved they could also cooperate.

The setup of her experiment was simple but very careful. She covered trees with plastic bags filled with labeled carbon dioxide. The trees took in this carbon during photosynthesis. Simard then used special tools to follow where the labeled carbon went. Within days, it showed up in nearby trees connected by the fungal network. This showed that trees share carbon through underground fungi.

In later studies, Simard discovered that this sharing changed with the seasons. In spring, Douglas fir trees helped birch trees before their leaves grew. In summer, birch trees helped fir trees when they were more active. This back-and-forth support showed trees work together, not just against each other. Simard's work laid the foundation for the idea of cooperative forests.

David Read's Pioneering Work

David Read's research at the University of Sheffield helped build the foundation for understanding how mycorrhizal networks work. In the 1980s and 1990s, he ran many lab and greenhouse experiments. These studies showed that mycorrhizal fungi connect plant roots and help move nutrients between different plant species.

What made Read's work important was the use of controlled environments. His team created special growth chambers. In these, they could control whether plants were connected through fungi or kept separate. They also changed the levels of nutrients to see how the plants responded.

The results showed that plants connected by mycorrhizal fungi grew better and absorbed more nutrients than plants that were isolated. He also found that different types of fungi form different types of networks. This showed that not all fungal connections work the same way.

Read's careful methods gave strong support to what scientists had seen in natural forests. His studies removed confusion caused by outside environmental factors. He also created standard procedures that other researchers later used and improved to study different kinds of forests.

Key Findings from International Forest Research

International research has helped scientists learn more about mycorrhizal networks in many types of forests. In Europe, studies in beech and oak forests found old and complex networks. In these systems, ancient trees act as central hubs. They support hundreds of smaller plants connected through underground fungi.

In Scandinavia, researchers studied boreal forests. These forests have very cold winters and short summers. Despite the harsh climate, the mycorrhizal networks remained strong and active. This showed their resilience in extreme environments (Berglund & Kuuluvainen, 2021).

In Australia, studies focused on eucalyptus forests. These areas often experience bushfires. The research showed that mycorrhizal networks help forests recover quickly after fires. The fungi support new plant growth by moving nutrients to young trees.

Tropical forests also showed interesting patterns. In some rainforests, researchers found over 100 different fungal species in a single area. These fungi form complex webs that connect many plant species and allow them to share resources.

In Japan, scientists studied how these networks change with the seasons. They observed specific patterns of communication between tree species. Their long-term research showed that as forests grow and change, the networks also adjust. The fungi adapt their connections as new plants appear and old ones die.

Criticism, Replication Challenges, and Alternative Theories

Even with strong evidence, research on mycorrhizal networks faces valid scientific criticism. Some experts say that earlier studies did not fully rule out other ways carbon might move between trees. For example, root grafting or the natural breakdown of plant matter could explain resource movement, without involving fungal networks.

There are also concerns about the methods used in these studies. It is hard to separate the effects of mycorrhizal networks from other soil-related processes. Results from lab experiments may not always apply to real forest settings, where many complex factors are at play.

Some scientists suggest that what looks like resource sharing may be caused by changes in soil chemistry near the roots, rather than direct transfer through fungal networks.

Statistical issues have also been raised. Critics point out that some studies use small sample sizes or run for short periods. Because forests are so complex, it's difficult to control every factor that could affect results.

Field-Based Versus Lab-Based Approaches

There is an ongoing debate between field-based and lab-based research in mycorrhizal network studies. Field experiments take place in natural settings, so they reflect real forest conditions. However, they often struggle to control variables or pinpoint exact causes. In contrast, laboratory studies allow for precise control of conditions. But they may not capture the full complexity of how forests work in nature.

Field studies are useful for showing how networks function in the wild. They help uncover large-scale patterns that might not appear in a lab. Still, it is hard to prove cause and effect in the field, and many environmental factors can influence results.

Laboratory studies help explain how specific processes work. They allow for accurate measurement and testing. But some researchers worry that these simple setups do not reflect the richness of natural forests.

The strongest results come from combining both methods. Lab work helps reveal how things function. Field work checks if those results apply to real ecosystems. Today, scientists use tools like molecular techniques, isotope tracing, and advanced imaging to connect lab findings with what happens in nature. This approach gives stronger and more reliable evidence for how mycorrhizal networks work.

ECOLOGICAL IMPACTS AND FUNCTIONS

Mycorrhizal networks play a key role in shaping forest ecosystems. They form subterranean systems that facilitate interspecific cooperation among trees and plants.. This cooperation improves survival, supports adaptation, and strengthens resilience. These networks influence forests at many levels. They affect how individual trees interact with each other. They also guide larger processes that shape the entire forest landscape. As a result, forests function not as groups of separate trees but as connected, cooperative systems.

Influence on Forest Regeneration and Succession

Mycorrhizal networks act like underground highways that speed up forest regrowth and support natural changes in forest structure. After disturbances like logging, fires, or storms, the remaining parts of these networks become essential for recovery. New plants use these existing connections to quickly access nutrients and information, improving their chances of survival.

As forests move through natural stages of growth, these networks help smooth the transition. Early-growing species build simple networks that support themselves and later species. Over time, as the forest matures, the networks become more complex. Older trees become central hubs that share resources with many smaller plants below.

These networks are especially important for helping young trees grow. Studies show that "mother trees" can pass up to 75% more resources to their seedlings through these connections. This helps not only their own offspring but also unrelated seedlings, creating a nursery effect that benefits the whole forest.

Enhancement of Biodiversity Through Cooperative Networks

Cooperative mycorrhizal networks help increase biodiversity by encouraging resource sharing among different plant species. Instead of allowing stronger species to push out weaker ones, these networks make it easier for many plants to live together. They do this by helping plants use different resources in different ways. For example, some plants are better at using sunlight, while others are better at absorbing nutrients from the soil. The networks allow these plants to share what they are best at, which supports more variety in the forest.

Rare and endangered plants benefit the most. These plants receive support through the network that helps them survive in tough conditions. Even if their main habitat changes or becomes less suitable, the network offers backup support to keep them alive.

Disease Resistance and Ecosystem Resilience

Mycorrhizal networks act like immune systems for forests, helping trees defend against disease and stress. When one tree is attacked by a pathogen, it sends out chemical warning signals through the network. Neighboring trees receive these signals and activate their own defenses early, which can cut disease spread by as much as 60% compared to trees without network connections.

These networks also offer support during tough conditions like drought or illness. If one tree is struggling, healthier trees connected to the network can share nutrients and water to help it survive. This teamwork helps keep the entire forest stable and reduces the risk of widespread damage.

Balancing Competition and Cooperation

Mycorrhizal networks help maintain a natural balance between competition and cooperation. Trees compete for resources like light, water, and nutrients when these are available in plenty. But during times of stress, such as drought or disease, they begin to support each other. This shift from competition to cooperation helps the forest survive as a whole. The system is flexible. It allows trees to grow and thrive when conditions are good. It also helps them stay strong and recover when the environment becomes difficult.

Forest-Wide Adaptation to Climate Change

Mycorrhizal networks help forests adjust quickly to climate change and environmental stress. These underground systems allow trees to share information about changing conditions. As a result, trees can adjust their internal functions and responses together, rather than acting alone. This teamwork helps the entire forest stay healthy and keep working even as the environment changes.

These networks also support long-term adaptation. They help climate-adapted trees move into new areas. They also allow trees with different traits to mix and produce stronger offspring. This genetic exchange helps forests evolve faster than if each tree adapted on its own.

MYSTERIES YET TO BE SOLVED

Even after many years of research showing how amazing mycorrhizal networks are, important questions still need answers. These unknowns show how complex forest communication really is. They also suggest that what we know now is just the start of a much bigger story.

Unanswered Questions About Signaling Speed and Specificity

We still do not fully understand how fast signals travel through mycorrhizal networks. Some studies show chemical signals reaching nearby trees in hours. Others find it takes days or weeks over similar distances. We need to study what controls this speed. It might depend on fungal species, signal types, environmental conditions, or how the network is built. Signal clarity is also a puzzle. How do these networks tell different chemical messages apart and send them to the right plants? Trees seem able to send messages to specific individuals or species. This suggests a complex system for addressing messages, but scientists have only just started to study this.

We also do not know how signal strength relates to accuracy over long distances. Do signals weaken as they travel? How do networks keep messages clear when they pass through many fungi and plants?

How Non-Mycorrhizal Plants Interact with the Network

We have limited knowledge about how plants without mycorrhizal partnerships take part in forest communication. Some plant groups, like many Brassicaceae and Proteaceae species, do not form these partnerships but still seem to send and receive signals connected to the network. These non-mycorrhizal plants might interact with the network by releasing chemicals from their roots that affect fungi. They could also pick up signals floating in the soil. Another possibility is that they connect indirectly by associating with nearby plants that do have mycorrhizal links.

Learning how these plants fit into the network is important to fully understand how communication works across the whole forest.

Network Boundaries and Limitations

We do not yet know the full physical and biological limits of mycorrhizal networks. How far can one network stretch? What stops it from growing larger? Things like geographic barriers, changes in soil chemistry, and shifts in plant species probably affect how networks spread. But no detailed mapping has been done so far.

Another question is how much these networks can handle. Can they get overwhelmed with too many signals or too many plants connected? How do they decide which messages are most important when many are sent at once? Also, what happens if the network gets damaged or broken into pieces?

Cross-Species Communication

How plants as different as trees, shrubs, ferns, and herbs manage to communicate is still not fully understood. Each group likely relies on its own set of chemical cues, yet the underground networks seem to carry messages across these differences. This raises important questions about the fungi involved. Do the fungi modify or "translate" these signals so that unrelated plants can interpret them? Or is there a shared chemical code that works for many species? More research is needed to understand how these signals move through such diverse plant communities.

Studying how forest understory plants like ferns, mosses, and wildflowers join these networks could reveal new sides of forest intelligence and teamwork. These findings could change how we think about how ecosystems and plant communities work.

COMMUNICATION VS COMPETITION DEBATE

The idea of the Wood Wide Web has led to major scientific and philosophical debates. It challenges long-held beliefs about how plants behave and interact with each other. The discussion goes beyond just mycorrhizal networks. It raises deeper questions about whether plants can cooperate, show intelligence, or act with purpose in nature.

Critics of the Wood Wide Web Concept

Some leading experts in plant competition believe that what looks like cooperation in mycorrhizal networks is actually a complex form of competition. Critics like James Cahill and André Kessler say that resource sharing mostly helps the donor plants. This benefits their own offspring or keeps good fungal partners for future returns.

Skeptics also question if the resource transfers are truly intentional. They think it might just be passive movement from plants with plenty of resources to those with less, following natural chemical flows. According to evolutionary theory, plants behave competitively rather than kindly. So, any cooperation likely serves selfish genetic goals.

Some researchers also doubt the experiments. They say that isotope-tracing methods might pick up chemicals from root leaks or decomposed matter, not actual network transfers. These critics call for stronger experimental controls and other explanations for the results seen.

Philosophical and Scientific Debates Around Plant Intelligence

The Wood Wide Web idea has sparked strong debates about whether plants have intelligence or consciousness. Traditional science sees plants as chemical systems that react to their environment without awareness or intention. However, plant neurobiologists like Stefano Mancuso believe these networks show information processing, memory, and decision-making. They say this means plants have a form of intelligence.

Critics disagree. They say calling plants intelligent is a mistake because it applies human traits to plants. They argue that complex behaviors can come from simple chemical reactions without needing awareness or choices.

This debate touches on bigger questions about what intelligence really means. If intelligence means processing information, learning, and adapting to challenges, then plant networks could be considered intelligent. But if intelligence requires conscious thought and intention, then plants are just automatic biological systems, no matter how complex their behavior seems.

Is Cooperation Intentional or Emergent?

A key question in the debate is whether mycorrhizal network cooperation is intentional teamwork or just a result of individual plants acting in their own interest.

Supporters say trees actively choose partners, decide how to share resources, and coordinate their actions. This shows they cooperate on purpose.

Critics think that what looks like cooperation actually comes from each plant trying to improve its own survival. They say natural selection makes plants behave in ways that seem cooperative but really serve their own reproduction.

This difference matters a lot for how we understand evolution. If cooperation is intentional, plants might have advanced thinking abilities. But if cooperation is just an outcome of individual actions, it shows how complex teamwork can come from simple self-interest.

Implications for Redefining Ecological Models

The Wood Wide Web debate challenges traditional ecological ideas about competition. Old models say species compete for limited resources, and stronger species push out weaker ones. This leads to simple communities with only a few dominant species. Newer network models show that sharing resources and information lets many species live together. These models predict more biodiversity and stronger ecosystems than competition alone.

This idea changes how we protect forests and manage nature. If networks create cooperation, then we should focus on keeping these connections strong and supporting teamwork among species. Instead of only protecting single species or favoring competition, management should help maintain these cooperative relationships.

HUMAN INFLUENCE AND TECHNOLOGICAL EXPLORATION

Effects of Deforestation and Fragmentation on Mycorrhizal Networks

Mycorrhizal networks are underground fungi that connect the roots of trees and plants. These networks help trees share water, nutrients, and warning signals when there is danger. They keep forests balanced, strong, and diverse. But human actions like cutting down forests and breaking up the land are harming these networks. When big parts of forests are cleared for farming, roads, or buildings, the networks get broken. Trees become separated and lose the help from these fungal connections. This makes trees weaker and less able to fight disease, drought, and pests. It also hurts young plants, soil health, and the movement of important nutrients between species. Broken forests cause stressed ecosystems that find it hard to heal. Because of this, forest life declines, and nature recovers more slowly. Protecting forest connections is key to saving mycorrhizal networks and the forests they support.

Use of Imaging, Sensors, and Machine Learning to Study Fungal Systems

Scientists use new technologies to study mycorrhizal networks without harming the forest floor. Tools like infrared cameras and ground-penetrating radar let them see underground fungal threads called hyphae that connect trees. Soil sensors measure temperature, moisture, and chemicals that show fungal activity. The large amount of data from these tools is analyzed with machine learning. This helps researchers find patterns, predict changes, and understand how fungi respond to stress from the environment. For example, they can link temperature or rainfall changes to fungal growth. Machine learning can also spot early signs of forest stress before it becomes visible. Together, these methods provide a safe way to watch forest health and track how fungal networks recover after human activity or natural disasters.

Citizen Science and Forest Monitoring

Citizen science is playing a bigger role in forest research. Around the world, regular people who are not scientists help collect data from forests. They use easy tools like smartphone apps to record tree growth, soil health, or signs of fungi. Some projects let people collect soil samples or upload photos of forest changes. This helps researchers get more data than they could alone. It also connects people more closely to nature. Citizen science lets students, farmers, hikers, and local communities learn about forests and join in protecting them. When people help monitor forests, they understand better how important forests are and how mycorrhizal networks work. This public participation raises awareness and builds support for conservation. In this way, citizen science is a strong tool for research and learning about the environment.

Ethical and Philosophical Implications of "Eavesdropping" on Nature

Studying how trees and fungi quietly communicate raises important ethical and philosophical questions. When we use advanced tools to watch how forests behave and react, we should ask if we are just learning or also intruding. Some scientists and thinkers believe forests show a kind of intelligence through their cooperation and support systems. For example, mycorrhizal networks help manage resources, protect weaker plants, and adjust to changes. If we accept this, then closely watching or recording their "conversations" might feel like eavesdropping. There is also a worry that this knowledge could be used for profit instead of protection. For example, companies might try to control these networks to grow trees faster or make more money, ignoring the forest's larger role in nature. Ethical research should focus on care, respect, and humility. By seeing forests as living communities, not just resources, we can study them in ways that protect their health and dignity.

CONSERVATION, RESTORATION, AND POLICY IMPLICATIONS

Role of the Wood Wide Web in Ecosystem Restoration

The underground fungal networks called the "Wood Wide Web" are very important for healing damaged ecosystems. These networks are made by mycorrhizal fungi and connect trees and plants below the ground. They let older, stronger trees help younger ones by sharing water, nutrients, and protective chemicals. When forests are damaged or cut down, rebuilding these fungal connections can help nature recover faster. Mycorrhizal fungi improve the soil, help more plants survive, and support forests to grow with more variety and balance. Restoration projects that bring back native fungi along with trees usually succeed better over time. These underground networks do more than help single trees—they rebuild whole communities and keep ecosystems in balance. Supporting the Wood Wide Web is not just useful but needed for healthy and strong forests.

Forest Management Informed by Fungal and Root Networking

Traditional forest management usually looks at things we can see, like tree height, timber quality, and surface plants. But new research shows that what happens underground is just as important. Fungal and root networks create a hidden web that connects and supports the forest. These systems help control tree growth, protect against pests, and keep the ecosystem healthy. Forest managers can use this knowledge to make better choices. For example, they can do selective logging that avoids harming important fungal hubs. They can also avoid disturbing soil where fungal connections are strong. When planting new forests, picking tree species that form good partnerships with fungi can improve success. Understanding underground networks helps managers focus on long-term health, not just quick resource gain. This way, the forest is seen as a connected living system, not just separate trees.

Policy Suggestions: Preserving Underground Biodiversity

Protecting underground biodiversity needs new policies that recognize the importance of fungal networks. Most current forest laws focus on trees, animals, or water, but often ignore the life living in the soil. Governments and conservation groups should include fungi when assessing environmental impacts and planning restoration. Policies could help fungi by protecting old forests that have strong fungal networks. They should also limit harmful actions like deep soil digging and heavy use of chemical fertilizers. More funding should go to studying soil life. Restoration work should include adding fungi along with planting trees. By making rules that value the hidden parts of the forest, we can create healthier and stronger ecosystems and prevent collapse caused by neglecting what lives underground.

Indigenous Knowledge and Traditional Forest Management Alignment

Many Indigenous communities have long seen forests as living systems. Their traditional ways often protect the balance of nature, including underground life. Practices like controlled burning, seasonal harvesting, and selective planting come from careful observation and respect for natural cycles. These methods help keep root systems and fungal networks healthy, even if not explained by science. Combining modern forest policies with Indigenous knowledge can lead to more sustainable and respectful management. When scientists and Indigenous groups work together, conservation efforts improve. These partnerships blend advanced technology with generations of local wisdom. Recognizing Indigenous rights and including their voices in decision-making not only ensures fairness but also deepens scientific understanding. The future of forest care may depend on this mix of tradition and technology, based on shared respect for the hidden systems that support forests.

CONCLUSION

The hidden world under the soil shows us how forests really work. The connections between trees and fungi, known as the Wood Wide Web, reveal that forests are not just separate trees but a shared community. These networks help plants pass resources and warnings, suggesting a cooperative side of nature that challenges the older idea that forests mainly operate through competition. This balance between cooperation and competition is still a major question in science, and researchers continue to learn more every year. These underground systems are important because they support biodiversity, help forests recover, and keep the environment stable. To protect them, we need experts from many fields to work together. Understanding the forest means paying attention to its quiet signals and acting with care. As science grows, we are only beginning to understand this hidden world, which reminds us to stay humble before nature's deep wisdom.

REFERENCES

Bas, T. G., Sáez, M. L., & Sáez, N. (2024). Sustainable Development versus Extractivist Deforestation in Tropical, Subtropical, and Boreal Forest Ecosystems: Repercussions and Controversies about the Mother Tree and the Mycorrhizal Network Hypothesis. *Plants*, *13*(9), 1231. DOI: 10.3390/plants13091231 PMID: 38732447

Berglund, H., & Kuuluvainen, T. (2021). Representative boreal forest habitats in northern Europe, and a revised model for ecosystem management and biodiversity conservation. *Ambio*, *50*(5), 1003–1017. DOI: 10.1007/s13280-020-01444-3 PMID: 33454914

Bhupenchandra, I., Chongtham, S. K., Devi, A. G., Dutta, P., Sahoo, M. R., Mohanty, S., & Swapnil, P. (2024). Unlocking the potential of arbuscular mycorrhizal fungi: Exploring role in plant growth promotion, nutrient uptake mechanisms, biotic stress alleviation, and sustaining agricultural production systems. *Journal of Plant Growth Regulation*, •••, 1–39.

Branco, S., Schauster, A., Liao, H. L., & Ruytinx, J. (2022). Mechanisms of stress tolerance and their effects on the ecology and evolution of mycorrhizal fungi. *The New Phytologist*, *235*(6), 2158–2175. DOI: 10.1111/nph.18308 PMID: 35713988

Cargill, R. I., Shimizu, T. S., Kiers, E. T., & Kokkoris, V. (2025). Cellular anatomy of arbuscular mycorrhizal fungi. *Current Biology*, *35*(11), R545–R562. DOI: 10.1016/j.cub.2025.03.053 PMID: 40494310

Chai, L., Zaburdaev, V., & Kolter, R. (2024). How bacteria actively use passive physics to make biofilms. *Proceedings of the National Academy of Sciences of the United States of America*, *121*(40), e2403842121. DOI: 10.1073/pnas.2403842121 PMID: 39264745

Chen, Y., Liang, M., Burslem, D. F., Johnson, D., Yu, S., & Liu, X. (2024). Contrasting response of root traits of arbuscular mycorrhizal and ectomycorrhizal trees to phosphorus availability in subtropical forests. *Plant and Soil*, •••, 1–13. DOI: 10.1007/s11104-024-07051-5

Dervash, M. A., Yousuf, A., Bhat, M. A., & Ozturk, M. (2024). Fungal Internet: The Natural Networking Systems. In *Soil Organisms: Deciphering the Life Beneath Our Feet* (pp. 77–83). Springer Nature Switzerland. DOI: 10.1007/978-3-031-66293-5_7

Figueiredo, A. F., Boy, J., & Guggenberger, G. (2021). Common mycorrhizae network: A review of the theories and mechanisms behind underground interactions. *Frontiers in Fungal Biology*, *2*, 735299. DOI: 10.3389/ffunb.2021.735299 PMID: 37744156

Fioroni, F., Carrón, A. I., Mancilla, M. A. S., Pastorino, M., & Fernández, N. V. (2025). Underground Neighbors: Shared Root-Associated Fungi Facilitating the Coexistence of *Nothofagus obliqua* and *Pinus ponderosa*. *Forest Science*, *71*(1), 97–122. DOI: 10.1007/s44391-024-00011-7

Goswami, R., Singh, V., Chauhan, A., Vamil, R., & Bhadauria, S. (2025). *The Hidden World Beneath: An Exploration of Soil Fungi*. Notion Press.

Govaert, S., Meeussen, C., Vanneste, T., Bollmann, K., Brunet, J., Calders, K., Cousins, S. A. O., De Pauw, K., Diekmann, M., Graae, B. J., Hedwall, P.-O., Iacopetti, G., Lenoir, J., Lindmo, S., Orczewska, A., Ponette, Q., Plue, J., Sanczuk, P., Selvi, F., & De Frenne, P. (2024). Trait–micro-environment relationships of forest herb communities across Europe. *Global Ecology and Biogeography*, *33*(2), 286–302. DOI: 10.1111/geb.13789

Kingsland, S. E., & Taiz, L. (2025). Plant "intelligence" and the misuse of historical sources as evidence. *Protoplasma*, *262*(2), 223–246. DOI: 10.1007/s00709-024-01988-1 PMID: 39276228

Merckx, V. S., Gomes, S. I., Wang, D., Verbeek, C., Jacquemyn, H., Zahn, F. E., Gebauer, G., & Bidartondo, M. I. (2024). Mycoheterotrophy in the wood-wide web. *Nature Plants*, *10*(5), 710–718. DOI: 10.1038/s41477-024-01677-0 PMID: 38641664

Moguel, R. G. (2022). *Using Radiocarbon to Trace Methane Emissions from Natural and Anthropogenic Sources in Two Canadian Ecosystems*. McGill University.

Mrak, T., Grebenc, T., Friedrich, S., & Münzenberger, B. (2024). Description, identification, and growth of *Tuber borchii Vittad.* mycorrhized *Pinus sylvestris L.* seedlings on different lime contents. *Mycorrhiza*, *34*(1), 85–94. DOI: 10.1007/s00572-023-01135-3 PMID: 38236414

Pollastrini, M., & Bussotti, F. (2023). Forest monitoring updates: physiological indicators of tree health for detecting and understanding a changing environment. In *118° Congresso della Società Botanica Italiana IX International Plant Science Conference (IPSC)*, PISA, 13–16 SEPTEMBER 2023 (pp. 1–252). Società Botanica Italiana.

Priyadarshani, T. D. C., Weerasinghe, P. A., Law, S., & Zakeel, M. C. M. (2025). Hidden Allies: Exploring the Cutting-Edge Insights into the Role of Mycorrhizal Fungi in Plant Health. In *Fungal Endophytes Volume II: Applications in Agroecosystems and Plant Protection* (pp. 551–585). Singapore: Springer Nature Singapore.

Rasheed, M. U., Brosset, A., & Blande, J. D. (2023). Tree communication: The effects of "wired" and "wireless" channels on interactions with herbivores. *Current Forestry Reports*, *9*(1), 33–47. DOI: 10.1007/s40725-022-00177-8

Robinson, D. G., Ammer, C., Polle, A., Bauhus, J., Aloni, R., Annighöfer, P., Baskin, T. I., Blatt, M. R., Bolte, A., Bugmann, H., Cohen, J. D., Davies, P. J., Draguhn, A., Hartmann, H., Hasenauer, H., Hepler, P. K., Kohnle, U., Lang, F., Löf, M., & Näsholm, T. (2024). Mother trees, altruistic fungi, and the perils of plant personification. *Trends in Plant Science*, *29*(1), 20–31. DOI: 10.1016/j.tplants.2023.08.010 PMID: 37735061

Simard, S. (2021). *Finding the mother tree: Uncovering the wisdom and intelligence of the forest.* Penguin UK.

Toca, A., Moler, E., Nelson, A., & Jacobs, D. F. (2022). Environmental conditions in the nursery regulate root system development and architecture of forest tree seedlings: A systematic review. *New Forests*, *53*(6), 1113–1143. DOI: 10.1007/s11056-022-09944-8

West, P. W. (2024). A review of the growth behaviour of stands and trees in even-aged, monospecific forest. *Annals of Forest Science*, *81*(1), 34. DOI: 10.1186/s13595-024-01250-x

Xiaofan, M. A., & LIMPENS, E. (2025). Networking via mycorrhizae. *Frontiers of Agricultural Science and Engineering*, *12*(1).

Zhang, Q., Wang, Z., Gao, R., & Jiang, Y. (2025). Sugars, lipids and more: New insights into Plant Carbon sources during plant–microbe interactions. *Plant, Cell & Environment*, *48*(2), 1656–1673. DOI: 10.1111/pce.15242 PMID: 39465686

KEY TERMS AND DEFINITIONS

Arbuscular Mycorrhizal Fungi (AMF): Fungi that enter plant root cells and form tree-shaped structures called arbuscules. These structures help plants absorb phosphorus and other minerals.

Carbon Transfer: Movement of carbon-rich sugars from one plant to another through mycorrhizal networks, often from larger trees to younger seedlings.

Cytoplasmic Streaming: Internal flow of fluid inside fungal hyphae that transports nutrients and chemical signals between connected plants.

Ecosystem Resilience: The capacity of a forest to recover from disturbances like fire, drought, or disease. Mycorrhizal networks help speed up this recovery.

Ectomycorrhizal Fungi (ECM): Fungi that cover plant root surfaces but do not enter the root cells. They help trees absorb nutrients from organic matter in the soil.

Forest Succession: The natural process of how a forest regrows and changes over time after disturbances. Mycorrhizal networks support the survival of early seedlings.

Fungal Mantle: A thick covering of fungal tissue surrounding plant roots in ectomycorrhizal associations, helping in nutrient absorption and protection.

Hartig Net: A network of fungal threads that form between root cells in ectomycorrhizal fungi, serving as the nutrient exchange zone.

Hyphae: Thin, hair-like fungal threads that spread through the soil, increasing the plant's ability to access water and minerals.

Kin Recognition: Ability of plants, especially trees, to detect their own relatives and share more nutrients with genetically related seedlings.

Mother Trees: Large, mature trees that act as central hubs in mycorrhizal networks. They support surrounding plants by sending extra carbon and nutrients.

Mycorrhizae: A symbiotic partnership between plant roots and fungi where plants give sugars to fungi, and fungi provide essential nutrients and water.

Mycorrhizal Network: An underground web of fungal hyphae connecting multiple plants, enabling the exchange of nutrients, water, and signals.

Nitrogen Transfer: Sharing of nitrogen nutrients from fungi or well-supplied plants to those with less access, helping maintain balance in the forest.

Resource Allocation: How trees and fungi distribute nutrients and water among connected plants based on need, health, or kinship.

Signaling Molecules: Chemical compounds such as VOCs and phenolics released by plants to communicate warnings or growth information through fungal networks.

Soil Metagenomics: A scientific method that studies genetic material in soil to understand fungal diversity and underground microbial communities.

Stress Signaling: Warning signals sent by stressed plants (due to drought, pests, or disease) to nearby plants through fungal networks.

Symbiosis: A close biological relationship where two different organisms live together and benefit each other, such as plants and fungi in mycorrhizae.

Wood Wide Web: A popular term describing the underground communication and resource-sharing system formed by plant roots and mycorrhizal fungi.

Chapter 6
Problems of Forest and Biodiversity Conservation in Developing Countries

Nofamataro Zebua
https://orcid.org/0009-0008-3902-2289
Universitas Negeri Malang, Indonesia

Laila Badriyatul Habibah
https://orcid.org/0009-0003-9094-6747
Universitas Negeri Malang, Indonesia

Ester Novi Kurnia Zebua
https://orcid.org/0009-0000-5879-9124
Universitas Negeri Malang, Indonesia

ABSTRACT

This chapter discusses governance and innovation issues related to forest conservation and biodiversity in developing countries, focusing on institutional, political, and technological aspects. It does not address biology or culture, but rather focuses on policy disjunction, poorly coordinated inter-sectoral collaboration, and institutional fragmentation. Challenges include scarcity of information and technological resources, complexity of regulatory frameworks, political myopia, and data scarcity. This chapter also discusses emerging governance innovations, including adaptive technologies, open-source monitoring, cross-border conservation diplomacy, and others. This multidisciplinary approach, complemented by case studies, informs the chapter's recommendations for building resilient, inclusive conservation systems that

DOI: 10.4018/979-8-3373-3992-4.ch006

are strategically supported by data. Ilsa emphasizes the need for active participation by policymakers, academics, and practitioners in the Global South working toward sustainable conservation strategies.

INTRODUCTION

Forest conservation is a vital pillar of sustainable development, serving as the foundation for maintaining biodiversity, protecting cultural heritage, and sustaining the livelihoods of millions of people who depend on natural ecosystems. Forests provide crucial ecosystem services such as carbon sequestration, water regulation, and soil fertility maintenance, which directly support global environmental stability and human well-being (Chazdon et al., 2017; WCED, 1987). In addition, forests are cultural and spiritual spaces for Indigenous peoples and local communities whose traditional ecological knowledge has sustained biodiversity for generations (Colchester, 2004; Dawson et al., 2021). Therefore, conserving forests is not solely an environmental obligation but also a social and cultural necessity aligned with the principles of sustainable development.

In developing countries, however, achieving effective conservation remains a complex and multidimensional challenge. Rich biodiversity often coexists with significant ecological degradation, weak governance, and competing economic interests (Bebbington et al., 2018). Institutional fragmentation, limited funding, and poor inter-sectoral coordination further undermine conservation effectiveness (Brockhaus et al., 2014). Ministries overseeing forestry, agriculture, mining, and energy frequently operate independently with overlapping mandates, resulting in policy inconsistencies and inefficient management of forest resources (Ardiansyah et al., 2015). These governance limitations are compounded by political dynamics that prioritize short-term economic gains over long-term ecological sustainability (Adams, 2019; Spash, 2022).

Forest conservation has also evolved conceptually—from a focus on strict protection of ecosystems to more inclusive approaches that integrate community participation, cultural preservation, and economic empowerment (Borrini-Feyerabend & Hill, 2015). This paradigm shift reflects a broader understanding of sustainability that seeks to harmonize environmental goals with socio-economic development. As a result, modern conservation strategies emphasize collaborative governance, multi-level policy integration, and adaptive management that respond to changing environmental and social contexts (Folke et al., 2005; Chaffin et al., 2014).

This chapter explores the multifaceted nature of forest conservation, emphasizing the interconnected roles of governance, institutions, technology, and community engagement in shaping conservation outcomes. It begins by tracing the historical evolution of conservation thought and practice, followed by an examination of governance structures, challenges, and innovations. It then discusses the ethical dimensions of conservation, offers policy recommendations for strengthening intersectoral coordination, and concludes with suggestions for future research directions. By framing forest conservation within the broader context of biodiversity, culture, and community well-being, this study underscores the need for integrated, participatory, and adaptive approaches that ensure the sustainability of both nature and society

HISTORICAL EVOLUTION

The historical trajectory of forest conservation reflects a gradual but profound transformation—from early protectionist paradigms focused solely on preserving nature, to more inclusive and participatory models that integrate human, cultural, and economic dimensions. Initially, conservation efforts in the late nineteenth and early twentieth centuries were grounded in a preservationist ideology, emphasizing the establishment of protected areas to shield natural landscapes from human exploitation (Adams, 2019). These early models, largely influenced by Western environmental thought, prioritized biodiversity protection but often excluded Indigenous peoples and local communities from decision-making processes, leading to displacement and social disintegration (Agrawal & Redford, 2009; Brockington et al., 2008).

By the mid-twentieth century, the limitations of strict preservationism became increasingly evident. Conservation policies that isolated people from nature failed to recognize the interdependence between ecosystems and human livelihoods. This period saw a shift toward the concept of "sustainable use," which sought to reconcile ecological protection with socio-economic needs (WCED, 1987). The emergence of international frameworks such as the World Conservation Strategy (1980) and the Brundtland Report (1987) marked a significant conceptual turning point—linking conservation directly with sustainable development and human welfare.

In the 1990s and early 2000s, conservation thought evolved further with the rise of *community-based natural resource management (CBNRM)* and participatory conservation approaches (Borrini-Feyerabend & Hill, 2015; Lemos & Agrawal, 2006). These models emphasized decentralization and local governance, granting communities greater authority in managing their natural resources. In countries like Indonesia, decentralization aimed to improve responsiveness to local ecological and cultural contexts, though it also introduced new governance complexities and policy fragmentation (Resosudarmo et al., 2014; Ardiansyah et al., 2015).

Simultaneously, the integration of indigenous and traditional ecological knowledge (TEK) into conservation policy gained momentum. Scholars and practitioners increasingly recognized that Indigenous peoples, as long-term stewards of biodiversity, possess ecological knowledge essential for sustainable forest management (Colchester, 2004; Dawson et al., 2021). This recognition spurred collaborative initiatives, such as co-management systems and participatory forest monitoring, that combined scientific and traditional knowledge frameworks to strengthen conservation outcomes.

The early twenty-first century ushered in a new phase characterized by adaptive governance and the use of technological innovations to enhance policy learning and monitoring. Tools such as satellite-based forest observation systems and artificial intelligence now enable near real-time detection of deforestation and ecological change (Hansen et al., 2013; Turubanova et al., 2018). These advances have facilitated more transparent and accountable governance structures, reinforcing evidence-based decision-making in conservation policy (Chaffin et al., 2014; Antonelli et al., 2023).

Overall, the historical evolution of forest conservation demonstrates an ongoing shift from exclusionary protectionism to integrated governance models that value participation, equity, and knowledge diversity. Contemporary conservation frameworks no longer view human presence as a threat to nature but as an essential component of sustainable ecosystems. This holistic approach—linking biodiversity protection, cultural continuity, and socio-economic development—provides a more just and resilient foundation for achieving long-term sustainability in forest conservation.

LITERATURE REVIEW

Forest conservation has long been positioned at the intersection of ecological sustainability, socio-economic development, and governance reform. Over the past three decades, scholarly literature has evolved to recognize that conservation success depends not only on ecological management but also on the political, institutional, and socio-cultural dimensions that shape implementation (Adams, 2019; Brockington et al., 2008). Research from developing countries consistently demonstrates that governance arrangements—particularly institutional coordination, decentralization, and stakeholder engagement—play decisive roles in determining conservation outcomes (Brockhaus et al., 2014; Lemos & Agrawal, 2006).

Early studies conceptualized conservation primarily as an environmental protection strategy. However, subsequent research revealed that conservation divorced from social realities is often ineffective and unjust, especially when it marginalizes Indigenous and local communities (Agrawal & Redford, 2009; Colchester, 2004). In Indonesia and across Southeast Asia, for instance, centralized conservation policies frequently resulted in community displacement and conflict over land tenure (Meilani et al., 2021). As a response, newer frameworks such as *community-based conservation* (CBC) and *co-management systems* emerged, emphasizing participatory decision-making and equitable benefit-sharing between the state and local populations (Borrini-Feyerabend & Hill, 2015; Dawson et al., 2021).

The governance perspective in conservation studies highlights the critical role of institutions, laws, and policies in structuring environmental outcomes. Scholars such as Young et al. (2013) and Ardiansyah et al. (2015) argue that fragmented institutional mandates, overlapping jurisdictions, and weak coordination mechanisms hinder effective implementation. This fragmentation is particularly pronounced in Indonesia, where multiple ministries—such as those responsible for forestry, agriculture, and mining—operate with conflicting priorities, leading to regulatory duplication and inconsistent enforcement. Consequently, policy incoherence becomes a structural barrier to sustainable forest management (Brockhaus et al., 2012).

Decentralization, widely adopted across developing nations since the early 2000s, was intended to enhance administrative efficiency and local participation (Resosudarmo et al., 2014). However, the literature reveals mixed outcomes. While decentralization promotes context-specific management and strengthens local accountability, it also risks policy fragmentation and elite capture if not accompanied by strong oversight and institutional capacity (Colfer & Capistrano, 2012; Wardhani et al., 2025). Comparative studies from Indonesia, Malaysia, and Tanzania indicate that decentralization is most effective when grounded in secure land tenure and transparent benefit-sharing mechanisms (Dunning, 2022; Kicheleri et al., 2021).

Recent scholarship underscores the growing integration of technology and data-driven tools in conservation governance. The advent of remote sensing, artificial intelligence (AI), and digital monitoring systems has transformed biodiversity management, allowing for real-time analysis and greater transparency (Hansen et al., 2013; Antonelli et al., 2023). Yet, gaps in data infrastructure and technical capacity remain major obstacles in many low-income nations (Xiao-hui et al., 2011; Cummings et al., 2023). The lack of interoperability between government and non-governmental databases often results in inefficiencies and duplication, undermining evidence-based policymaking (Brockhaus et al., 2012).

A growing body of literature also stresses the political economy of conservation. Policies are frequently shaped by elite interests, where conservation agendas may be used to consolidate power or secure funding rather than achieve ecological outcomes (Ghafran & Yasmin, 2025; Spash, 2022). These dynamics are reinforced by global funding structures and international diplomacy, which can impose external priorities on local contexts (Holmes, 2011; Bixler & Shmelev, 2015). Scholars advocate for *multi-level governance* frameworks that balance global commitments with local autonomy and indigenous rights (Ayana et al., 2018; Martin et al., 2016).

Overall, the literature converges on several key themes: the need for integrated governance, recognition of local and Indigenous knowledge systems, incorporation of advanced technologies, and alignment of conservation with socio-economic goals.

Modern forest conservation is no longer viewed as a static ecological activity but as a dynamic governance process requiring cross-sectoral collaboration, political inclusivity, and institutional learning (Folke et al., 2005; Chaffin et al., 2014). This synthesis underscores that successful conservation outcomes depend on coherent policies, strong institutions, equitable participation, and continuous adaptation to evolving ecological and social realities.

METHODOLOGY

This study employs a qualitative analytical approach designed to examine how institutional structures, governance mechanisms, and socio-political dynamics shape forest conservation in developing countries. The qualitative framework allows for an in-depth exploration of policies, institutions, and stakeholder relationships, emphasizing contextual interpretation rather than numerical generalization (Cook et al., 2013). Through this approach, the research aims to synthesize theoretical insights, empirical findings, and case studies drawn from both national and international contexts.

The analysis integrates policy review, document analysis, and comparative synthesis. Primary sources include official reports from environmental ministries, policy documents, and international agreements such as the *Convention on Biological Diversity (CBD)* and *REDD+* frameworks. Secondary sources consist of peer-reviewed articles, policy briefs, and institutional reviews published by recognized organizations such as CIFOR, UNEP, and WWF (Ardiansyah et al., 2015; Chazdon et al., 2017; World Resources Institute, 2021). The selection of literature prioritizes research that addresses governance challenges, decentralization effects, and conservation outcomes across tropical regions, particularly Southeast Asia, Africa, and Latin America (Murcia et al., 2016; Dunning, 2022; Ayambire et al., 2025).

To ensure conceptual coherence, this study employs a multi-level governance framework that evaluates conservation at three interrelated scales: national, subnational, and community levels (Lemos & Agrawal, 2006; Wardhani et al., 2025). This framework is particularly useful in examining the vertical and horizontal coordination among institutions responsible for environmental management. National-level analysis focuses on regulatory and policy coherence, while subnational analysis investigates decentralization outcomes, and community-level analysis explores participatory governance and indigenous knowledge integration (Colchester, 2004; Dawson et al., 2021).

Data interpretation follows a thematic content analysis approach, emphasizing the identification of recurring themes and patterns related to institutional fragmentation, political dynamics, and adaptive innovation (Brockhaus et al., 2014;

Borrini-Feyerabend & Hill, 2015). Thematic clustering helps reveal how institutional design and political incentives influence conservation outcomes across different governance contexts. Comparative insights from case studies—such as Indonesia's decentralized forest management, Tanzania's community-based wildlife areas, and Peru's indigenous reserves—provide empirical grounding for theoretical discussion (Kicheleri et al., 2021; Sarmiento Barletti et al., 2021).

In terms of data validation and triangulation, the study cross-references institutional documents, legal frameworks, and scholarly analyses to ensure the reliability of interpretations. Peer-reviewed studies from Scopus-indexed journals and CIFOR publications are used to verify key assumptions and reinforce analytical rigor (Angelsen et al., 2018; Ayana et al., 2018). Furthermore, the study incorporates insights from decision science and adaptive governance literature to enhance analytical consistency across the various dimensions of conservation governance (Hemming et al., 2022; Chaffin et al., 2014).

Ultimately, the methodological orientation of this research is interpretative and policy-focused. It seeks to uncover the underlying governance conditions that influence conservation success or failure, rather than to quantify environmental outcomes. This approach allows for a nuanced understanding of how conservation policies interact with political structures, community interests, and technological innovation—thereby providing a comprehensive analytical foundation for developing more integrated, equitable, and sustainable forest conservation strategies.

RESEARCH OBJECTIVES

The primary objective of this research is to critically examine how forest conservation policies can simultaneously protect biodiversity, sustain community livelihoods, and strengthen governance systems in developing countries. Conservation is not only an ecological necessity but also a socio-political and cultural process that requires the alignment of multiple actors, institutions, and knowledge systems (Adams, 2019; Borrini-Feyerabend & Hill, 2015). Therefore, this study seeks to explore the dynamic intersection between environmental sustainability and socio-economic development, emphasizing the crucial role of inclusive governance and adaptive institutional arrangements.

More specifically, the study pursues five interrelated objectives designed to address the complexity of conservation governance:

1) To analyze the evolution of forest conservation paradigms—from traditional protectionist approaches to participatory and adaptive governance models. This includes exploring how historical shifts in policy and practice have influenced

the integration of cultural, ecological, and economic dimensions within conservation strategies (Brockington et al., 2008; WCED, 1987).

2) To assess the institutional and governance structures that shape forest conservation outcomes in developing contexts. This objective focuses on understanding how inter-sectoral coordination, decentralization, and multi-level governance mechanisms contribute to—or hinder—the effectiveness of conservation policy (Brockhaus et al., 2014; Lemos & Agrawal, 2006).
3) To evaluate the role of local and Indigenous knowledge systems in sustaining biodiversity and enhancing conservation legitimacy. The study aims to understand how community participation and traditional ecological knowledge (TEK) can be integrated into formal conservation frameworks to ensure both ecological integrity and social justice (Colchester, 2004; Dawson et al., 2021).
4) To identify key challenges and ethical dilemmas in conservation policy implementation, such as elite capture, policy fragmentation, funding constraints, and the displacement of Indigenous communities (Agrawal & Redford, 2009; Ghafran & Yasmin, 2025). By highlighting these barriers, the study contributes to a more critical understanding of governance inequities in conservation.
5) To propose strategic recommendations for improving policy coherence, strengthening institutional capacity, and promoting collaboration among government agencies, civil society, and private stakeholders. This includes exploring technological innovations—such as remote sensing and AI—for improving monitoring, accountability, and evidence-based policymaking (Hansen et al., 2013; Antonelli et al., 2023).

Collectively, these objectives aim to build a comprehensive framework that links biodiversity conservation with socio-economic resilience and good governance. By achieving these aims, the study contributes to advancing integrated conservation approaches that align with the Sustainable Development Goals (SDGs) and promote equitable, culturally sensitive, and scientifically informed environmental policies (Chazdon et al., 2017; Seddon et al., 2021).

THEORETICAL FRAMEWORKS

The theoretical foundation of this study is grounded in three interrelated frameworks—sustainability science, political ecology, and governance theory—which together provide an analytical lens for examining the complexity of forest conservation in developing countries. Each framework contributes distinct yet complementary perspectives that illuminate how environmental, political, and institutional dimensions interact in shaping conservation outcomes.

The first framework, sustainability science, emphasizes the integration of ecological integrity, economic development, and social equity as the core pillars of sustainable development (WCED, 1987; Chazdon et al., 2017). This framework posits that forest ecosystems must be managed not as isolated biophysical entities but as socio-ecological systems where human well-being and environmental health are interdependent (Folke et al., 2005). Within this context, forest conservation is viewed as both a scientific and a governance challenge requiring adaptive management, multi-stakeholder participation, and iterative learning (Chaffin et al., 2014). Sustainability science provides the conceptual basis for linking conservation efforts to broader development agendas such as poverty alleviation, climate resilience, and cultural preservation (Seddon et al., 2021).

The second framework, political ecology, examines the power relations, political structures, and economic interests that underpin environmental governance (Robbins, 2012; Bixler & Shmelev, 2015). This perspective challenges the notion of conservation as a purely ecological activity, arguing instead that it is deeply political—shaped by competing interests among governments, corporations, and local communities (Adams, 2019; Spash, 2022). Political ecology explains how policy outcomes often reflect asymmetries of power and access to resources, resulting in social exclusion or elite capture (Ghafran & Yasmin, 2025). It also reveals the tensions between global environmental discourses and local realities, showing how external conservation agendas can undermine indigenous autonomy and traditional land tenure systems (Colchester, 2004; Dawson et al., 2021). Thus, political ecology serves as a critical lens for evaluating the fairness, legitimacy, and accountability of conservation governance.

The third framework, governance theory, focuses on the institutional arrangements and policy processes that structure conservation decision-making (Lemos & Agrawal, 2006; Young et al., 2013). Within this framework, governance is understood as a networked process involving state, market, and civil society actors working across multiple levels of authority (Borrini-Feyerabend & Hill, 2015). The concept of *multi-level governance* is particularly relevant, as it highlights the interactions between national ministries, regional administrations, and community institutions in managing forest resources (Ardiansyah et al., 2015; Wardhani et al., 2025). Effective governance requires coordination across sectors, transparency in decision-making, and the inclusion of marginalized stakeholders, especially Indigenous Peoples and Local Communities (IPLCs). Governance theory also underscores the importance of adaptive institutions that can respond flexibly to socio-ecological change and uncertainty (Folke et al., 2005; Chaffin et al., 2014).

Together, these three theoretical perspectives form an integrative framework that situates forest conservation within a broader socio-political and institutional landscape. Sustainability science provides the normative goal of balance between ecological and social systems; political ecology uncovers the power dynamics and socio-economic inequalities influencing conservation; and governance theory offers practical insights into institutional design and coordination. This integrated framework enables a holistic understanding of conservation as a multi-dimensional process—one that requires scientific knowledge, political negotiation, and institutional collaboration to achieve equitable and sustainable outcomes in forest management.

FOUNDATIONS

Forest conservation rests on the integration of ecological, social, and cultural dimensions that together sustain biodiversity and human well-being. While early conservation emphasized habitat and species protection, contemporary approaches recognize that ecosystems cannot be separated from the social and cultural contexts that shape them (Folke et al., 2005; Adams, 2019). Effective conservation, therefore, relies on three interdependent pillars: ecological integrity, social inclusion, and cultural continuity.

Ecological integrity forms the core of forest conservation. Forests function as biodiversity reservoirs, carbon sinks, and regulators of climate and water cycles (Chazdon et al., 2017; Seddon et al., 2021). Maintaining these processes ensures both species' survival and ecosystem resilience to climate change. Conservation policies must balance protection with restoration through reforestation, natural regeneration, and adaptive management (Murcia et al., 2016; Antonelli et al., 2023).

Social inclusion recognizes that conservation succeeds only when local and Indigenous communities are active partners. Their livelihoods, knowledge, and identities are deeply linked to forest ecosystems (Colchester, 2004; Dawson et al., 2021). Inclusive approaches—such as community-based forest management (CBFM)—strengthen sustainability by ensuring participation, secure land tenure, and equitable benefit-sharing (Borrini-Feyerabend & Hill, 2015; Resosudarmo et al., 2014). Exclusionary models, in contrast, often lead to conflict and resource degradation (Agrawal & Redford, 2009).

Cultural continuity emphasizes the role of traditional ecological knowledge (TEK) and indigenous worldviews that view forests as living systems requiring stewardship (Bixler & Shmelev, 2015; Dawson et al., 2021). Customary laws in regions like Indonesia, the Philippines, and the Amazon have effectively maintained forest cover and complemented formal governance (Ardiansyah et al., 2015; Ayana et al., 2018). Together, these three pillars reinforce one another: ecological resilience depends on social justice, and cultural respect strengthens conservation legitimacy. Integrating them establishes a holistic foundation for sustainable forest governance—where biodiversity protection, community empowerment, and cultural survival progress in harmony.

IMPORTANCE

Forest conservation is vital not only for protecting trees and wildlife but also for sustaining ecological stability, economic development, and cultural identity worldwide. Forests form the backbone of Earth's life-support systems, providing essential ecosystem services such as carbon sequestration, water regulation, soil protection, and air purification (Chazdon et al., 2017; Seddon et al., 2021). In tropical regions, where biodiversity is most concentrated, forests underpin global ecological resilience and climate balance (Antonelli et al., 2023; Murcia et al., 2016).

Ecologically, forests sustain nearly 80% of terrestrial biodiversity, safeguarding genetic diversity and evolutionary processes (WWF, 2020). Deforestation and fragmentation threaten species survival and ecosystem functionality, reducing resilience to environmental stressors such as drought and pests (Seddon et al., 2021). Maintaining large, connected forest landscapes is thus crucial for achieving global climate and biodiversity targets under the Paris Agreement and the Convention on Biological Diversity (CBD) (Ardiansyah et al., 2015; Ayana et al., 2018).

Economically, forests are lifelines for over 1.6 billion people who depend on them for food, energy, and income (World Bank, 2022). Sustainable forest management (SFM) promotes economic diversification through ecotourism, non-timber forest products, and carbon markets under REDD+ (Brockhaus et al., 2014; Dunning, 2022). Conversely, forest degradation results in economic losses from reduced ecosystem services and heightened climate risks (Hemming et al., 2022). Thus, investing in conservation is both an ecological and economic imperative.

Culturally, forests embody ancestral wisdom, spirituality, and community identity (Colchester, 2004; Dawson et al., 2021). Recognizing Indigenous Peoples and Local Communities (IPLCs) as co-managers ensures the preservation of traditional knowledge and promotes social justice (Borrini-Feyerabend & Hill, 2015; Resosudarmo et al., 2014). In sum, forest conservation is a multidimensional necessity that integrates environmental, economic, and cultural sustainability. It supports biodiversity, livelihoods, and heritage—making it indispensable for achieving global sustainable development goals (Adams, 2019; Chazdon et al., 2017).

GOVERNANCE

Governance fundamentally determines the success or failure of forest conservation. It encompasses the institutions, rules, and actor relationships that guide how power and responsibilities are exercised in managing natural resources (Lemos & Agrawal, 2006; Young et al., 2013). Effective governance extends beyond state control to include civil society, local communities, and the private sector, emphasizing coordination, transparency, accountability, and inclusiveness (Borrini-Feyerabend & Hill, 2015; Chaffin et al., 2014).

In many developing countries, forest governance suffers from institutional fragmentation and policy incoherence. Competing mandates among ministries—such as forestry, agriculture, and mining—produce overlapping jurisdictions and conflicting land-use decisions (Brockhaus et al., 2014; Ardiansyah et al., 2015). This misalignment weakens enforcement and creates disconnects between national conservation goals and local development agendas (Resosudarmo et al., 2014; Wardhani et al., 2025).

The multi-level governance framework offers a pathway to address these challenges by integrating decision-making across national, regional, and local levels (Lemos & Agrawal, 2006). While decentralization can enhance local responsiveness, it may also lead to elite capture and policy inconsistency without adequate oversight and capacity-building (Colfer & Capistrano, 2012; Kicheleri et al., 2021).

Inclusive participation is central to governance effectiveness. Collaborative models—such as community-based forest management (CBFM) and co-management systems—recognize Indigenous Peoples and Local Communities (IPLCs) as active partners, leveraging traditional knowledge for ecological stewardship (Colchester, 2004; Dawson et al., 2021). These participatory approaches align with global frameworks like REDD+ and the SDGs, promoting transparency and justice in environmental governance (Angelsen et al., 2018; Ayana et al., 2018).

Technological advances, including remote sensing, AI, and digital monitoring, strengthen transparency and enable real-time deforestation tracking (Hansen et al., 2013; Antonelli et al., 2023). Yet, limited digital infrastructure and data literacy in low-income regions remain barriers to effective implementation (Xiao-hui et al., 2011; Cummings et al., 2023). In essence, governance is the institutional backbone of forest conservation. Its success depends on coherent policy frameworks, participatory engagement, and adaptive institutions supported by technology. Weak governance fosters fragmentation, while inclusive and accountable systems ensure resilience, trust, and long-term sustainability.

ADVANTAGES

Effective forest conservation generates ecological, economic, social, and cultural benefits that collectively drive sustainable development. When governance is inclusive and well-coordinated, conservation safeguards biodiversity, strengthens community resilience, promotes livelihoods, and enhances institutional accountability (Borrini-Feyerabend & Hill, 2015; Chaffin et al., 2014).

Ecologically, conservation maintains biodiversity and ecosystem services vital for global stability. Forests act as carbon sinks, regulate water cycles, preserve soil fertility, and buffer against climate change impacts (Chazdon et al., 2017; Murcia et al., 2016; Antonelli et al., 2023). These functions contribute to achieving international environmental goals under the Paris Agreement and the Convention on Biological Diversity (Ayana et al., 2018).

Socio-economically, sustainable forest management (SFM) and community-based forest management (CBFM) provide jobs, diversify income, and reduce poverty through incentives like payment for ecosystem services (Borrini-Feyerabend & Hill, 2015; Resosudarmo et al., 2014). Locally managed forests in Indonesia, Tanzania, and Latin America show higher forest retention and fairer benefit distribution (Kicheleri et al., 2021; Sarmiento Barletti et al., 2021). Forests also act as natural infrastructure, reducing disaster risks and economic losses (Hemming et al., 2022).

Institutionally, conservation fosters policy coherence, collaboration, and accountability (Brockhaus et al., 2014; Ardiansyah et al., 2015). Digital monitoring and satellite systems enhance transparency and evidence-based decision-making (Hansen et al., 2013; Cummings et al., 2023). Culturally, forests sustain indigenous knowledge, identity, and spirituality, reinforcing social cohesion when local communities are treated as co-managers rather than beneficiaries (Colchester, 2004; Dawson et al., 2021). Ultimately, forest conservation creates a synergistic cycle—ecological stability strengthens socio-economic resilience, while inclusive governance sustains ecological integrity. Thus, conservation is not a developmental cost but a strategic investment in the shared future of people and the planet (Adams, 2019; Chazdon et al., 2017).

CHALLENGES

Despite growing global commitment, forest conservation remains constrained—especially in developing countries—by institutional, political, financial, and sociocultural barriers (Brockhaus et al., 2014; Ardiansyah et al., 2015). Understanding these obstacles is crucial for building adaptive and equitable governance systems that balance ecological and developmental goals.

A primary challenge is institutional fragmentation and weak inter-sectoral coordination. Overlapping mandates among forestry, agriculture, and mining ministries create policy contradictions, inefficiencies, and local confusion (Lemos & Agrawal, 2006; Resosudarmo et al., 2014). Without coherent frameworks, conservation objectives are often overshadowed by short-term economic agendas.

Financial limitations further undermine implementation. Many programs rely heavily on external donors, leaving them vulnerable to shifting priorities and donor dependency (Ghafran & Yasmin, 2025; Dunning, 2022). National budgets remain insufficient, while mechanisms like REDD+ and carbon markets have yet to provide fair benefits to local communities due to governance and tenure weaknesses (Angelsen et al., 2018; Ayana et al., 2018).

The political economy of conservation adds complexity. Elite influence, corruption, and weak accountability often redirect conservation policies toward private interests, eroding public trust and policy consistency (Adams, 2019; Holmes, 2011; Spash, 2022). Persistent social and ethical issues—including community exclusion, forced displacement, and marginalization of Indigenous Peoples and Local Communities (IPLCs)—undermine both social justice and ecological outcomes (Colchester, 2004; Dawson et al., 2021; Borrini-Feyerabend & Hill, 2015).

Lastly, technological and data constraints—limited expertise, infrastructure, and data interoperability—impede evidence-based policymaking (Xiao-hui et al., 2011; Cummings et al., 2023). In sum, overcoming these interlinked challenges requires systemic reform: coherent governance, sustainable financing, transparency, and inclusive participation. Only through such integrative approaches can forest conservation evolve from fragmented intent into transformative practice.

ETHICAL CONSIDERATIONS

Ethics form the foundation of forest conservation, linking environmental protection with justice, equity, and human rights. Conservation cannot be deemed successful if it safeguards biodiversity while undermining community welfare or cultural integrity. Ethical governance, therefore, requires balancing ecological goals with respect for the rights and agency of Indigenous and local peoples (Colchester, 2004; Dawson et al., 2021).

A major ethical challenge concerns the displacement of local communities under exclusionary "fortress conservation" models that historically restricted human activity in protected areas (Brockington et al., 2008; Agrawal & Redford, 2009). Such practices often violate human rights by forcing relocations and eroding cultural identity, contradicting principles like *Free, Prior, and Informed Consent (FPIC)* under the UNDRIP framework (Dawson et al., 2021).

Ethical issues also emerge from unequal benefit distribution in mechanisms such as REDD+ and carbon markets, which can privilege elites over local actors (Angelsen et al., 2018; Ghafran & Yasmin, 2025). Fair and transparent benefit-sharing is essential to ensure that communities gain real economic and social value from conservation (Borrini-Feyerabend & Hill, 2015). Inclusive participation represents another ethical imperative. The underrepresentation of Indigenous peoples, women, and youth in governance processes undermines both justice and conservation effectiveness (Colchester, 2004; Resosudarmo et al., 2014). Genuine participation enhances equity, accountability, and policy legitimacy.

Emerging technologies pose ethical concerns regarding surveillance and data governance. While digital monitoring improves transparency, it risks disempowering communities if used without consent or equitable access (Cummings et al., 2023). Responsible data practices must respect privacy, cultural values, and shared ownership. Ultimately, ethical forest conservation is rooted in relational responsibility—recognizing that ecological integrity and social justice are inseparable (Adams, 2019; Spash, 2022). Embedding ethics within governance ensures that conservation protects not only forests but also the dignity, culture, and rights of the people who depend on them.

FOSTERING COLLABORATION

Collaboration is fundamental to effective forest conservation, bridging institutional, disciplinary, and socio-political divides that often impede coordinated management. Given the complexity of forest ecosystems and stakeholder diversity, cooperative governance linking governments, local communities, NGOs, academia, and the private sector ensures inclusive and adaptive conservation (Borrini-Feyerabend & Hill, 2015; Chaffin et al., 2014). At the institutional level, collaboration enhances policy coherence and mitigates governance fragmentation, particularly in developing countries where sectoral silos create conflicting objectives (Ardiansyah et al., 2015; Brockhaus et al., 2014). Multi-sectoral coordination platforms—such as Indonesia's *Forest Management Units* and Tanzania's *Joint Forest Management*—demonstrate how shared authority improves transparency and accountability (Resosudarmo et al., 2014; Kicheleri et al., 2021).

At the community level, collaboration empowers Indigenous Peoples and Local Communities (IPLCs) as co-managers of forest resources. When local actors participate in decision-making and benefit-sharing, conservation becomes more equitable and sustainable (Colchester, 2004; Dawson et al., 2021). Recognizing Traditional

Ecological Knowledge (TEK) alongside science strengthens adaptive management and social trust (Borrini-Feyerabend & Hill, 2015). NGOs and academia serve as vital intermediaries, facilitating capacity-building, knowledge exchange, and evidence-based policymaking (Holmes, 2011; Ayana et al., 2018). Research organizations such as CIFOR and UNEP-WCMC advance monitoring and evaluation, ensuring conservation strategies remain scientifically sound.

The private sector contributes through sustainable certification, CSR, and public–private partnerships, aligning economic growth with environmental goals (Dunning, 2022). Yet, these collaborations require strong accountability to prevent *greenwashing* and ensure real ecological benefits (Spash, 2022; Ghafran & Yasmin, 2025). International cooperation—through initiatives like the *Heart of Borneo* and transboundary landscape management—strengthens regional connectivity and aligns national policies with global sustainability frameworks such as the SDGs and the Paris Agreement (Ayana et al., 2018). In essence, collaboration transforms forest conservation into a shared societal responsibility. It enhances legitimacy, fosters trust, and generates synergies across governance levels—making conservation not only technically effective but also socially just and sustainable.

IMPLEMENTATION OF REGULATORY FRAMEWORKS

The implementation of regulatory frameworks is central to translating conservation policies into tangible ecological and social outcomes. Regulations define how forest resources are managed, who holds authority, and how benefits and responsibilities are distributed among stakeholders. However, despite the proliferation of national and international policies, the effectiveness of forest conservation often depends on the coherence, enforcement, and adaptability of these frameworks (Brockhaus et al., 2014; Ardiansyah et al., 2015).

At the national level, regulatory frameworks serve as instruments for formalizing conservation objectives and integrating them into development planning. Many developing countries, including Indonesia, have enacted comprehensive laws to protect forests through mechanisms such as *Forest Management Units (FMUs)*, *Protected Areas (PAs)*, and *Ecosystem Restoration Concessions (ERCs)* (Ardiansyah et al., 2015; Resosudarmo et al., 2014). However, the implementation of these frameworks often faces challenges related to overlapping jurisdictions, unclear land tenure, and weak institutional coordination. The decentralization of forest governance, while intended to promote local accountability, has in some cases created inconsistencies between national directives and local enforcement (Colfer & Capistrano, 2012; Wardhani et al., 2025). Strengthening vertical coordination between ministries and regional governments remains essential to ensure policy alignment and compliance.

From an international perspective, global frameworks such as the *Convention on Biological Diversity (CBD)*, *Paris Agreement*, and *REDD+* have established guiding principles for sustainable forest governance and climate mitigation (Angelsen et al., 2018; Ayana et al., 2018). These frameworks encourage participating nations to integrate biodiversity conservation and carbon sequestration into national policies. Nevertheless, translating these commitments into effective local action remains a persistent challenge. International frameworks often depend on donor-driven funding and technical assistance, which can limit national autonomy and sustainability once external support declines (Dunning, 2022; Ghafran & Yasmin, 2025).

Another critical issue in regulatory implementation is policy fragmentation. The coexistence of multiple, sometimes conflicting, laws governing forestry, mining, and agriculture leads to administrative inefficiency and enforcement gaps (Brockhaus et al., 2012; Ardiansyah et al., 2015). Efforts to harmonize regulations, such as *One Map Policy* initiatives in Indonesia, have shown promise in improving transparency, reducing overlapping claims, and enhancing inter-agency coordination (Wardhani et al., 2025). However, the success of such reforms depends on strong political will, adequate funding, and technological capacity for spatial data integration (Cummings et al., 2023).

Figure 1. Limitations in Data and Information Flow for Forest and Biodiversity Conservation

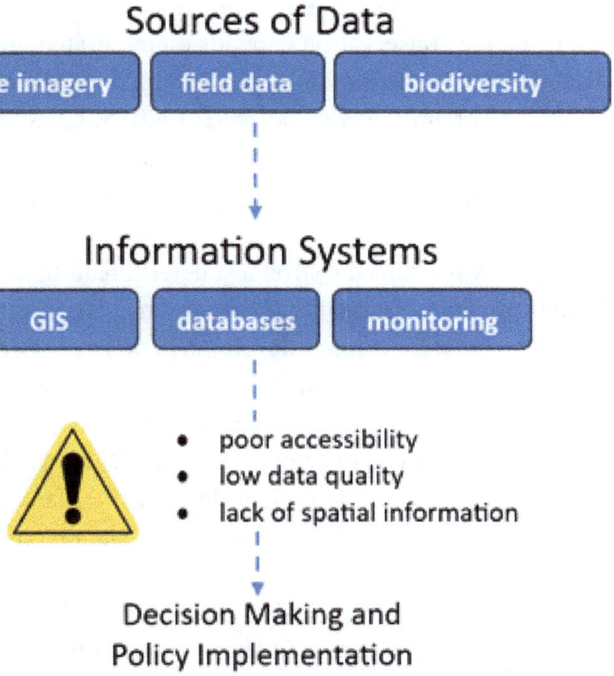

Enforcement capacity remains another weak link in many conservation frameworks. Limited resources, corruption, and insufficient monitoring tools hinder effective implementation (Holmes, 2011; Adams, 2019). The adoption of advanced technologies such as remote sensing, satellite imagery, and AI-based monitoring has improved surveillance and transparency (Hansen et al., 2013; Antonelli et al., 2023). Yet, the benefits of these innovations are unevenly distributed, with low-income countries often lacking the expertise or infrastructure to utilize them effectively (Xiao-hui et al., 2011).

To strengthen regulatory implementation, policymakers must prioritize policy coherence, participatory law enforcement, and adaptive regulation. This includes aligning legal instruments across sectors, integrating local and Indigenous rights into national frameworks, and promoting community-based monitoring to enhance compliance and legitimacy (Colchester, 2004; Dawson et al., 2021). Ultimately, the effectiveness of regulatory frameworks lies not merely in their existence but in their capacity to evolve with ecological realities and societal needs, ensuring that forest conservation remains both just and sustainable.

CONCLUSION

Forest conservation lies at the intersection of ecology, governance, and human development—representing both an environmental imperative and a socio-political responsibility. Effective conservation depends on integrated governance, inclusive participation, and adaptive institutions capable of balancing biodiversity protection with human needs (Borrini-Feyerabend & Hill, 2015; Chaffin et al., 2014). The shift from protectionist to participatory paradigms reflects a growing recognition of forests as socio-ecological systems rather than isolated natural assets (Adams, 2019; Folke et al., 2005).

Governance quality remains the strongest determinant of conservation success. Weak coordination, institutional fragmentation, and overlapping mandates continue to hinder policy implementation in many developing countries (Brockhaus et al., 2014; Ardiansyah et al., 2015). Strengthening multi-level and participatory governance—supported by capacity-building and transparent regulation—is vital to ensure policy coherence and accountability (Colfer & Capistrano, 2012; Kicheleri et al., 2021).

Indigenous Peoples and Local Communities (IPLCs) play a central role through their ecological knowledge and cultural stewardship. Their inclusion as co-managers strengthens conservation legitimacy and aligns strategies with justice, equity, and sustainability principles (Colchester, 2004; Dawson et al., 2021). Empowering these groups through secure land tenure, benefit-sharing, and participatory decision-making enhances both ecological resilience and social cohesion.

From an institutional perspective, adaptive and collaborative governance offers a critical pathway forward. Integrating scientific data, technological innovation, and local wisdom enables more responsive and evidence-based decision-making (Hansen et al., 2013; Antonelli et al., 2023). Digital monitoring and spatial mapping can improve transparency but must be governed ethically and equitably (Cummings et al., 2023).

Ultimately, forest conservation must be embedded within broader sustainable development goals—supporting climate mitigation, poverty reduction, food security, and cultural preservation (Chazdon et al., 2017; Seddon et al., 2021). Policies should evolve as living frameworks, adaptive to ecological realities and inclusive of diverse voices. In essence, conserving forests means sustaining life systems that unite nature, culture, and community. Its success depends on political will, institutional collaboration, and shared ethical commitment, ensuring forests remain foundations of ecological integrity, social justice, and global sustainability.

SUGGESTIONS

Strengthening forest conservation requires a holistic and adaptive approach that integrates governance reform, community empowerment, financial innovation, and technological advancement. Based on the preceding analysis, several strategic recommendations can be proposed to improve the effectiveness, equity, and sustainability of forest conservation in developing countries.

As conservation challenges become increasingly complex, governance systems must embrace flexibility, inclusivity, and innovation. Figure 2 below presents a conceptual model of *Innovation in Governance for Biodiversity Conservation*, emphasizing how adaptive governance integrates technology, community participation, and indigenous knowledge to promote continuous learning and resilience.

Figure 2. Conceptual Framework of Innovation in Biodiversity Governance

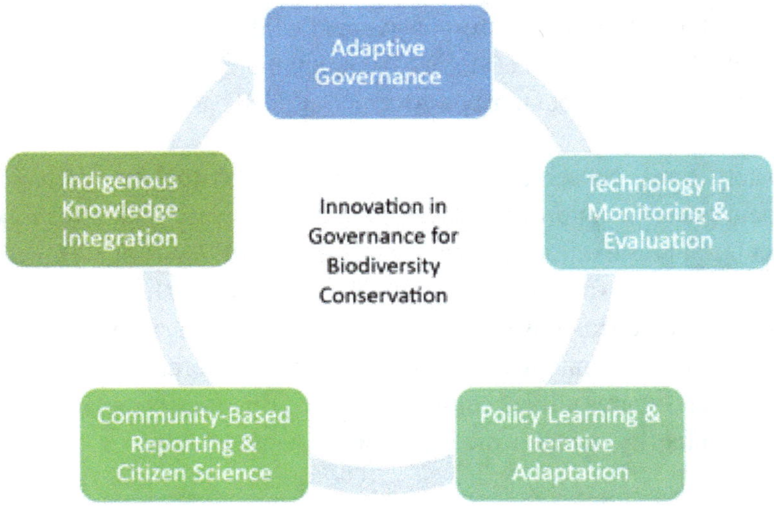

The model highlights the need for adaptive and participatory approaches that connect scientific innovation with local experience. Building on this framework, several strategic measures can be proposed to strengthen forest conservation:

1) Strengthen inter-sectoral coordination and policy coherence. One of the most urgent needs is to harmonize conservation policies across ministries and governance levels. Institutional fragmentation remains a persistent obstacle to effective implementation (Brockhaus et al., 2014; Ardiansyah et al., 2015). Establishing cross-sectoral coordination platforms—such as joint task forces or multi-agency councils—can align priorities among forestry, agriculture, mining, and energy sectors. The adoption of integrated planning frameworks, such as the *One Map Policy* in Indonesia, should be expanded to ensure transparency, reduce overlapping land claims, and enhance spatial governance (Wardhani et al., 2025).
2) Enhance institutional capacity and financial sustainability. Governments and local authorities need to invest in institutional strengthening, including training, monitoring systems, and anti-corruption mechanisms. Conservation budgets should be diversified beyond donor dependency through innovative financing, such as carbon markets, ecosystem service payments, and public–private partnerships (Angelsen et al., 2018; Dunning, 2022). These mechanisms can create long-term funding streams while incentivizing sustainable land use and reforestation efforts.

3) Promote inclusive governance and community participation. Forest conservation should be rooted in participatory processes that respect the rights, knowledge, and cultural values of Indigenous Peoples and Local Communities (IPLCs) (Colchester, 2004; Dawson et al., 2021). Governments must recognize customary land tenure and ensure equitable benefit-sharing from conservation programs. Strengthening community-based forest management (CBFM) and co-management schemes enhances legitimacy and fosters local stewardship, which has been proven to improve ecological outcomes (Borrini-Feyerabend & Hill, 2015; Resosudarmo et al., 2014).
4) Foster innovation and technological integration. Digital technologies, remote sensing, and artificial intelligence can significantly improve monitoring and enforcement (Hansen et al., 2013; Antonelli et al., 2023). To maximize these benefits, developing nations must invest in data infrastructure and build local technical expertise. Open-access platforms for forest data can increase transparency and allow civil society and researchers to contribute to monitoring and evaluation (Cummings et al., 2023).
5) Embed ethics and justice in conservation policy. Conservation should advance social equity and human rights alongside environmental goals. Ethical frameworks must guide project design and implementation to prevent displacement, ensure free and informed consent (FPIC), and prioritize vulnerable groups (Agrawal & Redford, 2009; Dawson et al., 2021). Embedding ethical standards strengthens legitimacy, trust, and long-term compliance among local communities.

In summary, improving forest conservation requires both structural and moral transformation—aligning ecological objectives with social justice and institutional accountability. By fostering coordination, innovation, and inclusion, conservation can evolve from a fragmented administrative task into an integrated pillar of sustainable development and climate resilience.

FUTURE RESEARCH DIRECTIONS

Future research on forest conservation must move beyond descriptive analyses of deforestation and policy design toward deeper, interdisciplinary investigations that explore the dynamic interactions between ecology, governance, and society. The complexity of modern conservation challenges—such as climate change, biodiversity loss, and sociopolitical tensions—requires innovative methods and multi-scalar frameworks that integrate ecological, technological, and socio-political dimensions (Chazdon et al., 2017; Folke et al., 2005).

A key direction is the integration of Indigenous and local knowledge (TEK) with modern conservation science. More empirical research is needed to understand how TEK can be institutionalized within national policies and adaptive management systems (Colchester, 2004; Dawson et al., 2021). Comparative studies across Southeast Asia, Africa, and Latin America may reveal institutional and cultural mechanisms that enable effective co-management between local communities and the state.

Another priority is evaluating multi-level and collaborative governance models. While decentralization and community-based approaches are widely promoted, their outcomes vary according to political context, institutional capacity, and socio-economic conditions (Brockhaus et al., 2014; Lemos & Agrawal, 2006). Longitudinal analyses are needed to assess how governance reforms affect forest conditions over time, including the roles of local leadership, gender inclusion, and inter-agency coordination (Resosudarmo et al., 2014; Wardhani et al., 2025).

Technological innovation and data-driven conservation also represent emerging frontiers. Although AI and remote sensing have enhanced forest monitoring, issues of accessibility, data governance, and ethics persist (Hansen et al., 2013; Cummings et al., 2023). Future research should examine how digital tools can be co-designed with communities to support participatory monitoring and ensure technology complements rather than replaces human decision-making. The potential of *blockchain* for traceability and accountability in forest supply chains also warrants empirical investigation (Antonelli et al., 2023).

In addition, the political economy of conservation finance deserves closer scrutiny. Studies should analyze how global mechanisms—such as REDD+ and carbon markets—affect national sovereignty, benefit distribution, and long-term sustainability (Angelsen et al., 2018; Dunning, 2022). Attention is also needed to the social implications of conservation financialization, where commodification may exacerbate inequalities (Ghafran & Yasmin, 2025; Spash, 2022).

Finally, future research must adopt transdisciplinary approaches that integrate natural and social sciences, economics, and ethics. Forest conservation should be viewed as a living system shaped by ecological processes, political structures, and cultural values (Adams, 2019; Seddon et al., 2021). By combining local wisdom with scientific rigor and linking ecological insights with governance reforms, future scholarship can guide a transformation toward more equitable, ethical, and sustainable forest conservation practices.

REFERENCES

Adams, W. M. (2019). *Green Development*. Routledge., DOI: 10.4324/9780203386033

Agrawal, A., & Redford, K. (2009). Conservation and Displacement: An Overview. *Conservation & Society*, *7*(1), 1. DOI: 10.4103/0972-4923.54790

Ardiansyah, F., Marthen, A. A., & Amalia, N. (2015). Forest and land-use governance in a decentralized Indonesia: A legal and policy review (Vol. 124). Center for International Forestry Research (CIFOR). DOI: 10.17528/cifor/005579

Ayambire, R. A., Rytwinski, T., Taylor, J. J., Luizza, M. W., Muir, M. J., Cadet, C., Armitage, D., Bennett, N. J., Brooks, J., Cheng, S. H., Martinez, J., Nagendran, M., Öckerman, S., Rivera, S. N., Savage, A., Wilkie, D. S., Cooke, S. J., & Bennett, J. R. (2025). Challenges in assessing the effects of environmental governance systems on conservation outcomes. *Conservation Biology*, *39*(1), 1–14. DOI: 10.1111/cobi.14392 PMID: 39417601

Ayana, A. N., Arts, B., & Wiersum, K. F. (2018). How environmental NGOs have influenced decision making in a 'semi-authoritarian' state: The case of forest policy in Ethiopia. *World Development*, *109*(621), 313–322. DOI: 10.1016/j.worlddev.2018.05.010

Bebbington, A. J., Humphreys Bebbington, D., Sauls, L. A., Rogan, J., Agrawal, S., Gamboa, C., Imhof, A., Johnson, K., Rosa, H., Royo, A., Toumbourou, T., & Verdum, R. (2018). Resource extraction and infrastructure threaten forest cover and community rights. *Proceedings of the National Academy of Sciences of the United States of America*, *115*(52), 13164–13173. DOI: 10.1073/pnas.1812505115 PMID: 30509982

Bixler, R., & Shmelev, S. (2015). The political ecology of participatory conservation: Institutions and discourse. *Journal of Political Ecology*, *22*(1), 164–182. DOI: 10.2458/v22i1.21083

Borrini-Feyerabend, G., & Hill, R. (2015). Governance for the Conservation of Nature. In Protected Area Governance and Management (pp. 169–206). ANU Press. DOI: 10.22459/PAGM.04.2015.07

Brockhaus, M., Di Gregorio, M., & Carmenta, R. (2014). REDD+ policy networks: Exploring actors and power structures in an emerging policy domain. *Ecology and Society*, *19*(4), art29. https://www.jstor.org/stable/26269649. DOI: 10.5751/ES-07098-190429

Brockhaus, M., Obidzinski, K., Dermawan, A., Laumonier, Y., & Luttrell, C. (2012). An overview of forest and land allocation policies in Indonesia: Is the current framework sufficient to meet the needs of REDD+? *Forest Policy and Economics, 18*, 30–37. DOI: 10.1016/j.forpol.2011.09.004

Brockington, D., Duffy, R., & J, I. (2008). Nature Unbound: Conservation, Capitalism and the Future of Protected Areas. In Paper Knowledge . Toward a Media History of Documents. earthscan.

Chaffin, B. C., Gosnell, H., & Cosens, B. A. (2014). A decade of adaptive governance scholarship: Synthesis and future directions. *Ecology and Society, 19*(3), art56. DOI: 10.5751/ES-06824-190356

Chazdon, R. L., Brancalion, P. H. S., Lamb, D., Laestadius, L., Calmon, M., & Kumar, C. (2017). A Policy-Driven Knowledge Agenda for Global Forest and Landscape Restoration. *Conservation Letters, 10*(1), 125–132. DOI: 10.1111/conl.12220

Colchester, M. (2004). Conservation policy and indigenous peoples. *Environmental Science & Policy, 7*(3), 145–153. DOI: 10.1016/j.envsci.2004.02.004

Colfer, C. J. P., & Capistrano, D. (2012). *The Politics of Decentralization.* Routledge., DOI: 10.4324/9781849773218

Colfer, C. J. P., & Capistrano, D. (2012). *The Politics of Decentralization.* Routledge., DOI: 10.4324/9781849773218

Cook, C. N., Mascia, M. B., Schwartz, M. W., Possingham, H. P., & Fuller, R. A. (2013). Achieving Conservation Science that Bridges the Knowledge–Action Boundary. *Conservation Biology, 27*(4), 669–678. DOI: 10.1111/cobi.12050 PMID: 23574343

Cummings, S., Dhewa, C., Kemboi, G., & Young, S. (2023). Doing epistemic justice in sustainable development: Applying the philosophical concept of epistemic injustice to the real world. *Sustainable Development (Bradford), 31*(3), 1965–1977. DOI: 10.1002/sd.2497

Dawson, N. M., Coolsaet, B., Sterling, E. J., Loveridge, R., Gross-Camp, N. D., Wongbusarakum, S., Sangha, K. K., Scherl, L. M., Phan, H. P., Zafra-Calvo, N., Lavey, W. G., Byakagaba, P., Idrobo, C. J., Chenet, A., Bennett, N. J., Mansourian, S., & Rosado-May, F. J. (2021). The role of Indigenous peoples and local communities in effective and equitable conservation. *Ecology and Society, 26*(3), 1–39. DOI: 10.5751/ES-12625-260319

Dunning, K. (2022). Biodiversity conservation policy in megadiverse countries: Comparing policy systems for 2020 targets to inform management in the coming decades. Journal of Environmental Management, 302(PA), 1–14. DOI: 10.1016/j.jenvman.2021.113815

Folke, C., Hahn, T., Olsson, P., & Norberg, J. (2005). ADAPTIVE GOVERNANCE OF SOCIAL-ECOLOGICAL SYSTEMS. *Annual Review of Environment and Resources*, *30*(1), 441–473. DOI: 10.1146/annurev.energy.30.050504.144511

Ghafran, C., & Yasmin, S. (2025). Participation Strategies and Ethical Considerations in NGO Led Community-Based Conservation Initiatives. *Journal of Business Ethics*, *196*(3), 659–675. DOI: 10.1007/s10551-024-05665-4

Hansen, M. C., Potapov, P. V., Moore, R., Hancher, M., Turubanova, S. A., Tyukavina, A., Thau, D., Stehman, S. V., Goetz, S. J., Loveland, T. R., Kommareddy, A., Egorov, A., Chini, L., Justice, C. O., & Townshend, J. R. G. (2013). High-Resolution Global Maps of 21st-Century Forest Cover Change. *Science*, *342*(6160), 850–853. DOI: 10.1126/science.1244693 PMID: 24233722

Hemming, V., Camaclang, A. E., Adams, M. S., Burgman, M., Carbeck, K., Carwardine, J., Chadès, I., Chalifour, L., Converse, S. J., Davidson, L. N. K., Garrard, G. E., Finn, R., Fleri, J. R., Huard, J., Mayfield, H. J., Madden, E. M., Naujokaitis-Lewis, I., Possingham, H. P., Rumpff, L., & Martin, T. G. (2022). An introduction to decision science for conservation. *Conservation Biology*, *36*(1), 1–16. DOI: 10.1111/cobi.13868 PMID: 34856010

Holmes, G. (2011). Conservation's Friends in High Places: Neoliberalism, Networks, and the Transnational Conservation Elite. *Global Environmental Politics*, *11*(4), 1–21. DOI: 10.1162/GLEP_a_00081

Kicheleri, R. P., Mangewa, L. J., Nielsen, M. R., Kajembe, G. C., & Treue, T. (2021). Designed for accumulation by dispossession: An analysis of Tanzania's Wildlife Management Areas through the case of Burunge. *Conservation Science and Practice*, *3*(1), e360. Advance online publication. DOI: 10.1111/csp2.360

Lemos, M. C., & Agrawal, A. (2006). Environmental Governance. *Annual Review of Environment and Resources*, *31*(1), 297–325. DOI: 10.1146/annurev.energy.31.042605.135621

Martin, A., Coolsaet, B., Corbera, E., Dawson, N. M., Fraser, J. A., Lehmann, I., & Rodriguez, I. (2016). Justice and conservation: The need to incorporate recognition. *Biological Conservation*, *197*, 254–261. DOI: 10.1016/j.biocon.2016.03.021

Meilani, M., Andayani, W., Faida, L. R. W., Susanti, F. D., Myers, R., & Maryudi, A. (2021). Symbolic consultation and cultural simplification in the establishment of an Indonesian national park and its impacts on local livelihoods. *Forest and Society, 5*(2), 494–505. DOI: 10.24259/fs.v5i2.11875

Murcia, C., Guariguata, M. R., Andrade, Á., Andrade, G. I., Aronson, J., Escobar, E. M., Etter, A., Moreno, F. H., Ramírez, W., & Montes, E. (2016). Challenges and Prospects for Scaling-up Ecological Restoration to Meet International Commitments: Colombia as a Case Study. *Conservation Letters, 9*(3), 213–220. DOI: 10.1111/conl.12199

Resosudarmo, I. A. P., Atmadja, S., Ekaputri, A. D., Intarini, D. Y., Indriatmoko, Y., & Astri, P. (2014). Does Tenure Security Lead to REDD+ Project Effectiveness? Reflections from Five Emerging Sites in Indonesia. *World Development, 55*, 68–83. DOI: 10.1016/j.worlddev.2013.01.015

Robbins, P. (2012). *Political Ecology* (2nd ed.). John Wiley & Sons Ltd.

Sarmiento Barletti, J. P., Begert, B., & Guerra Loza, M. A. (2021). Is the Formalization of Collective Tenure Rights Supporting Sustainable Indigenous Livelihoods? Insights from Comunidades Nativas in the Peruvian Amazon. *The International Journal of the Commons, 15*(1), 381–394. DOI: 10.5334/ijc.1126

Seddon, N., Smith, A., Smith, P., Key, I., Chausson, A., Girardin, C., House, J., Srivastava, S., & Turner, B. (2021). Getting the message right on nature-based solutions to climate change. *Global Change Biology, 27*(8), 1518–1546. DOI: 10.1111/gcb.15513 PMID: 33522071

Spash, C. L. (2022). Conservation in conflict: Corporations, capitalism and sustainable development. *Biological Conservation, 269*(March), 109528. DOI: 10.1016/j.biocon.2022.109528

Turubanova, S., Potapov, P. V., Tyukavina, A., & Hansen, M. C. (2018). Ongoing primary forest loss in Brazil, Democratic Republic of the Congo, and Indonesia. *Environmental Research Letters, 13*(7), 074028. DOI: 10.1088/1748-9326/aacd1c

Wardhani, L. T. A. L., Diamantina, A., & Rafsanjani, L. H. (2025). Framework for Biodiversity Conservation Governance in the Context of Regional Administration in Indonesia. *International Journal of Social Science and Human Research, 08*(01), 449–456. DOI: 10.47191/ijsshr/v8-i1-55

World Resources Institute. (2021). Global Forest Watch. https://www.globalforestwatch.org

Xiao-hui, W., Ke-he, W., & Xue-rong, X. (2011). Research of spatial index & caching for power GIS. 2011 6th International Conference on Computer Science & Education (ICCSE), 354–357. DOI: 10.1109/ICCSE.2011.6028653

Young, O. R., King, L. A., & Schroeder, H. (2013). *Institutions and Environmental Change: Principal Findings, Applications, and Research Frontiers*. MIT Press.

Chapter 7
Human-Animal Conflict:
A Case of Chinnathambi Elephant

K. P. Naachimuthu
https://orcid.org/0000-0002-2966-2089
PSG College of Arts and Science, India

S. B. Negha
https://orcid.org/0009-0005-1284-6542
PSG College of Arts and Science, India

ABSTRACT

This abstract offers a comprehensive analysis of Human-Elephant Conflict (HEC), a critical socio-ecological challenge in India. Using the compelling case of Chinnathambi, a translocated wild elephant, it critiques the limitations of reactive conservation strategies. The chapter contextualizes HEC within global and Indian frameworks, integrating quantitative data on conflict impacts with qualitative insights from diverse stakeholders. It examines the indispensable ecological roles of elephants as keystone species and scrutinizes existing policy frameworks, highlighting deficiencies in corridor protection, compensation mechanisms, and inter-sectoral coordination. A key focus is the justice dimension, addressing community perceptions of inequitable conservation policies. The chapter concludes by outlining forward-looking solutions, including corridor restoration, advanced early warning systems, community-based initiatives, and integrated land-use planning.

1. INTRODUCTION: CHINNATHAMBI – A SYMBOL OF CONFLICT AND COEXISTENCE

The boundary where human civilization meets the wild is a perpetual stage for narratives of both discord and harmony. In India, a nation teeming with life, both

DOI: 10.4018/979-8-3373-3992-4.ch007

human and wild, this intricate dance is particularly pronounced. Among the myriad stories that unfold on this stage, the saga of Chinnathambi, a wild elephant, emerged in January 2019, captivating national attention and igniting fervent debates across India. His journey became a powerful lens through which to examine the uneasy, often volatile, relationship between India's burgeoning human population and its iconic pachyderms.

Chinnathambi’s story began in the fertile agricultural fields of Coimbatore, a rapidly industrializing city nestled within the Western Ghats of Tamil Nadu. This region, a designated UNESCO World Heritage site and a global biodiversity hotspot, is characterized by a rich tapestry of dense forests, expansive plantations, and an ever-encroaching human footprint. Historically, these lands formed unhindered elephant habitats and vital migratory routes. However, decades of relentless land-use change—driven by the insatiable demands of agriculture, rapid urbanization, and extensive infrastructure development—have severely fragmented these pristine habitats, pushing elephants like Chinnathambi into frequent, often destructive, encounters with human communities.

Initially, Chinnathambi’s repeated forays into cultivated fields were perceived as a localized nuisance. Yet, as his presence became persistent, causing significant and recurrent damage to crops and property, he swiftly escalated from a mere nuisance to a “problem elephant” demanding state intervention. His subsequent capture, a dramatic and widely publicized operation, was envisioned as a definitive solution: translocate him to a distant, supposedly safer forest. However, Chinnathambi's persistent and almost defiant return to human settlements, despite these concerted efforts, transformed him from a local issue into a powerful protagonist in a national drama, rich with profound ecological, political, cultural, and ethical dimensions.

For dedicated conservationists, Chinnathambi’s unwavering homing instinct served as a stark testament to the futility of reactive management strategies. For beleaguered farmers, he represented a tangible menace. Animal rights activists passionately cast him as a victim of relentless human encroachment. The state, caught in the crossfire, viewed him as a complex “problem elephant” demanding effective management. Thus, Chinnathambi transcended his individual existence, becoming a living symbol of contradictions, systemic failures, and, crucially, the enduring possibilities of human-elephant coexistence in a rapidly changing India.

1.1. Why Chinnathambi Matters: Deconstructing a Symbol of Conflict

Selecting Chinnathambi’s compelling narrative as the central anchor for this study is a deliberate decision, acknowledging the inherent limitations of conventional academic discourse on Human-Elephant Conflict (HEC). Traditional scholarly discussions, while often statistically rigorous, frequently depersonalize the complex, visceral realities of HEC. Chinnathambi's widely documented and deeply personal story humanizes this conflict, compelling researchers, policymakers, and the public alike to confront the lived experiences of both elephants and the human communities who share these increasingly fragmented landscapes.

Chinnathambi's profound significance as a case study emanates from several critical points:

- **Limitations of Translocation:** His tenacious and repeated attempts to return to his original range illustrate the profound biological and psychological complexities that undermine translocation as a viable solution. Scientific literature consistently points to low success rates for relocating large, intelligent mammals like elephants, who possess sophisticated spatial memory, strong social bonds, and an intricate, inherited knowledge of their home ranges (Fernando et al., 2021; Kumar et al., 2020).
- **Emotional and Cultural Dimensions:** Elephants in India occupy a unique and often paradoxical status. They are revered as sacred beings, intrinsically linked with Lord Ganesha, yet simultaneously feared and resented as destructive forces. Chinnathambi's story starkly dramatized this dichotomy, forcing a national conversation about how deep-seated cultural reverence grapples with the harsh realities of daily conflict.
- **Policy Gaps and Reactive Conservation:** The governmental response to Chinnathambi—his capture and translocation—exposed the reactive, crisis-management orientation that pervades many conservation policies, particularly within Project Elephant. Instead of proactive measures, the focus remains on managing symptoms rather than addressing root causes (MoEFCC, 2023).
- **Ethical Dilemmas:** The intense public and policy conflict surrounding Chinnathambi prompted a profound national introspection on the ethical responsibilities owed to both wild animals and human communities. The case raised critical questions about animal welfare during capture, the fundamental rights of local communities to security, and the justice dimensions of conservation (Redpath et al., 2015).

By strategically centering Chinnathambi, this chapter aims to ground the abstract discourse of HEC in a tangible, relatable, and emotionally resonant narrative. His individual experience serves as a potent lens through which to systematically examine the intricate ecological drivers, deeply entrenched cultural perspectives, pervasive policy failures, and compelling ethical considerations that define human-elephant coexistence across India and globally.

1.2. Capture and Translocation: A State's Quick Fix and its Historical Roots

The decision to capture and translocate Chinnathambi was not an isolated incident but rather a reflection of a long-standing, deeply entrenched management paradigm in Indian wildlife conservation: the physical removal of "problem animals." This approach has historical roots stretching back to colonial-era practices, where wild animals were frequently viewed as mere obstacles to human development or as direct threats (Rangarajan, 2019; Sukumar, 1990).

Post-independence, the culling of protected species largely gave way to translocation, a practice often perceived as more humane, despite growing scientific skepticism regarding its efficacy.

Chinnathambi's capture in January 2019 became a widely publicized spectacle. It was a multi-agency, resource-intensive operation involving forest officials, skilled veterinarians, tranquilizer guns, and, crucially, kumki elephants—trained captive elephants used to control wild ones. This highly visible intervention was not merely a technical solution; it was also a powerful political performance to demonstrate governmental responsiveness and capacity to "restore order."

However, the inherent limitations and fundamental flaws of this strategy quickly became evident. Elephants are profoundly intelligent, highly social animals endowed with remarkable spatial memory, complex cognitive maps, and exceptionally strong homing instincts. These innate traits paradoxically make them exceptionally poor candidates for successful translocation. Studies consistently demonstrate that translocated elephants frequently attempt to return to their original ranges, sometimes traversing hundreds of kilometers through dangerous, unfamiliar, and human-dominated landscapes (Fernando et al., 2021; Kumar et al., 2020).

Translocation also imposes immense physiological and psychological stress. The combined trauma of capture, sedation, transport, and release into an unfamiliar environment can cause significantly elevated stress hormone levels, leading to disorientation, erratic movements, and increased aggression. Studies have documented translocated elephants exhibiting signs of trauma, moving unpredictably, and often gravitating towards forest edges or agricultural fields rather than retreating into dense forest interiors, leading to new conflicts (Kumar et al., 2020).

For Chinnathambi, the designated release forest was entirely unfamiliar, likely fragmented, and potentially already occupied by resident elephant herds. Uprooting him severed his vital connection to known food sources, water points, and escape routes. His post-release behavior mirrored documented patterns; he instinctively began to move back towards the familiar Coimbatore landscape—a terrain he knew intimately for sustenance and safety. This determined movement unequivocally signaled that translocation had fundamentally failed.

1.3. Return and Resistance: The Elephant Who Would Not Stay Away

Chinnathambi's arduous return journey, while deeply frustrating for officials, was, in ecological terms, entirely predictable. Elephants are creatures of profound habit, guided by exceptional memory and an innate understanding of resource availability (Sukumar, 2004). Their intelligence allows them to maintain detailed mental maps of seasonal resources, safe passageways, and potential threats. For Chinnathambi, Coimbatore represented familiarity; the alien release forest offered no such comfort.

Within weeks of his release, Chinnathambi's determined journey homeward began. He traversed bustling highways, crossed formidable rivers, and navigated through numerous villages and agricultural fields, displaying an extraordinary resolve driven by unwavering instinct. Each step highlighted the inherent vulnerability of elephants in human-dominated landscapes—exposed to electrocution, train collisions, and hostility. Yet, Chinnathambi persisted, a testament to his homing instinct and deep familiarity with his original range.

His reappearance near Coimbatore, once again raiding crops and clashing with residents, unequivocally confirmed the abject failure of the translocation strategy. For Chinnathambi, however, his return was not an act of defiance, but a profound matter of survival—a relentless, instinctual search for familiar landscapes that offered reliable food and water. His persistent "resistance" was a primal ecological imperative, a desperate return to known, resource-rich areas that offered the best chance for his continued existence.

This widely reported phase of Chinnathambi's journey highlighted several critical truths about Human-Elephant Conflict (HEC) in India:

- **Futility of Translocation:** Chinnathambi's case powerfully reinforced what numerous scientific studies consistently show: translocated elephants frequently attempt to return home, a phenomenon that often exacerbates conflict (Fernando et al., 2021; Kumar et al., 2020).
- **Limits of Human Control and Elephant Agency:** Despite significant deployment of state resources, Chinnathambi demonstrated profound and undeniable elephant agency. His determined return underscored that elephants are not passive objects to be managed solely at human whim.
- **Persistence and Structural Nature of Conflict:** Chinnathambi's return reignited tensions, demonstrating that HEC is not a temporary anomaly but a structural, persistent, and deeply contested feature of shared landscapes, demanding continuous negotiation.
- **Symbolic Power of Resistance:** In the public imagination, Chinnathambi rapidly evolved into a powerful symbol of resilience, embodying the fun-

damental ecological truth that elephants cannot simply be "removed" from human-dominated landscapes without first addressing root causes.

Chinnathambi's determined resistance thus mirrored broader HEC trends observed across India: recurring, unresolved conflicts, frequently subject to intense public and political contestation.

1.4. Public Debate: From Local Conflict to National Spectacle

Chinnathambi's unwavering return dramatically transformed a localized human-wildlife conflict into a national spectacle. His compelling story rapidly gained traction in media, fueled by his captivating appearances and the state's visible inability to contain him. Television crews followed his movements, broadcasting real-time images, while social media platforms propelled him into a household name.

For many urban Indians, far removed from the daily risks of living with wild elephants, Chinnathambi quickly became a tragic hero. He was widely perceived as a powerful symbol of the wild, a magnificent creature unjustly displaced, courageously fighting to reclaim his home. This narrative resonated deeply with a burgeoning urban environmental consciousness, evoking widespread sympathy and calls for his protection. Animal rights activists rapidly mobilized, launching petitions and legal actions to prevent his recapture, framing him as a victim of habitat loss.

Simultaneously, the voices of affected farming communities grew louder. For them, Chinnathambi was not a romantic symbol, but a tangible, daily threat—a destroyer of harvests and a pervasive source of fear. Local political leaders, responding to the distress, organized protests, demanding stronger government action. Their frustration stemmed from a perception that elephant welfare was being prioritized over human survival.

Caught between these powerful pressures, the state government oscillated, promising farmers compensation while facing legal and moral pressure from activists. This tension starkly exposed a deeper policy paralysis endemic to HEC management: a persistent reliance on reactive, piecemeal measures that ultimately satisfied no one.

Media narratives throughout this period vividly reflected this ambivalence. Some outlets portrayed Chinnathambi as a "gentle giant" wronged by expansion, highlighting the ecological tragedy. Others, closer to affected communities, emphasized his destructive actions, framing him as a menace. This duality mirrored India's complex relationship with elephants—a blend of deep reverence and simmering resentment.

Chinnathambi thus transcended being merely an individual elephant; he transformed into a symbolic battleground where competing visions of conservation, social justice, and economic development fiercely collided:

- **For farming communities:** He represented the injustice of bearing direct costs of conservation without adequate state support.
- **For animal rights activists:** He embodied the rights of non-human beings facing ecological crisis, serving as a call for ethical conservation.
- **For policymakers:** He became a critical test case: how to balance electoral pressures with constitutional mandates for wildlife conservation?
- **For the broader Indian public:** His story served as a potent mirror reflecting India's contradictions: worshipping elephants yet struggling to coexist with them.

This widely publicized stage vividly illustrated that HEC is not merely an ecological issue, but a profound political, cultural, and socio-economic spectacle, forcing an uncomfortable, yet necessary, national conversation.

1.5. The Symbol: Chinnathambi as the "Elephant in the Room"

Chinnathambi's individual story, rich with incident and imbued with deep symbolic meaning, transcended its specific boundaries to become a powerful and enduring symbol of India's broader Human-Elephant Conflict (HEC) crisis. His journey, meticulously marked by repeated incursions, dramatic capture, failed translocation, and determined return, illuminated several fundamental truths often unacknowledged in conventional conservation discourse. He effectively became the quintessential "elephant in the room"—a living metaphor for uncomfortable realities that policymakers, conservationists, and communities alike must confront directly.

Chinnathambi's narrative crystallized these uncomfortable truths with striking clarity:

- **Elephants Cannot Simply Be Removed:** His unwavering homing instinct powerfully demonstrated the inherent futility of simplistic removal strategies. Intelligent, wide-ranging, and deeply connected to their environments, elephants possess ingrained ecological needs that ensure their persistent return to landscapes they perceive as home (Kumar et al., 2020; Fernando et al., 2021).
- **Humans Cannot Ignore Elephants:** The pervasive presence of elephants in shared landscapes, their undeniable capacity for destruction, and profound cultural significance mean their presence cannot be dismissed. Their impact demands constant attention and a fundamental paradigm shift from viewing them as external "problems" to recognizing them as permanent, integral components of India's living landscapes.

- **Conflict is Recurring, Contested, and Systemic:** Chinnathambi's cyclical narrative of conflict, intervention, and renewed conflict underscored that HEC is not a temporary anomaly but a structural, persistent, and deeply contested feature of shared landscapes, demanding continuous negotiation and adaptive management.

Chinnathambi's journey, therefore, became a resonant mirror reflecting the broader trajectory of HEC across the Indian subcontinent. He embodies the profound paradox of elephants in India: simultaneously revered as gods, yet reviled as pests; fiercely protected, yet tragically killed. This inherent ambivalence lies at the heart of the HEC challenge.

Recounting Chinnathambi's journey in such granular detail is a deliberate methodological choice. It aims to reveal how one well-documented elephant's life can powerfully encapsulate the intricate ecological, profound cultural, and complex political dimensions of HEC. His story speaks directly to the urgent future of elephants and humans throughout India and to the global challenge of biodiversity conservation.

- **For Conservationists:** Chinnathambi serves as a compelling case study illustrating the widespread failure of translocation for large, intelligent species (Kumar et al., 2020; Fernando et al., 2021).
- **For Policymakers:** His story is an urgent reminder of the limitations of reactive governance models, advocating for a shift towards proactive, integrated frameworks (MoEFCC, 2023).
- **For Affected Communities:** Chinnathambi embodies both a formidable threat and a revered deity, powerfully reflecting the deep ambivalence in human attitudes (Rangarajan, 2019; Menon et al., 2017).
- **For Scholars:** His case provides a potent lens to analyze the intricate justice dimensions of conservation (Redpath et al., 2015).

While Chinnathambi's story is unique in its dramatic unfolding, it is also profoundly archetypal, mirroring countless similar conflicts daily across India, invariably involving human fear, insufficient official interventions, and persistent tensions.

1.6. Chinnathambi in the Public Eye: A Deeper Dive

Chinnathambi's initial capture might have remained a local news item, but his tenacious return swiftly elevated him to a resonant figure across Tamil Nadu and far beyond. Continuous television coverage, impassioned newspaper editorials, and fervent social media discussions propelled a localized conflict into a national spectacle. This unprecedented public engagement was driven not just by his actions, but by the powerful symbolic weight society attached to his plight.

For many urban viewers, Chinnathambi became a tragic hero—a magnificent wild creature unjustly displaced. For rural farmers, the narrative was starkly different: Chinnathambi was unequivocally a destructive force. For policymakers, his enduring presence became a critical test case—a challenge to balance constitutional conservation mandates with political pressures for public safety. This multiplicity of meanings reveals a crucial point: elephants in India are deeply interwoven into the cultural fabric and act as significant political figures.

Media played a crucial and often complex role in shaping public perception, frequently oscillating between two dominant narratives:

- **The Victim Narrative:** Many prominent headlines depicted Chinnathambi as a "gentle giant" forced into conflict by habitat loss. Animal rights groups amplified this, portraying him as an innocent casualty. Poignant images evoked widespread sympathy and calls for humane treatment.
- **The Menace Narrative:** Local and regional news outlets highlighted devastating crop destruction, palpable fear among villagers, and economic losses. Here, Chinnathambi was unequivocally framed as a threat.

These coexisting narratives vividly reflect Indian society's profound ambivalence towards elephants: sacred yet dangerous, revered yet resented.

To fully grasp Chinnathambi's national resonance, a nuanced understanding of the pervasive cultural symbolism of elephants in India is crucial:

- **Religious Icons:** Elephants are intimately associated with Lord Ganesha, symbolizing wisdom and prosperity. They are central to Hindu rituals and festivals, fostering a cultural ethos of reverence (Lahiri-Choudhury, 2020).
- **Royal Symbols and Historical Significance:** Historically, elephants were indispensable to Indian royalty and warfare, symbolizing power and strength. Ancient texts and art are replete with elephant depictions (Rangarajan, 2019).
- **Everyday Presences and Economic Utility:** Beyond sacred associations, elephants have historically been utilitarian animals, employed in logging, transport, and tourism, maintaining a tangible connection.

This profound cultural embedding creates a paradox. While harming elephants is often considered taboo, this reverence reaches its limits when human livelihoods or lives are directly threatened. Chinnathambi's case dramatically dramatized this tension.

Chinnathambi's prolonged saga ignited critical policy debates across various governance levels. His actions became a potent catalyst for scrutinizing existing HEC management approaches:

- **Translocation Under Scrutiny:** His determined return empirically challenged the very premise of translocation. Conservation biologists argued it was ineffective and harmful (Fernando et al., 2021; Kumar et al., 2020), calling for holistic solutions prioritizing habitat restoration.
- **Critique of Compensation Mechanisms:** Farmer complaints brought compensation mechanisms under a harsh spotlight. Farmers demanded quicker, fairer compensation. Bureaucratic delays, inadequate payouts, and allegations of corruption fueled dissatisfaction (Menon et al., 2017).
- **Legal and Ethical Questions:** Animal rights groups' petitions thrust the judiciary into the HEC debate, raising complex legal and ethical questions about animal welfare and the interpretation of wildlife protection laws.
- **Project Elephant Under Scrutiny:** Chinnathambi's case starkly exposed operational limitations of Project Elephant. Critics argued it had devolved into reactive "firefighting" rather than comprehensive planning, highlighting weak enforcement of corridor protection and disproportionate allocation of funds (MoEFCC, 2023).

1.7. Chinnathambi as Metaphor: Unveiling Systemic Truths

Beyond his specific story, Chinnathambi functions as a powerful and enduring metaphor for India's broader Human-Elephant Conflict crisis. He embodies a profound convergence of ecological, social, economic, and governance challenges defining the future of human-wildlife coexistence in the Anthropocene. His journey represents several profound systemic truths:

- **The Persistence of Elephants:** Despite centuries of human expansion, elephants stubbornly assert their presence. Chinnathambi's unyielding resistance to relocation signifies that these creatures are not easily eradicated or confined. Resourceful, intelligent, and deeply connected, they continually remind humanity that nature persists and demands recognition.
- **The Limits of Human Control in the Anthropocene:** His remarkable ability to evade recapture and his determined return demonstrated a profound limitation in the modern state's capacity for absolute control over natural processes. Even with sophisticated technology, managing a single elephant in dynamic landscapes proves to be an immense challenge.

- **The Interconnectedness of Issues:** Chinnathambi's story intricately links disparate domains: ecology, economy, culture, and politics. It defies simplistic compartmentalization, arguing forcefully that HEC cannot be resolved by isolated interventions. It is a systemic issue demanding systemic solutions.

By consciously treating Chinnathambi as the "elephant in the room," this study underscores the urgent need to confront uncomfortable truths: that development without ecological foresight is unsustainable; that conservation without social justice is untenable; and that genuine coexistence is impossible without robust community participation. His story serves as a constant reminder that ignoring these fundamental realities will only perpetuate conflict.

1.8. Chinnathambi in a Global Frame: Universal Dilemmas, Shared Lessons

While Chinnathambi's narrative unfolded in Tamil Nadu, its profound resonance extends globally. Across Asia and Africa, elephants are central to numerous human-wildlife conflicts that mirror India's challenges. These global parallels underscore universal ecological and socio-economic dilemmas, making the lessons gleaned from Chinnathambi's experience broadly applicable.

- **Sri Lanka: Microcosm of Asian HEC:** Sri Lanka faces exceptionally high incidences of HEC. Lone bull elephants raiding paddy fields are common, leading to hundreds of human and elephant fatalities annually (Fernando et al., 2021). Translocation efforts have also yielded disappointing results, reinforcing the limitations of this approach.
- **East Africa (Kenya and Tanzania): Competing for Resources:** In East African nations, elephants frequently clash with pastoralist and agricultural communities over scarce resources. Expansion of large-scale commercial agriculture creates significant flashpoints (Hoare, 2015; Shaffer et al., 2019). Successful models often lean towards community conservancies and benefit-sharing schemes.
- **Southeast Asia (Thailand and Malaysia): The Palm Oil Frontier:** In countries like Thailand and Malaysia, rapid conversion of tropical forest into monoculture plantations, notably oil palm, is a primary driver of HEC. This large-scale habitat destruction displaces elephants into human settlements (Campos-Arceiz & Blake, 2011).

Chinnathambi's tenacious resistance mirrors global cases where elephants consistently demonstrate profound homing instincts and innate resistance to human-imposed boundaries. His story exemplifies a universal dilemma: how to reconcile the vital ecological needs of wide-ranging megafauna with the pressing socio-economic realities of human communities. Lessons from India, particularly the failure of translocation, are broadly applicable. Conversely, successful models from other nations offer adaptable paradigms for India's HEC mitigation.

Figure 1.

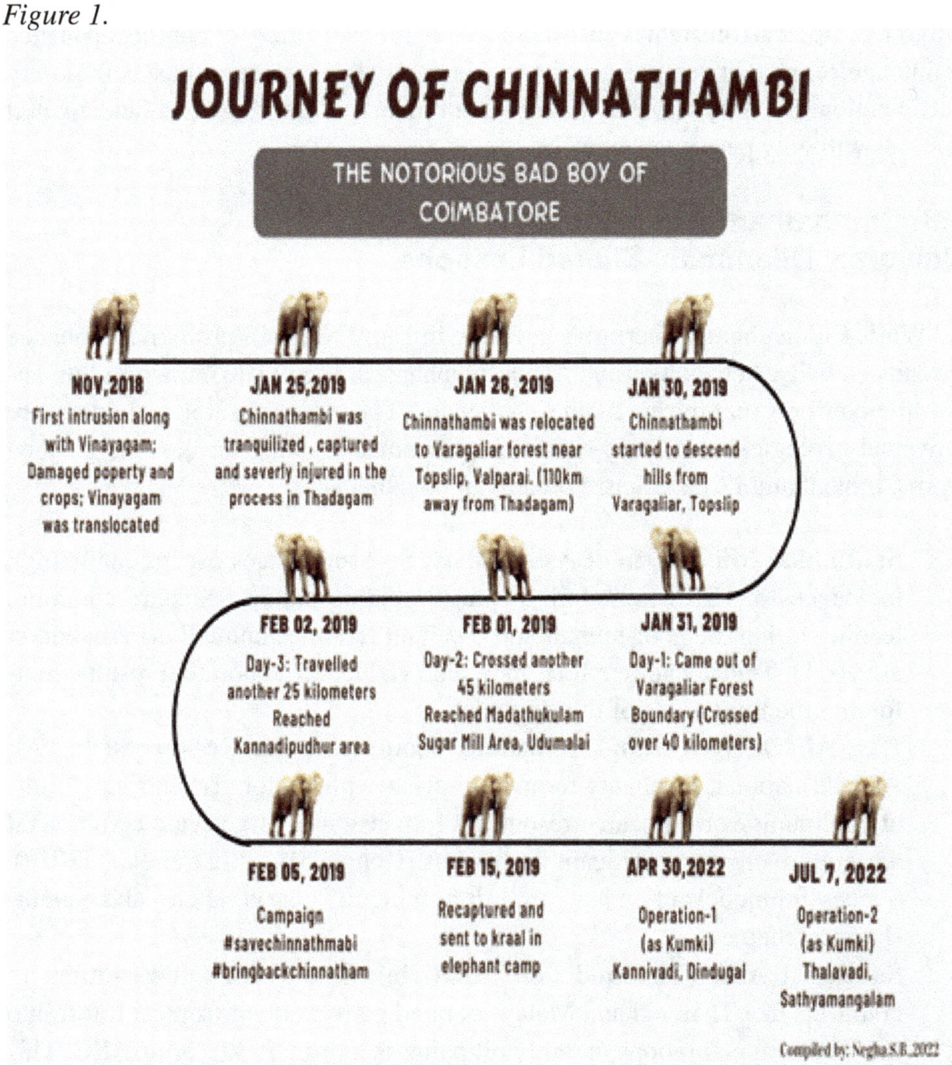

1.10. Research Questions Emerging from Chinnathambi's Case

Chinnathambi's case exposes limitations and highlights intricate factors driving HEC, giving rise to critical research questions aiming to delve into underlying causes, community perceptions, ecological impacts, and pathways for sustainable coexistence:

Policy Effectiveness and Governance Deficiencies:

- Why do translocation and the kumki system persist despite scientific evidence of low efficacy and ethical concerns? What institutional, political, and socio-cultural factors perpetuate reactive over proactive measures?
- How can policy instruments (Project Elephant, state wildlife acts) be reformed to genuinely prioritize landscape-level connectivity, habitat restoration, and integrated land-use planning?

Community Perspectives and Psycho-Social Dimensions:

- How do diverse stakeholders (farmers, forest officials, NGOs) perceive elephants, HEC, and current mitigation strategies? How do these divergent perceptions shape conflict dynamics?
- What narratives of justice, injustice, and equity emerge from lived experiences of affected communities? What psychological impacts does chronic HEC have on human well-being?

Ecological Dimensions and Landscape Dynamics:

- How do elephants' ecological roles (keystone species, ecosystem engineers) intersect with human land-use? What are the long-term implications for forest regeneration, biodiversity, and ecosystem health in fragmented landscapes?
- What are the precise ecological consequences of HEC management strategies, particularly on habitat fragmentation and the long-term viability of elephant populations? How does climate change modify these dynamics?

Future Pathways and Integrated Solutions:

- What portfolio of integrated solutions—combining legal and ecological protection of corridors, advanced early warning systems, genuine community-based mitigation, and comprehensive land-use planning—offers the most promise for sustainable and equitable HEC mitigation in India?

- Given its dual identity as a biodiversity hotspot and rapidly developing economy, how can India strategically balance conservation mandates with developmental aspirations, ensuring equitable coexistence globally?

1.11. Objectives of the Study

Building on Chinnathambi's case and the research questions, this study aims to achieve four overarching objectives:

1. **Contextualize HEC within Global and National Patterns:** Provide a robust understanding of India's HEC challenges by positioning them within global trends, identifying unique drivers, spatial/temporal patterns, and socio-ecological contexts in different Indian regions, drawing comparative lessons internationally.
2. **Critically Analyze Existing Policy Gaps and Reactive Measures:** Scrutinize the design, implementation, and effectiveness of current conservation policies (Project Elephant, state forest acts, compensation schemes). Highlight their limitations, bureaucratic inefficiencies, and the prevalence of reactive, symptomatic measures over proactive, holistic strategies.
3. **Foreground Community Narratives and Diverse Stakeholder Perspectives:** Move beyond scientific/policy-centric analyses by emphasizing lived experiences, perceptions, fears, and adaptive strategies of diverse stakeholders directly impacted by HEC. Capture qualitative narratives for a human-centered understanding of conflict complexities.
4. **Propose Integrated and Sustainable Solutions for Coexistence:** Develop and advocate for a comprehensive portfolio of holistic solutions grounded in ecological science, leveraging technological innovation, ensuring equitable community participation, and integrated within broader land-use planning.

Chinnathambi's story is more than a localized incident; it is a profound reflection on the future of elephants and humans in India and a significant contribution to the global challenge of coexistence.

2 THE GLOBAL AND INDIAN CONTEXT OF HUMAN-ELEPHANT CONFLICT (HEC)

2.1. Global Overview of Human-Elephant Conflict

Human-Elephant Conflict (HEC) is a pervasive and escalating global conservation and socio-economic challenge across Africa and Asia. Its fundamental driver is the inexorable expansion of human populations and associated land-use activities (agriculture, infrastructure, settlements) into, or adjacent to, traditional elephant habitats and migratory routes. As wilderness shrinks and fragments, elephants are increasingly forced into direct contact with human communities, leading to resource competition.

2.1.1. HEC in Africa: Varied Landscapes, Shared Challenges

In African nations, HEC manifests with distinct regional characteristics yet shares common underlying drivers. Kenya, Tanzania, and Botswana, with large and often recovering elephant populations, experience frequent crop raiding, property damage, and human fatalities. Expansion of large-scale commercial agriculture powerfully attracts elephants, creating flashpoints with commercial entities and smallholder farmers (Hoare, 2015; Shaffer et al., 2019).

Beyond agriculture, burgeoning human settlements encroaching on forest edges significantly contribute to conflict. Elephants moving between fragmented forest patches often traverse villages, leading to accidental encounters and retaliatory killings. In parts of West and Central Africa, illegal logging and mining intensify pressures, further fragmenting habitats.

Globally, HEC is framed as a conservation-development dilemma, pitting the imperative of protecting keystone species against the equally pressing need to safeguard human livelihoods (Redpath et al., 2015). Recent literature increasingly emphasizes HEC as a profoundly socio-political challenge, shaped by governance structures, land tenure systems, and socio-economic inequalities (Treves, 2009; Guarnieri et al., 2023).

2.1.2. HEC in Southeast Asia: The Palm Oil Frontier and Beyond

In Southeast Asian nations like Sri Lanka, Thailand, Malaysia, and Indonesia, the primary HEC driver has been rapid, uncontrolled conversion of tropical forest into monoculture plantations, notably oil palm. This large-scale habitat destruction and fragmentation directly displaces elephants into human settlements (Fernando et al., 2021). Sri Lanka faces one of the highest global HEC mortality rates (Fernando et al., 2021). Relentless plantation expansion severs crucial migratory corridors, forcing elephants into desperate foraging, leading to intensified crop raiding and retaliatory killings (Campos-Arceiz & Blake, 2011).

In other Southeast Asian areas, infrastructure development and illegal logging further exacerbate fragmentation. The resulting "edge effect" leads to higher encounter frequency. Similar to Africa, cultural reverence often coexists with acute resentment when elephants threaten human safety and economic well-being (Rangarajan, 2019).

Thus, HEC emerges globally as a multidimensional challenge: ecological in its drivers, economic in its impacts, and political in its governance. Global evidence consistently underscores that effective coexistence strategies must move beyond purely biological interventions, integrating ecological science with principles of social justice.

2.2. The Indian Context: A Unique Tapestry of Conflict and Coexistence

India, a nation of unparalleled biodiversity and demographic scale, is undeniably the most crucial range state for the long-term survival of the Asian elephant. Home to an estimated 27,000 wild Asian elephants, India shelters over 60% of the global population of this endangered species (MoEFCC, 2017; Sukumar, 2004). Unlike many parts of Africa, Indian elephants roam across an intricate and often overlapping mosaic of designated forests, private agricultural lands, plantations, and densely populated human settlements (Menon et al., 2017). This unique pattern is both a strength (contributing to genetic diversity) and a profound vulnerability, leading to alarmingly frequent and often fatal human conflict (Fernando et al., 2021; Natarajan etal., 2025).

Several interconnected factors render the Indian context of HEC particularly complex:

- **Deep Cultural Reverence and Spiritual Integration:** Elephants are profoundly embedded in Indian culture and religion. They are revered as sacred animals, intrinsically linked to Lord Ganesha. This deep-seated cultural reverence has historically fostered a unique degree of tolerance (Rangarajan, 2019). However, this spiritual regard coexists uneasily with acute resentment when elephant actions threaten human lives or economic stability.
- **Immense Demographic Pressure and High Population Density:** India is one of the most densely populated nations. Rural communities are typically situated adjacent to, or even within, historical elephant habitats. This unprecedented proximity of millions of people to thousands of wild elephants dramatically increases the frequency and intensity of encounters (Shaffer et al., 2019).
- **Highly Fragmented and Anthropogenically Modified Landscapes:** Rapid urbanization, extensive infrastructure development, and relentless agricultural expansion have severely fragmented India's elephant habitats, creating isolated “islands” of forest within human-dominated matrices (Menon et al., 2017; MoEFCC, 2023). This forces elephants into constricted corridors, often funneling them directly through villages and farms, intensifying conflict.

A recent study in Tamil Nadu found that proximity to forest edges, attractive crops, and human population density were strong predictors of conflict intensity, highlighting the central role of land-use change (Shameer et al., 2024).

This confluence of factors creates a profound paradox in India: elephants are simultaneously worshipped in temples yet resented as pests in fields; fiercely protected yet tragically killed. This intrinsic ambivalence underscores the complexity of HEC in India, where ecological realities are inextricably intertwined with cultural symbolism, socio-economic concerns, and governance challenges. As Natarajan et al. (2025) argue, elephant range expansion into human-dominated landscapes indicates that conflict is a structural and inherent feature of coexistence in the Anthropocene.

2.3. Policy Gaps and Ineffective Management Strategies: The Case of Translocation and the Kumki System

Despite decades of conservation efforts and policy frameworks like Project Elephant, India continues to grapple with persistent HEC. A critical analysis reveals that many prevalent management strategies suffer from fundamental policy gaps, being predominantly reactive, symptomatic, and often ineffective. Among these, translocation and the kumki system stand out as prime examples.

2.3.1. Translocation: A Flawed Strategy in the Anthropocene

Translocation, the capture and relocation of "problem elephants," has been and continues to be a widely used management strategy. The rationale is to mitigate immediate conflict and protect human lives. However, vast and growing research consistently demonstrates the inherent ineffectiveness and often counterproductive nature of this approach for large, intelligent, and socially complex species like elephants.

- **The Homing Instinct and Spatial Memory:** Elephants possess extraordinary spatial memory and strong homing instincts. Studies have documented countless cases of translocated elephants embarking on arduous journeys to return to their original ranges (Fernando et al., 2021; Kumar et al., 2020), placing them at increased risk of electrocution, train collisions, and retaliatory attacks.
- **Physiological and Psychological Stress:** The entire translocation process imposes immense physiological and psychological stress. Elevated cortisol levels are commonly observed, leading to behavioral changes such as increased aggression and disorientation (Kumar et al., 2020). These stressed

animals may struggle to integrate into new herds, making them more prone to conflict.

- **Displacement of Conflict, Not Resolution:** Rather than solving HEC, translocation often merely displaces the problem. A translocated elephant may create new conflicts in the release zone. Chinnathambi's determined return was a stark demonstration of this flaw.
- **Ethical Concerns:** The practice raises significant ethical questions regarding animal welfare. Critics argue that translocation disregards their individuality, social structures, and deep connection to native landscapes.

2.3.2. The Kumki System: A Tool with Limited Scope and Ethical Considerations

The use of trained captive elephants (kumkis) to manage conflicts with wild elephants is unique to India and parts of Southeast Asia. Kumkis are deployed to herd wild elephants, assist in capture, or provide physical barriers. While effective in specific contexts, their broader utility as a systemic HEC mitigation tool is limited, and their use raises ethical considerations.

- **Limited Scale and Scope:** The number of trained kumkis is limited, making it logistically impossible to deploy them as a primary solution across all affected landscapes.
- **Temporary Relief, Not Systemic Solution:** The kumki system primarily offers temporary relief. It is a reactive measure that does not address underlying causes of conflict. Once kumkis are removed, wild elephants are likely to return.
- **Welfare Concerns for Kumkis:** The welfare of kumkis is a growing ethical concern. These elephants often undergo harsh training regimes, and their living conditions and working hours are subjects of scrutiny (Lahiri-Choudhury, 2020).

Both translocation and the kumki system represent reactive approaches that tackle symptoms rather than fundamental causes. This highlights an urgent need for a paradigm shift towards more sustainable, ecologically informed, socially equitable, and landscape-level strategies.

2.4. Data on Human-Elephant Conflict in India: A Sobering Reality

Human-Elephant Conflict (HEC) in India represents one of the most pressing socio-ecological challenges of the 21st century. India is both a crucial stronghold for the species and an undeniable hotspot for conflict (MoEFCC, 2017; Sukumar, 2004). The coexistence of elephants and humans across increasingly shared landscapes has produced a complex tapestry of interactions, ranging from cultural reverence to acute economic resentment and tragic violence. A comprehensive analysis of quantitative data reveals a sobering reality of escalating casualties and economic losses.

2.4.1. Human Fatalities: A Chronic Crisis

Scale of the Problem: India consistently records an alarming average of 400-500 human deaths annually due to HEC (MoEFCC, 2023; Shaffer et al., 2019). These figures represent a significant toll on human lives. Between 2019 and 2022, over 1,700 people were tragically killed by elephants, indicating a persistent and unabated rate of fatalities (India Today, 2025).

Geographical Concentration: Human fatalities are disproportionately concentrated in states that harbor large elephant populations alongside high rural population densities, creating frequent opportunities for negative encounters. Key hotspots include Eastern India (Assam, West Bengal, Odisha, Jharkhand) and Southern India (Karnataka, Kerala, Tamil Nadu).

Victim Profiles and Circumstances: The majority of victims are smallholder farmers, agricultural laborers, or tribal communities. Many deaths occur when individuals encounter elephants while guarding fields at night, during accidental encounters on footpaths, or during retaliatory harassment. A study in Chhattisgarh found a strong correlation between human fatalities and specific cropping cycles, highlighting the link between agricultural practices and conflict risk (Nature, 2025).

2.4.2. Elephant Fatalities: The Cost of Coexistence

Elephants also suffer heavy losses annually. Approximately 100-150 elephants die each year due to conflict-related causes, including electrocution, train collisions, poaching, and retaliatory killings (MoEFCC, 2023).

- **Electrocution:** The Silent Killer: Electrocution is a tragically prevalent cause of elephant mortality, accounting for nearly 60% of elephant deaths in some states (Ghosh et al., 2025; Lahiri-Choudhury, 2020). Primary culprits are illegal electric fences erected by desperate farmers and poorly maintained power lines.

- **Train Collisions:** A Peril of Linear Infrastructure: Train collisions constitute another major source of elephant mortality, particularly where railway lines bisect critical elephant habitats and migratory corridors. States like Assam, West Bengal, and Odisha witness frequent fatalities from train strikes. Despite calls for underpasses, speed restrictions, and early warning systems, effective implementation remains patchy (Ahmed & Saikia, 2022).
- **Poaching:** A Lingering Threat: While large-scale poaching for ivory dramatically reduced after the 1980s, it remains a lingering concern in some regions, with male elephants still vulnerable for their tusks. Local poaching for bushmeat also persists (Menon et al., 2017).
- **Retaliatory Killings and Other Forms of Violence:** In extreme cases of prolonged conflict, frustrated communities resort to retaliatory killings, involving poisoning, spearing, or explosives. Such acts, though illegal, are often driven by desperation and perceived abandonment by the state (Shaffer et al., 2019).

2.4.3. Economic Losses: The Burden on Rural Livelihoods

The economic impacts of HEC are profound and disproportionately borne by rural communities, creating a significant impediment to livelihoods.

- **Crop Damage:** A Devastating Blow: Crop raiding is the most widespread and economically impactful aspect of HEC. Elephants are naturally attracted to calorie-rich crops cultivated on forest fringes. Annual crop losses run into hundreds of millions of rupees (Menon et al., 2017; Shaffer et al., 2019). For subsistence farmers, a single raid can wipe out an entire season's income.
- **Property Damage:** Beyond the Fields: Elephants frequently cause extensive property damage, destroying houses, fences, and grain storage facilities in search of food and water. Such damage imposes significant financial costs and creates a pervasive sense of insecurity and fear (Kumar et al., 2020).
- **Impact on Human Labor and Opportunity Cost:** HEC incurs significant indirect economic costs. Farmers spend sleepless nights guarding fields, diverting labor from other activities. Children may miss school, and land values can depress in conflict-prone areas.
- **Inadequate Compensation Schemes:** Fueling Resentment: Most Indian states have compensation programs, but these are universally criticized for being slow, bureaucratic, and inadequate (MoEFCC, 2023; Shaffer et al., 2019). Lengthy delays, complex paperwork, insufficient payouts, and allega-

tions of corruption fuel resentment, undermining cooperation with conservation efforts.

2.4.4. Trends and Future Outlook

The data from the last two decades paints a concerning picture of intensifying HEC across India. The period between 2019 and 2022, with 1,783 human deaths and 341 elephant mortalities, indicates rising casualties for both species (India Today, 2025). Long-term studies confirm an increasing frequency and severity of conflicts, directly correlated with shrinking elephant habitats and expanding human populations (Nature, 2025; MoEFCC, 2023).

- **Climate Change as an Exacerbating Factor:** The looming threat of climate change is expected to further exacerbate HEC. Altered rainfall patterns and more intense droughts will impact natural vegetation and crop cycles (Guarnieri et al., 2023). Droughts may compel elephants further into agricultural fields, intensifying competition for diminishing resources.
- **The Urgency of Systemic Intervention:** Without a fundamental shift towards systemic, proactive, and justice-oriented interventions, both human and elephant casualties are likely to continue their upward trajectory. The MoEFCC's 2023 framework for Elephant Conservation Plans (ECPs) emphasizes integrated approaches focusing on corridor protection, habitat restoration, and community-based mitigation. However, effective implementation remains a significant challenge.

In conclusion, quantitative and qualitative data on HEC in India reveal a sobering and unsustainable reality: hundreds of humans and elephants are dying annually, and thousands of livelihoods are disrupted. Conflict is a chronic, structural feature of coexistence. Addressing this multifaceted challenge demands a strategic departure from reactive, piecemeal measures towards comprehensive, landscape-level, and justice-oriented solutions.

3. ECOLOGICAL ROLE OF ELEPHANTS: KEYSTONE SPECIES AND ECOSYSTEM ENGINEERS

Elephants are not merely large, charismatic mammals; they are, in ecological terms, quintessential ecosystem engineers and keystone species. This designation reflects their profound and often irreplaceable impact on the structure, composition, and resilience of the landscapes they inhabit across both Asia and Africa (Campos-

Arceiz & Blake, 2011; Sukumar, 2004). Their behaviors, ranging from feeding and foraging to movement and waste deposition, trigger cascading effects throughout the ecosystem. Consequently, the decline of elephant populations—a direct outcome of habitat loss, fragmentation, and Human-Elephant Conflict—threatens not only the survival of a single iconic species but fundamentally undermines the ecological integrity, biodiversity, and functional health of entire landscapes. Their conservation is, therefore, a conservation of the entire ecosystem.

3.1. Seed Dispersal: The Forest's Gardeners

Elephants are among the most important large-bodied seed dispersers in tropical and subtropical forests. They consume vast quantities and varieties of fruits, nuts, and seeds, which pass through their digestive tracts largely intact. Critically, they then disperse these seeds over exceptionally long distances through their copious dung. This process is absolutely vital for forest regeneration, maintaining plant diversity, and ensuring the genetic connectivity of plant populations (Campos-Arceiz & Blake, 2011).

- **Dispersal Distance and Scale:** Unlike smaller frugivores, elephants can carry seeds for many kilometers, effectively facilitating gene flow between isolated forest patches. Their large gut size and slow digestion allow for long-distance transport, increasing the chances of seeds being deposited in suitable microhabitats away from the parent plant.
- **Large-Seeded Species Reliance:** Many large-seeded tree species are almost entirely reliant on megaherbivores like elephants for dispersal. Without elephants, these species would face severe recruitment limitation, leading to significant shifts in forest composition. In South and Southeast Asia, elephants disperse seeds of over 80 plant species, many ecologically and economically important (Sekar et al., 2015).
- **Fertilization and Germination Enhancement:** Elephant dung acts as an excellent organic fertilizer, providing an ideal substrate for seed germination. Passage through the digestive tract can also scarify hard seed coats, enhancing germination rates.

3.2. Habitat Modification: Landscape Architects

Through their distinctive feeding and movement behaviors, elephants act as powerful landscape architects and habitat modifiers. They exert immense physical force on their environment, fundamentally altering vegetation structure and creating a mosaic of different habitats. While sometimes perceived as destructive, these actions are essential for maintaining ecosystem balance and heterogeneity (Sukumar, 2004).

- **Forest Opening and Grassland Maintenance:** Elephants browse on saplings, strip bark, and push over trees, creating clearings within dense forests and maintaining open grasslands or savannas. In India, elephant browsing helps maintain open glades and meadows, crucial habitats for other herbivores (MoEFCC, 2023).
- **Creation of Microhabitats:** Physical disturbances by elephants create a variety of microhabitats. Uprooted trees create deadwood, supporting invertebrate communities. Openings in the canopy promote understory growth. Wallowing sites create temporary pools for amphibians and insects.
- **Fire Regime Modulation:** In some savanna ecosystems, by reducing woody biomass and promoting grass growth, elephants can influence fire regimes, making fires more frequent but less intense.

3.3. Water Access: Hydrological Engineers

During arid seasons or in water-scarce environments, elephants use their powerful trunks and feet to dig waterholes in sandy riverbeds. These "hydrological engineers" access subsurface groundwater, making this critical resource available not only to themselves but also to a wide array of other species (Redpath et al., 2015).

- **Lifelines in Droughts:** In regions experiencing prolonged droughts, these elephant-dug wells can become literal lifelines for entire faunal communities, sustaining biodiversity through challenging periods (Sukumar, 2004).
- **Groundwater Recharge:** Their digging activities can also contribute to localized groundwater recharge by breaking up compacted soil layers.

3.4. Corridor Maintenance: Guardians of Connectivity

Elephants are inherently wide-ranging species, requiring vast territories. Their movements follow established pathways that are critical for maintaining biodiversity corridors—linear strips of habitat linking fragmented forest patches, enabling gene flow and facilitating wildlife movement (Menon et al., 2017).

- **Facilitating Gene Flow:** By moving between fragmented habitats, elephants help maintain genetic connectivity among plant and animal populations, preventing genetic isolation and inbreeding.

- **Benefiting Other Species:** Paths created by elephants serve as vital movement routes for other large and medium-sized mammals, including tigers, leopards, and various ungulates. Protecting elephants directly contributes to the conservation of a broader spectrum of species (Krishnan et al., 2023).

3.5. Cascading Ecological Consequences of Elephant Decline

The decline or local extinction of elephant populations would trigger profound and often irreversible cascading ecological consequences, undermining the resilience of ecosystems. The removal of such a powerful ecological force would lead to dramatic shifts in vegetation structure, species composition, and ecological processes:

- **Altered Forest Composition and Reduced Diversity:** Without elephants, large-seeded trees would experience severe recruitment failure, leading to a significant decline in these species and altering forest composition (Campos-Arceiz & Blake, 2011).
- **Habitat Homogenization and Loss of Grasslands:** In the absence of elephant browsing, woody encroachment would proceed unchecked, converting open grasslands into denser scrub or closed-canopy forest. This would reduce habitat heterogeneity and lead to a decline in species dependent on open habitats.
- **Diminished Water Availability:** During dry seasons, the absence of elephant-dug waterholes would significantly diminish water availability for other species, potentially leading to localized extinctions (Redpath et al., 2015).
- **Collapse of Habitat Connectivity:** The disruption or loss of elephant movements would lead to the collapse of crucial biodiversity corridors, isolating wildlife populations. This isolation would impede gene flow and increase vulnerability to disease and climate change (Menon et al., 2017).

Thus, elephants are not only keystone species but also umbrella species, as their conservation requires the protection of vast, connected landscapes that benefit a multitude of other species. Their survival is inextricably linked to the health and resilience of countless other species and the critical ecological services upon which human societies ultimately depend. Protecting elephants is about safeguarding the fundamental ecological integrity and long-term sustainability of entire landscapes.

REFERENCES

Ahmed, R., & Saikia, A. (2022). Pandora's Box: A spatiotemporal assessment of elephant–train casualties in Assam, India. *PLoS One*, *17*(7), e0271416. journals.plos.org. DOI: 10.1371/journal.pone.0271416

Campos-Arceiz, A., & Blake, S. (2011). Megagardeners of the forest — The role of elephants in seed dispersal. *Acta Oecologica*, *37*(6), 542–553. DOI: 10.1016/j.actao.2011.01.014

Fernando, P., Jayewardene, J., Prasad, T., Hendavitharana, W., & Pastorini, J. (2021). Current status of Asian elephants and lessons for conservation. *Gajah*, *54*, 14–25.

Ghosh, S., Mandal, M., Vijayaprakash, V., Pandey, D., & Gayen, S. K. (2025). Geospatial analysis of elephant mortalities by electrocution in West Bengal, India. *Environmental Monitoring and Assessment*, *197*, 1124. DOI: 10.1007/s10661-025-14492-7

Guarnieri, M., Kumaishi, G., Brock, C., Chatterjee, M., Fabiano, E., Katrak-Adefowora, R., Larsen, A., Lockmann, T. M., & Roehrdanz, P. R. (2023). Effects of climate, land use, and human population change on human–elephant conflict risk in Africa and Asia. *Proceedings of the National Academy of Sciences of the United States of America*, *120*(12), e2209463120. DOI: 10.1073/pnas.2209463120

Hoare, R. (2015). Lessons from 20 years of human–elephant conflict mitigation in Africa. *Pachyderm*, *55*, 60–74.

Human-wildlife conflict. (2022, June). IUCN. https://www.iucn.org/resources/issues-brief/human-wildlife-conflict

Krishnan, S., Warren, N., Ganguly, U., Ranganathan, E., & Kumar, D. (2023). Road for elephants: Use of the Karadikkal–Madeshwara corridor in southern India. *Gajah*, *56*(56), 34–42. DOI: 10.2305/DIHX2111

Kumar, M. A.. (2020). Translocation as a management strategy: Failures and alternatives. *The Journal of Wildlife Management*, *84*(5), 1024–1035.

Menon, V., Tiwari, S. K., Easa, P. S., & Sukumar, R. (2017). *Right of Passage: Elephant Corridors of India* (2nd ed.). Wildlife Trust of India.

Ministry of Environment. Forest and Climate Change (MoEFCC). (2017). *Synchronized Elephant Population Estimation of India*. Government of India.

Natarajan, L., Nigam, P., & Pandav, B. (2025). Human–elephant conflict in expanding Asian elephant range in east-central India: Implications for conservation and management. *Oryx, 59*(2), 234–245. DOI: 10.1017/S0030605324000930

Project Elephant. (2021, September 10). Ministry of Environment, Forest and Climate Change: Government of India. https://moef.gov.in/en/division/forest-divisions-2/project-elephant-pe/introduction/

Rangarajan, M. (2019). *Nature and Nation: Essays on Environmental History.* Ranikhet: Permanent Black.

Redpath, S. M., Young, J., Evely, A., Adams, W. M., Sutherland, W. J., Whitehouse, A., Amar, A., Lambert, R. A., Linnell, J. D. C., Watt, A., & Gutiérrez, R. J. (2015). Understanding and managing conservation conflicts. *Trends in Ecology & Evolution, 28*(2), 100–109. DOI: 10.1016/j.tree.2012.08.021

Sekar, N., Lee, C., & Sukumar, R. (2015). Asian elephants as seed dispersers of large-seeded plants in a fragmented landscape. *Biotropica, 47*(4), 465–473.

Shaffer, L. J., Khadka, K. K., Van Den Hoek, J., & Naithani, K. J. (2019). Human–elephant conflict: A review of current management strategies and future directions. *Frontiers in Ecology and Evolution, 6*, 235. DOI: 10.3389/fevo.2018.00235

Shameer, T. T., Routray, P., Udhayan, A., Ranjan, N., Ganesan, M. G., Manimozhi, A., & Vasanthakumari, D. (2024). Understanding the patterns and predictors of human-elephant conflict in Tamil Nadu, India. *European Journal of Wildlife Research, 70*(95), 95. Advance online publication. DOI: 10.1007/s10344-024-01848-6

Sukumar, R. (1990). Ecology of the Asian elephant in southern India. II. Feeding habits and crop raiding patterns. *Journal of Tropical Ecology, 6*(1), 33–53. DOI: 10.1017/S0266467400004004

Sukumar, R. (2004). *The Living Elephants: Evolutionary Ecology, Behaviour, and Conservation.* Oxford University Press.

Today, I. (2025). *Deepening human–wildlife crisis killed 1,783 Indians, 341 elephants in 3 years. India Today Insight.* India Today.

Treves, A. (2009). The social dimensions of human–elephant conflict in Africa: A literature review and case studies from Uganda and Cameroon. *Human Dimensions of Wildlife, 14*(5), 315–326.

Chapter 8
Strategies for Biodiversity Conservation

G. R. Manjunatha Reddy
https://orcid.org/0009-0002-4476-5710
Bangalore University, India

ABSTRACT

Biodiversity conservation is a global imperative in the face of accelerating habitat degradation, climate change, and unsustainable development. This chapter delves into comprehensive strategies to safeguard biodiversity—from local grassroots actions to global frameworks, while integrating scientific innovation, traditional ecological knowledge, policy, and community engagement. The chapter explores in situ and ex-situ conservation methods, ecosystem-based approaches, and restoration initiatives, alongside an in-depth review of enabling legal, economic, and technological tools. Drawing on interdisciplinary insights and real-world case studies, the chapter emphasizes adaptive, inclusive, and resilient models for conserving the Earth's biological richness in the 21st century.

1. INTRODUCTION

The richness of life on Earth—its biodiversity—underpins the stability and functioning of ecosystems as well as the well-being of human societies (Xu et al., 2025). However, we stand at a crossroads. Accelerated biodiversity loss, primarily driven by human activities, threatens wildlife, plants, and the ecological systems that sustain our agriculture, climate, water resources, and health (Wang et al., 2024).

According to the Intergovernmental Science-Policy Platform on Biodiversity and Ecosystem Services (IPBES), an estimated one million species are expected to face extinction within the coming decades (Tollefson, 2019). This staggering figure

DOI: 10.4018/979-8-3373-3992-4.ch008

is symptomatic of larger ecological imbalances precipitated by land-use change, pollution, overexploitation, invasive species, and climate change. The erosion of biodiversity compromises the very systems that enable life to flourish.

Biodiversity conservation, therefore, is not just about saving species. It is about maintaining the integrity, productivity, and resilience of ecosystems to ensure a sustainable future for all. It involves a complex interplay of science, policy, community values, cultural beliefs, and economic structures (Sintayehu, 2018).

This chapter presents a holistic analysis of strategies for biodiversity conservation, examining both time-tested and emerging approaches through an interdisciplinary lens. It aims to deepen our understanding of conservation strategies by weaving together ecological theory, practical tools, socio-political context, and lived experiences from around the world.

2. FOUNDATIONS OF BIODIVERSITY CONSERVATION

2.1 Historical Perspectives on Conservation

The idea of conservation is not a new invention—it is deeply rooted in human history, although the reasons behind it have changed over time and across cultures. In many traditional societies, protecting nature was closely tied to spiritual beliefs and cultural practices (Taylor et al., 2020). For example, certain forests in India are considered sacred spaces protected from exploitation due to their significant religious value. Similarly, in Polynesia, cultural taboos and customs often restricted activities like overfishing or excessive hunting, ensuring that natural resources are balanced and sustainable.

Moving into the 19th century, conservation began to take on a more formalized and scientific tone, particularly in Western countries such as the United States and Europe. This period marked the beginning of organized conservation efforts, often focused on preserving vast, untouched landscapes. The creation of Yellowstone National Park in 1872 was a landmark moment—it became the world's first national park and symbolized a growing desire to protect "wilderness" areas. However, this early version of conservation typically saw nature as something separate from humanity. Indigenous and local communities were frequently excluded from these protected areas (PAs), sometimes even displaced, under the belief that humans harmed nature simply by being present.

By the mid-to-late 20th century, a significant shift had occurred in how conservation was understood and practiced. Scientists, activists, and policymakers are beginning to recognize that successful conservation must involve the people who are closest to nature. New strategies emerged that prioritized collaboration with local communities, recognizing their knowledge and role in stewardship. Approaches such as community-based conservation, integrated ecosystem management, and the concept of sustainable development became increasingly prominent (Dwivedi et al., 2022). These ideas reflect a more nuanced understanding that people and nature are interconnected and that long-term conservation efforts must address environmental and social dimensions.

2.2 *In-situ* and *Ex-situ* Approaches

Conserving biodiversity—the incredible variety of life on Earth—requires a multifaceted strategy. Two key worldwide methods are *in situ* (on-site) and *ex-situ* (off-site) conservation. Each plays a vital role in protecting species, ecosystems, and genetic diversity, often working most effectively when used in conjunction with one another.

2.2.1 In-situ Conservation: Protecting Nature Where It Lives

In situ conservation refers to efforts to protect species within their natural environments. This approach helps preserve not just the species themselves but also the intricate relationships they share with other organisms and their ecosystems. It enables the continuation of natural processes, such as pollination, predation, migration, and evolution, which are essential for long-term survival.

Common examples of *in situ* conservation include:

- National parks and wildlife sanctuaries, such as Kaziranga National Park in India, protect the endangered one-horned rhinoceros in its natural habitat within the floodplain (Sarmah, 2023).
- Biosphere reserves, such as the Serengeti in Tanzania, safeguard ecosystems while allowing for sustainable human activities in surrounding areas (Vallin et al., 2025).
- Community reserves and indigenous-managed lands are places where local people actively participate in conservation, blending traditional knowledge with modern ecological practices. For instance, many Indigenous communities in the Amazon rainforest manage land sustainably, preserving biodiversity and cultural heritage (Bontempi et al., 2023).

This method is ideal whenever possible, as it supports the full complexity of ecosystems and allows species to adapt naturally to environmental changes.

2.2.2 Ex-situ Conservation: A Safety Net Beyond the Wild

On the other hand, *ex-situ* conservation involves protecting species outside of their natural habitats. This approach is critical when a species' natural habitat is affected by deforestation, climate change, pollution, or human encroachment (Kasso & Balakrishnan, 2013).

Examples of *ex-situ* conservation include:

- Seed banks, such as the Svalbard Global Seed Vault in Norway, store seeds worldwide as a backup in case of agricultural crises or species loss.
- Zoos and aquariums protect endangered animals, such as the Amur leopard and the Axolotl, and serve as research centers and educational platforms.
- Botanical gardens, such as the Royal Botanic Gardens at Kew in the UK, are where rare and endangered plant species are cultivated and studied.
- Gene banks store genetic material, such as DNA, sperm, eggs, and embryos, to help preserve the genetic diversity of threatened species.

Ex-situ efforts are crucial when a species is on the verge of extinction or its habitat is no longer viable (Jaisankar et al., 2018). They also provide opportunities for scientific research, public awareness, and even the future reintroduction of these species into the wild.

2.2.3 A Complementary Strategy

While *in-situ* conservation is generally preferred because it maintains species within their natural contexts, *ex-situ* conservation is a critical backup. Together, they offer a holistic approach to biodiversity protection. For example, the California condor was brought back from the brink of extinction through *ex-situ* breeding programs and later reintroduced into protected wild areas, demonstrating how both approaches can work in tandem.

Ultimately, using both *in-situ* and *ex-situ* methods strengthens global conservation efforts by addressing the diverse and urgent challenges facing wildlife today.

2.3 The Ecosystem and Landscape Approach

In today's world, where environmental challenges are increasingly complex and interconnected, conservation thinking has evolved beyond protecting isolated patches of nature. Modern strategies now emphasize a broader, more integrated view that acknowledges the interdependence of ecosystems, wildlife, and human communities. Two such approaches gaining ground globally are the ecosystem and landscape approaches.

2.3.1 The Ecosystem Approach: Nature as a Living System

The ecosystem approach is a comprehensive method of conservation that focuses on managing ecosystems as dynamic, living systems rather than isolated components. This strategy, strongly promoted by the Convention on Biological Diversity (CBD), encompasses the full range of ecological processes, including nutrient cycling, water flow, species interactions, and climate regulation (Scarano, 2017). It also acknowledges the critical role people play in shaping and maintaining these systems.

The ecosystem approach aims to sustain the health, productivity, and diversity of entire ecosystems rather than preserving individual species or specific habitats (Table 1). It encourages flexible, adaptive management that can respond to changing environmental conditions, scientific knowledge, and community needs (Morand & Lajaunie, 2018).

Example: In the Great Barrier Reef Marine Park in Australia, the ecosystem approach is employed to manage a vast marine environment that faces threats from coral bleaching, tourism, and fishing. Managers collaborate with local communities, industries, and scientists to strike a balance between conservation and sustainable use, thereby ensuring that the reef supports both biodiversity and livelihoods (Pascoe et al., 2024).

This approach also values the participation of local communities, understanding that long-term conservation success depends on including those who live within the ecosystem. In this way, it blends ecological science with traditional knowledge and social realities.

2.3.2 The Landscape Approach: Balancing Nature and Human Use

The landscape approach is closely related to the ecosystem approach, which looks at conservation through the lens of land use and spatial planning. Landscapes are not just wilderness or farmland—they are mosaics that include forests, farms, villages, towns, rivers, and PAs, all coexisting in varying degrees of harmony (Khoroshev, 2020).

The landscape approach is beneficial in areas where strict conservation, creating national parks, or banning human activities is neither practical nor fair to the communities. Instead, this method balances multiple land uses to support biodiversity, food production, economic development, and cultural values (Adams et al., 2023).

In the Satoyama landscapes of Japan, traditional agricultural practices coexist with forests and wetlands. Local communities and conservation managers actively manage these areas to enhance biodiversity while supporting sustainable farming, fishing, and forestry practices. The Satoyama model has become a global symbol of the landscape approach, demonstrating how people and nature can thrive together (Jiao et al., 2019).

In parts of East Africa, conservationists and local communities use a landscape approach to integrate wildlife conservation with pastoralism and small-scale farming. Here, conservationists collaborate with local communities to develop land-use plans that preserve wildlife corridors, safeguard key habitats, and foster sustainable livelihoods (Bollig, 2024).

2.3.3 A Shift Toward Integration and Sustainability

The ecosystem and landscape approaches reflect a growing understanding that conservation cannot be achieved solely through human development. They emphasize integration rather than isolation, encouraging collaboration among conservationists, farmers, policymakers, Indigenous peoples, and private landowners (Cazzolla Gatti, 2025).

These approaches are not about returning to a pristine, untouched wilderness. Instead, they are about managing change wisely—so ecosystems remain resilient, people can thrive, and future generations inherit a planet rich in natural and cultural heritage.

Adopting ecosystem—and landscape-level thinking offers one of the most promising paths toward sustainable conservation in the face of environmental, social, and economic pressures.

2.4 Philosophical and Indigenous Foundations

Conservation is not solely a scientific or policy-driven endeavor; it is deeply intertwined with human values, cultural worldviews, and ethical considerations. At the heart of every conservation initiative lies a profound and often challenging question: *What do we value in nature, and what are our responsibilities to it?* These are not merely technical concerns but philosophical ones, prompting reflection on our moral obligations to other species, future generations, and the planet.

Understanding and addressing these questions requires engaging with diverse philosophical traditions and ethical frameworks. Environmental philosophy, for example, explores concepts such as intrinsic value in nature, ecological justice, and the rights of non-human beings. It challenges anthropocentric worldviews prioritizing human interests above all else and promotes more inclusive perspectives on the interconnectedness of all life forms.

Indigenous knowledge systems, too, offer rich, time-tested perspectives on the human-nature relationship. Rooted in deep ecological understanding and cultural traditions, Indigenous worldviews often emphasize harmony with nature, stewardship, reciprocity, and the sacredness of all life. These frameworks do not separate humans from nature but view people as integral parts of ecological systems responsible for caring for the land and all living beings. As Piccolo et al. (2022) noted, these perspectives provide essential insights that are often undervalued in mainstream conservation discourse but are crucial for fostering more holistic, ethical, and sustainable approaches.

By integrating philosophical inquiry and Indigenous wisdom into conservation strategies, we can transcend narrow, utilitarian views of nature and cultivate a more compassionate, respectful, and resilient relationship with the natural world. This broader, values-based foundation is essential for achieving ecological sustainability and social and environmental justice.

2.4.1 Philosophical Traditions: Beyond Human-Centered Thinking

Much of modern society operates from an anthropocentric (human-centered) perspective, where nature is often seen primarily in terms of its usefulness to people—resources to extract, land to develop, or scenery to admire. However, several philosophical traditions challenge this mindset, offering alternative ways of thinking about our place in the world (Heikkinen et al., 2024).

- Deep ecology, a movement developed in the 1970s by Norwegian philosopher Arne Næss, argues that all living beings, regardless of their utility to humans, have inherent worth. According to deep ecology, humans are not superior to other species but are part of a complex, interconnected web of life (Glasser, 2011). This view calls for a radical shift in how we live, consume, and make decisions.
- Eco-centrism takes this further by placing the ecosystem, not just individual species, at the center of ethical concern. Under this framework, rivers, forests, and even entire biomes deserve respect and protection, not just because they benefit people but because they are valuable in their own right (Kopnina et al., 2022).
- Bio-cultural ethics bridges ecological and cultural dimensions, recognizing that the well-being of human communities and ecosystems is deeply intertwined. This approach emphasizes the importance of preserving biological and cultural diversity, especially where traditional lifestyles are tied to the land (Hong, 2013).

These philosophical foundations challenge us to rethink the dominant narratives of control and consumption, urging a humbler and respectful relationship with nature.

2.4.2 Indigenous Knowledge Systems: Living in Relationship with the Land

While philosophical movements in the West have begun to question anthropocentrism in recent decades, Indigenous communities worldwide have long held relational worldviews that view humans as an integral part of nature, not separate from or above it (Pelizzon, 2025).

For many Indigenous cultures, the land is not just a resource; it is sacred. Mountains, rivers, animals, and plants are often regarded as relatives, each with their roles, rights, and responsibilities within the greater whole. These knowledge systems are intensely local, built over generations through direct, sustained interaction with specific ecosystems (Martinez et al., 2023).

Example: In the Amazon Basin, Indigenous communities such as the Asháninka and the Kayapo have developed sophisticated knowledge of medicinal plants, forest ecology, and seasonal cycles—knowledge that has been crucial in conserving vast tracts of rainforest that might otherwise have been destroyed (Virtanen, 2012).

These knowledge systems are not only spiritually rich but ecologically compelling. Studies increasingly show that biodiversity is often higher in Indigenous-managed territories than in state PAs. This demonstrates that Indigenous conservation practices are not just cultural expressions—they are vital tools in the global effort to safeguard nature.

2.4.3 Integration: A Moral and Practical Imperative

Recognizing and respecting Indigenous and local knowledge is an ethical responsibility and a strategic necessity. Conservation that excludes local voices can lead to conflict, failure, and injustice (Cairney et al., 2023). By contrast, inclusive, culturally grounded conservation is more likely to be resilient, equitable, and prosperous.

Efforts such as co-management agreements, Indigenous Protected and Conserved Areas (IPAs), and the recognition of bio-cultural heritage sites demonstrate that modern conservation is beginning to acknowledge these diverse foundations (Bishop et al., 2025).

Building a sustainable future requires more than protecting nature—it requires reimagining our place within it. Drawing on philosophical inquiry and Indigenous wisdom, conservation can evolve into a deeper practice of reciprocity, respect, and responsibility that honors the land and the people who know it best (Olivadese & Dindo, 2025).

3. KEY STRATEGIES FOR BIODIVERSITY CONSERVATION

Conserving biodiversity—the variety of life on Earth—is a multifaceted challenge that requires coordinated action at global, national, and local levels. This chapter examines the practical and policy tools currently employed to protect, restore, and sustain ecosystems and the numerous species they support (Wang et al., 2024). These strategies range from PAs and community-led initiatives to advanced technologies and economic incentives.

Table 1. Key Strategies for Biodiversity Conservation

Strategy	Description	Application Level	Expected Impact	References
In-situ Conservation	Conservation of species in their natural habitats	National Parks, Reserves	Preserves entire ecosystems and species interactions	Offord (2017).
Ex-situ Conservation	Conservation outside natural habitats (zoos, gene banks, botanical gardens)	Global/National	Protects endangered species and allows research and breeding	Murray (2017).
Habitat Restoration	Rehabilitating degraded ecosystems	Local/Regional	Revives native flora and fauna	(UNEP 2021).
Community-based Conservation	Engaging local communities in biodiversity protection	Local	Sustainable use and protection with socio-economic benefit	Koricha, & Adem (2024).
Legislation and Policy Frameworks	Legal protection through national/international laws	National/ International	Regulatory enforcement of conservation goals	Lausche (2011).
Biodiversity Monitoring and Research	Use of technologies (e.g., GIS, AI) to track species and ecosystems	All levels	Informs policy and intervention strategies	(Stefanni et al., 2022).
Education and Awareness	Spreading knowledge on the importance of biodiversity	Schools, NGOs, Public	Enhances public participation and support	Børresen et al. (2022).

3.1 PAs and Conservation Planning

3.1.1 National Parks and Reserves

PAs remain the cornerstone of global biodiversity conservation, serving as vital refuges for countless species and ecosystems under increasing threat from human activities (Figure 1). As of 2024, over 17% of the world's terrestrial and more than 8% of marine environments are under formal protection through various designations, including national parks, wildlife sanctuaries, marine reserves, and community-managed conservation areas (International Union for Conservation of Nature [IUCN], 2024). These protected spaces differ in their governance structures and permitted land uses, ranging from strictly protected nature reserves, which limit human intervention, to sustainable use areas, which allow for regulated resource use in harmony with conservation goals.

Figure 1.

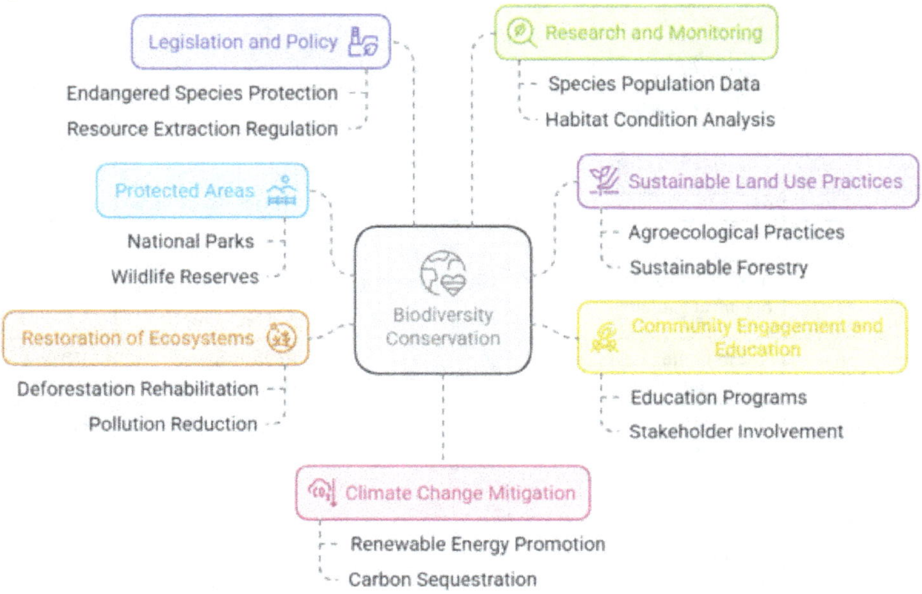

However, designation alone does not guarantee conservation success. The effectiveness of PAs depends on several critical factors determining whether biodiversity is genuinely safeguarded or merely symbolically protected on paper.

1. Adequate funding and human capacity are essential for managing PAs. Without proper resources, managers are unable to effectively monitor wildlife populations, enforce regulations, or address emerging threats such as invasive species or climate change.
2. Robust legal frameworks and enforcement mechanisms are crucial for deterring and penalizing illegal activities, such as poaching, unregulated tourism, logging, and land encroachment. Clear policies and penalties enhance compliance and legitimacy.

3. Community engagement and local participation are fundamental for long-term conservation success (Figure 1). When local communities are included as stakeholders and beneficiaries, rather than being excluded from traditional lands, they are more likely to support and uphold conservation goals. Participatory management also integrates traditional ecological knowledge into stewardship practices.
4. Ecological representation ensures that PAs encompass a wide range of ecosystems, not just aesthetically appealing or charismatic landscapes. This principle covers biodiversity across various biomes, including deserts, grasslands, wetlands, and lesser-known yet equally important ecosystems.
5. Climate resilience planning is becoming increasingly important as climate change alters species distributions, disrupts seasonal cycles, and increases the frequency of extreme weather events. PAs must be designed and managed with adaptive strategies that anticipate such shifts, including establishing buffer zones and connectivity corridors.

Kruger National Park in South Africa is a prime example of effective protected area management. While renowned for its biodiversity and ecotourism, Kruger's success has been strengthened by collaborative governance with local communities and transboundary cooperation through the Great Limpopo Transfrontier Park, which links conservation efforts across South Africa, Mozambique, and Zimbabwe. This cross-border initiative enhances ecological connectivity, enables species migration, and promotes peace and shared responsibility for environmental stewardship (Martinez et al., 2023).

PAs remain indispensable for biodiversity conservation, but their success hinges on holistic, inclusive, and adaptive management. As global environmental pressures intensify, there is an urgent need to scale up and strengthen protected area networks in a manner that is both scientifically sound and socially just, as well as climate-resilient.

3.1.2 Marine Protected Areas (MPAs)

MPAs play a pivotal role in maintaining the health and resilience of ocean ecosystems. These designated zones conserve biodiversity by safeguarding ecologically significant habitats such as coral reefs, seagrass beds, mangrove forests, and spawning and nursery grounds. When effectively managed, MPAs support biodiversity conservation, enhance fisheries productivity, and secure coastal livelihoods by maintaining ecosystem services and ensuring the availability of resources (Hoppit et al., 2022).

Well-designed and effectively enforced MPAs must deliver a range of ecological and socio-economic benefits. They can replenish overfished populations by providing safe havens for species to breed and grow, thereby increasing biomass within and beyond their boundaries through the spill-over effect. They also help protect endangered and vulnerable marine species by reducing threats from human activities, such as overfishing, habitat destruction, and pollution. Moreover, MPAs enhance the resilience of marine ecosystems to climate-related stresses, such as ocean acidification and coral bleaching, by preserving ecological complexity and promoting adaptive capacity (Hoppit et al., 2022).

However, the potential of MPAs is hindered by significant implementation and management challenges. Many MPAs suffer from weak enforcement, insufficient funding, unclear legal frameworks, and a lack of stakeholder engagement, particularly with local fishing communities. As a result, some MPAs remain "paper parks"—protected in legislation (Table 1) but ineffective in practice due to minimal on-the-ground conservation action or monitoring (Chukwuka et al., 2025). Thus, conservation goals require inclusive governance, participatory management, community buy-in, and adequate financial and institutional support.

A compelling example is the Great Blue Hole Marine Reserve in Belize, which demonstrates how a well-managed MPA can conserve biodiversity and support the national economy. This iconic site protects rich marine ecosystems, including coral reefs and endemic species, while serving as a major attraction for ecotourism. The revenue generated from sustainable tourism contributes significantly to local livelihoods and national income, creating a positive feedback loop between conservation and economic development (Adams et al., 2023).

MPAs are vital for ocean conservation, but their effectiveness hinges on robust design, equitable governance, and sustained investment. Integrating ecological science with community participation and policy enforcement is essential to ensure that MPAs fulfill their intended role in securing marine biodiversity for future generations.

3.1.3 Ecological Corridors and Connectivity

Conservation does not—and cannot—stop at the borders of national parks or wildlife reserves. As natural habitats become increasingly fragmented due to urbanization, agriculture, infrastructure development, and climate change, ecological connectivity emerges as a critical component of effective biodiversity conservation. Ecological corridors, also known as wildlife corridors or greenways, are strategically preserved or restored stretches of natural or semi-natural habitat—such as riparian zones, hedgerows, forest strips, or even engineered structures like green bridges and overpasses—that facilitate the movement of species between otherwise isolated patches of habitat (Cazzolla Gatti, 2025).

These corridors serve multiple vital functions in landscape-level conservation. Firstly, they help maintain genetic diversity by enabling individuals from different populations to interact, mate, and exchange genetic material, thus preventing inbreeding and genetic bottlenecks. Secondly, corridors allow species to migrate or shift their ranges in response to climate change, enabling adaptation by accessing new suitable habitats (Christie & Knowles, 2015). This is especially important for species with large home ranges or specific climatic needs. Thirdly, ecological corridors mitigate human-wildlife conflict by providing designated, safer routes for animal movement, reducing the likelihood of encounters in human-dominated areas.

A compelling example comes from India, where elephant corridors have been identified and maintained to facilitate the seasonal migration of elephants across fragmented forest landscapes. These corridors are crucial for reducing deadly confrontations between humans and elephants, especially in agricultural zones where crop raiding and property damage can lead to retaliation. Similarly, in North America, the construction of wildlife overpasses and underpasses—such as those on the Trans-Canada Highway in Banff National Park—has proven highly effective in reducing vehicle-wildlife collisions while restoring connectivity for species like bears, deer, and mountain lions.

Ecological corridors enhance the resilience of ecosystems and species, enabling them to survive and thrive in dynamic and human-altered environments. Their inclusion in conservation planning reflects a shift from isolated, site-specific strategies toward integrated landscape and seascape-level approaches that recognize the importance of movement, adaptation, and ecological processes across space and time.

3.2 Community-Based and Indigenous Conservation

Local and Indigenous communities have long acted as guardians of biodiversity. Modern conservation increasingly recognizes the importance of sustainable stewardship by involving the people closest to nature (Artelle et al., 2019).

Community-based conservation includes a range of practices:

- Sacred natural sites are characterized by their cultural or spiritual significance.
- Community-managed forests, such as those in Nepal or Mexico, where locals maintain forest health while benefiting economically.
- Participatory mapping and monitoring enable communities to define and protect traditional territories.

Research shows that biodiversity outcomes in Indigenous and community-managed areas often equal or surpass those in government-run parks (Gerstner & Zarnetske, 2025). However, for these initiatives to work, several conditions must be met:

- Secure land and resource rights,

- Fair distribution of benefits,
- Genuine involvement in decision-making, not just token participation.

Example: In the Philippines, Indigenous Peoples and Community Conserved Territories and Areas (ICCAs) have become a formal part of national conservation strategies.

3.3 Restoration and Rewilding

3.3.1 Restoration: Repairing Damaged Ecosystems

Ecological restoration is the process of restoring degraded or destroyed ecosystems to their original state. This might involve:

- Planting native trees in deforested areas,
- Removing invasive species that crowd out local flora and fauna,
- Restoring wetlands to improve water quality and flood control.

Sometimes, nature is given a helping hand through assisted natural regeneration, which supports the ecosystem's ability to recover (Edwards et al., 2021).

3.3.2 Rewilding: Letting Nature Lead

Rewilding is a bold approach that seeks to reintroduce species and restore natural processes with minimal human interference. The goal is not to return to a specific historical state but to allow nature to evolve freely and dynamically (Gorghiu et al., 2024).

Example: The reintroduction of wolves in Yellowstone National Park is a landmark case. Their return altered deer behavior, allowed vegetation to recover, and even changed the course of rivers—a powerful example of how top predators influence entire ecosystems.

3.4 Legal and Policy Frameworks

3.4.1 International Agreements

Global environmental treaties establish the legal framework for coordinated conservation efforts across international borders. Some of the key agreements include:

- The CBD promotes sustainable use and equitable sharing of genetic resources.

- CITES regulates international trade in endangered species.
- The Ramsar Convention focused on wetland protection.

The Kunming-Montreal Global Biodiversity Framework, adopted in 2022, sets ambitious goals to halt biodiversity loss by 2030, including the protection of 30% of land and sea areas and the restoration of 30% of degraded ecosystems (Hughes & Grumbine, 2023).

3.4.2 National and Regional Laws

While international treaties provide the framework, real action happens at the national and local levels. Countries must implement:

- Biodiversity action plans,
- Environmental impact assessments (EIAs) before significant developments,
- Protected area legislation,
- Moreover, legal recognition of nature's rights is increasingly being granted, with ecosystems being granted legal personhood.

3.5 Economic Tools and Incentives

3.5.1 Payments for Ecosystem Services (PES)

PES programs compensate landowners or communities for managing their land in ways that provide public ecological benefits, such as clean water, carbon storage, or biodiversity protection.

Example: Costa Rica's national PES scheme has paid farmers to conserve forests, resulting in a dramatic recovery of forest cover since the 1990s (Tafoya et al., 2020).

3.5.2 Ecotourism and Conservation Finance

Eco-friendly tourism, conservation trust funds, biodiversity credits, and green bonds are innovative methods for funding conservation. When designed ethically, these tools:

- Generate income for local communities,
- Reduce reliance on extractive industries,
- Raise awareness about the value of biodiversity.

However, they must be carefully regulated to avoid greenwashing, exploitation, or the marginalization of local communities.

3.6 Technological and Scientific Innovations

Science and technology are revolutionizing our understanding and protection of biodiversity. New tools allow faster, more precise, cost-effective conservation efforts (Reynolds et al., 2025).

Key innovations include:

- Remote sensing and satellite imagery to track land use, deforestation, and ecosystem health (Gu & Zeng, 2024).
- Environmental DNA (eDNA) can identify species from a single drop of water or soil (Rishan et al., 2023).
- Bioacoustic monitoring uses microphones to track bird calls, frog songs, or illegal logging (Oestreich et al., 2024).
- Artificial Intelligence (AI) and machine learning are utilized in various applications, ranging from species recognition to anti-poaching surveillance (Fergus et al., 2024).
- Biotechnology, including gene editing and seed banks that preserve genetic diversity for future use (Salgotra & Chauhan, 2023).

Together, these strategies represent a dynamic and interconnected approach to biodiversity conservation that blends tradition, innovation, local wisdom, and global action. As threats to the natural world intensify, so must our efforts to protect it through policy, partnerships, science, and a deep respect for the life around us.

4. SOCIAL AND POLITICAL DIMENSIONS OF BIODIVERSITY CONSERVATION

4.1 The Role of Governance

Effective governance is essential for biodiversity conservation. It encompasses laws, institutions, policies, power dynamics, and decision-making processes (Sterling et al., 2017). Key governance principles include:

- Transparency and accountability
- Stakeholder inclusion and participation
- Equity and justice

- Rule of law and enforcement

Multilevel governance—linking global, national, and local actors—is increasingly important in addressing transboundary and systemic biodiversity threats (Kok & Ludwig, 2022).

4.2 Environmental Justice, Indigenous Rights, and Conservation Displacement

While conservation is widely seen as a positive force, it has not always been implemented fairly. In some cases, Indigenous and local communities have been forcibly removed from their ancestral lands to make way for PAs — creating what are now known as "conservation refugees." These actions raise serious ethical questions and often backfire by disrupting long-standing stewardship systems that have effectively managed ecosystems for generations.

A rights-based conservation model challenges this history by centering the rights of Indigenous peoples and local communities. It emphasizes:

- Legal recognition of customary land ownership
- Upholding the right to Free, Prior, and Informed Consent (FPIC)
- Equitable sharing of conservation benefits

Incorporating environmental justice into conservation planning ensures that marginalized voices are included, cultural heritage is respected, and local livelihoods are supported. Ultimately, addressing these issues is not just about fairness — it is essential for making conservation efforts more effective, legitimate, and resilient in the long term.

4.3 Education, Awareness, and Behavior Change

Conservation is also a cultural and behavioral challenge. Public education campaigns, citizen science initiatives, environmental storytelling, and eco-literacy programs can help shape attitudes and foster stewardship (Srisathan et al., 2024).

Examples include:

- The Jane Goodall Institute's Roots & Shoots program
- School garden biodiversity clubs
- Media campaigns for wildlife protection

Changing behavior requires understanding psychological, cultural, and economic drivers—linking knowledge to action.

4.4 Forest Governance and Community Forestry

Forests remain central to global biodiversity conservation, yet they are also contested spaces where ecological priorities intersect with economic pressures and local rights. Traditional top-down protection strategies have often proven inadequate, sometimes excluding the very communities that sustain these ecosystems. In response, innovative governance models—such as community forestry, co-management arrangements, and international frameworks like REDD+ (Reducing Emissions from Deforestation and Degradation)—have emerged as more inclusive alternatives.

For example, Nepal's community forestry programs have enabled local groups to regenerate degraded hillsides while deriving livelihoods from forest products. Similarly, Indigenous territories in the Amazon consistently record lower rates of deforestation than surrounding lands, demonstrating the effectiveness of customary governance. These cases highlight that participatory forest governance is not only ecologically effective but also socially just, ensuring that conservation strategies align with local realities while contributing to global sustainability goals.

5. CHALLENGES AND LIMITATIONS IN BIODIVERSITY CONSERVATION

Despite growing awareness and a wide range of strategies, biodiversity conservation continues to face significant and persistent challenges. These obstacles are complex, interlinked, and often driven by broader economic, social, and political forces. Understanding the limitations and threats to biodiversity is essential, not to discourage action but to strengthen and adapt conservation efforts in a rapidly changing world (Hoffmann, 2022).

5.1 Habitat Fragmentation and Land-Use Change

One of the most pressing threats to biodiversity is the fragmentation and conversion of natural habitats. Expanding cities, intensive agriculture, mining, and infrastructure, such as highways and dams, divide once-continuous ecosystems into smaller, isolated patches. This disrupts animal migration, gene flow, and ecosystem processes, making species more vulnerable to extinction.

To address this, several approaches are being employed:

- Integrated land-use planning considers ecological needs alongside development goals to strike a balance between conservation and human activity.
- Conservation easements—legal agreements that limit land use while maintaining private ownership—are used in countries like the United States to protect vital habitats on working lands.
- Agro ecology and wildlife-friendly farming promote agroforestry, organic farming, and the preservation of hedgerows and wetlands to maintain biodiversity on agricultural lands.

Habitat connectivity and more innovative land use are essential to ensure species survive and adapt as their surroundings change.

5.2 Climate Change

Climate change adds another layer of complexity to biodiversity conservation. Shifts in temperature, precipitation, and seasonal cycles disrupt the delicate timing of ecological processes, such as flowering, migration, and breeding. Species may be forced to shift their ranges uphill or toward the poles in search of suitable conditions (Rubenstein et al., 2023). Others may not move fast enough or find themselves trapped in fragmented habitats with nowhere to go.

Addressing this requires climate-adaptive conservation, which includes:

- Resilient, protected area networks that are large, connected, and span different climate zones to allow species movement.
- Assisted migration, the intentional relocation of species to more suitable habitats, is being debated as a last-resort measure for critically endangered species facing climate threats.
- Nature-based solutions provide dual benefits by mitigating climate change and protecting biodiversity.

Integrating climate and conservation policy is no longer optional—it is essential for the survival of countless species and ecosystems.

5.3 Invasive Species

Invasive species are non-native organisms that, when introduced—often through human activities—outcompete or displace native species, thereby disrupting the balance of ecosystems. They can alter food webs, change fire regimes, and even modify entire landscapes. Islands and freshwater ecosystems are especially vulnerable (Mayfield et al., 2021).

Effective invasive species management includes:

- Prevention and strict biosecurity measures at borders, ports, and shipping hubs to stop introductions before they happen.
- Early detection and rapid response systems are needed to identify and contain new invasions before they spread.
- Biological control is when carefully selected predators or diseases are used to manage invasives (though this must be done cautiously to avoid unintended impacts).
- Ecosystem restoration can help reestablish native species and stabilize ecosystems following the removal of invasive species.

Fighting invasive species is an ongoing battle that requires vigilance, public awareness, and scientific innovation.

5.4 Unsustainable Exploitation and Illegal Trade

Overusing natural resources—whether through unsustainable logging, overfishing, hunting, or mining—continues to be a significant contributor to biodiversity loss. The illegal wildlife trade, often driven by international demand for exotic pets, traditional medicine, or luxury goods, has pushed many species to the brink (Mozer & Prost, 2023).

Example: Elephants and rhinos have been heavily targeted for ivory and horn, while pangolins have become the most trafficked mammals worldwide due to demand for their scales.

Tackling this issue demands a multi-pronged approach:

- Strengthened law enforcement, including ranger training, better surveillance technologies, and judicial support to prosecute offenders.
- Alternative livelihoods, such as ecotourism, sustainable agriculture, or handicrafts, can reduce dependence on exploitative activities.
- As seen in partnerships like the INTERPOL Wildlife Crime Group, global cooperation is crucial for tracking and dismantling international trafficking networks.

The fight against illegal exploitation must also address poverty, inequality, and demand-side behaviors to be genuinely effective.

5.5 Funding Gaps and Economic Pressures

Despite the scale of the biodiversity crisis, conservation remains severely underfunded. According to the United Nations, an estimated $700 billion annual gap exists between current biodiversity funding and what is needed to halt biodiversity loss (United Nations Environment Program. (2025).

This shortfall hampers efforts to manage PAs, invest in restoration, and support community conservation initiatives.

To bridge the gap, new financial models are being explored:

- Mainstreaming biodiversity into national budgets, treating conservation as an essential public service rather than a luxury.
- Public-private partnerships that bring together governments, NGOs, and businesses to invest in sustainable landscape management.
- Market-based instruments, such as biodiversity credits, green bonds, and PES, offer innovative ways to mobilize funding. However, they must be carefully designed to avoid exploitation or unintended harm.

Example: The Global Environment Facility (GEF) and the UNDP's BIOFIN initiative support countries in developing financing strategies that align with national biodiversity goals.

Ultimately, securing the future of biodiversity will depend on political will and societal commitment to investing in the planet's health, not just for the sake of nature but also for our own.

5.6 Agro-Biodiversity and Food Systems

Agro-biodiversity—the diversity of crops, livestock, soil organisms, and associated ecosystems—represents an often-overlooked but vital dimension of conservation. Beyond its ecological value, agro-biodiversity underpins food security, cultural heritage, and climate resilience. Diverse farming systems support ecosystem services, such as pollination, pest regulation, and soil fertility, while providing dietary diversity and reducing dependence on monocultures that are vulnerable to shocks.

Examples abound: The Satoyama landscapes of Japan illustrate how traditional land-use mosaics integrate forests, paddy fields, and wetlands to sustain both biodiversity and local livelihoods. Similarly, community seed banks in South Asia and Africa preserve traditional crop varieties, ensuring resilience to pests, diseases, and climatic extremes. By linking conservation to food systems, agro-biodiversity underscores the principle that protecting biodiversity is not only about wilderness but also about the working landscapes that nourish human societies.

6. CASE STUDIES IN BIODIVERSITY CONSERVATION

Learning from real-world examples provides powerful insights into the successes and ongoing challenges of biodiversity conservation. Although each case study represents a different geography, culture, and strategy, they all highlight a common truth: Conservation is most effective when it integrates ecology with community values, policy innovation, and long-term vision.

6.1 The Western Ghats, India: Community Forest Stewardship

The Western Ghats stretch along the western edge of India, sheltering some of the planet's most biologically rich ecosystems. This UNESCO World Heritage Site harbors thousands of plant and animal species—many of which are found nowhere else—and is deeply intertwined with the lives and traditions of local Indigenous communities.

In the Indian state of Kerala, local participation in conservation has taken on innovative forms. One key initiative is the People's Biodiversity Register (PBR), a community-driven effort to document local biodiversity and traditional ecological knowledge. These registers recognize community expertise and serve as tools for planning, education, and legal protection.

Sacred groves—forest patches preserved for spiritual reasons—are another example of deeply rooted conservation ethics. Traditionally managed by local communities, these groves often contain rare and endemic species, serving as reservoirs of genetic diversity (Wadhwa, 2023).

Moreover, Joint Forest Management (JFM) programs have enabled communities to co-manage forest areas with government agencies. This shared responsibility model fosters local stewardship while providing sustainable livelihoods, such as harvesting non-timber forest products (Thoker et al., 2024).

6.2 Costa Rica: A National Conservation Model

Once facing some of the highest deforestation rates in Latin America, Costa Rica has become a global leader in environmental restoration and sustainability. Beginning in the 1980s, the country reversed environmental decline through a blend of policy innovation, ecological awareness, and economic incentives.

A cornerstone of Costa Rica's strategy is its extensive network of PAs, which now covers approximately 25% of the national territory. These protected zones, ranging from lush rainforests to mangrove estuaries, are managed in conjunction with the achievement of the Sustainable Development Goals.

Another game-changing policy has been the PES program, which offers financial compensation to landowners who conserve forests, protect watersheds, or restore degraded lands. This improvement in biodiversity and boost to rural economies (Mamedes et al., 2023).

Ecotourism plays a significant role in the nation's conservation economy. With millions of visitors each year attracted to its natural beauty and wildlife, tourism revenues support conservation and local livelihoods (Samal & Dash, 2023).

Costa Rica's political commitment distinguishes it. The constitution enshrines the right to a healthy environment and biodiversity in national development planning.

6.3 Yellowstone National Park, USA: Rewilding with Wolves

In the vast landscapes of Yellowstone National Park, a bold ecological experiment unfolded in the 1990s: the reintroduction of gray wolves, which had been absent from the region for nearly 70 years due to eradication efforts. Their return had far-reaching consequences, revealing the power of trophic cascades—ecological chains triggered by apex predators.

With wolves hunting elk (wapiti) again, overgrazing pressures on vegetation, such as willow and aspen, were reduced. This allowed riparian habitats to recover, benefiting other species such as songbirds, beavers, and amphibians. Beavers, in turn, helped reshape river systems by building dams and ponds, thereby supporting aquatic biodiversity and regulating water flow.

This case demonstrated that rewilding—allowing nature to restore its balance—can lead to systemic changes when keystone species are reintroduced.

Lesson: Apex predators play a critical role in maintaining ecosystem health, and their recovery can trigger cascading ecological benefits that go far beyond individual species (Ordiz et al., 2021).

6.4 Brazil's Amazon: Indigenous Territories and Biodiversity

The Amazon Rainforest, often referred to as the "lungs of the planet," is a vital carbon sink and home to an extraordinary array of biodiversity, as well as hundreds of Indigenous communities. Surprisingly, research has shown that Indigenous-managed lands often outperform government-designated parks in curbing deforestation and protecting biodiversity.

These territories, governed under customary practices and traditional knowledge, support sustainable livelihoods while preserving large tracts of intact forest. However, this success is now under threat.

Recent years have seen policy rollbacks, increased illegal logging, and land invasions, putting both Indigenous rights and biodiversity at risk. Despite legal protections under Brazil's constitution and international agreements, enforcement remains weak, and political pressures persist.

The solution lies in law, recognizing and strengthening Indigenous leadership, securing land rights, and supporting territorial governance with technical and legal assistance.

Lesson: Indigenous stewardship is not just a moral imperative but an ecological necessity for preserving the Amazon and stabilizing the global climate.

6.5 Rwanda: Conservation and Economic Renewal

Few countries illustrate the link between biodiversity and national development as powerfully as Rwanda. After the devastation of the 1994 genocide, Rwanda made a remarkable recovery—not just economically but also in its relationship with nature.

At the heart of this renewal is Volcanoes National Park, home to the endangered mountain gorilla. Rwanda's government prioritized conservation as a pillar of economic growth, using gorilla tourism as a key strategy. A portion of tourism revenues is returned to surrounding communities to fund schools, healthcare, and infrastructure, building public support for conservation (Cohen-Shacham et al., 2016).

The country also adopted a "Green Growth" strategy, integrating environmental protection into climate adaptation, agriculture, and poverty reduction. Investments in reforestation, clean energy, and sustainable land use have helped balance human development with ecological integrity.

7. EMERGING TRENDS AND FUTURE DIRECTIONS IN BIODIVERSITY CONSERVATION

As the world confronts intensifying environmental crises, from climate change to mass species extinction, biodiversity conservation is undergoing a dynamic transformation. New approaches are emerging—more integrated, inclusive, and technology-driven—responding to ecological imperatives and societal needs. This chapter examines the innovative trends shaping the future of conservation and outlines where we are headed next.

7.1 Nature-Based Solutions (NbS): Working with Nature for a Sustainable Future

NbS are becoming a cornerstone of global sustainability efforts (Figure 2). Instead of relying solely on engineered or technological fixes, NbS harnesses the power of healthy ecosystems to address human challenges such as climate change, disaster risk, food security, and public health (Debele et al., 2023).

Figure 2.

These solutions recognize that thriving ecosystems are beneficial not only for wildlife but also for human well-being.

Examples of Nature-Based Solutions include:

- Mangrove restoration to buffer coastlines from storm surges and erosion while providing nursery habitats for fish.
- Urban green spaces, such as parks and green roofs, reduce heat in cities, improve air quality, and support mental and physical health (Figure 2).
- Wetland conservation for natural flood control and water purification is often more cost-effective than artificial flood control measures.

NbS are now being actively promoted in international frameworks, including the UN Framework Convention on Climate Change (UNFCCC) and the CBD. However, for NbS to succeed, it must be designed with equity and ecological integrity in mind, not just as green window dressing.

7.2 Urban Biodiversity and Green Infrastructure: Greening the Cities of Tomorrow

As more people live in cities than ever before, urban areas have become crucial for biodiversity conservation. Traditionally seen as ecological deserts, cities are reimagined as spaces where nature and urban life coexist.

Urban biodiversity contributes to resilience, liveability, and justice, offering green refuges in dense areas, supporting pollinators, and reducing the urban heat island effect (Wang et al., 2024).

Innovative urban conservation strategies include:

- Green roofs and vertical gardens provide habitat while insulating buildings and lowering energy use.
- Native landscaping and wildlife corridors connect fragmented habitats, encouraging urban species such as birds, butterflies, and small mammals to thrive.
- Biodiverse architecture, where buildings themselves become hosts for life, incorporates features such as bird-friendly glass or bat boxes.

7.3 Digital Technologies and Conservation Intelligence

The digital revolution is rapidly reshaping the tools available for conservation. From satellite imaging to smartphone apps (Figure 3), technology is enhancing our ability to monitor, protect, and understand biodiversity in unprecedented ways (Hariyani et al., 2025).

Figure 3.

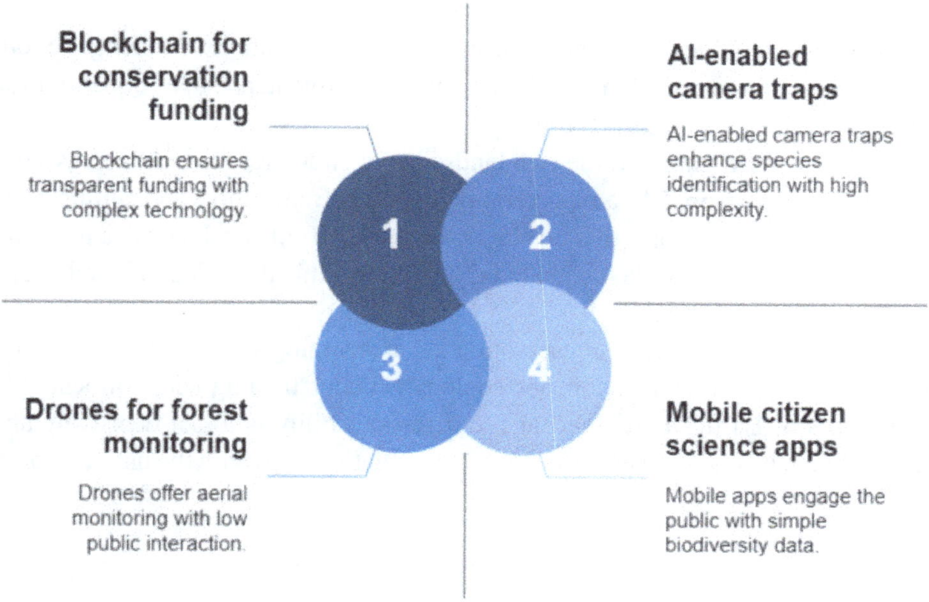

These advances fall under a growing field known as "conservation intelligence." Emerging technologies include:

- Drones that provide aerial imagery for forest health monitoring, anti-poaching patrols, and habitat mapping (Figure 3).
- AI-enabled camera traps are capable of identifying species automatically from thousands of images.
- Mobile citizen science apps, such as iNaturalist and eBird, empower the public to record and share biodiversity data.
- Blockchain for transparent and traceable conservation funding, ensuring accountability in complex projects.

These tools enhance efficiency, scalability, and public engagement, but they also raise important ethical and equity concerns, including who controls the data and whose knowledge is prioritized.

8. CONCLUSION

Conserving biodiversity is one of the most urgent and complex challenges of our time. It demands systemic thinking, long-term commitment, and deep collaboration across disciplines, sectors, and cultures.

This chapter has explored the rich landscape of strategies for biodiversity conservation, ranging from PAs and restoration to policy frameworks and cutting-edge technologies. It underscores that no single solution will suffice. Instead, a portfolio of approaches tailored to the context and guided by ethical, ecological, and social considerations is essential.

The future of biodiversity conservation lies in bridging science with values, local knowledge with global action, and ecological understanding with political will. As stewards of Earth's living heritage, our responsibility is to act decisively and compassionately, restoring ecosystems and our relationships with the natural world.

REFERENCES

Adams, V. M., Chauvenet, A. L. M., Stoudmann, N., Gurney, G. G., Brockington, D., & Kuempel, C. D. (2023). Multiple-use PAs are critical to equitable and effective conservation. *One Earth*, *6*(9), 1173–1189. DOI: 10.1016/j.oneear.2023.08.011

Artelle, K. A., Zurba, M., Bhattacharyya, J., Chan, D. E., Brown, K., Housty, J., & Moola, F. (2019). Supporting resurgent Indigenous-led governance: A nascent mechanism for just and effective conservation. *Biological Conservation*, *240*, 108284. DOI: 10.1016/j.biocon.2019.108284

Bishop, A., Roth, R., McGregor, D., Moola, F., & Nitah, S. (2025). Catalyzing transformative change in conservation: Lessons learned from a decolonial conservation partnership. *Conservation & Society*, *23*(1), 24–36. DOI: 10.4103/cs.cs_9_24

Bollig, M. (2024). Wildlife corridors in a Southern African conservation landscape: The political ecology of multispecies mobilities along the arteries of anthropogenic conservation. *Anthropology Southern Africa*, *47*(2), 216–235. DOI: 10.1080/23323256.2024.2327467 PMID: 39262499

Bontempi, A., Venturi, P., Del Bene, D., Scheidel, A., Zaldo-Aubanell, Q., & Maneja Zaragoza, R. (2023). Conflict and conservation: On the role of PAs for environmental justice. *Global Environmental Change*, *82*, 102740. DOI: 10.1016/j.gloenvcha.2023.102740

Børresen, S. T., Ulimboka, R., Nyahongo, J., Ranke, P. S., Skjaervø, G. R., & Røskaft, E. (2022). The role of education in biodiversity conservation: Can knowledge and understanding alter locals' views and attitudes towards ecosystem services? *Environmental Education Research*, *29*(1), 148–163. DOI: 10.1080/13504622.2022.2117796

Cairney, P., Timonina, I., & Stephan, H. (2023). How can policy and policymaking foster climate justice? A qualitative systematic review. *Open Research Europe*, *3*, 51. DOI: 10.12688/openreseurope.15719.2 PMID: 38106639

Cazzolla Gatti, R. (2025). Ecological peace corridors: A new conservation strategy to protect human and biological diversity. *Biological Conservation*, *302*, 110947. DOI: 10.1016/j.biocon.2024.110947

Christie, M. R., & Knowles, L. L. (2015). Habitat corridors facilitate genetic resilience irrespective of species dispersal abilities or population sizes. *Evolutionary Applications*, *8*(5), 454–463. doi: 12255DOI: 10.1111/Eva

Chukwuka, A. V., Adegboyegun, A. D., & Adeogun, A. O. (2025). Marine PAs of Africa: Current challenges and future opportunities in conservation. *Aquatic Sciences*, *87*(3), 61. DOI: 10.1007/s00027-025-01188-w

Cohen-Shacham, E., Walters, G., Janzen, C., & Maginnis, S. (Eds.). (2016). *Nature-based solutions to address global societal challenges*. IUCN., DOI: 10.2305/IUCN.CH.2016.13.en

Debele, S. E., Leo, L. S., Kumar, P., Sahani, J., Ommer, J., Bucchignani, E., Vranić, S., Kalas, M., Amirzada, Z., Pavlova, I., Shah, M. A. R., Gonzalez-Ollauri, A., & Di Sabatino, S. (2023). Nature-based solutions can help reduce the impact of natural hazards: A global analysis of NBS case studies. *The Science of the Total Environment*, *902*, 165824. DOI: 10.1016/j.scitotenv.2023.165824 PMID: 37527720

Dwivedi, Y. K., Hughes, L., Kar, A. K., Baabdullah, A. M., Grover, P., Abbas, R., Andreini, D., Abumoghli, I., Barlette, Y., Bunker, D., Kruse, L. C., Constantiou, I., Davison, R. M., De', R., Dubey, R., Fenby-Taylor, H., Gupta, B., He, W., Kodama, M., & Wade, M. (2022). Climate change and COP26: Are digital technologies and information management part of the problem or the solution? An editorial reflection and call to action. *International Journal of Information Management*, *63*, 102456. DOI: 10.1016/j.ijinfomgt.2021.102456

Edwards, D. P., Cerullo, G. R., Chomba, S., Worthington, T. A., Balmford, A. P., Chazdon, R. L., & Harrison, R. D. (2021). Upscaling tropical restoration to deliver environmental benefits and socially equitable outcomes. *Current Biology*, *31*(19), R1326–R1341. DOI: 10.1016/j.cub.2021.08.058 PMID: 34637743

Fergus, P., Chalmers, C., Longmore, S., & Wich, S. (2024). Harnessing artificial intelligence for wildlife conservation. *Conservation*, *4*(4), 685–702. DOI: 10.3390/conservation4040041

Gerstner, B. E., & Zarnetske, P. L. (2025). Evaluating the effectiveness of PAs and community-managed lands in capturing multiple dimensions of frugivorous biodiversity in the Tropical Andes. *Biological Conservation*, *302*, 110904. DOI: 10.1016/j.biocon.2024.110904

Glasser, H. (2011). Naess's deep ecology: Implications for the human prospect and challenges for the future. *Inquiry (Oslo, Norway)*, *54*(1), 52–77. DOI: 10.1080/0020174X.2011.542943

Gorghiu, G., Bîzoi, M., Gorghiu, L. M., Buruleanu, C. L., & Suduc, A.-M. (2024). Rewilding as a multifaceted concept and emerging approach: The Romanian experience. *Sustainability (Basel)*, *16*(4), 1645. DOI: 10.3390/su16041645

Gu, Z., & Zeng, M. (2024). The use of artificial intelligence and satellite remote sensing in land cover change detection: Review and perspectives. *Sustainability (Basel), 16*(1), 274. DOI: 10.3390/su16010274

Hariyani, D., Hariyani, P., & Mishra, S. (2025). Digital technologies for the Sustainable Development Goals. *Green Technologies and Sustainability, 3*(3), 100202. DOI: 10.1016/j.grets.2025.100202

Heikkinen, H. L. T., Huttunen, R., Mahon, K., & Kemmis, S. (2024). Beyond an anthropocentric view of praxis: Towards education for planetary well-being. *Environmental Education Research, 30*(7), 1147–1160. DOI: 10.1080/13504622.2024.2326460

Hoffmann, S. (2022). Challenges and Opportunities of Area-Based Conservation in Achieving Biodiversity and Sustainability Goals. *Biodiversity and Conservation, 31*(2), 325–352. DOI: 10.1007/s10531-021-02340-2

Hong, S. K. (2013). Biocultural diversity conservation for island and islanders: Necessity, goal, and activity. *Journal of Marine and Island Cultures, 2*(2), 102–106. DOI: 10.1016/j.imic.2013.11.004

Hoppit, G., Schmidt, D. N., Brazier, P., Mieszkowska, N., & Pieraccini, M. (2022). Are marine PAs an adaptation measure against climate change impacts on coastal ecosystems? A UK case study. *Nature-Based Solutions, 2*, 100030. DOI: 10.1016/j.nbsj.2022.100030

Hughes, A. C., & Grumbine, R. E. (2023). The Kunming-Montreal Global Biodiversity Framework: What's next? *Nature Ecology & Evolution, 7*(1), 10–12. DOI: 10.1038/s41559-022-01927-9 PMID: 36396970

Jaisankar, I., Velmurugan, A., & Sivaperuman, C. (2018). Biodiversity conservation: Issues and strategies for the tropical islands. In Sivaperuman, C., Velmurugan, A., Singh, A. K., & Jaisankar, I. (Eds.), *Biodiversity and climate change adaptation in tropical islands* (pp. 525–552). Academic Press., DOI: 10.1016/B978-0-12-813064-3.00019-3

Jiao, Y., Ding, Y., Zha, Z., & Okuro, T. (2019). Crises of biodiversity and ecosystem services in Satoyama landscape of Japan: A review on the role of management. *Sustainability (Basel), 11*(2), 454. DOI: 10.3390/su11020454

Kasso, M., & Balakrishnan, M. (2013). *Ex situ* conservation of biodiversity with particular emphasis to Ethiopia. *International Scholarly Research Notices, 985037*, 1–11. Advance online publication. DOI: 10.1155/2013/985037

Khoroshev, A. (2020). Landscape-ecological approach to spatial planning as a tool to minimize socio-ecological conflicts: Case study of agro landscape in the taiga zone of Russia. *Land (Basel)*, *9*(6), 192. DOI: 10.3390/land9060192

Kok, M. T. J., & Ludwig, K. (2022). Understanding international non-state and subnational actors for biodiversity and their possible contributions to the post-2020 CBD global biodiversity framework: Insights from six international cooperative initiatives. *International Environmental Agreement: Politics, Law and Economics*, *22*(1), 1–25. DOI: 10.1007/s10784-021-09547-2

Kopnina, H., Gray, J., Lynn, W., Heister, A., & Srivastava, R. (2022). Uniting ecocentric and animal ethics: Combining non-anthropocentric approaches in conservation and the care of domestic animals. *Ethics, Policy & Environment*, *26*(2), 265–286. DOI: 10.1080/21550085.2022.2127295

Koricha, H. G., & Adem, M. J. (2024). Investigated the role of community-based approaches for biodiversity conservation and socio-economic development in Bale Mountains National Park, Southeast Ethiopia. *Scientific Reports*, *14*(1), 12241. DOI: 10.1038/s41598-024-60177-5 PMID: 38806458

Lausche, B. (2011). *Guidelines for PAs legislation*. IUCN.

Mamedes, I., Guerra, A., Rodrigues, D. B. B., Garcia, L. C., Godoi, R. de F., & Oliveira, P. T. S. (2023). Brazilian payment for environmental services programs emphasize water-related services. *International Soil and Water Conservation Research*, *11*(2), 276–289. DOI: 10.1016/j.iswcr.2023.01.001

Martinez, D. J., Cannon, C. E. B., McInturff, A., Alagona, P. S., & Pellow, D. N. (2023). Back to the future: Indigenous relationality, kincentricity and the North American model of wildlife management. *Environmental Science & Policy*, *140*, 202–207. DOI: 10.1016/j.envsci.2022.12.010

Mayfield, A. E.. (2021). Impacts of invasive species in terrestrial and aquatic systems in the United States. In Poland, T. M., Patel-Weynand, T., Finch, D. M., Miniat, C. F., Hayes, D. C., & Lopez, V. M. (Eds.), *Invasive species in forests and rangelands of the United States* (pp. 5–39). Springer., DOI: 10.1007/978-3-030-45367-1_2

Morand, S., & Lajaunie, C. (2018). A brief history of the links between health and biodiversity. In Morand, S., & Lajaunie, C. (Eds.), *Biodiversity and health* (pp. 1–14). Elsevier., DOI: 10.1016/B978-1-78548-115-4.50001-9

Mozer, A., & Prost, S. (2023). An introduction to illegal wildlife trade and its effects on biodiversity and society. *Forensic Science International. Animals and Environments*, *3*, 100064. DOI: 10.1016/j.fsiae.2023.100064

Murray, B. G. (2017). Plant Diversity, Conservation, and Use. In Thomas, B., Murray, B. G., & Murphy, D. J. (Eds.), *Encyclopedia of applied plant sciences* (2nd ed., pp. 289–308). Academic Press., DOI: 10.1016/B978-0-12-394807-6.00047-2

Oestreich, W. K., Mercier, E., Holt, D. E., Duarte, C. M., Klinck, H., Parsons, M. J. G., Schlosser, E., & Širović, A. (2024). Listening to animal behavior to understand changing ecosystems. *Trends in Ecology & Evolution*, *39*(10), 961–973. DOI: 10.1016/j.tree.2024.06.007 PMID: 38972787

Offord, C. A. (2017). Germplasm conservation. In Thomas, B., Murray, B. G., & Murphy, D. J. (Eds.), *Encyclopedia of applied plant sciences* (2nd ed., pp. 281–288). Academic Press., DOI: 10.1016/B978-0-12-394807-6.00046-0

Olivadese, M., & Dindo, M. L. (2025). Ancient wisdom in modern tourism: Sustainable solutions from Greek and Roman literature. *Land (Basel)*, *14*(1), 109. DOI: 10.3390/land14010109

Ordiz, A., Aronsson, M., Persson, J., Støen, O.-G., Swenson, J. E., & Kindberg, J. (2021). Effects of human disturbance on terrestrial apex predators. *Diversity*, *13*(2), 68. DOI: 10.3390/d13020068

Pascoe, S., Anthony, K., Scheufele, G., & Pears, R. J. (2024). Identifying coral reef restoration objectives: A framework. *Ocean and Coastal Management*, *251*, 107081. DOI: 10.1016/j.ocecoaman.2024.107081

Pelizzon, A. (2025). Anthropocentrism redefined. In *Ecological jurisprudence: Contemporary environmental law and policy* (pp. 61–107). Springer., DOI: 10.1007/978-981-96-0173-8_3

Piccolo, J. J., Taylor, B., Washington, H., Kopnina, H., Gray, J., Alberro, H., & Orlikowska, E. (2022). "Nature's contributions to people" and people's moral obligations to nature. *Biological Conservation*, *270*, 109572. DOI: 10.1016/j.biocon.2022.109572

Reynolds, S. A., Beery, S., Burgess, N., Burgman, M., Butchart, S. H. M., Cooke, S. J., Coomes, D., Danielsen, F., Di Minin, E., Durán, A. P., Gassert, F., Hinsley, A., Jaffer, S., Jones, J. P. G., Li, B. V., Mac Aodha, O., Madhavapeddy, A., O'Donnell, S. A. L., Oxbury, W. M., & Sutherland, W. J. (2025). The potential for AI to revolutionize conservation: A horizon scan. *Trends in Ecology & Evolution*, *40*(2), 191–207. DOI: 10.1016/j.tree.2024.11.013 PMID: 39694720

Rishan, S. T., Kline, R. J., & Rahman, M. S. (2023). Applications of environmental DNA (eDNA) to detect subterranean and aquatic invasive species: A critical review on the challenges and limitations of eDNA metabarcoding. *Environmental Advances*, *12*, 100370. DOI: 10.1016/j.envadv.2023.100370

Rubenstein, M. A., Weiskopf, S. R., Bertrand, R., Merow, C., Cook-Patton, S. C., & Trisos, C. H. (2023). Climate change and the global redistribution of biodiversity: Substantial variation in empirical support for expected range shifts. *Environmental Evidence*, *12*(1), 7. DOI: 10.1186/s13750-023-00296-0 PMID: 39294691

Salgotra, R. K., & Chauhan, B. S. (2023). Genetic diversity, conservation, and utilization of plant genetic resources. *Genes*, *14*(1), 174. DOI: 10.3390/genes14010174 PMID: 36672915

Samal, R., & Dash, M. (2023). Ecotourism, biodiversity conservation, and livelihoods: Understanding the convergence and divergence. *International Journal of Geoheritage and Parks*, *11*(1), 1–20. DOI: 10.1016/j.ijgeop.2022.11.001

Sarmah, B. (2023). Reinventing nature in Kaziranga National Park: A history of rhinoceros' conservation in Assam, 1948–1974. *Conservation & Society*, *21*(2), 133–143. DOI: 10.4103/cs.cs_143_21

Scarano, F. R. (2017). Ecosystem-based adaptation to climate change: Concept, scalability and a role for conservation science. *Perspectives in Ecology and Conservation*, *15*(2), 65–73. DOI: 10.1016/j.pecon.2017.05.003

Sintayehu, D. W. (2018). Impact of climate change on biodiversity and associated key ecosystem services in Africa: A systematic review. *Ecosystem Health and Sustainability*, *4*(9), 225–239. DOI: 10.1080/20964129.2018.1530054

Srisathan, W. A., Malai, K., Narathawaranan, N., Coochampoo, K., & Naruetharadhol, P. (2024). The impact of citizen science on environmental attitudes, environmental knowledge, environmental awareness, to pro-environmental citizenship behavior. *International Journal of Sustainable Engineering*, *17*(1), 360–378. DOI: 10.1080/19397038.2024.2354269

Stefanni, S., Mirimin, L., Stanković, D., Chatzievangelou, D., Bongiorni, L., Marini, S., Modica, M. V., Manea, E., Bonofiglio, F., del Río Fernández, J., Cukrov, N., Gavrilović, A., De Leo, F. C., & Aguzzi, J. (2022). Framing cutting-edge integrative deep-sea biodiversity monitoring via environmental DNA and optoacoustic augmented infrastructures. *Frontiers in Marine Science*, *8*, 806060. DOI: 10.3389/fmars.2021.797140

Sterling, E. J., Betley, E., Sigouin, A., Gomez, A., Toomey, A., Cullman, G., Malone, C., Pekor, A., Arengo, F., Blair, M., Filardi, C., Landrigan, K., & Porzecanski, A. L. (2017). Assessing the Evidence for stakeholder engagement in biodiversity conservation. *Biological Conservation*, *209*, 159–171. DOI: 10.1016/j.biocon.2017.02.008

Tafoya, K. A., Brondizio, E. S., Johnson, C. E., Beck, P., Wallace, M., Quirós, R., & Wasserman, M. D. (2020). Costa Rica's national PES scheme: Perspectives on their emergence, evolution, and effectiveness. *Frontiers in Environmental Science*, *8*, 580724. DOI: 10.3389/fenvs.2020.580724

Taylor, B., Chapron, G., Kopnina, H., Orlikowska, E., Gray, J., & Piccolo, J. J. (2020). The need for ecocentrism in biodiversity conservation. *Conservation Biology*, *34*(5), 1089–1096. DOI: 10.1111/cobi.13541 PMID: 32424955

Thoker, I. A., Bhat, M. S., Shah, S. A., Lone, F. A., & Jeelani, P. (2024). An appraisal of people's participation in the joint forest management program in the Kashmir Himalayas. *Forest Policy and Economics*, *166*, 103265. DOI: 10.1016/j.forpol.2024.103265

Tollefson, J. (2019, May 6). Humans are driving one million species to extinction: Landmark United Nations-backed report finds that agriculture is one of the biggest threats to Earth's ecosystems. *Nature*. https://www.nature.com/articles/d41586-019-01448-4

United Nations Environment Program. (2021, April 30). *A beginner's guide to ecosystem restoration*. Nature Action. https://www.unep.org/news-and-stories/story/beginners-guide-ecosystem-restoration

United Nations Environment Program. (2025, March 3). Governments adopt first global strategy to finance biodiversity: Implications for financial institutions. [News]. *Nature*.

Vallin, J., Raycraft, J., Bettermann, D., Kioko, J., Kissui, B. M., Koester, S., Lindsay, K., Ole, E. M., Ramirez, E., Spizuco, B., Loos, J., & Kiffner, C. (2025). Variable social-ecological indicators across a Tanzanian protected area network. *Biological Conservation*, *308*, 111214. DOI: 10.1016/j.biocon.2025.111214

Virtanen, P. K. (2012). *Indigenous youth in Brazilian Amazonia: Changing lived worlds*. Palgrave Macmillan. DOI: 10.1057/9781137266514

Wadhwa, A. (2023, August 28). Sacred groves, the secret wizards of conservation. *IUCN Blog*.

Wang, D., Xu, P.-Y., An, B.-W., & Guo, Q.-P. (2024). Urban green infrastructure: Bridging biodiversity conservation and sustainable urban development through adaptive management approach. *Frontiers in Ecology and Evolution*, *12*, 1440477. Advance online publication. DOI: 10.3389/fevo.2024.1440477

Wang, Z., Wang, T., Zhang, X., Wang, J., Yang, Y., Sun, Y., Guo, X., Wu, Q., Nepovimova, E., Watson, A. E., & Kuca, K. (2024). Biodiversity conservation in the context of climate change: Facing challenges and management strategies. *The Science of the Total Environment*, *937*, 173377. DOI: 10.1016/j.scitotenv.2024.173377 PMID: 38796025

Xu, Q., Qu, B., Zhao, P., Clement, S., Shi, Y., Jordan, K. P., Li, L., & Chen, Y. (2025). Temporal species stability underpins the link between biodiversity and human health. *Environmental Research*, *278*, 121728. DOI: 10.1016/j.envres.2025.121728 PMID: 40306449

Chapter 9
Addressing the Problems of Forest and Biodiversity in Developing Nations: A Conservation Conundrum

Ananya Pandey
https://orcid.org/0000-0002-2419-6314
Christ University, Bangalore, India

ABSTRACT

Forests and biodiversity are essential for ecological stability and human well-being, but they confront unprecedented threats in developing countries. This chapter investigates the multifaceted interactions among environmental decline, socio-economic stresses, inadequate governance, and global market forces that drive biodiversity loss and deforestation. It underscores the difficulties of reconciling development with conservation in areas characterized by poverty, land tenure disputes, and weak institutions. The chapter emphasizes the necessity for methods that are integrated, participatory, and equitable through an examination of local and global frameworks, such as community-based conservation, indigenous knowledge, and international agreements. Case studies demonstrate conservation initiatives that succeeded and others that did not, providing vital perspectives on the future of biodiversity governance. In conclusion, the chapter calls for global cooperation, inclusive policy frameworks, and context-specific strategies to ensure sustainable management of forests and biodiversity.

DOI: 10.4018/979-8-3373-3992-4.ch009

INTRODUCTION

Forests and biodiversity are at the base of life on Earth, providing vital life support systems that provide ecological equilibrium, climate stabilization and support livelihoods, particularly for people in developing countries. Approximately 31% of the world's land area is covered by forests, and these land areas produce critical ecosystem services that include carbon sequestration, water filtration, climate stabilization, and soil stabilization (FAO, 2020). Moreover, forests consist of more than 80% of all terrestrial species of animals, plants, and fungi, producing complex ecological relationships on these life support systems. Biodiversity refers to the variability of life across all levels of ecological organization and is the foundation of these life support systems (Cardinale et al., 2012). Biodiversity is essential to maintaining agricultural productivity, health resilience, and cultural identities. Biodiversity and forests are incredibly valuable global assets currently confronted with unprecedented pressure, especially in the Global South. Developing countries, many within the climate-rich tropical belt with high biodiversity, grapple with dual responsibilities: conserving their natural ecosystem services while pursuing economic development and poverty alleviation. Competing demands create a conservation dilemma, often forcing short-term decisions that undermine long-term ecological stability. According to FAO (2020), there is a cumulative loss globally of approximately 420 million hectares of forest between 1990 and 2020, primarily in developing tropical regions like the Amazon, Southeast Asia, and Central Africa. The 2019 IPBES report similarly estimated that about one million multimodal plant and animal species are now threatened with extinction, many to occur within decades, without transformative change (IPBES, 2019).

Over the last several decades, conservation initiatives have grown considerably due to global commitments, scientific understanding, and the escalating urgency of climate change. International initiatives like the Convention on Biological Diversity (CBD), the Aichi Biodiversity Targets, and the newly established Kunming-Montreal Global Biodiversity Framework have established ambitious objectives for protecting ecosystems, ensuring sustainable use, and achieving fair benefit-sharing. At the same time, initiatives such as REDD+ (Reducing Emissions from Deforestation and Forest Degradation) have sought to connect forest conservation efforts with climate finance and development priorities. While these frameworks indicate a shared international intent, the execution of conservation initiatives frequently does not meet expectations in developing countries because of governance shortcomings, financial limitations, and socio-political intricacies (Chhatre & Agrawal, 2009). A significant difference in global conservation efforts is the lack of uniformity in financial and institutional resources. Developed countries usually have more resources, cutting-edge technologies, and regulatory systems to safeguard their environments.

Conversely, developing nations frequently struggle with inadequate enforcement, a lack of technical know-how, and clashing socio-economic priorities. The World Bank (2021) points out that low-income countries dedicate only a tiny fraction of their GDP to environmental conservation, even though they contain some of the world's most vulnerable ecosystems. Additionally, conservation strategies in these contexts are often imposed from above, leaving out the communities whose livelihoods are most connected to natural ecosystems. This exclusion frequently results in conflicts, mismanagement, and unintended ecological repercussions. Transboundary factors like the worldwide demand for timber, palm oil, soy, and other natural commodities exacerbate the issue by driving deforestation and habitat destruction in developing countries. As an example, deforestation due to trade in nations such as Indonesia and Brazil demonstrates the degradation of local ecosystems to meet the demands of global markets, frequently involving little reinvestment in local conservation efforts or community development (Curtis et al., 2018). Moreover, the illegal wildlife trade, frequently organized via international networks, has led to a dramatic decline in animal populations and has hindered efforts to preserve biodiversity (Warchol, 2004). Simultaneously, numerous developing countries house dynamic traditional knowledge systems and community-driven conservation methods that present encouraging alternatives to conventional conservation models. Indigenous peoples and local communities (IPLCs) manage approximately 40% of the world's protected and ecologically intact regions, and evidence indicates that their territories often demonstrate better conservation outcomes than those of state-managed protected areas (Garnett et al., 2018). Nonetheless, these practices are often sidelined in national policy frameworks and omitted from decision-making processes.

The increasing acknowledgment of the relationship between biodiversity loss and climate change has heightened global discussions on nature-based solutions. These solutions promote the restoration and safeguarding of natural ecosystems as a sustainable and economical means to tackle various environmental challenges at once. However, to deploy such solutions effectively in developing countries, it is necessary to develop context-sensitive strategies that reconcile conservation goals with the needs of local development. This chapter aims to elucidate the complex challenges of conserving forests and biodiversity in developing countries. The main aim is to deliver a thorough examination of the systemic and structural barriers that impede conservation efforts in the Global South, while also pointing out avenues for sustainable and inclusive solutions. This chapter takes an interdisciplinary approach, incorporating perspectives from environmental science, development studies, law, economics, and political ecology to create a nuanced understanding of the conservation crisis in developing countries. It stresses the necessity of moving away from conservation models that are imposed from outside and driven by donors, toward those that are based in the local context, attuned to cultural sensitivities, and informed by ecological considerations. To sum up, as the world advances toward realizing the Sustainable Development Goals (SDGs)—especially SDG 15 (Life on Land) and SDG 13 (Climate Action)—it is urgent to reconceptualize conservation so that it harmonizes ecological integrity with human well-being. For developing nations, this entails formulating policies and practices that safeguard biodiversity while also empowering communities, honoring indigenous rights, and promoting sustainable development.

THE ECOLOGICAL SIGNIFICANCE OF FORESTS AND BIODIVERSITY

Forests and biodiversity are crucial for sustaining ecological balance, particularly in developing countries where natural resources are abundant and human pressures are high. These nations frequently host some of the planet's most affluent ecosystems, such as tropical rainforests and mangrove forests, which act as biodiversity hotspots. The ecological significance of these ecosystems transcends national boundaries, impacting global climate, water cycles, and species survival. It is essential for the development of successful conservation strategies to comprehend their ecological functions and the dangers they encounter.

Forest Coverage and Biodiversity Hotspots in Developing Nations

A considerable part of the world's biological diversity is found in developing countries, which are mainly situated in the Global South. As stated by Myers et al. (2000), biodiversity hotspots are areas that harbor a large number of endemic species but face considerable danger from human actions. Out of the 36 acknowledged global biodiversity hotspots, most are located in Africa, Asia, and Latin America—regions generally classified as developing areas. These comprise the Amazon Basin, the Congo Rainforest, the Indo-Burma area, and the Sundaland forests. The most biologically diverse terrestrial ecosystems are tropical rainforests. Although they account for only around 6% of the Earth's surface, they are home to over 50% of its species (Barlow et al., 2016). Among the most crucial areas for biodiversity conservation are the Amazon rainforest in South America, the Congo Basin in Africa, and the Southeast Asian rainforests, including those found in Indonesia and Malaysia. A wide variety of flora and fauna, much of it not yet scientifically documented, is housed in these forests.

Mangrove forests, located along the coasts of tropical and subtropical regions, are equally vital. Though frequently underappreciated, they function as nurseries for many marine species, help avert coastal erosion, and are vital carbon sinks. According to Alongi (2012), mangrove ecosystems in nations such as Bangladesh, Indonesia, and Nigeria are vital for maintaining biodiversity and supporting the livelihoods of coastal communities.

Forest cover in developing countries, despite its importance, faces severe threats from deforestation, infrastructure development, and agricultural expansion. The Food and Agriculture Organization (FAO, 2020) reports that the area of global forests is still decreasing, with South America and Africa experiencing the greatest loss rates. This trend jeopardizes not just biodiversity but also the ecosystem services crucial for human survival. Ecosystem Services Forests and biodiversity offer a multitude of ecosystem services essential for sustaining life on Earth. According to the Millennium Ecosystem Assessment (2005), these services are typically divided into provisioning, regulating, supporting, and cultural services. Considering their rich natural resources and ecological functions, developing countries play a crucial role in sustaining these services. A vital regulating service that forests provide is carbon sequestration. Through photosynthesis, forests absorb atmospheric carbon dioxide and store it in biomass and soils, functioning as carbon sinks. According to Pan et al. (2011), tropical forests located in developing nations hold about 25% of the Earth's terrestrial carbon. Not only does their destruction liberate carbon that was previously stored, it also diminishes the Earth's ability to counteract climate change.

Forest ecosystems play a vital role in conserving soil and water. Forests play a crucial role in regulating hydrological cycles, sustaining groundwater levels, and averting soil erosion. Forests in mountainous areas like the Himalayas and Andes help avert landslides and floods by stabilizing soil and absorbing excess rainwater (Bruijnzeel, 2004). The disruption of these ecological functions due to the loss of forest cover can result in a rise in natural disasters and a decline in agricultural productivity. Healthy ecosystems are largely responsible for pollination, a supporting service that is essential for food security. A variety of pollinator species, such as bees, birds, and bats, depend on forest habitats for nesting and foraging. They guarantee the reproduction of more than 75% of global food crops (IPBES, 2016). With forest degradation, there is a reduction in pollinator populations, which has a direct impact on agricultural yields and the food supply in urban and rural settings alike. Forest-derived medicinal resources hold great worth, especially regarding traditional medicine and the development of pharmaceuticals. According to estimates by the World Health Organization (WHO, 2013), around 80% of people in developing countries depend on traditional herbal medicine for their primary healthcare needs. Moreover, numerous contemporary medications are sourced from forest species, including quinine (derived from Cinchona trees) and paclitaxel (from yew trees). The reduction of biodiversity diminishes the genetic pool available for future medical discoveries. Another aspect of forest ecosystems that is often neglected is their cultural and spiritual value. In numerous indigenous and local communities, forests serve not only as economic resources but also as sacred spaces that embody spiritual beliefs and cultural identity. This non-physical connection nurtures a feeling of custodianship and enduring accountability for conservation efforts. Local ecological knowledge and cosmologies, as emphasized by Escobar (2008), are essential for sustainable environmental management.

KEY DRIVERS OF FOREST AND BIODIVERSITY LOSS

In developing countries, biodiversity and forest ecosystems are increasingly stressed by a variety of human-induced and systemic factors. These drivers are intricately linked, and their combined impact threatens ecological resilience and sustainable development. The subsequent sections examine critically the four main drivers: deforestation and land use change, illegal wildlife trade and poaching, climate change, and governance and policy gaps.

Deforestation and Land Use Change

In many developing countries, one of the greatest dangers to forests and biodiversity is land use change-induced deforestation. Forests are being rapidly cleared to accommodate agricultural expansion, industrial activities, and infrastructure development. In tropical regions of Africa, Asia, and Latin America, large-scale commercial agriculture—mainly the cultivation of commodities like palm oil, soy, and cattle—is the leading cause of deforestation (Curtis et al., 2018). Furthermore, although it is frequently neglected, smallholder agriculture plays a significant role in forest degradation, especially in subsistence economies where local populations rely directly on land resources for their survival.

Logging, whether legal or illegal, is a key factor in forest degradation. In areas like the Congo Basin and Amazon, logging is frequently conducted with little supervision, leading to habitat fragmentation and the loss of species reliant on undisturbed forest canopies. In addition, mining activities have caused considerable deforestation in areas rich in biodiversity, such as the Andes and Southeast Asia. This has resulted in the displacement of plant and animal species and the contamination of water bodies through runoff (Sonter et al., 2017). Land use change is aggravated further by urbanization. The swift urban growth in developing countries results in the encroachment upon peri-urban forests and wetlands. This process displaces biodiversity and eliminates essential ecosystem services like water filtration and climate regulation, which are vital for urban resilience (Seto et al., 2012). These processes have a cumulative effect that compromises the integrity of forests and disrupts ecological corridors, resulting in cascading losses of biodiversity.

Illegal Wildlife Trade and Poaching

In numerous developing countries, illegal wildlife trade and poaching are widespread, mainly fueled by significant global demand and ineffective enforcement measures. Wildlife trafficking, a multi-billion-dollar illegal industry, preys on endangered and at-risk species such as elephants, rhinos, pangolins, and exotic birds. This trade flourishes in areas where poverty, absence of alternative means of making a living, and corruption generate conducive circumstances for poaching networks (Nijman, 2010). In Africa, the illegal hunting of bushmeat and ivory has led to the destruction of megafauna populations. Meanwhile, in Southeast Asia, species like the orangutan and tiger face imminent extinction as a result of habitat destruction and illegal trapping. Besides its direct impact on species that are targeted, the wildlife trade frequently causes ripple effects throughout entire ecosystems by disturbing predator-prey dynamics and genetic diversity (Ripple et al., 2016). In developing countries, enforcement continues to be one of the weakest aspects of biodiversity conservation. Conservation officers often face issues of inadequate funding, lack of proper training, and exposure to hazardous situations. In addition, traffickers may act without fear of consequences due to inadequately prepared or compromised judicial systems.

Climate Change

Climate change is a widespread force that intensifies current dangers to biodiversity and forest ecosystems, especially in the Global South. As temperatures rise, precipitation patterns shift, and extreme weather becomes more common, habitats are being transformed. Species must migrate, adapt, or risk extinction (Pecl et al., 2017). In tropical rainforests, droughts driven by climate change have resulted in higher tree mortality rates and increased susceptibility to fires. Meanwhile, montane species are experiencing habitat loss as they move to higher altitudes in search of viable ecosystems that are diminishing. In coral reef ecosystems like the Coral Triangle, ocean acidification and warming are causing extensive bleaching and a collapse of biodiversity. Likewise, salinization, erosion, and loss of productivity are affecting wetlands, mangroves, and estuaries in developing countries, jeopardizing the food security of local communities and the reproductive cycles of aquatic species (Díaz et al., 2019). Natural disasters like floods and cyclones, which wreak havoc on biodiversity and forest cover, become more frequent and severe due to climate change. Environmental stress that results from this interacts with other dangers like deforestation, increasing the vulnerability of ecosystems to degradation and diminishing their recovery capacity. Therefore, climate change functions not only as a driver but also as a risk multiplier across ecological and socio-economic systems.

Governance and Policy Gaps

The enduring issue of weak governance and fragmented policy frameworks in developing countries underlies many of the direct drivers of biodiversity loss. A number of countries do not have comprehensive and enforceable environmental laws. Even where such laws are in place, their implementation is frequently obstructed by factors such as inadequate institutional capacity, insufficient political commitment, and corruption (Kashwan, 2017). A frequent problem is the overlap or contradiction between development policies and environmental laws. As an example, plans for developing infrastructure could go ahead without sufficient evaluations of environmental impact or consultations with the community. Furthermore, the lack of security regarding land tenure acts as a deterrent to long-term conservation investments and contributes to an uptick in land grabbing, especially in regions of significant ecological importance (Larson et al., 2013). Another significant obstacle to effective conservation is corruption. In multiple instances, logging or mining permits are issued unlawfully or due to political coercion, frequently aimed at protected regions. The effectiveness of conservation programs is further constrained by inadequate coordination among government agencies, as well as a deficiency in transparency and accountability. Moreover, international funding for conservation is marked by inconsistency and a frequent misalignment with local needs.

SOCIO-ECONOMIC PRESSURES IN DEVELOPING COUNTRIES

In their quest for socio-economic development, developing nations encounter a paradox regarding the necessity of conserving forests and biodiversity. Due to limited resources and growing populations, these countries frequently place a higher priority on economic development than on ecological sustainability. The pressures stemming from poverty, dependence on resources, demographic expansion, and the conflict between conservation and development all contribute to a complex set of challenges for safeguarding the environment.

Dependence on Resources and Poverty

In developing countries, poverty is a key factor behind environmental degradation. Often, communities that rely on forests utilize natural resources due to necessity rather than carelessness. In rural and marginalized regions where other means of livelihood are scarce or absent, forests provide essential fuelwood, food, medicinal resources, and income (Angelsen et al., 2014). Consequently, poverty fosters a reliance on forests that frequently results in overexploitation, particularly when sustainable extraction methods or sufficient regulatory systems are lacking. As an example, in Sub-Saharan Africa and certain regions of South Asia, households depend largely on fuelwood for cooking, which plays a major role in deforestation (FAO, 2020). Likewise, the expansion of smallholder agriculture into forested regions is frequently a means of survival when land is scarce, food insecurity is present, and agricultural productivity is low. Though these actions are economically rational for individual households, they cumulatively degrade ecosystems and threaten biodiversity.

The issue is further exacerbated by government responses that penalize local forest use without offering viable alternatives, leading to conflict and resentment. Unless there is significant investment in sustainable livelihoods—like agroforestry, ecotourism, or value-added non-timber forest products—poor communities are unlikely to adopt conservation objectives (Shackleton et al., 2007). Therefore, alleviating the pressure on forest ecosystems requires that poverty be tackled with inclusive and diversified economic strategies.

Population Growth

The rapid increase in population puts more pressure on natural ecosystems, especially in areas where governance and infrastructure do not keep pace with demographic changes. Numerous developing nations contend with elevated fertility rates, influenced by elements like restricted availability of reproductive health services, early matrimonial practices, and socio-cultural values that promote large families (UNFPA, 2023). The demographic pressure that results from this increases the demand for arable land, water, timber, and energy—resources that are frequently obtained from forested and biodiverse areas. A primary driver of deforestation is the growth of agriculture to satisfy the dietary requirements of increasing population numbers. As per the Intergovernmental Science-Policy Platform on Biodiversity and Ecosystem Services (IPBES, 2019), alterations in agricultural land use have been the main factor behind biodiversity decline in tropical areas.In nations including Indonesia and Brazil, the clearing of forests for cash crops like palm oil and soybeans has transformed ecologically rich forests into monoculture plantations, leading to a significant reduction in species richness.

Additionally, the urbanization linked to population growth plays a role in habitat fragmentation and degradation. When forests are transformed into roads, residential areas, and industrial zones, it undermines ecological connectivity and brings about pollution and the introduction of invasive species. Without robust urban planning and environmental protections, cities in developing countries often grow randomly, invading peri-urban forests and designated protected areas (Seto et al., 2012). Thus, population growth management via education, empowering women, and providing reproductive health services can serve as a crucial means of conserving biodiversity.

Development vs. Conservation Dilemma

Nations in the developing world often face a dilemma between pursuing their development objectives and fulfilling conservation obligations. Infrastructure endeavors like roadways, dams, mining activities, and industrial corridors are essential for economic development and alleviating poverty. Nevertheless, such advancements frequently come at the cost of forests and biodiversity hotspots, especially when environmental impact assessments are insufficient or neglected for the sake of economic expediency (Laurance et al., 2014). The extractive industries, which include mining, oil and gas exploration, and logging, pose especially significant issues. In areas such as the Congo Basin and the Amazon, the extraction of minerals and hydrocarbons has caused direct deforestation and created a network of access roads that expose previously unreachable regions to illegal logging and hunting. Foreign investment and government incentives often support these industries, leading to a political economy that prioritizes short-term revenue over long-term ecological health. Moreover, the development of infrastructure can lead to the displacement of local and indigenous populations, whose traditional knowledge and management of land are vital for conservation efforts. As an illustration, the building of dams can entail moving communities, which interrupts social ties and ecological connections. This has a detrimental effect on resilience in both human and environmental spheres (Brockington & Igoe, 2006). In such contexts, conservation is viewed not as a collaborative effort but as a top-down imposition, which diminishes its effectiveness and legitimacy. Addressing this dilemma requires development planning that integrates economic activities with ecological limits. This encompasses implementing strategic environmental assessments, funding green infrastructure, and advocating for ecosystem-based development methods. Payments for ecosystem services and climate adaptation funds are examples of international cooperation and financing that can assist developing countries in pursuing low-carbon and nature-positive growth trajectories (World Bank, 2021).

INDIGENOUS AND LOCAL COMMUNITY PERSPECTIVE

Indigenous and local communities (ILCs) have historically acted as guardians of biodiversity and ecosystems, managing forest resources with generational knowledge systems based on sustainable practices. Their profound interdependence with the natural environment has influenced the development of traditional ecological knowledge (TEK), which provides adaptive strategies for conservation, resilience, and restoration. Nevertheless, contemporary conservation policies—frequently motivated by top-down models that exclude certain groups—have overlooked these contributions in many instances. This has led to socio-environmental conflicts and displacement.

Role of Traditional Knowledge in Conservation

Traditional ecological knowledge includes practices, beliefs, and understandings that have been transmitted across generations, based on observations of local ecosystems. TEK integrates the spiritual, cultural, and ecological dimensions of the natural world, in contrast to Western science, which often compartmentalizes nature (Berkes, 2018). As an illustration, rotational farming employed by numerous indigenous communities boosts soil fertility and biodiversity, whereas sacred groves safeguarded for spiritual purposes act as vital biodiversity sanctuaries (Gadgil, Berkes, & Folke, 1993).

Indigenous communities in nations like India, Brazil, and Kenya have cultivated complex knowledge systems about medicinal plants, the seasonal behavior of animals, and the management of forests. This knowledge supports their livelihoods and complements scientific conservation methods. Research indicates that in the Amazon, indigenous territories experience lower deforestation rates compared to state-managed protected areas (Nepstad et al., 2006). In a similar vein, the Zabo system in northeast India—an indigenous method for water harvesting and forest protection—exemplifies nature-based solutions that adhere to sustainability and biodiversity conservation principles (Pretty et al., 2009). Initiatives for community-based conservation (CBC) that focus on traditional knowledge have demonstrated potential. These models view local communities as stakeholders and rights-holders, not just as beneficiaries. Such methods have been embraced in Namibia's communal conservancies, the ejido forestry system in Mexico, and India's Joint Forest Management (JFM) program. These initiatives integrate traditional governance structures and often yield improved ecological and social outcomes (Ostrom, 1990). The potential of TEK is not fully utilized in many national conservation strategies, which are largely shaped by Western scientific paradigms and development priorities. Traditional knowledge systems are often marginalized by formal education and state policies, risking their erosion (Brosius, Tsing, & Zerner, 2005). To incorporate TEK into formal conservation, it is essential to acknowledge indigenous ways of knowing, establish co-management frameworks, and ensure fair involvement in decisions.

Community-Based Stewardship and Ecological Responsibility

Numerous ILCs see themselves as embedded within a larger ecological continuum, where achieving harmony with nature is both a cultural standard and a moral necessity. Therefore, stewardship practices are integrated into daily life, with traditional rituals, taboos, and community sanctions playing crucial roles in resource governance. As an example, the Maasai of East Africa manage grazing patterns using traditional knowledge that helps to prevent overuse and degradation (Goldman, 2003). The indigenous "muyong" system in the Philippines combines agroforestry with forest conservation, ensuring biodiversity and subsistence needs are met (Pulhin et al., 2007). These practices embody an ecological ethic that emphasizes sustainability over extraction in the long term. There is a growing body of research confirming that when ILCs possess legal land tenure and resource control, biodiversity outcomes are enhanced. The World Resources Institute reported in 2016 that ensuring indigenous land rights results in enhanced carbon storage and reduced deforestation, highlighting the importance of local governance for conservation initiatives.

Conflict and Displacement: The Dark Side of Conservation

Despite these contributions, ILCs often suffer the consequences of conservation policies that are created and enforced without their consent. Historically, the establishment of national parks and protected areas has adhered to a "fortress conservation" model that excludes human presence in order to preserve "pristine" nature (Adams & Hutton, 2007). This method has resulted in the forced removal of millions of indigenous peoples, compromising their rights and breaking their cultural and spiritual connections to the land. For instance, the eviction of the Sengwer people from the Embobut Forest in Kenya, purportedly to safeguard the watershed, has faced global condemnation for breaching human rights and not acknowledging community stewardship (Human Rights Watch, 2018). Similar patterns have been observed in Asia, Latin America, and Africa, where conservation initiatives financed by international donors have led to community displacement, the criminalization of traditional practices, and worsened poverty (Dowie, 2009). Displacement often leads to physical dislocation as well as the erosion of social and cultural identity. The loss of access to forests jeopardizes food security, spiritual practices, and the transmission of traditional knowledge (Pandey & Joseph, 2025). Furthermore, displaced populations can be moved to degraded areas or the outskirts of cities, leading to their marginalization and restricted economic opportunities (Colchester, 2004). The clash between worldwide conservation aims and local human rights has led to a rising push for "rights-based conservation." This change in model stresses the inclusion of indigenous viewpoints, safeguarding customary land rights, and adhering to the principles of Free, Prior and Informed Consent (FPIC). Tools like the United Nations Declaration on the Rights of Indigenous Peoples (UNDRIP) and the Aichi Biodiversity Targets promote conservation strategies that include communities. Nonetheless, the implementation of these principles is inconsistent and complicated by political difficulties. Recognizing the conservation contributions of local and indigenous communities is essential for both ethical and ecological reasons. Their traditional knowledge systems provide methods for sustainable resource use and biodiversity management that have been proven effective over time. Simultaneously, excluding them from conservation planning perpetuates injustices and diminishes the effectiveness of environmental protection. For developing countries facing ecological degradation and social inequality, it is essential to move toward conservation models that are inclusive, participatory, and culturally respectful. This necessitates the legal acknowledgment of indigenous land rights, fair governance structures, and authentic collaborations that respect local expertise and leadership. Only then can conservation efforts be genuinely just, sustainable, and transformative.

POLICY AND LEGAL INSTRUMENTS

In developing nations, the preservation of forests and biodiversity depends on ecological awareness and community involvement, as well as the robustness, coherence, and enforceability of policy and legal frameworks. Legal instruments at both the national and international levels are crucial in determining conservation priorities, implementing regulatory mechanisms, and facilitating sustainable development. Nevertheless, the lack of sufficient institutional capacity and difficulties in execution persistently weaken conservation initiatives throughout the Global South, along with legislative gaps.

National Conservation Policies

In light of escalating ecological crises and global commitments, numerous developing nations have established national forest laws, biodiversity acts, and environmental protection regulations. The objective of these legal frameworks is to regulate land use, protect endangered species, promote sustainable resource management, and maintain ecological integrity. Examples include India's Forest Conservation Act (1980) and Biological Diversity Act (2002), Brazil's Forest Code, and Kenya's Forest Conservation and Management Act (2016), which establish the legal framework for forest and biodiversity governance in their respective jurisdictions (Pandey, 2025; Pandey, 2026). These laws often require the creation of protected areas, promote reforestation and afforestation initiatives, and stipulate penalties for illegal logging, poaching, and habitat destruction.

Several countries have notably included provisions for community participation and benefit-sharing, exemplified by India's recognition of community forest rights through the Scheduled Tribes and Other Traditional Forest Dwellers (Recognition of Forest Rights) Act, 2006, which empowers forest-dwelling communities. Nonetheless, in spite of these forward-thinking actions, serious gaps persist. Environmental laws often lack enforceability because of ambiguous wording, overlapping jurisdictions, or a lack of sufficient decentralization of authority (Phelps, Carrasco, Webb, Koh, & Pascual, 2014). In addition, these laws frequently overlook the socio-economic circumstances surrounding conservation efforts, which can result in tension between conservation objectives and development needs. The lack of integration among forestry, wildlife, and environmental departments contributes to a lack of policy coherence. At the same time, poor coordination between central and local governments worsens inefficiencies.

International Agreements

Numerous conservation initiatives in developing nations have been catalyzed by international legal instruments that offer normative frameworks, technical support, and financial incentives. The CBD, which was adopted in 1992, is still a cornerstone of global biodiversity governance. It calls on countries to create National Biodiversity Strategies and Action Plans (NBSAPs), incorporate biodiversity into wider policies, and encourage fair benefit-sharing. Developing nations especially gain from the Global Environment Facility (GEF), which funds biodiversity initiatives that align with CBD objectives (UNEP, 2020). A further important mechanism is REDD+ (Reducing Emissions from Deforestation and Forest Degradation), which promotes forest conservation in developing nations by associating emission reductions with monetary incentives. REDD+ initiatives aimed at forest protection, community engagement, and governance improvement have been adopted by nations such as Indonesia, Colombia, and the Democratic Republic of Congo (Angelsen, 2017). The Convention on International Trade in Endangered Species of Wild Fauna and Flora (CITES) offers a legal structure to address the illegal wildlife trade, which poses a significant threat to biodiversity in numerous developing countries. CITES manages the international movement of threatened species through a system of permits and trade controls, but its effectiveness often relies on the domestic enforcement capabilities of member states. Although these instruments are important, there are many challenges to their implementation. Numerous developing nations find it challenging to synchronize national policies with international obligations because of a lack of technical know-how, bureaucratic resistance, and conflicting policy priorities. Furthermore, some treaties (for example, the CBD) do not have binding enforcement mechanisms. This reduces their practical effectiveness. It is common for countries to submit national reports and strategies without guaranteeing compliance on the ground or measurable impact (Morgera & Tsioumani, 2010).

Implementation Challenges

A country's ability to administer, finance, and engage politically has a significant impact on how effective conservation policy is. In numerous developing countries, the lack of funding is one of the most urgent limitations. Due to limited budgetary allocations for forest departments and biodiversity conservation, authorities often rely on foreign aid or NGO support. Long-term planning and monitoring are complicated by this financial fragility (Watson et al., 2014). Absence of monitoring and evaluation systems further hinders implementation. Even when policies are in place, a lack of strong mechanisms for tracking biodiversity loss, assessing forest degradation, or ensuring compliance leads to policy failure. Due to a lack of technical skills and insufficient infrastructure, digital monitoring tools such as satellite surveillance and geographic information systems (GIS) are not used to their full potential. Weaknesses within institutions also contribute significantly. Corruption, patronage networks, and inadequate staffing in environmental agencies undermine the effectiveness of regulations. Moreover, the emphasis on short-term economic growth—often driven by external pressures such as debt servicing or global trade—results in the weakening or reversal of conservation efforts (Geist & Lambin, 2002). Due to discretionary powers and legal loopholes, extractive industries are frequently able to bypass environmental regulations, particularly in forest areas that are politically marginalized. In order to tackle these challenges, developing nations require not just more robust legislation, but also enhanced institutional capacity, international collaboration, and creative financing solutions. To convert legal norms into effective conservation outcomes, it is vital to enhance environmental governance through transparency, capacity-building, and participatory frameworks.

STRATEGIES FOR MOVING FORWARD

Forests and biodiversity in developing countries can only be preserved through strategies that take into account ecological and socio-economic realities on multiple levels. This section suggests a future-oriented roadmap that incorporates nature-based solutions, empowers communities, utilizes technological innovations, and taps into financial mechanisms. These strategies are interdependent pillars that work together and are essential for fostering resilient and inclusive conservation models.

Integrative Approaches: Nature-Based Solutions and Landscape-Level Conservation

Nature-based solutions (NbS) are gaining recognition as effective strategies for combating biodiversity loss while bolstering climate resilience and promoting sustainable livelihoods. NbS encompasses the sustainable management and restoration of ecosystems to tackle societal challenges such as climate change, food security, and disaster risk (Seddon et al., 2021). In developing countries, agroforestry — which involves incorporating trees into agricultural landscapes — acts as a vital nature-based solution (NbS) that reconciles productivity with ecological responsibility. According to the FAO (2019), it improves soil fertility, manages erosion, and creates habitats for indigenous species, all while giving farmers access to various sources of income. Conservation at the landscape level provides a comprehensive view that goes beyond individual protected areas, advocating for the management of whole ecosystems along with their socio-ecological contexts. This method takes into account ecological connectivity, watershed integrity, and transboundary conservation initiatives. As an example, the Central African Forest Initiative (CAFI) encourages various countries to work together in order to conserve one of the largest carbon sinks globally via sustainable land-use planning (UNDP, 2020). Landscape-level conservation promotes consistency between environmental safeguarding and socio-economic progress by synchronizing conservation initiatives with wider development objectives.

Community Empowerment: Legal Recognition and Participatory Governance

The involvement of local and indigenous communities is crucial for the success of forest and biodiversity conservation. These communities frequently hold profound ecological insights and have traditionally served as guardians of biodiversity. However, conservation policies have often excluded them, leading to conflict, displacement, and ineffective outcomes (Rights and Resources Initiative [RRI], 2020). The first step to empowering communities is the legal acknowledgment of their land rights and traditional knowledge. Secure land tenure has been associated with reduced deforestation and improved resource management outcomes, especially in indigenous territories (Blackman et al., 2017). In Brazil and Colombia, for instance, the rates of deforestation on legally recognized indigenous lands are significantly lower than those in adjacent areas that lack such protections. Participatory governance mechanisms, such as community forest management, joint management of protected areas, and co-creation of conservation plans, guarantee the inclusion of local voices in decision-making processes. Models like ecotourism revenue-sharing and PES that ensure fair benefit-sharing increase local involvement and offer economic incentives for conservation efforts. These models turn local communities from passive subjects of conservation policy into active partners and beneficiaries.

Technology and Innovation: Digital Tools for Conservation

New technologies are changing the methods used to monitor, manage, and safeguard biodiversity and forest resources. Near real-time data on changes in forest cover, illegal logging, and habitat fragmentation are provided by remote sensing and satellite imagery. Instruments such as Global Forest Watch allow civil society and governments to monitor deforestation worldwide with a level of accuracy that has never been seen before (Hansen et al., 2013). Artificial intelligence (AI) and machine learning models are being utilized more and more for predicting biodiversity hotspots, detecting poaching activities via pattern recognition, and automating image classification in camera trap data (Lamba et al., 2019). In India, tools that utilize AI are being used to keep track of tiger populations and avert human-wildlife clashes. Mobile apps also contribute to the decentralization of conservation monitoring. Applications such as iNaturalist and Forest Watcher enable local communities, students, and citizen scientists to record species, report unlawful actions, and add information to biodiversity databases. This democratization of data collection enhances local capacity and fosters community involvement in conservation initiatives. To ensure respect for local knowledge systems and data sovereignty, the effective deployment of technology must be accompanied by capacity-building, inclusive access, and ethical safeguards.

Funding Mechanisms: Green Funds and Market-Based Solutions

In developing countries, sustainable financing continues to be a key challenge for conservation endeavors. Public funding is often restricted, and reliance on donor aid is not a sustainable solution over the long term. To close the conservation finance gap, innovative financing mechanisms are crucial. Forest and biodiversity projects receive vital assistance from green climate funds like the Global Environment Facility (GEF) and the Green Climate Fund (GCF). These funds help nations fulfill their obligations under international accords such as the Convention on Biological Diversity and the Paris Agreement. Financial incentives for the conservation of carbon-rich ecosystems are provided by market-based mechanisms like carbon credits, particularly through initiatives such as REDD+ (Reducing Emissions from Deforestation and Forest Degradation). Despite facing criticism regarding implementation gaps and equity concerns, REDD+ remains a promising tool if adapted to local contexts and governed transparently (Angelsen et al., 2018). Scalable solutions are also provided by public-private partnerships. Private sector involvement can enhance public initiatives by incorporating businesses into sustainable sourcing, supply chain certification (such as the Forest Stewardship Council), and conservation financing. In the end, a combination of blended finance, regulatory backing, and performance-based payments can drive sustainable and fair conservation results.

CONCLUSION

The conservation crisis affecting forests and biodiversity in developing countries is intricately linked to socio-economic, political, and ecological factors. This chapter has shown that although forests and biodiversity are essential ecosystems that support climate regulation, water cycles, cultural heritage, and livelihoods (MEA, 2005), they continue to degrade due to unsustainable development paths, weak governance, and global market pressures. The interconnected challenges—spanning deforestation, habitat fragmentation, illegal trade, climate change, and the marginalization of indigenous communities—highlight the complexity of achieving effective conservation in low- and middle-income countries. A major understanding that arises is that conservation problems in the Global South are inherently multidimensional. According to Geist & Lambin (2002), environmental degradation is not solely caused by ecological mismanagement; rather, it is closely linked to factors such as poverty, insecurity in land tenure, population growth, and conflicting development priorities. In numerous cases, conservation efforts are compromised by inconsistent policies, a lack of financial resources, and exclusionary top-down methods that neglect local conditions and indigenous knowledge systems (Chomba et al., 2016). Thus, safeguarding biodiversity needs to be integrated into wider frameworks of sustainable development, human rights, and social equity.

The case studies and analyses presented in this chapter further underscore that community-based approaches, bolstered by inclusive governance and benefit-sharing mechanisms, can lead to significant improvements in conservation outcomes. IPLCs frequently act as custodians of landscapes teeming with biodiversity and hold crucial ecological knowledge that can enhance scientific conservation strategies (Sobrevila, 2008). To achieve lasting conservation, it is essential to acknowledge and empower these stakeholders through legal reforms, participatory planning, and decentralized resource management. Internationally, frameworks like the Convention on Biological Diversity (CBD), REDD+, and the Sustainable Development Goals (especially SDG 15) provide normative guidance and financial support for conservation efforts (UNEP, 2020). The effectiveness of these frameworks, however, depends on their successful localization and incorporation into national development agendas. Furthermore, the integration of technological advancements—such as remote sensing, AI for biodiversity tracking, and mobile-based monitoring—provides promising resources for enhancing transparency, data collection, and enforcement in biodiversity management (Turner, Nakamura, & Dinetti, 2004). This moment calls for action. Solving the conservation dilemma in developing countries necessitates worldwide unity, with affluent nations backing ecological stewardship in areas rich in biodiversity but limited by economic constraints. This encompasses fair climate finance, transfers of technology, and trade mechanisms that do not sustain extractive dependencies. Meanwhile, local participation must be bolstered through democratic environmental governance, rights-based strategies, and grassroots education that fosters environmental literacy. Ultimately, comprehensive frameworks that integrate ecology, economy, culture, and justice should direct both policy and practice. We can only guarantee the conservation of forest ecosystems and biodiversity as vital foundations for human and planetary well-being, rather than just natural resources, through such integrative, multiscalar initiatives.

REFERENCES

Adams, W. M., & Hutton, J. (2007). People, parks and poverty: Political ecology and biodiversity conservation. *Conservation & Society*, *5*(2), 147–183.

Alongi, D. M. (2012). Carbon sequestration in mangrove forests. *Carbon Management*, *3*(3), 313–322. DOI: 10.4155/cmt.12.20

Angelsen, A., Jagger, P., Babigumira, R., Belcher, B., Hogarth, N. J., Bauch, S., Börner, J., Smith-Hall, C., & Wunder, S. (2014). Environmental income and rural livelihoods: A global-comparative analysis. *World Development*, *64*, S12–S28. DOI: 10.1016/j.worlddev.2014.03.006 PMID: 32405139

Angelsen, A., Martius, C., De Sy, V., Duchelle, A. E., Larson, A. M., & Pham, T. T. (2018). *Transforming REDD+: Lessons and new directions*. Center for International Forestry Research (CIFOR).

Barlow, J., França, F., Gardner, T. A., Hicks, C. C., Lennox, G. D., Berenguer, E., & Parry, L. (2016). The future of hyperdiverse tropical ecosystems. *Nature*, *559*(7715), 517–526. DOI: 10.1038/s41586-018-0301-1 PMID: 30046075

Berkes, F. (2018). *Sacred Ecology* (4th ed.). Routledge.

Blackman, A., Corral, L., Santos Lima, E., & Asner, G. (2017). Titling indigenous communities protects forests in the Peruvian Amazon. *Proceedings of the National Academy of Sciences of the United States of America*, *114*(16), 4123–4128. DOI: 10.1073/pnas.1603290114 PMID: 28373565

Brockington, D., & Igoe, J. (2006). Eviction for conservation: A global overview. *Conservation & Society*, *4*(3), 424–470.

Brosius, J. P., Tsing, A. L., & Zerner, C. (2005). *Communities and Conservation: Histories and Politics of Community-Based Natural Resource Management*. AltaMira Press.

Bruijnzeel, L. A. (2004). Hydrological functions of tropical forests: Not seeing the soil for the trees? *Agriculture, Ecosystems & Environment*, *104*(1), 185–228. DOI: 10.1016/j.agee.2004.01.015

Cardinale, B. J., Duffy, J. E., Gonzalez, A., Hooper, D. U., Perrings, C., Venail, P., Narwani, A., Mace, G. M., Tilman, D., Wardle, D. A., Kinzig, A. P., Daily, G. C., Loreau, M., Grace, J. B., Larigauderie, A., Srivastava, D. S., & Naeem, S. (2012). Biodiversity loss and its impact on humanity. *Nature*, *486*(7401), 59–67. DOI: 10.1038/nature11148 PMID: 22678280

Chhatre, A., & Agrawal, A. (2009). Trade-offs and synergies between carbon storage and livelihood benefits from forest commons. *Proceedings of the National Academy of Sciences of the United States of America, 106*(42), 17667–17670. DOI: 10.1073/pnas.0905308106 PMID: 19815522

Chomba, S., Nathan, I., Minang, P. A., & van Noordwijk, M. (2016). Illusions of empowerment? Questioning policy and practice of community forestry in Kenya. *Ecology and Society, 21*(3). Advance online publication. DOI: 10.5751/ES-07741-200302

Colchester, M. (2004). Conservation policy and indigenous peoples. *Environmental Science & Policy, 7*(3), 145–153. DOI: 10.1016/j.envsci.2004.02.004

Curtis, P. G., Slay, C. M., Harris, N. L., Tyukavina, A., & Hansen, M. C. (2018). Classifying drivers of global forest loss. *Science, 361*(6407), 1108–1111. DOI: 10.1126/science.aau3445 PMID: 30213911

Díaz, S., Settele, J., Brondízio, E. S., Ngo, H. T., Guèze, M., Agard, J., & Zayas, C. N. (2019). *Summary for policymakers of the global assessment report on biodiversity and ecosystem services of the Intergovernmental Science-Policy Platform on Biodiversity and Ecosystem Services*. IPBES.

Dowie, M. (2009). *Conservation Refugees: The Hundred-Year Conflict Between Global Conservation and Native Peoples*. MIT Press. DOI: 10.7551/mitpress/7532.001.0001

Escobar, A. (2008). *Territories of difference: Place, movements, life, redes*. Duke University Press.

FAO. (2019). *Agroforestry and climate change: Understanding the linkages*. Food and Agriculture Organization of the United Nations.

FAO. (2020). *Global Forest Resources Assessment 2020*. Food and Agriculture Organization of the United Nations. https://www.fao.org/forest-resources-assessment

Gadgil, M., Berkes, F., & Folke, C. (1993). Indigenous knowledge for biodiversity conservation. *Ambio, 22*(2-3), 151–156. PMID: 33566330

Garnett, S. T., Burgess, N. D., Fa, J. E., Fernández-Llamazares, Á., Molnár, Z., Robinson, C. J., & Watson, J. E. (2018). A spatial overview of the global importance of Indigenous lands for conservation. *Nature Sustainability, 1*(7), 369–374. DOI: 10.1038/s41893-018-0100-6

Geist, H. J., & Lambin, E. F. (2002). Proximate causes and underlying driving forces of tropical deforestation. *Bioscience, 52*(2), 143–150. DOI: 10.1641/0006-3568(2002)052[0143:PCAUDF]2.0.CO;2

Goldman, M. J. (2003). Partitioned nature, privileged knowledge: Community-based conservation in Tanzania. *Development and Change, 34*(5), 833–862. DOI: 10.1111/j.1467-7660.2003.00331.x

Hansen, M. C., Potapov, P. V., Moore, R., Hancher, M., Turubanova, S. A., Tyukavina, A., Thau, D., Stehman, S. V., Goetz, S. J., Loveland, T. R., Kommareddy, A., Egorov, A., Chini, L., Justice, C. O., & Townshend, J. R. (2013). High-resolution global maps of 21st-century forest cover change. *Science, 342*(6160), 850–853. DOI: 10.1126/science.1244693 PMID: 24233722

Human Rights Watch. (2018). *Kenya: Forest evictions violate rights*. Retrieved from https://www.hrw.org

IPBES. (2019). *Global assessment report on biodiversity and ecosystem services*. Intergovernmental Science-Policy Platform on Biodiversity and Ecosystem Services. https://ipbes.net/global-assessment

Kashwan, P. (2017). *Democracy in the Woods: Environmental Conservation and Social Justice in India, Tanzania, and Mexico*. Oxford University Press. DOI: 10.1093/acprof:oso/9780190637385.001.0001

Lamba, A., Chettri, N., & Sharma, E. (2019). Mapping the research knowledge for managing biodiversity in the Hindu Kush Himalaya. *Environmental Science & Policy, 94*, 21–31.

Larson, A. M., Barry, D., & Ram Dahal, G. (2013). *Tenure rights and access to forests: A training manual for research. Center for International Forestry Research.* CIFOR.

Laurance, W. F., Sayer, J., & Cassman, K. G. (2014). Agricultural expansion and its impacts on tropical nature. *Trends in Ecology & Evolution, 29*(2), 107–116. DOI: 10.1016/j.tree.2013.12.001 PMID: 24388286

Millennium Ecosystem Assessment (MEA). (2005). *Ecosystems and Human Well-being: Biodiversity Synthesis*. World Resources Institute.

Myers, N., Mittermeier, R. A., Mittermeier, C. G., da Fonseca, G. A., & Kent, J. (2000). Biodiversity hotspots for conservation priorities. *Nature, 403*(6772), 853–858. DOI: 10.1038/35002501 PMID: 10706275

Nepstad, D. C., Schwartzman, S., Bamberger, B., Santilli, M., Ray, D., Schlesinger, P., Lefebvre, P., Alencar, A., Prinz, E., Fiske, G., & Rolla, A. (2006). Inhibition of Amazon deforestation and fire by parks and indigenous lands. *Conservation Biology, 20*(1), 65–73. DOI: 10.1111/j.1523-1739.2006.00351.x PMID: 16909660

Nijman, V. (2010). An overview of international wildlife trade from Southeast Asia. *Biodiversity and Conservation, 19*(4), 1101–1114. DOI: 10.1007/s10531-009-9758-4

Ostrom, E. (1990). *Governing the Commons: The Evolution of Institutions for Collective Action*. Cambridge University Press. DOI: 10.1017/CBO9780511807763

Pan, Y., Birdsey, R. A., Fang, J., Houghton, R., Kauppi, P. E., Kurz, W. A., Phillips, O. L., Shvidenko, A., Lewis, S. L., Canadell, J. G., Ciais, P., Jackson, R. B., Pacala, S. W., McGuire, A. D., Piao, S., Rautiainen, A., Sitch, S., & Hayes, D. (2011). A large and persistent carbon sink in the world's forests. *Science, 333*(6045), 988–993. DOI: 10.1126/science.1201609 PMID: 21764754

Pandey, A. (2025). The Transformative Potential of Artificial Intelligence in Building Sustainable Agricultural Systems: Innovations, Challenges, and Future Directions. In *AI and Ecological Change for Sustainable Development* (pp. 161-182). IGI Global Scientific Publishing.

Pandey, A. (2026). Technological Innovation and Intellectual Property in Agriculture: A Critical Study With Reference to Farmers' Rights. In *Precision and Intelligence in Agriculture: Advanced Technologies for Sustainable Farming* (pp. 29-50). IGI Global Scientific Publishing.

Pandey, A., & Joseph, J. (2025). Leveraging Artificial Intelligence for Sustainable Development in Agriculture: An Exploratory Study. In *Artificial Intelligence and Data Science for Sustainability: Applications and Methods* (pp. 187-212). IGI Global Scientific Publishing.

Pecl, G. T., Araújo, M. B., Bell, J. D., Blanchard, J., Bonebrake, T. C., Chen, I. C., Clark, T. D., Colwell, R. K., Danielsen, F., Evengård, B., Falconi, L., Ferrier, S., Frusher, S., Garcia, R. A., Griffis, R. B., Hobday, A. J., Janion-Scheepers, C., Jarzyna, M. A., Jennings, S., & Williams, S. E. (2017). Biodiversity redistribution under climate change: Impacts on ecosystems and human well-being. *Science, 355*(6332), eaai9214. Advance online publication. DOI: 10.1126/science.aai9214 PMID: 28360268

Pretty, J., Adams, B., & Berkes, F.. (2009). The intersections of biological diversity and cultural diversity: Towards integration. *Conservation & Society*, *7*(2), 100–112. DOI: 10.4103/0972-4923.58642

Pulhin, J. M., Gevana, D. T., & Dizon, J. T. (2007). Integrating community-based forest management and protected area management in the Philippines. *Journal of Environmental Science and Management, 10*(1), 1–14.

Rights and Resources Initiative (RRI). (2020). *The state of Indigenous Peoples' and Local Communities' lands and territories: A technical review of a global dataset.*

Ripple, W. J., Abernethy, K., Betts, M. G., Chapron, G., Dirzo, R., Galetti, M., Levi, T., Lindsey, P. A., Macdonald, D. W., Machovina, B., Newsome, T. M., Peres, C. A., Wallach, A. D., Wolf, C., & Young, H. (2016). Bushmeat hunting and extinction risk to the world's mammals. *Royal Society Open Science, 3*(10), 160498. DOI: 10.1098/rsos.160498 PMID: 27853564

Seddon, N., Chausson, A., Berry, P., Girardin, C. A., Smith, A., & Turner, B. (2021). Understanding the value and limits of nature-based solutions to climate change and other global challenges. *Philosophical Transactions of the Royal Society B*, 375(1794), 20190120.

Seto, K. C., Güneralp, B., & Hutyra, L. R. (2012). Global forecasts of urban expansion to 2030 and direct impacts on biodiversity and carbon pools. *Proceedings of the National Academy of Sciences of the United States of America, 109*(40), 16083–16088. DOI: 10.1073/pnas.1211658109 PMID: 22988086

Shackleton, S., Shackleton, C., & Cousins, B. (2007). The role of land-based strategies in rural livelihoods: The contribution of arable production, animal husbandry and natural resource harvesting in communal areas in South Africa. *Development Southern Africa, 18*(5), 581–604. DOI: 10.1080/03768350120097441

Sobrevila, C. (2008). *The Role of Indigenous Peoples in Biodiversity Conservation.* World Bank.

Sonter, L. J., Ali, S. H., & Watson, J. E. M. (2017). Mining and biodiversity: Key issues and research needs in conservation science. *Proceedings of the Royal Society B: Biological Sciences*, 284(1861), 20162527. https://doi.org/DOI: 10.1098/rspb.2016.2527

Turner, W., Nakamura, T., & Dinetti, M. (2004). Global urbanization and the separation of humans from nature. *Bioscience, 54*(6), 585–590. DOI: 10.1641/0006-3568(2004)054[0585:GUATSO]2.0.CO;2

UNDP. (2020). *Central African Forest Initiative: Annual Progress Report.* United Nations Development Programme.

UNFPA. (2023). *State of World Population 2023: 8 billion lives, infinite possibilities.* United Nations Population Fund.

United Nations Environment Programme (UNEP). (2020). *Global Biodiversity Outlook 5.* https://www.cbd.int/gbo/gbo5/publication/gbo-5-en.pdf

Warchol, G. L. (2004). The transnational illegal wildlife trade. *Criminal Justice Studies, 17*(1), 57–73. DOI: 10.1080/0888431042000167933

WHO. (2013). *WHO traditional medicine strategy: 2014–2023*. World Health Organization. https://apps.who.int/iris/handle/10665/92455

World Bank. (2021). *Mobilizing Private Finance for Nature: Turning a Billion into a Trillion*. World Bank Group. https://www.worldbank.org/en/topic/environment/publication/mobilizing-private-finance-for-nature

World Bank. (2021). *The changing wealth of nations 2021: Managing assets for the future*. World Bank Group.

World Resources Institute. (2016). *Climate Benefits, Tenure Costs: The Economic Case for Securing Indigenous Land Rights*. WRI.

Chapter 10
Forest Microbiomes as Hidden Pillars of Biodiversity:
Roles, Threats, and Conservation Strategies

Anurag Yadav
https://orcid.org/0000-0001-6791-8869
Sardarkrushinagar Dantiawada Agricultural University, India

Kusum Yadav
University of Lucknow, India

ABSTRACT

Forest biodiversity conservation has largely overlooked the functionally paramount microbial communities, the forest microbiome—that underpin ecosystem health. This "invisible majority" of bacteria, fungi, and other microbes drives fundamental processes like nutrient cycling, soil formation, and plant growth, determining overall ecosystem functioning. However, these vital communities are threatened by anthropogenic pressures such as deforestation, land-use change, and climate change, leading to microbial diversity loss with cascading consequences for ecosystem stability. A paradigm shift is urgently needed to integrate microbial conservation into forest management and restoration, as ignoring this dimension limits the efficacy of current strategies. This review synthesizes our understanding of forest microbiome structure and function, evaluates major threats, and outlines technologies and policies for their conservation. We argue that stewarding forest microbiomes is a prerequisite for achieving global biodiversity and climate goals.

DOI: 10.4018/979-8-3373-3992-4.ch010

1. INTRODUCTION

Forest ecosystems have long been cherished and conserved primarily for their visible biodiversity, the majestic trees, charismatic megafauna, and vibrant understory that capture the public imagination and dominate conservation agendas. This perception, however, overlooks the vast and functionally paramount world of invisible biodiversity that resides within and around every plant, in every gram of soil, and on every leaf surface. This hidden dimension is known as the forest microbiome, an intricate assemblage of bacteria, archaea, fungi, protists, and viruses that collectively comprise the vast majority of life forms and genetic diversity within forest ecosystems (Bardgett & Van Der Putten, 2014). For decades, the limitations of culture-based methods rendered this world largely inaccessible, but the advent of high-throughput sequencing has unveiled a microbial universe of staggering complexity, forcing a fundamental re-evaluation of what constitutes biodiversity and how ecosystems function.

The ecological relevance of the forest microbiome cannot be overstated. These microorganisms are not merely passive inhabitants but are the primary engines of nearly all critical ecosystem processes. They are the master decomposers, breaking down complex organic matter and driving the biogeochemical cycling of essential elements like carbon, nitrogen, and phosphorus, thereby regulating soil fertility and nutrient availability for the entire plant community (Crowther et al., 2019). As ecosystem engineers, they create and stabilize soil structure, influencing water retention and aeration. As intimate partners with plants, they form symbioses, such as mycorrhizal networks, that are essential for nutrient uptake, stress mitigation, and inter-plant communication. The functional traits encoded within the collective microbial metagenome represent a massive reservoir of adaptive potential that underpins the resistance and resilience of the entire forest to environmental change (Van Der Heijden et al., 2008).

Despite this foundational importance, the microbial world has been largely ignored by mainstream conservation science, policy, and practice. A profound "microbial myopia" has led to a situation where management, restoration, and conservation strategies are designed and implemented with little to no consideration for the microscopic life that underpins their success (Averill et al., 2022). Restoration projects may fail because the soil lacks the specific microbial symbionts required by planted trees; the effectiveness of a protected area is judged by its charismatic fauna, while its soil health degrades unnoticed. This oversight is no longer a benign neglect but a critical flaw that jeopardizes the long-term success of conservation investments and misinterprets the very nature of ecosystem resilience. As forests face unprecedented threats from climate change, habitat loss, and pollution, a paradigm shift is necessary. The effective stewardship of our planet's forests requires a holistic perspective that explicitly recognizes and integrates the "invisible majority" into our conservation calculus.

This chapter aims to synthesize the current state of knowledge on forest microbiomes and advocate for their central placement within conservation and restoration frameworks. We begin by outlining the structure and immense diversity of microbial communities across global forest biomes and provide a detailed account of their indispensable ecological roles. We then critically evaluate the major anthropogenic threats from deforestation and pollution to climate change and biological invasions that are rapidly eroding this hidden biodiversity and highlight the cascading ecological consequences of its loss. Following a review of the powerful technological advances that now allow for unprecedented insight into these complex communities, we conclude by proposing a series of actionable strategies for integrating microbial ecology into research, management, policy, and practice. Ultimately, we argue that the long-term sustainability of forests is inextricably linked to the health and diversity of their microbial inhabitants, and that embracing this microbial perspective is essential for the future of biodiversity conservation.

2. STRUCTURE AND DIVERSITY OF FOREST MICROBIOMES

Forest ecosystems host a staggering diversity of microorganisms, which are organized into complex, structured communities from the soil to the canopy. The specific composition and richness of these microbial assemblages are determined by both the immediate physical environment of their niche and overarching global factors like climate and geography. This section explores the key constituents of these microbiomes and the large-scale patterns that govern their diversity across different forest types.

2.1 Overview of Forest Microbiome Constituents

The forest microbiome comprises a taxonomically vast and functionally critical assemblage of microorganisms, including bacteria, archaea, fungi, protists, and their associated viruses, which are organized into distinct communities across a range of ecological niches. The bulk soil and the rhizosphere, the zone of soil directly influenced by plant roots, represent the most concentrated hubs of microbial biomass and activity. Within these subterranean habitats, bacterial communities exhibit immense taxonomic diversity, while the mycobiome, particularly symbiotic arbuscular and ectomycorrhizal fungi, often dominates in terms of biomass and direct functional influence on plant nutrition and carbon cycling (Vasar et al., 2022). In contrast, the phyllosphere (leaf surfaces) and endosphere (internal plant tissues) constitute more specialized niches, harboring communities adapted to unique selective pressures such as high ultraviolet radiation, fluctuating moisture, and the host plant's immune system (Morella et al., 2020). Beyond the dominant bacteria and fungi, other constituents play pivotal roles. Archaea are key players in specific biogeochemical pathways like nitrification, especially in oligotrophic or acidic soils. Protists act as major consumers, shaping bacterial and fungal community structures through predation, while viruses, the most abundant biological entities, modulate host populations and serve as vectors for horizontal gene transfer, constantly introducing new genetic potential into the system (Zhang et al., 2012). **Table 1** provides a structured overview of the key microbial players across these habitats and their principal ecological roles.

Table 1. Key microbial taxa in different forest niches and their primary ecological functions

Forest Niche	Dominant Microbial Group	Key Phyla & Genera	Primary Ecological Function	Source(s)
Phyllosphere	Bacteria, Fungi	*Proteobacteria, Actinobacteria, Bacteroidota, Firmicutes, Ascomycota, Basidiomycota* Genera: *Pseudomonas, Bacillus, Methylobacterium*	Pathogen suppression, Nutrient acquisition (e.g., N-fixation), Phytohormone production, Initial decomposition of senesced leaves	(Shi et al., 2024; Thapa & Prasanna, 2018)
Litter	Fungi, Bacteria	*Ascomycota, Basidiomycota, Proteobacteria, Bacteroidetes* Genera: *Trichoderma, Aspergillus, Penicillium, Mucor*	Primary decomposition of cellulose, hemicellulose, and other labile organic matter; Nutrient mineralization	(Meng et al., 2024; Song et al., 2010; Ye et al., 2024)
Deadwood	Fungi	*Basidiomycota, Ascomycota* Genera: *Fomes, Ganoderma, Trametes* (White-rot); *Fomitopsis* (Brown-rot)	Decomposition of recalcitrant biopolymers (lignin, cellulose); Creation of habitat for other organisms; N-fixation by associated bacteria	(K et al., 2015; Tláskal et al., 2021)
Soil (Bulk)	Bacteria, Fungi, Archaea	*Acidobacteriota, Proteobacteria, Actinobacteria, Ascomycota, Basidiomycota, Thaumarchaeota* Genera: *Candidatus Solibacter, Streptomyces*	Decomposition of complex organic matter (humus); Soil aggregation and structure formation; Nutrient cycling (Nitrification, Denitrification)	(Huang et al., 2023; Singavarapu et al., 2023)
Rhizosphere	Bacteria, Fungi	*Proteobacteria, Firmicutes, Glomeromycota, Basidiomycota, Ascomycota* Genera: *Rhizobium, Bradyrhizobium, Burkholderia, Pseudomonas, Funneliformis, Cortinarius*	Symbiotic N-fixation, Phosphate solubilization, Mycorrhizal symbiosis (nutrient/water uptake), Production of plant growth-promoting substances, Protection from root pathogens	(Guo et al., 2020; Mendes et al., 2013)

2.2 Comparative Microbial Diversity Across Global Forests

Forest microbiome structure and diversity are strongly patterned by biogeography, exhibiting predictable shifts along latitudinal and climatic gradients that reflect co-evolved relationships with host plant communities and adaptation to local soil properties. Tropical forests, sustained by high temperatures and precipitation, are generally considered hotspots of microbial diversity. Their soils, though often highly weathered and acidic, host a vast array of bacterial taxa and are typically dominated by arbuscular mycorrhizal (AM) fungi. These AM fungi form symbioses with the immense diversity of tree species and are thought to facilitate rapid nutrient cycling, which supports high primary productivity (Delgado-Baquerizo et al., 2018; Vasar et al., 2022). In stark contrast, high-latitude boreal forests are characterized by lower overall alpha-diversity. These ecosystems are dominated by cold-adapted specialists, particularly ectomycorrhizal (EcM) fungi associated with a less diverse, coniferous-dominated plant community. EcM fungi are critical for acquiring nitrogen and phosphorus by "mining" the thick, recalcitrant organic horizons typical of these acidic, nutrient-limited environments (Bahram et al., 2018).

Temperate forests represent a dynamic transition zone between these extremes, where the balance between AM and EcM fungal guilds shifts in response to changes in dominant vegetation (e.g., broadleaf vs. conifer), soil type, and climate. This predictable global shift from AM-dominated tropics to EcM-dominated boreal zones, with temperate forests as the crossover region, is a fundamental rule in forest microbial biogeography, reflecting co-evolutionary histories between plants and their fungal symbionts and having profound implications for ecosystem carbon and nutrient cycling (Steidinger et al., 2019). The pronounced seasonality of temperate biomes also imposes strong temporal dynamics on microbial activity and composition, with distinct microbial assemblages flourishing under different seasonal conditions.

The Indian subcontinent serves as a powerful case study for these global patterns at a regional scale. Its dramatic eco-climatic gradient, which spans from tropical rainforests and coastal mangroves in the south to temperate broadleaf and alpine systems in the Himalayas, encapsulates a remarkable range of forest types in close proximity. This environmental heterogeneity fosters exceptionally high beta-diversity (the turnover of species between habitats), creating a rich mosaic of distinct soil microbial communities that likely harbor unique, endemic microbial lineages (Zhang et al., 2020). This uniqueness makes the region a natural laboratory for studying microbial adaptation and the response of forest microbiomes to diverse environmental drivers. Another study examined soil microbial diversity along a 2,500 km forest latitudinal gradient in China and found that environmental heterogeneity across the gradient significantly influenced soil microbial beta-diversity and community structure. These findings strongly support your point about eco-climatic gradients fostering high beta-diversity and endemic microbial lineages, making it the most appropriate citation for your passage.

3. ECOLOGICAL ROLES OF FOREST MICROBIOMES

The microbial communities within forests are not merely passive inhabitants but are master regulators of essential ecosystem processes. They drive foundational activities such as nutrient cycling and soil formation, directly support plant health and protection, and are ultimately integral to the overall resilience and stability of the forest. These multifaceted contributions underscore the microbiome's role as a cornerstone of ecosystem function.

3.1 Nutrient Cycling and Soil Fertility

The forest microbiome functions as the primary engine driving biogeochemical cycling, a role that directly underpins soil fertility and sustains the productivity of the entire ecosystem. Microorganisms are the principal agents of decomposition, responsible for the catabolism of complex organic matter—such as leaf litter, dead wood, and root exudates—into simpler, bioavailable inorganic forms. This process is a collaborative effort, often initiated by fungi whose powerful extracellular enzymes and hyphal networks allow them to break down recalcitrant polymers like cellulose and lignin. Bacteria subsequently metabolize the simpler organic byproducts, a division of labor that ensures the complete breakdown of organic inputs (Jansson & Hofmockel, 2020). This microbial-led mineralization is the fundamental process that liberates essential elements like carbon, nitrogen, and phosphorus from dead organic matter, making them available for uptake by plants and other microbes, thereby closing the ecosystem's nutrient loops (Crowther et al., 2019; Trivedi et al., 2013).

Beyond the immediate release of inorganic nutrients, the microbiome is integral to the formation and stabilization of soil organic matter (SOM), the long-term reservoir of soil fertility. Current understanding, articulated through concepts like the "microbial carbon pump," emphasizes that microbial biomass itself is a primary precursor to stable SOM. As microbes rapidly process labile carbon from plant inputs and subsequently die, their cellular remains or necromass, which are rich in nitrogen and other nutrients- become a major source of organic matter (Liang et al., 2017). This microbial necromass, through interactions with soil minerals and aggregation processes, can be stabilized and protected from further decomposition, forming a persistent repository of both carbon and nutrients. Therefore, soil fertility is not merely a static chemical property but an emergent, dynamic feature of a living system, governed by the continuous turnover of microbial life and its profound influence on the storage and flow of energy and matter (Liang et al., 2017). These interconnected microbial processes, ranging from decomposition and nutrient mineralization to root-microbe interactions and soil aggregation, are conceptually illustrated in **Fig. 1**.

Figure 1. Microbial mediation of nutrient cycling in forest ecosystems.

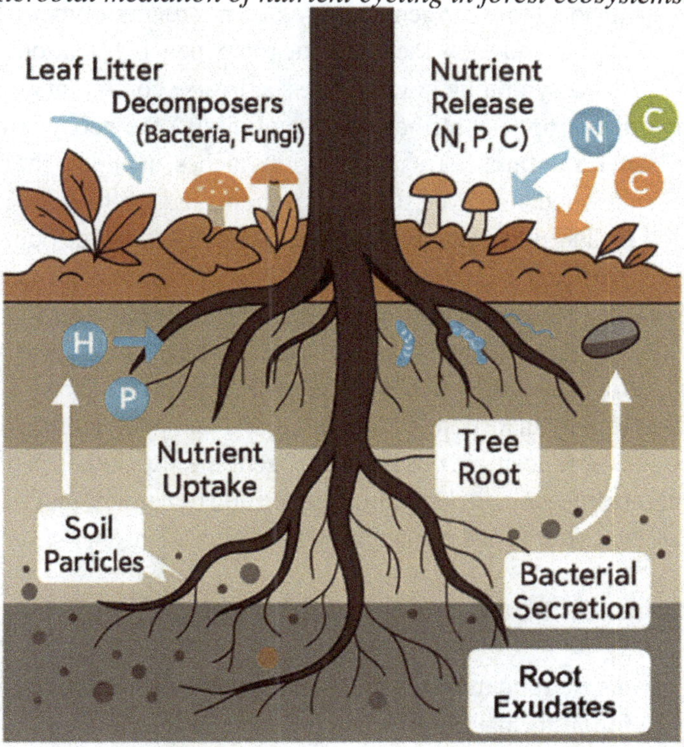

3.2 Nitrogen Fixation and Nutrient Availability

Biological nitrogen fixation (BNF) is a paramount microbial process that serves as the primary pathway for introducing new reactive nitrogen into many forest ecosystems, thereby alleviating the nitrogen limitation that frequently constrains net primary productivity. This essential function is carried out by a specialized group of microorganisms known as diazotrophs, which possess the oxygen-sensitive nitrogenase enzyme complex capable of converting inert atmospheric dinitrogen (N_2) into biologically available ammonia (NH_3). The most prolific contributors are often symbiotic bacteria that form intimate relationships with host plants. These include rhizobia forming root nodules on leguminous trees and shrubs, and actinobacteria of the genus *Frankia* forming similar structures on a diverse group of actinorhizal plants, such as *Alnus* spp. (alders), which are common in temperate and boreal zones (Taylor, 2025). Cyanobacteria also form critical symbioses with lichens, bryophytes, and some higher plants, contributing significant nitrogen inputs, particularly in nutrient-poor or early successional environments (Chen et al., 2025). These tight symbioses create an ideal low-oxygen, high-carbon microenvironment that protects the nitrogenase enzyme and fuels the energetically expensive fixation process.

In addition to these symbiotic powerhouses, a diverse community of free-living and associative diazotrophs contributes substantially to the ecosystem's nitrogen budget. These bacteria and archaea reside in various niches, including the rhizosphere, bulk soil, decaying wood, and even the phyllosphere, and while their fixation rates are typically lower than symbiotic systems, their ubiquity and large population sizes can result in a significant cumulative input of new nitrogen, especially in forests lacking dominant N-fixing plants (Xu et al., 2024). The diversity and distribution of these crucial microorganisms can be studied by targeting the *nifH* gene, a conserved marker for the nitrogenase enzyme. Ultimately, the nitrogen fixed by this wide array of diazotrophs is rapidly incorporated into microbial and plant biomass. Upon their turnover through decomposition, this fixed nitrogen enriches the soil N pool, directly enhancing soil fertility, influencing plant competition dynamics, and sustaining the productivity of the entire forest ecosystem (Hu et al., 2024).

3.3 Soil formation and structure stabilization

Microorganisms are fundamental geological agents, acting as the primary architects of both the initial formation of soil and the ongoing stabilization of its physical structure. In the earliest stages of pedogenesis, lithobiontic and soil-dwelling microbes initiate the weathering of parent rock material through the secretion of a suite of organic acids, protons, and chelating ligands. This biogeochemical weathering dissolves minerals, releasing essential nutrients and creating the foundational mineral-organic matrix that distinguishes soil from mere regolith. Beyond this initial formation, the microbiome is continuously responsible for developing and maintaining soil structure through the process of aggregation. Bacteria contribute by producing sticky extracellular polymeric substances (EPS), which act as a glue to bind fine soil particles into microaggregates. Fungal hyphae, particularly those of arbuscular mycorrhizal fungi (AMF), are arguably the most critical structural engineers, physically enmeshing microaggregates and soil particles into larger, more stable macroaggregates (Lehmann & Rillig, 2024). Furthermore, AMF secrete glomalin-related soil proteins (GRSP), highly persistent glycoproteins that coat aggregates, enhancing their water stability and contributing significantly to long-term carbon sequestration (Wang et al., 2017). This microbially-engineered architecture is critical for soil health, as it enhances porosity, aeration, water infiltration, and root penetration, while physically protecting organic matter from rapid decomposition.

3.4 Microbial influence on plant growth and productivity

Forest microbiomes directly enhance plant growth and ecosystem-level productivity through a diverse suite of mechanisms that extend far beyond nutrient cycling alone. These plant growth-promoting (PGP) activities are central to a plant's ability to establish, compete, and thrive. A primary contribution is the direct enhancement of nutrient acquisition, effectively augmenting the plant's root system. This function includes the critical process of biological nitrogen fixation by both symbiotic and free-living diazotrophs, which converts atmospheric N_2 into bioavailable ammonia. Furthermore, a vast number of soil bacteria and fungi enhance phosphorus availability through the solubilization of mineral phosphates via the secretion of organic acids, and the mineralization of organic phosphorus via the production of phosphatase enzymes (Alori et al., 2017). Mycorrhizal fungi are particularly crucial, as their extensive hyphal networks act as a physical extension of the root system, foraging for and transporting water and nutrients like phosphorus and zinc from soil volumes far beyond the reach of the roots themselves.

In addition to functioning as biofertilizers, forest microbes actively modulate plant growth and development through the synthesis and regulation of phytohormones. Many soil and endophytic microbes produce their auxins, cytokinins, and gibberellins, which are absorbed by the plant and can directly stimulate root initiation, cell division, and shoot elongation (Fadiji & Babalola, 2020). Perhaps one of the most well-documented mechanisms is the microbial production of the enzyme ACC (1-aminocyclopropane-1-carboxylate) deaminase. This enzyme cleaves ACC, the immediate precursor to the plant stress hormone ethylene, within the plant. By lowering ethylene levels, these microbes mitigate the inhibitory effect of stress on root growth, promoting a more robust root architecture that is better equipped to acquire resources, particularly under challenging environmental conditions (Glick, 2014).

The cumulative result of these microbially-mediated activities is a profound enhancement of net primary productivity at both the individual plant and ecosystem scales. By alleviating nutrient limitations and optimizing hormonal signaling for growth, the microbiome allows plants to allocate more resources to biomass accumulation. This direct influence on plant vigor and carbon assimilation capacity means that the composition and functional traits of the forest microbiome are a key determinant of the forest's overall productivity and its ability to act as a global carbon sink.

3.5 Microbe-mediated plant protection and disease suppression

The forest microbiome provides a critical ecosystem service by actively suppressing soil-borne and foliar diseases, functioning as a dynamic and responsive protective layer that regulates pathogen populations. This natural biological control is achieved through a sophisticated suite of direct and indirect mechanisms enacted by diverse microbial consortia. Direct antagonism is a primary mode of action, encompassing intense competition for limiting resources and overt chemical warfare. Beneficial microbes effectively engage in niche preemption by rapidly colonizing root surfaces (the rhizoplane), thereby physically excluding pathogens from infection courts (Xiong et al., 2023). This exclusion is often coupled with superior nutrient scavenging, such as the production of high-affinity siderophores that sequester iron, making it unavailable to invading pathogens. Furthermore, many forest microbes engage in antibiosis, synthesizing a vast arsenal of antimicrobial secondary metabolites. These include diffusible compounds like phenazines and 2,4-diacetylphloroglucinol (DAPG), volatile organic compounds that can act over a distance, and lytic enzymes such as chitinases and glucanases that directly degrade the structural integrity of fungal cell walls (Haas & Défago, 2005). Some beneficial microbes are also hyperparasites, exemplified by fungal genera like *Trichoderma*, which actively prey on and consume pathogenic fungi.

Beyond these confrontations, the microbiome orchestrates host-mediated defense through the phenomenon of induced systemic resistance (ISR). In contrast to pathogen-triggered systemic acquired resistance (SAR), which is typically dependent on the salicylic acid pathway, ISR is activated by beneficial microbes and relies on jasmonic acid and ethylene signaling pathways (Pieterse et al., 2014). This interaction primes the plant's innate immune system throughout its tissues, enabling a faster and more robust defense response against a broad spectrum of subsequent attackers without the significant fitness costs associated with constitutive high-level defense expression. The cumulative effect of these competitive, antagonistic, and host-priming interactions gives rise to the emergent ecosystem property of "disease-suppressive soils." This suppression can be general, resulting from the total microbial biomass and activity creating an inhospitable environment for pathogens, or specific, where a particular species or functional guild is primarily responsible for controlling a certain disease (Spooren et al., 2024). This microbially driven disease suppression is fundamental to the health, stability, and regenerative capacity of forest ecosystems worldwide.

3.6 Role in maintaining ecosystem resilience and stability

Forest microbiomes are fundamental architects of ecosystem resilience and stability, providing a crucial buffer against environmental perturbations and facilitating recovery from disturbances. This phenomenon of stabilizing influence stems mainly from two intrinsic properties of complex microbial communities: functional redundancy and response diversity. Functional redundancy, a hallmark of high biodiversity, ensures that multiple, taxonomically distinct microorganisms can perform the same key ecological functions, such as decomposition, nitrification, or methane oxidation (Nyström, 2006). This mechanism provides a form of biological insurance, as the loss of a few sensitive species following a disturbance (e.g., a drought or pollution event) does not lead to the collapse of the associated ecosystem process, because other, more tolerant taxa can compensate and maintain functional continuity. Complementing this is response diversity, the variation in how different species respond to environmental fluctuations. A community rich in response diversity will contain taxa with varying tolerances and optima for temperature, moisture, and pH, ensuring that at least some components of the community can remain active or recover quickly regardless of the nature of the stressor, thereby stabilizing aggregate ecosystem functions over time (Elmqvist et al., 2003).

These microbial community properties translate directly into whole-ecosystem stability. By underpinning the continuous cycling of essential nutrients, microbes regulate the resource base upon which the entire plant community depends, thereby mediating the forest's resistance to stress and its capacity for regrowth post-disturbance. For instance, after a fire, the rapid recovery of nitrogen-cycling microbes is a rate-limiting step for the re-establishment of vegetation. Furthermore, the web of interactions within the soil, including predation by protists and viral lysis, prevents the unchecked dominance of any single microbial group, contributing to the stability of the community structure itself (Bardgett & Van Der Putten, 2014). The resilience of the microbiome is therefore not merely a reflection of ecosystem health but is a primary determinant of it, providing the foundation for forests to absorb shocks and persist through environmental change. This "microbial memory" and adaptive capacity are critical for navigating both abrupt pulse disturbances and long-term press disturbances associated with global change (Shade et al., 2012).

4. MICROBIAL NETWORKS AND FOREST RESILIENCE

4.1 Mycorrhizal Networks: Communication, Nutrient Transfer, and Stress Mitigation

Common Mycorrhizal Networks (CMNs), colloquially known as the "Wood-Wide Web," are extensive belowground hyphal networks formed by mycorrhizal fungi that simultaneously colonize and connect the root systems of multiple plants, often spanning different species. These networks create a critical infrastructure within forest soils, fundamentally altering plant-plant interactions and resource dynamics. The most well-established function of CMNs is the facilitation of interplant nutrient transfer. Carbon, fixed by photosynthesizing plants, is allocated belowground to the fungal symbiont and can be transported throughout the hyphal network. In return, the fungus explores a vast volume of soil, absorbing and delivering limiting nutrients, such as nitrogen and phosphorus, to all connected plants (Van Der Heijden et al., 2008). This mechanism creates a dynamic system for resource redistribution, where nutrients can move from areas of high availability to areas of low availability, potentially buffering the entire plant community against resource patchiness.

Beyond resource sharing, emerging evidence suggests that CMNs function as a conduit for interplant communication and defense signaling. When a plant is subjected to biotic stress, such as an insect herbivore attack or infection by a pathogen, it can produce chemical defense signals that are transported through the network to neighboring, unattacked plants. Upon receiving these infochemicals, the neighboring plants can preemptively activate their defense pathways, effectively "eavesdropping" on the stress of their neighbors and preparing for a potential attack (Babikova et al., 2013)This rapid, belowground transfer of information can enhance the defensive posture of the entire plant community, demonstrating that the network mediates not just resource flows but also risk perception.

The integrated functions of CMNs are crucial for mitigating environmental stress and promoting forest resilience and regeneration. By extending the effective root system, the network significantly enhances a plant's capacity for water uptake, thereby improving drought tolerance for the connected community. Perhaps most critically, CMNs play a vital role in the establishment and survival of seedlings in the shaded forest understory. Large, established "hub" trees can shuttle significant amounts of carbon and other essential nutrients through the network to support young, light-limited seedlings, increasing their survival rates until they can reach the canopy (Pickles et al., 2015). This process facilitates natural regeneration and helps maintain the structure and diversity of the forest community over time.

The accompanying diagram (**Fig. 2**) succinctly visualizes the functional architecture of Common Mycorrhizal Networks (CMNs), depicting two interconnected plants engaged in belowground symbiosis through fungal hyphae. It illustrates key processes such as nutrient exchange, carbon flow, microbial interactions with PGPR, and the transmission of defense signals—visually reinforcing CMNs as dynamic networks for both resource sharing and interplant communication.

Figure 2. Functional dynamics of mycorrhizal networks and microbial symbiosis in the rhizosphere

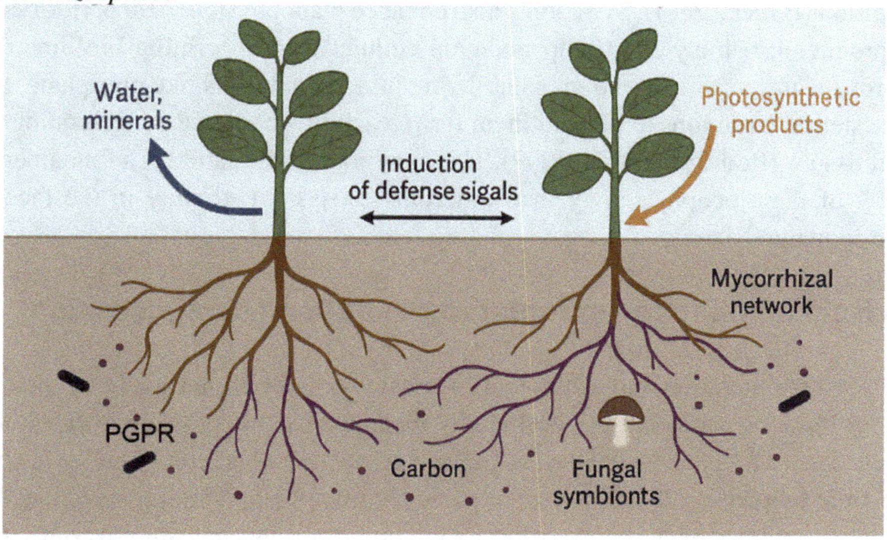

4.2 Microbial symbiosis and plant stress resistance

Beyond the well-documented role of mycorrhizal networks, a diverse assemblage of microbial symbionts, particularly endophytic fungi and bacteria residing within plant tissues, confers critical stress resistance to host plants, thereby enhancing forest resilience. These symbionts bolster defenses against a range of biotic stressors through multifaceted mechanisms. They can directly produce a formidable arsenal of anti-fungal, antibacterial, or anti-herbivore secondary metabolites, such as alkaloids and polyketides, which deter pathogens and consumers (Qin et al., 2024). Additionally, they can outcompete pathogens for niche space and essential nutrients within the plant's internal environment. Perhaps most significantly, these microbes can prime the plant's systemic immune system through induced systemic resistance (ISR), a state of heightened defensive readiness that provides broad-spectrum protection against future attacks. This microbe-mediated defense reduces disease incidence and herbivory, conferring a direct survival and competitive advantage to the host plant in complex forest communities. Furthermore, these symbiotic partners are instrumental in mitigating a suite of abiotic stresses, including drought, salinity, and heavy metal toxicity, effectively expanding the host plant's fundamental niche. The mechanisms are varied and sophisticated. A key strategy is the microbial production of the enzyme 1-aminocyclopropane-1-carboxylate (ACC) deaminase, which cleaves the precursor of the plant stress hormone ethylene. By downregulating ethylene synthesis, these microbes prevent stress-induced root growth inhibition, allowing the plant to continue exploring the soil for water and nutrients even under adverse conditions (Glick, 2014). Symbionts also enhance plant physiological performance by producing osmolytes to help maintain cellular turgor, forming biofilms that improve soil moisture retention, solubilizing mineral nutrients like phosphate, and sequestering toxic ions to prevent them from reaching damaging concentrations in plant tissues (Rodriguez et al., 2009). This symbiotic fortification is a fundamental aspect of plant ecophysiology that underpins ecosystem stability in the face of environmental change.

4.3 Role of microbes in forest regeneration and succession

Forest regeneration and ecological succession are not driven solely by the life-history traits of plants but are profoundly mediated by the dynamic interplay between plants and their associated soil microbiomes. As plant communities change over time following a disturbance, so too do the microbial communities in the soil, creating a coupled successional trajectory. The central mechanism governing this process is the phenomenon of plant-soil feedbacks (PSFs), wherein a plant species alters the biotic and abiotic properties of the soil it occupies, which in turn influences the performance of the next generation of plants, including its offspring and those of other species (Liu & Zhao, 2023). This feedback can be positive, where a plant cultivates microbes that benefit its growth, or negative, where it accumulates species-specific pathogens or depletes essential resources, thereby inhibiting its seedlings and facilitating the establishment of competitors.

The balance and direction of these plant-soil feedbacks are critical in driving the transition from early- to late-successional forest communities. Early-successional or pioneer plant species often accumulate soil-borne pathogens and create negative feedbacks that limit their dominance over time, opening niches for the establishment of later-successional species (Liu & Zhao, 2023). Furthermore, early-successional plants typically associate with generalist AM fungi. In contrast, many late-successional tree species depend on more specific ectomycorrhizal (EcM) fungi. The establishment and proliferation of these EcM fungi, which are superior at acquiring nutrients from the complex organic matter that accumulates in mature forests, can create a positive feedback loop that reinforces the dominance of the late-successional plant community (Kadowaki et al., 2018). Thus, the soil microbiome is not a passive respondent to plant succession but an active participant that shapes the trajectory, pace, and ultimate composition of the regenerating forest.

4.4 Microbial response to environmental disturbances and ecological stability

The capacity of a forest to persist through and recover from environmental disturbances, its ecological stability, is fundamentally linked to the dynamic responses of its microbial communities. The stability of a microbiome is characterized by two key components: resistance, the ability to withstand a perturbation with little change, and resilience, the capacity to recover its structure and function after being altered (Shade et al., 2012). Intrinsic properties of the community govern the degree of resistance and resilience. High taxonomic and functional diversity often confers greater resilience, as the loss of sensitive species can be compensated for by functionally redundant taxa that perform similar roles (the "insurance hypothesis"). The architecture of the microbial interaction network also plays a critical role; highly connected and modular networks may be more robust to the loss of individual nodes (De Vries et al., 2018). Furthermore, many microbes can persist through disturbances by entering dormant states (e.g., forming spores or cysts), creating a microbial "seed bank" in the soil that facilitates the community's rapid recovery once favorable conditions return (Lennon & Jones, 2011).

The specific nature of the microbial response is critical for the trajectory of the entire ecosystem's recovery. Following a significant disturbance like a fire or insect outbreak that deposits a large pulse of organic matter, the rapid functional response of the decomposer community dictates the rate of nutrient release and subsequent plant regrowth. A resilient microbial community can quickly re-establish nutrient cycling pathways, preventing nutrient loss from the system and supporting the regeneration of the plant community (Griffiths & Philippot, 2013). Moreover, the rapid re-colonization of soil niches by native, beneficial microbes can prevent the establishment of opportunistic pathogens or invasive species in the post-disturbance environment. In this way, the "microbial memory" of the pre-disturbance state, encoded in the surviving taxa and the dormant seed bank, not only enables the microbiome's recovery but actively steers the recovery of the entire visible ecosystem, highlighting that forest stability is an emergent property of its belowground resilience.

5. THREATS TO FOREST MICROBIAL DIVERSITY

Despite their fundamental importance and inherent resilience, forest microbial communities are under unprecedented and escalating pressure from a suite of anthropogenic drivers. These threats, operating at local to global scales, are profoundly altering the composition, structure, and function of this invisible biodiversity. This section will detail the primary stressors, including direct habitat loss and fragmentation from deforestation, intensive land-use change, the pervasive impacts of chemical pollution and industrial contaminants, the overarching influence of climate change, and the disruptive effects of biological invasions and pathogen spread. Understanding the mechanisms by which these drivers impact the microbiome is critical, as the erosion of this foundational biodiversity has severe cascading consequences for the stability and functioning of the entire forest ecosystem (Cavicchioli et al., 2019). **Table 2** provides a consolidated overview of the major anthropogenic stressors discussed in this section. It summarizes their primary mechanisms of impact on forest microbial communities and highlights the key documented ecological consequences of these disruptions, offering a comprehensive snapshot of the threats facing this critical component of forest biodiversity.

Table 2. Summary of anthropogenic stressors and their documented impacts on forest microbiomes

Stressor	Impact on Microbial Diversity (e.g., Shannon Index)	Impact on Microbial Biomass	Impact on Key Enzyme Activities (e.g., Dehydrogenase, Phosphatase)	Impact on Community Structure	Source(s)
Deforestation / Logging	Generally reduced, especially specialist taxa (e.g., ECM fungi)	Often reduced due to loss of organic inputs and soil compaction	Variable; can decrease due to loss of microbial biomass	Major shift from fungal-dominated (ECM) to bacterial- or AM-dominated communities; Loss of network connectivity	(Holden & Treseder, 2013; S. J. B. Robinson et al., 2024)
Heavy Metals (e.g., Cd, Pb, Zn)	Significantly reduced	Reduced	Strongly inhibited	Shift towards metal-tolerant species; Reduced network stability; Co-selection for antibiotic resistance genes	(Campillo-Cora et al., 2025; Li et al., 2025)
Pesticides (Fungicides)	Reduced, especially for fungi; Shannon & Simpson indices decrease	Variable; can be reduced	Inhibited, particularly dehydrogenase, phosphatase, and urease activity	Strong negative effect on fungal abundance; Bacterial communities less affected or may increase	(Baćmaga et al., 2016; Wang et al., 2025)

continued on following page

Table 2. Continued

Stressor	Impact on Microbial Diversity (e.g., Shannon Index)	Impact on Microbial Biomass	Impact on Key Enzyme Activities (e.g., Dehydrogenase, Phosphatase)	Impact on Community Structure	Source(s)
Pesticides (Glyphosate)	Variable and context-dependent; can decrease, increase, or have no effect	Variable; can be inhibitory or stimulatory (as a nutrient source)	Variable effects reported	Shift in community composition, favoring glyphosate-degrading or resistant taxa; Can reduce AMF colonization	(Meena et al., 2020; Sibalekile et al., 2025a)
Microplastics	Can alter diversity	Can be negatively impacted	Can be altered	Changes in community structure and function; Act as vectors for other pollutants and microbes	(Yan et al., 2025)
Climate Change (Warming, Altered Precipitation)	Shifts in community composition; potential loss of sensitive taxa	Variable; can alter microbial activity and biomass	Enzyme activities can be altered, with responses varying by enzyme and stressor	Can shift C balance, potentially turning forests into C sources	(Cuartero et al., 2024; Ruan et al., 2024)
Biological Invasions (Invasive Plants)	Can be altered	Can be altered	Can be altered	Invasive plants recruit distinct microbial communities, disrupting native plant-microbe interactions	(Bell et al., 2020; Torres et al., 2021)

5.1 Habitat Loss and Fragmentation Due To Deforestation

Habitat loss and fragmentation driven by deforestation represent one of the most severe and direct threats to forest microbiomes. The removal of forest cover triggers a cascade of changes in the physical and chemical environment of the soil, profoundly disrupting the microhabitats that support diverse microbial life. The loss of canopy results in elevated soil temperatures, increased desiccation, and greater fluctuations in moisture levels, creating harsh selective pressures that are detrimental to many sensitive, forest-specialist microbes (Onet et al., 2025). This physical disruption is coupled with the cessation of critical resource inputs—the diverse leaf litter and root exudates from the native plant community—leading to the local extinction of host-specific symbionts. The overall effect is a rapid "biotic homogenization" of the soil microbial community, where a few tolerant generalists replace a rich assemblage of endemic specialists, as has been demonstrated in the conversion of Amazonian rainforest to pastureland (Rodrigues et al., 2013).

Fragmentation of the remaining forest exacerbates this vulnerability by creating extensive edge habitats. These edges are characterized by altered microclimates that penetrate deep into forest remnants, changing the selective pressures on soil biota and favoring different functional guilds. For example, ectomycorrhizal fungi, which are often dependent on stable host associations and intact forest floor conditions, typically decline sharply in fragmented landscapes, while more generalist saprotrophic fungi may proliferate (Haddad et al., 2015). Compounding this, the physical isolation of forest patches restricts microbial dispersal and gene flow. This phenomenon mirrors the genetic bottlenecks observed in macro-organisms, limiting the potential for re-colonization after local disturbances and potentially reducing the adaptive capacity of microbial communities to ongoing environmental change (Raimbault et al., 2024).

The consequences of these changes are not merely taxonomic but are deeply functional and persistent. The disruption of crucial plant-microbe symbioses, such as those involving mycorrhizal fungi and nitrogen-fixing bacteria, directly impairs nutrient acquisition for remaining plants and impedes the natural regeneration of the forest (Van Der Heijden et al., 2008). Long-term studies confirm that microbial communities in deforested and fragmented landscapes show lasting alterations in both diversity and function, with significant shifts in carbon and nitrogen cycling processes that can persist for decades after the initial disturbance (Jangid et al., 2011).

5.2 Land-Use Changes (Agriculture, Monoculture Plantations, Urbanization)

The conversion of complex forest ecosystems into simplified, human-dominated landscapes represents a profound threat to microbial diversity. Transforming forests into agricultural land imposes a suite of intense and simultaneous disturbances on the soil microbiome. The initial act of deforestation removes the primary source of diverse organic inputs, while subsequent tillage physically decimates fungal hyphal networks and destroys soil aggregates that serve as critical microbial habitats. This transition is often followed by the application of synthetic fertilizers and pesticides, which drastically alter soil chemistry and can be toxic to non-target, beneficial microorganisms. The cumulative effect is a significant loss of microbial diversity and a functional shift from complex, fungal-dominated communities to simplified assemblages of fast-growing, disturbance-tolerant bacteria, a process often termed microbial homogenization (Peng et al., 2024).

Even when forest cover is maintained, its replacement with monoculture plantations (e.g., pine, oil palm, or eucalyptus) drives a significant, albeit different, loss of microbial diversity. By replacing a rich, multi-species plant community with a single species, the variety of resources entering the soil via litter and root exudates is dramatically reduced. This biotic homogenization of the plant community inevitably leads to a corresponding simplification of the soil microbiome, selecting for a narrow subset of microbes adapted to the specific host tree (Widyati et al., 2025). This loss of diversity erodes the functional redundancy of the soil community, potentially impairing its resilience and increasing the vulnerability of the entire plantation to outbreaks of host-specific pathogens.

Urbanization represents one of the most extreme forms of land-use change, creating novel ecosystems with profound impacts on soil life. The conversion of forests to urban and suburban landscapes leads to soil sealing by impervious surfaces, severe compaction, and chronic contamination with a unique suite of pollutants, including heavy metals, hydrocarbons, and road salts. These pressures, combined with the "urban heat island" effect, eliminate most native forest soil taxa and select for a distinct urban microbiome. Recent studies have shown that these urban soils can become hotspots for antibiotic resistance genes and potential human pathogens, representing a complete and often detrimental transformation of the original microbial community (Jacques et al., 2025).

5.3 Pollution Impacts on Microbial Communities

Pollution is a pervasive and multifactorial threat to forest microbial communities, with profound implications for soil health, nutrient cycling, and overall ecosystem functioning. Forest microbiomes are highly sensitive to both acute and chronic pollution, including chemical pesticides, heavy metals, and industrial effluents. These contaminants can alter microbial diversity, reduce functional redundancy, and disrupt vital plant-microbe interactions that underpin forest productivity and resilience.

5.3.1 Chemical pesticides and heavy metals

Chemical pollutants, even when originating from adjacent agricultural or industrial landscapes, can infiltrate forest ecosystems through atmospheric deposition and runoff, posing a significant threat to the soil microbiome. Pesticides, including fungicides, herbicides, and insecticides, are of particular concern due to their potential for non-target effects. Fungicides, by design, can have a devastating impact on fungal diversity, indiscriminately harming beneficial, non-pathogenic fungi alongside their intended targets. This disruption is especially detrimental to mycorrhizal fungi, whose suppression can decouple nutrient transfer pathways to host trees and impair forest health (Ni et al., 2025). Herbicides, such as the widely used glyphosate, can indirectly alter microbial community structure by changing plant root exudation patterns or by exhibiting direct toxicity to certain microbial groups, thereby disrupting the delicate balance of the rhizosphere (Sibalekile et al., 2025b). The overall consequence of pesticide contamination is often a reduction in microbial functional diversity and the impairment of critical ecosystem services like decomposition and nutrient cycling.

Heavy metal contamination from sources such as industrial emissions, mining activities, and contaminated amendments presents another severe and persistent threat. Metals like lead (Pb), cadmium (Cd), mercury (Hg), and copper (Cu) are toxic to most microorganisms at elevated concentrations, as they can disrupt cell membrane integrity, denature essential enzymes, and generate oxidative stress (Campillo-Cora et al., 2025). This toxicity imposes a strong selective pressure on the soil community, leading to a sharp decline in overall microbial biomass and diversity and favoring the proliferation of a few metal-resistant taxa (Campillo-Cora et al., 2025). A particularly alarming consequence of heavy metal pollution is the co-selection for antibiotic resistance. The genes conferring resistance to heavy metals are often located on the same mobile genetic elements (e.g., plasmids) as antibiotic resistance genes (ARGs). Therefore, the presence of heavy metals in the soil can inadvertently promote the persistence and spread of ARGs in the environment, linking industrial pollution to a global health threat (Gillieatt & Coleman, 2024).

5.3.2 Industrial effluents and contaminants

Forest ecosystems, particularly those near industrial centers or downstream of urban areas, are increasingly exposed to a complex cocktail of industrial effluents and contaminants delivered through atmospheric deposition, wastewater discharge, and land application of sludges. This chemical milieu includes a wide range of persistent and emerging pollutants that can profoundly restructure soil microbial communities. Classic contaminants like polycyclic aromatic hydrocarbons (PAHs) from incomplete combustion and petroleum hydrocarbons from spills are directly toxic to many microorganisms, yet they also serve as a carbon source for a specialized subset of hydrocarbon-degrading bacteria and fungi (Das & Chandran, 2011). The introduction of these substances typically leads to a decrease in overall microbial diversity and a strong selective enrichment for catabolically-specialized genera such as *Pseudomonas*, *Rhodococcus*, and *Mycobacterium*, often at the expense of key symbiotic fungi and other sensitive taxa (Gkorezis et al., 2016). While this enriches the community's bioremediation potential, it often comes at the cost of disrupting other essential ecosystem functions, such as nutrient cycling. More recently, concern has grown over "emerging contaminants," including pharmaceuticals, personal care products (PPCPs), and per- and polyfluoroalkyl substances (PFAS), which are now detected globally (F. Wang et al., 2024). The long-term impacts of these bioactive and highly persistent compounds on the forest microbiome's structure and function are still poorly understood but represent a significant and growing threat to forest soil health (Cycoń et al., 2019).

5.4 Climate change and altered microbial community dynamics

Climate change acts as a pervasive and multifaceted stressor on forest microbiomes, directly altering community structure and function with significant feedbacks to the global carbon cycle. Rising temperatures directly affect microbial physiology by increasing metabolic and enzymatic rates up to a thermal optimum. While this can initially boost productivity, sustained warming can restructure communities, often selecting for faster-growing, copiotrophic bacteria over slower-growing, oligotrophic taxa and fungi (Jansson & Hofmockel, 2020). This shift can alter the very fabric of microbial food webs and may also facilitate the poleward expansion of temperature-sensitive plant and animal pathogens, increasing disease risks in forests.

Altered precipitation patterns, characterized by both prolonged droughts and more frequent extreme rainfall events, impose further selective pressures. Drought induces severe osmotic stress, which can lead to a decline in overall microbial biomass and activity. It particularly disadvantages fungi, whose extensive hyphal networks are dependent on soil water continuity for nutrient transport, potentially decoupling critical plant-mycorrhizal symbioses (De Vries et al., 2018). Conversely, intense precipitation can saturate soils, creating anoxic conditions that suppress aerobic microbes and favor anaerobic functional guilds, such as denitrifying bacteria and methanogenic archaea, thereby altering the flux of potent greenhouse gases like nitrous oxide (N_2O) and methane (CH_4).

The most critical consequence of these climate-driven microbial shifts is their effect on soil carbon cycling, which can create a powerful positive feedback to climate change. Increased temperatures generally accelerate the microbial decomposition of soil organic matter, leading to a greater efflux of carbon dioxide (CO_2) into the atmosphere and further amplifying warming (Crowther et al., 2016). This process is metabolically complex and is influenced by the community's overall carbon use efficiency (CUE), the fraction of carbon taken up that is allocated to biomass versus respired as CO_2. Climate-induced shifts in community composition or physiology that lower CUE can result in a greater proportion of carbon being lost from the soil, weakening the capacity of forest ecosystems to act as long-term carbon sinks (Malik et al., 2020).

5.5 Biological invasions and pathogen spread

Biological invasions, involving both macroscopic and microscopic organisms, can fundamentally disrupt the structure and function of native forest microbiomes. Invasive plant species, in particular, can act as ecosystem engineers, actively altering the soil environment to their benefit, often at the expense of the resident microbial community. They can achieve this by releasing novel allelopathic chemicals that inhibit native microbes, producing litter with different chemistry that alters decomposition rates, or by associating with a different type of mycorrhizal fungi than the native flora, thereby disrupting established nutrient-sharing networks (Zhao et al., 2024). Invasive animals can also have profound impacts; for example, the invasion of previously earthworm-free North American forests by European earthworm species has led to the rapid consumption of the entire forest floor organic layer, destroying the primary habitat for countless fungi and other soil organisms (Frelich et al., 2006). These invasions by macro-organisms create a cascade of belowground changes that can permanently alter the microbial community and the ecosystem services it provides.

The direct introduction and spread of invasive microbial pathogens represents an even more acute threat, responsible for some of the most devastating declines of keystone forest tree species. Global trade and travel have dramatically increased the long-distance dispersal of virulent fungi, oomycetes, and viruses into naive ecosystems. Classic examples include the introductions of *Cryphonectria parasitica* (chestnut blight) and *Ophiostoma ulmi* (Dutch elm disease) into North America, and *Phytophthora cinnamomi* and *P. ramorum* (sudden oak death) into various ecosystems worldwide, each causing landscape-scale mortality of their host trees (LeBoldus et al., 2024). These pathogens are often so destructive due to the "enemy release hypothesis," as the native host plants lack the co-evolved resistance mechanisms necessary to combat the novel threat. The loss of a dominant tree species not only impacts the forest canopy but triggers a complete restructuring of the associated soil microbiome, from the loss of specific mycorrhizal partners to altered nutrient inputs, with long-lasting consequences for ecosystem recovery and function (Gomez-Aparicio et al., 2022).

5.6 Consequences of microbial diversity loss: cascading ecological effects

The erosion of forest microbial diversity, driven by the cumulative pressures of habitat loss, pollution, and climate change, precipitates a cascade of deleterious effects that destabilize entire ecosystems. The loss of microbial taxa is not merely a reduction in species counts but a critical depletion of the ecosystem's functional repertoire and its inherent "biological insurance" (Delgado-Baquerizo et al., 2020). This functional decay first manifests as a weakening of ecosystem resilience; with diminished functional redundancy, the ability of the soil community to buffer against disturbances and maintain process stability is compromised. This impairment quickly cascades to core biogeochemical cycles. The precise enzymatic and metabolic pathways governing nutrient turnover become disrupted, leading to inefficient cycling, nutrient leakage from the system, and a decline in soil fertility. This belowground failure propagates upward, directly impacting plant communities through reduced nutrient availability and the loss of beneficial symbioses, resulting in lower primary productivity and increased vulnerability to stress and pathogens (Delgado-Baquerizo et al., 2025). Ultimately, this cascade diminishes ecosystem multifunctionality, the simultaneous provisioning of multiple ecosystem services, from carbon sequestration to water purification—propelling the forest toward a simplified, less resilient, and functionally impoverished state (Duffy et al., 2017).

6. MICROBES AS BIOINDICATORS OF FOREST ECOSYSTEM HEALTH

As forests face mounting pressures from climate change, land-use alterations, and pollution, there is an urgent need for sensitive and integrative tools to monitor ecosystem health. Microbial communities, owing to their rapid responsiveness, ecological ubiquity, and functional centrality, have emerged as powerful bioindicators capable of revealing subtle shifts in environmental quality well before visible signs of degradation appear. This section explores the conceptual foundation, methodological advances, and practical applications of using forest microbes as biological sentinels for ecosystem monitoring and management.

6.1 Concept and Importance of Microbial Bioindicators

Microbial bioindicators are specific taxa, functional guilds, or emergent community properties whose presence, abundance, or activity reflects the integrated health and functioning of an ecosystem. Unlike direct chemical or physical measurements, which quantify individual parameters, microbial indicators provide a holistic biological assessment, signaling how environmental stressors are impacting life at the foundational trophic level (Semenov et al., 2025). An indicator could be a single sensitive species that disappears under pollution stress, the ratio of major phyla (e.g., Firmicutes to Bacteroidetes), the activity rate of a key enzyme, or a structural attribute of the entire ecological network, such as its connectivity and modularity. The core concept rests on the premise that because microbial communities are intimately linked to and rapidly responsive to their environment, their composition and function serve as a highly sensitive and early-warning signal of ecological change.

The importance of leveraging these "living sensors" in forest management and conservation is threefold. First, microbes exhibit rapid turnover and high sensitivity, allowing them to respond to environmental shifts—such as soil compaction, heavy metal contamination, or altered moisture regimes—far more quickly than the overlying plant community. This capacity for early detection enables proactive management and intervention before degradation becomes severe and potentially irreversible (Vivian et al., 2025). Second, because microbes drive fundamental ecosystem processes like decomposition and nutrient cycling, their status is a direct proxy for the functional integrity of the entire forest. A change in the abundance of nitrifying bacteria, for example, has direct consequences for nitrogen availability to plants. Third, the advent of high-throughput molecular tools has made microbial monitoring more accessible and data-rich than ever, providing a powerful, evidence-based approach to quantify the success of restoration efforts or the impact of land-use change, thereby informing adaptive management strategies (Yang et al., 2024).

6.2 Methods for Assessing Microbial Indicators

The assessment of microbial indicators in forest ecosystems relies on an integrated suite of methods, from classical microbiological techniques to modern molecular and biochemical approaches. These methods collectively enable the evaluation of both taxonomic composition and the functional potential or activity of microbial communities, thereby offering a multidimensional understanding of forest health and soil quality. While traditional culturing on selective media can provide valuable insights into the abundance of specific viable groups like nitrogen-fixers or phosphate-solubilizers, it is severely limited by the "great plate count anomaly"—the fact that less than 1% of soil microbes are readily culturable under standard laboratory conditions. Therefore, culture-independent approaches have become essential for any comprehensive microbiome analysis (Torsvik & Øvreås, 2002).

Among culture-independent techniques, amplicon-based sequencing of taxonomic marker genes—such as the 16S rRNA gene for bacteria and archaea, and the Internal Transcribed Spacer (ITS) regions for fungi—is the most widely used molecular method. High-throughput sequencing platforms allow for large-scale profiling of microbial communities, revealing shifts in diversity, richness, and composition across environmental gradients or response to disturbance (Nilsson et al., 2019). For direct functional assessment, shotgun metagenomic sequencing provides access to the collective genomic content of the community. This approach allows researchers to identify and quantify genes associated with key ecosystem functions, such as nitrogen fixation (*nifH*), ammonia oxidation (*amoA*), and lignin degradation, which serve as robust biomarkers for nutrient cycling potential (H. Wang et al., 2024). High-throughput functional gene arrays (e.g., GeoChip) further enable the simultaneous detection of thousands of functional genes, explicitly linking microbial community structure to ecosystem function (He et al., 2007).

Biochemical indicators provide sensitive, integrated measures of microbial function and metabolic potential. Assays for soil microbial biomass carbon (MBC) and nitrogen (MBN), as well as the activity of extracellular enzymes like dehydrogenase, phosphatase, and β-glucosidase, respond rapidly to environmental changes and have been validated as reliable proxies for soil quality and nutrient availability in diverse forest types (Nannipieri et al., 2017). To complement these broad measures, quantitative PCR (qPCR) is increasingly employed to precisely quantify the abundance of specific microbial taxa or functional genes, offering high sensitivity for monitoring key microbial guilds or detecting pathogens (Smith & Osborn, 2009).

Emerging "meta-omics" technologies provide an even more dynamic view by moving beyond genetic potential to measure active processes. Metatranscriptomics (gene expression), metaproteomics (protein synthesis), and metabolomics (metabolic byproducts) offer a snapshot of what a community is doing under specific environmental conditions, which is crucial for understanding the functional resilience of microbial communities under stress (Jansson & Hofmockel, 2020). The vast datasets generated by these methods are interpreted using advanced bioinformatics tools such as LEfSe, for identifying differentially abundant features, and PICRUSt or FUNGuild, for predicting community function from marker gene data, allowing for a more complete, systems-level assessment of the microbiome's role in forest health (Nguyen et al., 2016; Segata et al., 2011).

7. TECHNOLOGICAL ADVANCES IN FOREST MICROBIAL ECOLOGY

The field of forest microbial ecology has entered a new era, propelled by remarkable technological advancements that have revolutionized our ability to observe, quantify, and model microbial life. These innovations have not only enhanced the resolution at which microbial communities can be studied but have also expanded our understanding from static taxonomic inventories to dynamic, functionally-informed ecosystems. In the following subsections, we explore the most transformative tools, ranging from high-throughput sequencing to systems-level multi-omics and predictive modeling, that are reshaping our grasp of the forest microbiome.

7.1 High-Throughput Sequencing and Metagenomics

The study of forest microbiomes has been fundamentally transformed by the advent of high-throughput sequencing (HTS) technologies, which have circumvented the severe limitations of traditional cultivation-based methods and provided unprecedented access to the "unculturable majority." The most widely adopted HTS approach is marker gene amplicon sequencing, which provides a taxonomic census of the community by targeting and sequencing hypervariable regions of conserved genes, typically the 16S rRNA gene for bacteria and archaea, and the Internal Transcribed Spacer (ITS) region for fungi (Quince et al., 2017). This metabarcoding approach is highly scalable and cost-effective, enabling large-scale ecological studies that have been instrumental in mapping the biogeography of forest microbes and identifying community-level responses to environmental drivers. While powerful for determining "who is there," amplicon sequencing provides limited direct information about the functional capabilities of the community.

To address this limitation, shotgun metagenomics has become an indispensable tool. This approach involves sequencing the entire collective pool of DNA from an environmental sample, thereby capturing a snapshot of the community's total genetic potential and answering the question "what could they do?"(Sharpton, 2014). By analyzing the millions of short sequence reads, researchers can identify genes encoding specific metabolic pathways, such as those for cellulose degradation, nitrogen fixation, or antibiotic resistance, and assess their relative abundance under different environmental conditions. Furthermore, advances in bioinformatics have enabled the reconstruction of near-complete genomes from metagenomic data, known as Metagenome-Assembled Genomes (MAGs) (Parks et al., 2017; Tyson et al., 2004). This breakthrough allows for the characterization of the metabolic capabilities of novel, uncultured microorganisms and the linking of key functions to specific lineages, thus bridging the critical gap between community structure and ecosystem function.

7.2 Environmental DNA (eDNA) Monitoring

Environmental DNA (eDNA) metabarcoding has emerged as a revolutionary, non-invasive technology for monitoring biodiversity, offering unprecedented insights into the composition of forest microbial communities. This approach relies on the analysis of genetic material extracted directly from environmental samples such as soil, water, or even air, bypassing the need for organismal isolation or cultivation, which is a significant bottleneck for studying microbes (Thomsen & Willerslev, 2015). The typical workflow involves amplifying specific taxonomic marker genes (e.g., the 16S rRNA gene for bacteria and archaea, the ITS region for fungi) from the total extracted eDNA, followed by high-throughput sequencing. The resulting sequence data provide a comprehensive snapshot of the community's taxonomic composition. The primary advantages of this method are its high sensitivity, which allows for the detection of rare and cryptic taxa, and its scalability, which makes it feasible to conduct large-scale biodiversity assessments across vast and remote forest landscapes efficiently (Taberlet et al., 2018).

The applications of eDNA monitoring in forest ecology and conservation are rapidly expanding. It is now routinely used to generate baseline inventories of soil microbial diversity, which are crucial for understanding biogeographical patterns and for tracking community responses to environmental change. One of its most powerful applications is in the early detection of invasive species and pathogens. For instance, eDNA from soil or water samples can reveal the presence of a pathogenic fungus like *Batrachochytrium dendrobatidis* or an invasive plant pathogen long before visible signs of disease emerge, enabling rapid management responses. Furthermore, eDNA provides a highly sensitive tool for evaluating the efficacy of forest restoration projects. By tracking the recovery of key microbial functional guilds, such as mycorrhizal fungi or nitrogen-fixers, managers can gain a much deeper understanding of ecological recovery than from plant surveys alone (Cristescu & Hebert, 2018). This capacity to provide rapid, comprehensive, and non-invasive biodiversity data is fundamentally changing how we observe and manage the invisible microbial dimension of forest ecosystems.

7.3 Meta-Omics Approaches (Transcriptomics, Proteomics, Metabolomics)

While metagenomics reveals the functional *potential* of a microbial community ("what could they do?"), a deeper understanding of ecosystem processes requires knowing which genes are actively being expressed and which functions are being performed at a given moment. To this end, a suite of "meta-omics" approaches, including metatranscriptomics, metaproteomics, and metabolomics, has become essential for moving from a static catalog of genes to a dynamic picture of microbial activity. These methods provide layered insights into the flow of biological information, from gene transcription to protein synthesis and metabolic output, allowing researchers to address the critical question of "who is doing what, and when?" within the forest microbiome (Aguiar-Pulido et al., 2016).

Metatranscriptomics involves the large-scale sequencing of all messenger RNA (mRNA) from an environmental sample, providing a snapshot of the community's collective gene expression profile. Because mRNA molecules are short-lived, the metatranscriptome reflects the community's immediate response to prevailing environmental conditions. This approach can reveal which metabolic pathways are actively "switched on"—for example, identifying the upregulation of genes for cellulose degradation in response to fresh litter input, or stress response genes during a drought. By linking active transcripts back to specific genomes (either from isolates or MAGs), it becomes possible to identify the specific microbial taxa that are driving key ecosystem processes under particular circumstances. However, the inherent instability of RNA makes sample collection, preservation, and analysis technically challenging.

Metaproteomics provides an even more direct measure of function by identifying and quantifying the entire complement of proteins produced by the microbial community, typically using high-resolution mass spectrometry. As proteins are the actual molecular machinery—the enzymes, transporters, and structural components—that carry out cellular functions, metaproteomics offers a view that is one step closer to the active phenotype of the community (Pan et al., 2024). This technique can confirm whether transcribed genes are translated into functional proteins and can reveal post-translational modifications that regulate protein activity. Like metatranscriptomics, metaproteomics is technically demanding, requiring sophisticated protein extraction protocols and complex bioinformatic analyses to match peptide fragments to a reference sequence database.

The ultimate functional readout of a microbial community is provided by metabolomics, the comprehensive analysis of the small-molecule metabolites present in a sample. These metabolites—including amino acids, organic acids, lipids, and signaling molecules—are the substrates, intermediates, and end-products of the community's collective metabolic activity. Metabolomics can therefore identify the specific biochemical compounds being exchanged between microbes or between microbes and their plant hosts, providing direct evidence of cross-feeding, competition, and symbiotic interactions (Bhattacharjya et al., 2024). The true power of these approaches lies in their integration. By combining metagenomics, metatranscriptomics, metaproteomics, and metabolomics, researchers can build a systems-level understanding of the forest microbiome, creating a holistic model that links genetic potential to gene expression, protein synthesis, and ultimate ecological impact.

7.4 Ecological Network Analysis and Microbiome Modeling

To move beyond simple taxonomic inventories, researchers are increasingly employing ecological network analysis to decipher the complex web of interactions governing forest microbiome structure and function. These networks are typically inferred from high-throughput sequencing data by identifying statistically significant patterns of co-occurrence or co-abundance among microbial taxa across multiple samples. A positive correlation may suggest cooperation, cross-feeding, or shared niche preferences, while a negative correlation can indicate competition, amensalism, or predator-prey relationships (Kitagami & Matsuda, 2024). This approach provides powerful insights into community organization, allowing for the identification of “keystone taxa”, highly connected nodes whose influence is disproportionately large relative to their abundance and whose loss could precipitate a cascade of secondary extinctions (Banerjee et al., 2018). Network analysis also reveals the modular structure of communities, where densely connected clusters of microbes may represent distinct functional guilds or synergistic consortia. However, it is critical to recognize that these inferred networks are correlational and represent hypotheses of interaction that require experimental validation to confirm causal mechanisms (Kajihara & Hynson, 2024).

While network analysis provides a powerful static snapshot of potential interactions, the next frontier lies in dynamic microbiome modeling to predict community behavior over time. These models aim to simulate how microbial communities assemble, evolve, and respond to specific environmental drivers, such as changes in temperature, moisture, or substrate availability. Approaches range from implementing classic ecological models, like generalized Lotka-Volterra equations to simulate population dynamics, to more complex trait-based and metabolic models (Blonder et al., 2023). Genome-scale metabolic models (GEMs), for instance, can leverage genomic information to simulate the flow of energy and nutrients through entire communities, predicting metabolic handoffs and competitive outcomes (Wu et al., 2024). The ultimate goal is to integrate these modeling frameworks with empirical data from multi-omics and network analyses, enabling a shift from descriptive to predictive microbial ecology. Such predictive power is essential for forecasting the fate of forest ecosystems under future global change scenarios.

8. CONSERVATION AND RESTORATION STRATEGIES INTEGRATING FOREST MICROBIOMES

The integration of microbiome science into forest conservation marks a paradigm shift in how we understand, restore, and manage ecosystem integrity. As the foundational drivers of nutrient cycling, plant health, and resilience to stressors, microbes are not merely passive components but active engineers of ecological recovery. This section critically examines the current gaps in conservation frameworks regarding microbial inclusion and highlights innovative strategies to incorporate microbiome knowledge into restoration practices and policy development.

8.1 Current Shortcomings of Microbial Awareness in Conservation

Despite their incontrovertible role as engines of planetary biogeochemistry and determinants of host health, microorganisms remain conspicuously absent from the vast majority of conservation frameworks, policies, and on-the-ground practices. This pervasive "microbial myopia" stems from a combination of historical momentum, perceptual biases, and logistical challenges (Junker & Farwig, 2024). Conservation as a discipline was built around the charisma of visible, often vertebrate, species and the tangible structure of plant communities. The unseen nature of the microbial world, coupled with the abstract and usually complex task of defining microbial "species" or conservation units, has created a persistent barrier to their inclusion in mainstream conservation planning. Consequently, flagship initiatives from global protected area networks to national endangered species acts are designed and implemented with a profound blindness to the foundational biodiversity that underpins the survival of the macro-organisms they aim to protect.

This conceptual oversight translates into significant practical failings that can compromise the long-term success of conservation interventions. For example, habitat restoration and reforestation projects are frequently planned based on plant selection and soil chemistry alone, without assessing or actively managing the soil microbiome. Such projects risk failure if the soil lacks the specific mycorrhizal symbionts or nitrogen-fixing bacteria essential for the target tree species to thrive (Peddle et al., 2025). Similarly, captive breeding and reintroduction programs for endangered animals often overlook the critical importance of maintaining a healthy gut microbiome, potentially releasing animals that are immunologically compromised or unable to efficiently digest their natural diet (Trevelline et al., 2019). The failure to incorporate a microbial perspective means that current conservation strategies are often addressing only the symptoms of ecosystem decline, rather than the root causes, thereby risking inefficient allocation of resources and suboptimal outcomes.

8.2 Microbiome-Informed Forest Restoration Techniques

Integrating microbial ecology into forest restoration offers a robust suite of tools to accelerate ecosystem recovery and improve long-term success, moving beyond a sole focus on plant-centric metrics. These techniques can be broadly categorized into direct microbial introductions and indirect habitat manipulations that facilitate the recovery of native microbial communities. Direct approaches often involve the inoculation of degraded soils to re-establish key microbial players. This practice can range from the application of "whole soil" inocula, where a small amount of soil from a healthy reference ecosystem is transferred to the restoration site, bringing with it a complex and co-adapted community of bacteria, fungi, and other organisms (Peddle et al., 2025). While effective, this method can be logistically challenging and carries a risk of unintentionally introducing pathogens. A more targeted approach involves the use of specific microbial inoculants, such as commercially produced arbuscular mycorrhizal (AM) fungi or nitrogen-fixing bacteria, which can be applied to nursery seedlings to ensure they enter the field with critical symbionts already established (Koziol et al., 2024). Additionally, biotic amendments like high-quality compost and biochar can serve as both a microbial inoculum and a habitat conditioner, improving soil structure and nutrient content to support the establishment of both the introduced and resident microbes (Koziol et al., 2024).

Indirect strategies focus on creating favorable environmental conditions to steer the natural recovery of the soil microbiome. A key aspect of this is managing plant-soil feedback. The selection of plant species for revegetation is critical, as different plants recruit distinct microbial communities. Utilizing a diverse mix of native, early-successional species, or employing "nurse plants," can help to break negative soil legacies left by invasive species and rebuild a diverse, functional soil community over time (Peddle et al., 2025). This approach recognizes that the plant community is a primary driver of the microbiome it supports. Furthermore, physical and chemical habitat restoration, such as soil decompaction, re-establishing natural hydrological flows, and adding complex organic matter like wood chips or mulch, can recreate the structural and chemical environment necessary for the recovery of sensitive native microbes, particularly fungi. Ultimately, the most robust restoration frameworks will likely be those that synergistically combine direct inoculation with thoughtful, indirect habitat management, tailored to the specific ecological context of the degraded site (J. M. Robinson et al., 2024).

8.3 Strategies for Mitigating Threats

Integrating microbial ecology into forest restoration offers a robust suite of tools to accelerate ecosystem recovery and improve long-term success, moving beyond a sole focus on plant-centric metrics. These techniques can be broadly categorized into direct microbial introductions and indirect habitat manipulations that facilitate the recovery of native microbial communities. Direct approaches often involve the inoculation of degraded soils to re-establish key microbial players. This strategy can range from the application of "whole soil" inocula, where a small amount of soil from a healthy reference ecosystem is transferred to the restoration site, bringing with it a complex and co-adapted community of bacteria, fungi, and other organisms (Peddle et al., 2025). While effective, this method can be logistically challenging and carries a risk of unintentionally introducing pathogens. A more targeted approach involves the use of specific microbial inoculants, such as commercially produced arbuscular mycorrhizal (AM) fungi or nitrogen-fixing bacteria, which can be applied to nursery seedlings to ensure they enter the field with critical symbionts already established (Koziol et al., 2024). Additionally, biotic amendments like high-quality compost and biochar can serve as both a microbial inoculum and a habitat conditioner, improving soil structure and nutrient content to support the establishment of both the introduced and resident microbes (Barbosa et al., 2024).

Indirect strategies focus on creating favorable environmental conditions to steer the natural recovery of the soil microbiome. A key aspect of this is managing plant-soil feedback. The selection of plant species for revegetation is critical, as different plants recruit distinct microbial communities. Utilizing a diverse mix of native, early-successional species, or employing "nurse plants," can help to break negative soil legacies left by invasive species and rebuild a diverse, functional soil community over time (Peddle et al., 2025). This approach recognizes that the plant community is a primary driver of the microbiome it supports. Furthermore, physical and chemical habitat restoration, such as soil decompaction, re-establishing natural hydrological flows, and adding complex organic matter like wood chips or mulch, can recreate the structural and chemical environment necessary for the recovery of sensitive native microbes, particularly fungi. Ultimately, the most robust restoration frameworks will likely be those that synergistically combine direct inoculation with thoughtful, indirect habitat management, tailored to the specific ecological context of the degraded site (Neuenkamp et al., 2024).

8.4 Integration of Microbial Ecology into Policy and Management Frameworks

A critical disconnect persists between the advancing scientific understanding of microbial ecosystems and their near-total invisibility within existing environmental policy and forest management frameworks. This "great microbial oversight" means that major international agreements, national laws, and industry standards designed to protect biodiversity and ecosystem services are operating without acknowledging the foundational layer of life that underpins them. To achieve meaningful and lasting conservation outcomes, this gap must be bridged by formally integrating microbial ecology into decision-making processes at all scales.

This integration must begin at the highest levels of global governance. Landmark agreements like the Convention on Biological Diversity (CBD) and the Kunming-Montreal Global Biodiversity Framework require explicit targets that move beyond the traditional focus on charismatic megafauna and flora. As leading scientists have argued, a failure to incorporate the microbial world into our planetary stewardship goals is a failure of the entire endeavor (Cavicchioli et al., 2019). This obligation requires signatory nations to translate these global goals into their National Biodiversity Strategies and Action Plans (NBSAPs) with specific, measurable, and relevant indicators for microbial communities. Such indicators could include targets for maintaining soil organic carbon levels, preserving microbial functional diversity within protected area management plans, and establishing clear regulatory limits on pollutants known to be detrimental to key microbial guilds, such as mycorrhizal fungi or nitrogen-fixing bacteria.

Translating high-level policy into on-the-ground action necessitates leveraging operational frameworks like forest certification and climate mitigation schemes. Market-based certification programs, including the Forest Stewardship Council (FSC), offer a direct mechanism to incentivize sustainable practices, yet their criteria for soil health are often underdeveloped. These standards could be significantly enhanced by mandating practices that protect microbial habitats, such as restricting clear-felling, preserving legacy trees, and limiting soil compaction and the use of harmful agrochemicals (Pezdevšek Malovrh et al., 2019). In parallel, climate initiatives such as REDD+ must evolve to recognize that the permanence of forest carbon sinks is not guaranteed by tree biomass alone; it is fundamentally governed by the microbial processes that regulate soil carbon sequestration and stability. Incorporating microbial indicators into the monitoring, reporting, and verification (MRV) systems of REDD+ would provide a more holistic and scientifically robust assessment of a forest's true climate mitigation value (Visca et al., 2025).

9. INTEGRATING TRADITIONAL AND INDIGENOUS KNOWLEDGE IN MICROBIAL CONSERVATION

Efforts to conserve forest microbiomes must extend beyond laboratory techniques and molecular tools to include the accumulated wisdom of traditional and Indigenous communities. These knowledge systems, deeply rooted in local ecologies and sustained through generations, offer ecologically sound and culturally grounded approaches to land stewardship. This section explores how integrating Traditional Ecological Knowledge (TEK) and Indigenous Knowledge (IK) can enrich microbial conservation, reveal hidden ecological insights, and foster more inclusive, resilient strategies for managing forest ecosystems.

9.1 Importance of Local and Indigenous Knowledge Systems

A truly holistic approach to forest microbiome conservation necessitates moving beyond a purely Western scientific paradigm to respectfully integrate the deep, place-based insights of Traditional Ecological Knowledge (TEK) and Indigenous Knowledge (IK) systems. These knowledge systems, developed through millennia of direct observation, empirical learning, and cultural transmission, represent a rich and often-overlooked repository of information on sustainable ecosystem management (Reyes-García et al., 2022). While not framed in the language of microbial ecology, many traditional land management practices—such as specific polyculture cropping systems, the use of biochar, prescribed low-intensity burning, and the selection of particular landscape mosaics for cultivation—implicitly foster soil health and conserve beneficial microbial communities. This knowledge is not a static collection of facts but a dynamic and adaptive system that is holistic, qualitative, and diachronic, offering a powerful complement to the often reductionist and synchronic nature of conventional scientific inquiry (Berkes, 2017). Recognizing the validity of TEK and IK is, therefore, not only an ethical imperative for decolonizing conservation science but also a pragmatic necessity. These systems can provide novel hypotheses, guide research toward locally relevant questions, and offer time-tested models for resilient land stewardship that can inform and enhance modern, science-based conservation strategies (Garnett et al., 2018).

9.2 Traditional Forest Management and Microbial Stewardship

Traditional forest management systems, guided by Indigenous and local ecological knowledge, are typically founded on a holistic worldview that views humans as integrated components of, rather than separate from, the ecosystem. This perspective fosters management principles that stand in stark contrast to the resource-extraction focus of many conventional forestry models. These principles often include a deep sense of reciprocity, respect for the agency of non-human entities, an emphasis on long-term, multi-generational sustainability, and a dynamic, adaptive approach to management based on continuous observation and feedback (Berkes, 2017; Turner et al., 2000). Practices emerging from this philosophy—such as selective harvesting, maintaining complex polycultures, seasonal resource rotation, and the retention of legacy trees and coarse woody debris—are designed to maintain the overall structure and integrity of the forest social-ecological system.

While not explicitly framed in the language of microbiology, these traditional principles and practices result in a form of "unintentional microbial stewardship." The emphasis on minimizing soil disturbance and maintaining continuous canopy cover protects the physical architecture of the soil, preserving the habitats of sensitive microbial communities, particularly fungal networks. By fostering a diverse, multi-species, and multi-aged plant community, these systems ensure a varied and continuous supply of resources to the belowground food web through litterfall and root exudates, which in turn support a more diverse and functionally resilient microbiome (Parrotta & Agnoletti, 2011; Poe et al., 2016). Consequently, traditional forest management often leads to the de facto conservation of microbial diversity and the robust maintenance of critical ecosystem functions like nutrient cycling and soil organic matter formation, demonstrating a time-tested model of sustainable coexistence that modern scientific approaches are only beginning to appreciate fully.

9.3 Sacred Groves and Community-Conserved Areas in India

Sacred groves—forest fragments conserved on religious or cultural grounds—are among India's oldest and most ecologically significant traditional conservation practices. Scattered across diverse biogeographical regions, from the Western Ghats to Northeast India, these groves serve as living repositories of both visible and invisible biodiversity. The protection of these patches is rooted in centuries-old beliefs that often prohibit extraction, disturbance, or desecration, thus offering an enduring model of community-based conservation that has persisted through periods of widespread land-use change (Bhagwat & Rutte, 2006; Malhotra, 2007).

While celebrated for conserving rare and endemic plant and animal species, recent research highlights the equally profound role of sacred groves in preserving soil microbial diversity. Because they are protected from the anthropogenic disturbances affecting the surrounding landscape, these groves act as invaluable *in situ* refugia. Studies consistently reveal that soils from sacred groves harbor higher abundance and richness of beneficial microbes—such as arbuscular mycorrhizal fungi, phosphate-solubilizing bacteria, and nitrogen-fixers—compared to adjacent disturbed or cultivated lands (Lyngwi & Joshi, 2015). These enriched microbial communities are vital for maintaining nutrient cycling, organic matter decomposition, and disease-suppressive properties, functions central to long-term ecosystem health. This tradition extends to the broader category of Community-Conserved Areas (CCAs), which include sacred landscapes, village woodlots, and community forests governed under customary law. Evidence suggests that CCAs maintain higher ecosystem integrity and microbial function compared to state-managed forests, as community norms typically encourage the retention of litter, fallen wood, and minimal soil disturbance—practices that directly support microbial habitats (Bag et al., 2022). Their role in landscape connectivity further amplifies the ecological significance of both sacred groves and CCAs; they can serve as genetic and functional corridors, supporting microbial fluxes between fragmented habitats.

In recent years, recognition of these traditional conservation systems has grown within both national policy and microbial ecology. They are now seen not only as bastions of cultural heritage but also as crucial, time-tested empirical models for *in situ* microbial conservation. Their integration into larger conservation frameworks offers a blueprint for synergizing Indigenous knowledge, community stewardship, and scientific understanding to safeguard forest biodiversity both above and below ground.

9.4 Global Examples of Indigenous Practices Benefiting Microbial Diversity

Across the world, Indigenous and local communities have evolved land management practices that promote not only aboveground biodiversity but also the resilience and diversity of soil microbial communities. These culturally rooted practices, grounded in Traditional Ecological Knowledge (TEK), shape soil structure, enhance organic matter, and sustain intricate belowground food webs, thus ensuring continued nutrient cycling and ecosystem stability. The convergence of TEK and scientific research underscores the critical value of Indigenous management for sustaining the invisible foundation of ecosystem health: the soil microbiome (Garnett et al., 2018; ITPS, 2021).

Perhaps the most celebrated case is the creation of *terra preta do índio* (Amazonian Dark Earths). In the Amazon Basin, Indigenous peoples intentionally enriched the notoriously infertile soils through the long-term addition of charred biomass (biochar), animal manure, pottery shards, and other organic refuse. These anthropogenic soils are globally renowned for their high microbial biomass, elevated enzymatic activities, and the exceptional persistence of beneficial microbial taxa, including nitrogen fixers and mycorrhizal fungi, centuries after their creation (Lehmann et al., 2006; O'neill et al., 2009). This example demonstrates that Indigenous soil stewardship can dramatically and permanently enhance soil fertility and microbial diversity.

In Australian landscapes, traditional fire management or "cultural burning" by Aboriginal peoples creates patchy mosaics of vegetation at different successional stages. These frequent, low-intensity fires are distinct from destructive, large-scale wildfires and function to maintain higher microbial diversity, stimulate the decomposition of organic matter, and foster the persistence of fire-adapted fungal communities, thereby supporting broader ecosystem regeneration (Johnston et al., 2024). Similarly, in the mountain agroecosystems of the Andes and Himalayas, Indigenous communities implement terracing, rotational fallows, and organic amendments methods that minimize soil erosion and maintain high organic carbon levels. The integration of traditional composting and livestock manure in Andean "chacras" or Himalayan homegardens results in soils with greater microbial biomass and disease-suppressive properties (Hadar & Papadopoulou, 2012; Jayaraman et al., 2021).

In East Africa, the continued practice of maintaining community-managed sacred "kaya" forests by the Mijikenda people preserves undisturbed soil profiles that harbor high levels of fungal and bacterial diversity. These sites are recognized for their unique and intact mycorrhizal communities, which are crucial for the regeneration and carbon cycling of the surrounding forest ecosystems (Onyango et al., 2023). These examples, from boreal forests managed by Canadian First Nations to Pacific

Islands agroforestry, empirically link Indigenous stewardship to enhanced soil health and microbial functional diversity, confirming their vital role in global conservation.

10. FUTURE PERSPECTIVES AND RESEARCH DIRECTIONS

The study of forest microbiomes is rapidly advancing from descriptive cataloging to mechanistic and predictive science, yet critical knowledge gaps remain. Future research must prioritize moving beyond "who is there" to understand "what are they doing, and why." Key research needs include elucidating the specific links between taxonomic diversity and functional resilience, identifying the microbial traits that confer resistance to climate change stressors like drought and heatwaves, and untangling the complex trophic interactions involving viruses, protists, and other microfauna that regulate bacterial and fungal populations (Jansson & Hofmockel, 2020). Furthermore, there is an urgent need for long-term, large-scale experimental manipulations that can causally link microbial community shifts to changes in ecosystem-level processes, providing the robust data needed for model parameterization and validation.

Addressing these complex questions will necessitate the integration of cutting-edge technologies and a deeply interdisciplinary approach. The continued development and application of multi-omics—combining metagenomics with metatranscriptomics, proteomics, and metabolomics—will be essential for linking genetic potential to *in situ* activity. Emerging technologies such as single-cell genomics, stable isotope probing (SIP) combined with sequencing, and advanced imaging techniques (e.g., nanoscale secondary ion mass spectrometry, NanoSIMS) will allow researchers to trace nutrient flows through specific taxa and visualize interactions within intact soil microstructures (Kleiner et al., 2023). However, technology alone is insufficient. Progress will depend on fostering collaboration between microbial ecologists, soil scientists, plant physiologists, remote sensing experts, climate modelers, and social scientists to integrate data across scales and disciplines, from molecules to entire biomes.

The profound implications of microbial science for forest health and conservation will only be realized if they are effectively communicated beyond academia. A concerted effort is needed to enhance "microbial literacy" among all stakeholders, including policymakers, land managers, conservation practitioners, and the public Dubinsky et al., 2022. This effort involves translating complex microbial data into accessible and actionable concepts, such as soil health indices or functional potential scores. Developing clear communication strategies and educational materials that illustrate the tangible benefits of a healthy microbiome, such as improved tree growth, disease resistance, and carbon storage, is crucial for garnering public support and driving the inclusion of microbial metrics into policy and management frameworks.

Finally, the potential role of microbial biotechnology in proactive forest conservation and restoration is a promising, albeit nascent, frontier. This innovation includes the development of next-generation "probiotics for ecosystems" scientifically designed microbial inoculants tailored to specific restoration goals, such as enhancing drought tolerance, degrading specific pollutants, or outcompeting invasive species. The principles of synthetic ecology could one day be used to construct robust microbial consortia with bespoke functions to bolster forest resilience (Toju et al., 2018). While this field holds immense potential, it must proceed with caution, requiring rigorous ecological risk assessments to understand the potential unintended consequences of releasing engineered or non-native microbes into the environment.

11. CONCLUSION

The vast and complex microbial communities inhabiting forest ecosystems are not merely passive residents but are fundamental architects of forest structure, function, and resilience. This review has synthesized the extensive evidence demonstrating that, from cycling essential nutrients and forming the bedrock of soil fertility to directly promoting plant growth and defending against pathogens, the forest microbiome is an indispensable pillar of biodiversity. Historically, conservation science and policy have operated with a profound "microbial myopia," focusing almost exclusively on the visible world of plants and animals. We argue that this oversight is no longer tenable; it has led to an incomplete understanding of ecosystem dynamics and has fundamentally limited the efficacy of management and restoration strategies. Recognizing forests as complex holobionts—superorganisms composed of macroscopic hosts and their myriad microscopic partners—is essential for advancing our approach to their stewardship.

This recognition must now translate into a tangible paradigm shift in conservation action and policy. We call for the explicit integration of the microbial dimension into all facets of forest management, from local restoration projects to global biodiversity frameworks. This shift requires moving beyond traditional metrics to incorporate microbial indicators into routine monitoring programs, thereby providing a more sensitive and early warning signal of ecosystem health. It demands that international agreements, such as the Convention on Biological Diversity, and certification schemes, such as the FSC, formally acknowledge and create targets for the protection of soil biodiversity and function. It also necessitates a transdisciplinary approach that values and integrates Traditional Ecological Knowledge systems, which have often implicitly managed for microbial health for millennia.

Ultimately, the long-term resilience and sustainability of the world's forests in an era of unprecedented global change will depend on the health and stability of their microbial inhabitants. The capacity of a forest to withstand and recover from disturbances like drought, fire, and invasive species is inextricably linked to the functional redundancy and adaptive potential encoded within its microbiome. Therefore, protecting the "invisible majority" is not a niche concern but a central and urgent priority. The future of our forests rests upon the stewardship of this hidden world, and a commitment to integrating microbial ecology into conservation is a commitment to a more resilient and sustainable planet for all.

REFERENCES

Aguiar-Pulido, V., Huang, W., Suarez-Ulloa, V., Cickovski, T., Mathee, K., & Narasimhan, G. (2016). Metagenomics, metatranscriptomics, and metabolomics approaches for microbiome analysis: supplementary issue: bioinformatics methods and applications for big metagenomics data. *Evolutionary bioinformatics, 12*, EBO. S36436.

Alori, E. T., Glick, B. R., & Babalola, O. O. (2017). Microbial phosphorus solubilization and its potential for use in sustainable agriculture. *Frontiers in Microbiology, 8*, 971. DOI: 10.3389/fmicb.2017.00971 PMID: 28626450

Averill, C., Anthony, M. A., Baldrian, P., Finkbeiner, F., van den Hoogen, J., Kiers, T., Kohout, P., Hirt, E., Smith, G. R., & Crowther, T. W. (2022). Defending Earth's terrestrial microbiome. *Nature Microbiology, 7*(11), 1717–1725. DOI: 10.1038/s41564-022-01228-3 PMID: 36192539

Babikova, Z., Gilbert, L., Bruce, T. J., Birkett, M., Caulfield, J. C., Woodcock, C., Pickett, J. A., & Johnson, D. (2013). Underground signals carried through common mycelial networks warn neighbouring plants of aphid attack. *Ecology Letters, 16*(7), 835–843. DOI: 10.1111/ele.12115 PMID: 23656527

Baćmaga, M., Wyszkowska, J., & Kucharski, J. (2016). The effect of the Falcon 460 EC fungicide on soil microbial communities, enzyme activities and plant growth. *Ecotoxicology (London, England), 25*(8), 1575–1587. DOI: 10.1007/s10646-016-1713-z PMID: 27631526

Bag, S., Biswal, D., Roy, A., Sarkar, B., Mondal, A., & Chatterjee, S. (2022). Diversity of soil bacteria in some sacred patches of Purba Bardhaman District, West Bengal, India. *Ecology, environment and conservation*, 269-279.

Bahram, M., Hildebrand, F., Forslund, S. K., Anderson, J. L., Soudzilovskaia, N. A., Bodegom, P. M., & Harend, H. (2018). Structure and function of the global topsoil microbiome. *Nature, 560*(7717), 233–237. DOI: 10.1038/s41586-018-0386-6 PMID: 30069051

Banerjee, S., Schlaeppi, K., & Van Der Heijden, M. G. (2018). Keystone taxa as drivers of microbiome structure and functioning. *Nature Reviews. Microbiology, 16*(9), 567–576. DOI: 10.1038/s41579-018-0024-1 PMID: 29789680

Barbosa, F. L., Santos, J. M., Mota, J. C., Costa, M. C., Araujo, A. S., Garcia, K. G., & Ferreira, O. P. (2024). Potential of biochar to restoration of microbial biomass and enzymatic activity in a highly degraded semiarid soil. *Scientific Reports, 14*(1), 26065. DOI: 10.1038/s41598-024-77368-9 PMID: 39478100

Bardgett, R. D., & Van Der Putten, W. H. (2014). Belowground biodiversity and ecosystem functioning. *Nature, 515*(7528), 505–511. DOI: 10.1038/nature13855 PMID: 25428498

Bell, J. K., Siciliano, S. D., & Lamb, E. G. (2020). A survey of invasive plants on grassland soil microbial communities and ecosystem services. *Scientific Data*, *7*(1), 86. DOI: 10.1038/s41597-020-0422-x PMID: 32152302

Berkes, F. (2017). *Sacred ecology*. Routledge. DOI: 10.4324/9781315114644

Bhagwat, S. A., & Rutte, C. (2006). Sacred groves: Potential for biodiversity management. *Frontiers in Ecology and the Environment*, *4*(10), 519–524. DOI: 10.1890/1540-9295(2006)4[519:SGPFBM]2.0.CO;2

Bhattacharjya, S., Ghosh, A., Sahu, A., Agnihotri, R., Pal, N., Sharma, P., Manna, M. C., Sharma, M. P., & Singh, A. (2024). Utilizing soil metabolomics to investigate the untapped metabolic potential of soil microbial communities and their role in driving soil ecosystem processes: A review. *Applied Soil Ecology*, *195*, 105238. DOI: 10.1016/j.apsoil.2023.105238

Blonder, B. W., Gaüzère, P., Iversen, L. L., Ke, P. J., Petry, W. K., Ray, C. A., Salguero-Gómez, R., Sharpless, W., & Violle, C. (2023). Predicting and controlling ecological communities via trait and environment mediated parameterizations of dynamical models. *Oikos*, *2023*(6), e09415. DOI: 10.1111/oik.09415

Campillo-Cora, C., Rodríguez-Seijo, A., Pérez-Rodríguez, P., Fernández-Calviño, D., & Santás-Miguel, V. (2025). Effect of heavy metal pollution on soil microorganisms: Influence of soil physicochemical properties. A systematic review. *European Journal of Soil Biology*, *124*, 103706. https://doi.org/https://doi.org/10.1016/j.ejsobi.2024.103706. DOI: 10.1016/j.ejsobi.2024.103706

Cavicchioli, R., Ripple, W. J., Timmis, K. N., Azam, F., Bakken, L. R., Baylis, M., & Classen, A. T. (2019). Scientists' warning to humanity: Microorganisms and climate change. *Nature Reviews. Microbiology*, *17*(9), 569–586. DOI: 10.1038/s41579-019-0222-5 PMID: 31213707

Chen, C.-C., Xie, Q.-Y., Chuang, P.-S., Darnajoux, R., Chien, Y.-Y., Wang, W.-H., & Tang, S.-L. (2025). A thallus-forming N-fixing fungus-cyanobacterium symbiosis from subtropical forests. *Science Advances*, *11*(7), eadt4093. DOI: 10.1126/sciadv.adt4093 PMID: 39937913

Cristescu, M. E., & Hebert, P. D. (2018). Uses and misuses of environmental DNA in biodiversity science and conservation. *Annual Review of Ecology, Evolution, and Systematics*, *49*(1), 209–230. DOI: 10.1146/annurev-ecolsys-110617-062306

Crowther, T. W., Todd-Brown, K. E., Rowe, C. W., Wieder, W. R., Carey, J. C., Machmuller, M. B., & Allison, S. D. (2016). Quantifying global soil carbon losses in response to warming. *Nature, 540*(7631), 104–108. DOI: 10.1038/nature20150 PMID: 27905442

Crowther, T. W., Van den Hoogen, J., Wan, J., Mayes, M. A., Keiser, A., Mo, L., Averill, C., & Maynard, D. S. (2019). The global soil community and its influence on biogeochemistry. *Science, 365*(6455), eaav0550. DOI: 10.1126/science.aav0550 PMID: 31439761

Cuartero, J., Querejeta, J. I., Prieto, I., Frey, B., & Alguacil, M. M. (2024). Warming and rainfall reduction alter soil microbial diversity and co-occurrence networks and enhance pathogenic fungi in dryland soils. *The Science of the Total Environment, 949*, 175006. https://doi.org/https://doi.org/10.1016/j.scitotenv.2024.175006. DOI: 10.1016/j.scitotenv.2024.175006 PMID: 39069184

Cycoń, M., Mrozik, A., & Piotrowska-Seget, Z. (2019). Antibiotics in the soil environment—Degradation and their impact on microbial activity and diversity. *Frontiers in Microbiology, 10*, 338. DOI: 10.3389/fmicb.2019.00338 PMID: 30906284

Das, N., & Chandran, P. (2011). Microbial degradation of petroleum hydrocarbon contaminants: An overview. *Biotechnology Research International, 2011*(1), 941810. DOI: 10.4061/2011/941810 PMID: 21350672

De Vries, F. T., Griffiths, R. I., Bailey, M., Craig, H., Girlanda, M., Gweon, H. S., & Kretzschmar, M. (2018). Soil bacterial networks are less stable under drought than fungal networks. *Nature Communications, 9*(1), 3033. DOI: 10.1038/s41467-018-05516-7 PMID: 30072764

Delgado-Baquerizo, M., Eldridge, D. J., Liu, Y.-R., Liu, Z.-W., Coleine, C., & Trivedi, P. (2025). Soil biodiversity and function under global change. *PLoS Biology, 23*(3), e3003093. DOI: 10.1371/journal.pbio.3003093 PMID: 40146744

Delgado-Baquerizo, M., Oliverio, A. M., Brewer, T. E., Benavent-González, A., Eldridge, D. J., Bardgett, R. D., Maestre, F. T., Singh, B. K., & Fierer, N. (2018). A global atlas of the dominant bacteria found in soil. *Science, 359*(6373), 320–325. DOI: 10.1126/science.aap9516 PMID: 29348236

Delgado-Baquerizo, M., Reich, P. B., Trivedi, C., Eldridge, D. J., Abades, S., Alfaro, F. D., & Gallardo, A. (2020). Multiple elements of soil biodiversity drive ecosystem functions across biomes. *Nature Ecology & Evolution, 4*(2), 210–220. DOI: 10.1038/s41559-019-1084-y PMID: 32015427

Duffy, J. E., Godwin, C. M., & Cardinale, B. J. (2017). Biodiversity effects in the wild are common and as strong as key drivers of productivity. *Nature, 549*(7671), 261–264. DOI: 10.1038/nature23886 PMID: 28869964

Elmqvist, T., Folke, C., Nyström, M., Peterson, G., Bengtsson, J., Walker, B., & Norberg, J. (2003). Response diversity, ecosystem change, and resilience. *Frontiers in Ecology and the Environment, 1*(9), 488–494. DOI: 10.1890/1540-9295(2003)001[0488:RDECAR]2.0.CO;2

Fadiji, A. E., & Babalola, O. O. (2020). Elucidating mechanisms of endophytes used in plant protection and other bioactivities with multifunctional prospects. *Frontiers in Bioengineering and Biotechnology, 8*, 467. DOI: 10.3389/fbioe.2020.00467 PMID: 32500068

Frelich, L. E., Hale, C. M., Reich, P. B., Holdsworth, A. R., Scheu, S., Heneghan, L., & Bohlen, P. J. (2006). Earthworm invasion into previously earthworm-free temperate and boreal forests. In *Biological invasions belowground: earthworms as invasive species* (pp. 35–45). Springer. DOI: 10.1007/978-1-4020-5429-7_5

Garnett, S. T., Burgess, N. D., Fa, J. E., Fernández-Llamazares, Á., Molnár, Z., Robinson, C. J., & Brondizio, E. S. (2018). A spatial overview of the global importance of Indigenous lands for conservation. *Nature Sustainability, 1*(7), 369–374. DOI: 10.1038/s41893-018-0100-6

Gillieatt, B. F., & Coleman, N. V. (2024). Unravelling the mechanisms of antibiotic and heavy metal resistance co-selection in environmental bacteria. *FEMS Microbiology Reviews, 48*(4), fuae017. DOI: 10.1093/femsre/fuae017 PMID: 38897736

Gkorezis, P., Daghio, M., Franzetti, A., Van Hamme, J. D., Sillen, W., & Vangronsveld, J. (2016). The interaction between plants and bacteria in the remediation of petroleum hydrocarbons: An environmental perspective. *Frontiers in Microbiology, 7*, 1836. DOI: 10.3389/fmicb.2016.01836 PMID: 27917161

Glick, B. R. (2014). Bacteria with ACC deaminase can promote plant growth and help to feed the world. *Microbiological Research, 169*(1), 30–39. DOI: 10.1016/j.micres.2013.09.009 PMID: 24095256

Gomez-Aparicio, L., Dominguez-Begines, J., Villa-Sanabria, E., García, L. V., & Munoz-Pajares, A. J. (2022). Tree decline and mortality following pathogen invasion alters the diversity, composition and network structure of the soil microbiome. *Soil Biology & Biochemistry, 166*, 108560. DOI: 10.1016/j.soilbio.2022.108560

Griffiths, B. S., & Philippot, L. (2013). Insights into the resistance and resilience of the soil microbial community. *FEMS Microbiology Reviews*, *37*(2), 112–129. DOI: 10.1111/j.1574-6976.2012.00343.x PMID: 22568555

Guo, Y., Ren, C., Yi, J., Doughty, R., & Zhao, F. (2020). Contrasting Responses of Rhizosphere Bacteria, Fungi and Arbuscular Mycorrhizal Fungi Along an Elevational Gradient in a Temperate Montane Forest of China [Original Research]. *Frontiers in microbiology, Volume 11 - 2020*. DOI: 10.3389/fmicb.2020.02042

Haas, D., & Défago, G. (2005). Biological control of soil-borne pathogens by fluorescent pseudomonads. *Nature Reviews. Microbiology*, *3*(4), 307–319. DOI: 10.1038/nrmicro1129 PMID: 15759041

Hadar, Y., & Papadopoulou, K. K. (2012). Suppressive composts: Microbial ecology links between abiotic environments and healthy plants. *Annual Review of Phytopathology*, *50*(1), 133–153. DOI: 10.1146/annurev-phyto-081211-172914 PMID: 22920558

Haddad, N. M., Brudvig, L. A., Clobert, J., Davies, K. F., Gonzalez, A., Holt, R. D., & Collins, C. D. (2015). Habitat fragmentation and its lasting impact on Earth's ecosystems. *Science Advances*, *1*(2), e1500052. DOI: 10.1126/sciadv.1500052 PMID: 26601154

He, Z., Gentry, T. J., Schadt, C. W., Wu, L., Liebich, J., Chong, S. C., & Jardine, P. (2007). GeoChip: A comprehensive microarray for investigating biogeochemical, ecological and environmental processes. *The ISME Journal*, *1*(1), 67–77. DOI: 10.1038/ismej.2007.2 PMID: 18043615

Holden, S. R., & Treseder, K. K. (2013). A meta-analysis of soil microbial biomass responses to forest disturbances [Original Research]. *Frontiers in microbiology, Volume 4 - 2013*. DOI: 10.3389/fmicb.2013.00163

Hu, W., Wang, X., Xu, Y., Wang, X., Guo, Z., Pan, X., Dai, S., Luo, Y., & Teng, Y. (2024). Biological nitrogen fixation and the role of soil diazotroph niche breadth in representative terrestrial ecosystems. *Soil Biology & Biochemistry*, *189*, 109261. DOI: 10.1016/j.soilbio.2023.109261

Huang, Z., He, X., Zhang, C., Zhang, M., Wang, J., Hou, Y., . . . Ji, K. (2023). Microbial communities and functions changed in rhizosphere soil of Pinus massoniana provenances with different carbon storage [Original Research]. *Frontiers in microbiology, Volume 14 - 2023*. DOI: 10.3389/fmicb.2023.1264670

ITPS. F. (2021). Recarbonizing global soils—a technical manual of recommended management practices. In: Volume.

Jacques, K. J., Xiang, Q., Hao, Y., Sun, A., Lu, C., & Chen, Q.-L. (2025). Urbanization increases high-risk antibiotic resistance genes and pathogenic bacteria in soil and phyllosphere microbiomes. *Journal of Hazardous Materials*, *495*, 139025. DOI: 10.1016/j.jhazmat.2025.139025 PMID: 40555022

Jangid, K., Williams, M. A., Franzluebbers, A. J., Schmidt, T. M., Coleman, D. C., & Whitman, W. B. (2011). Land-use history has a stronger impact on soil microbial community composition than aboveground vegetation and soil properties. *Soil Biology & Biochemistry*, *43*(10), 2184–2193. DOI: 10.1016/j.soilbio.2011.06.022

Jansson, J. K., & Hofmockel, K. S. (2020). Soil microbiomes and climate change. *Nature Reviews. Microbiology*, *18*(1), 35–46. DOI: 10.1038/s41579-019-0265-7 PMID: 31586158

Jayaraman, S., Naorem, A. K., Lal, R., Dalal, R. C., Sinha, N. K., Patra, A. K., & Chaudhari, S. K. (2021). Disease-Suppressive Soils-Beyond Food Production: A Critical Review. *Journal of Soil Science and Plant Nutrition*, *21*(2), 1437–1465. DOI: 10.1007/s42729-021-00451-x PMID: 33746349

Johnston, M. T., Miller, A. D., Holland, O. J., Clark, Z., & Wevill, T. (2024). Impacts of planned burning on the fungal diversity and vascular plants in a heathy woodland. *Forest Ecology and Management*, *563*, 121958. DOI: 10.1016/j.foreco.2024.121958

Junker, R. R., & Farwig, N. (2024). Microbes as conservation targets.

K, C., Kumar, V., Kattany, V., & Ganesh. (2015). Decomposition of Wood by Polypore Fungi in Tropics -Biological, Ecological and Environmental Factors-A Case Study. *Research Journal of Agriculture and Forestry Sciences ISSN 2320-6063, 3*, 1-23.

Kadowaki, K., Yamamoto, S., Sato, H., Tanabe, A. S., Hidaka, A., & Toju, H. (2018). Mycorrhizal fungi mediate the direction and strength of plant–soil feedbacks differently between arbuscular mycorrhizal and ectomycorrhizal communities. *Communications Biology*, *1*(1), 196. DOI: 10.1038/s42003-018-0201-9 PMID: 30480098

Kajihara, K. T., & Hynson, N. A. (2024). Networks as tools for defining emergent properties of microbiomes and their stability. *Microbiome*, *12*(1), 184. DOI: 10.1186/s40168-024-01868-z PMID: 39342398

Kitagami, Y., & Matsuda, Y. (2024). Forest types matter for the community and co-occurrence network patterns of soil bacteria, fungi, and nematodes. *Pedobiologia*, *107*, 151004. DOI: 10.1016/j.pedobi.2024.151004

Kleiner, M., Kouris, A., Violette, M., D'Angelo, G., Liu, Y., Korenek, A., & Lipton, M. S. (2023). Ultra-sensitive isotope probing to quantify activity and substrate assimilation in microbiomes. *Microbiome, 11*(1), 24. DOI: 10.1186/s40168-022-01454-1 PMID: 36755313

Koziol, L., Lubin, T., & Bever, J. D. (2024). An assessment of twenty-three mycorrhizal inoculants reveals limited viability of AM fungi, pathogen contamination, and negative microbial effect on crop growth for commercial products. *Applied Soil Ecology*, *202*, 105559. DOI: 10.1016/j.apsoil.2024.105559

LeBoldus, J. M., Lynch, S. C., Newhouse, A. E., Søndreli, K. L., Newcombe, G., Bennett, P. I., & Gordon, M. (2024). Biotechnology and genomic approaches to mitigating disease impacts on forest health. *Annual Review of Phytopathology*, *62*(1), 309–335. DOI: 10.1146/annurev-phyto-021622-114434 PMID: 39251210

Lehmann, A., & Rillig, M. C. (2024). Systematic mapping of experimental approaches to studying common mycorrhizal networks.

Lehmann, J., Kern, D. C., Glaser, B., & Woods, W. I. (2006). Amazonian dark earths: origin properties management.

Lennon, J. T., & Jones, S. E. (2011). Microbial seed banks: The ecological and evolutionary implications of dormancy. *Nature Reviews. Microbiology*, *9*(2), 119–130. DOI: 10.1038/nrmicro2504 PMID: 21233850

Li, D., Zhang, X., Zhang, H., Fan, Q., Guo, B., & Li, J. (2025). A global meta-analysis reveals effects of heavy metals on soil microorganisms. *Journal of Hazardous Materials*, *491*, 138018. https://doi.org/https://doi.org/10.1016/j.jhazmat.2025.138018. DOI: 10.1016/j.jhazmat.2025.138018 PMID: 40138950

Liang, C., Schimel, J. P., & Jastrow, J. D. (2017). The importance of anabolism in microbial control over soil carbon storage. *Nature Microbiology*, *2*(8), 1–6. DOI: 10.1038/nmicrobiol.2017.105 PMID: 28741607

Liu, Q., & Zhao, W. (2023). *Plant–soil microbe feedbacks drive seedling establishment during secondary forest succession: the 'successional stage hypothesis'* (Vol. 16). Oxford University Press UK.

Lyngwi, N., & Joshi, S. (2015). 'Traditional Sacred Groves', an ethnic strategy for conservation of microbial diversity. *Indian Journal of Traditional Knowledge*, *14*, 474–480.

Malhotra, K. C. (2007). *Sacred groves in India: An overview*. Indira Gandhi Rashtriya Manav Sangrahalaya.

Malik, A. A., Martiny, J. B., Brodie, E. L., Martiny, A. C., Treseder, K. K., & Allison, S. D. (2020). Defining trait-based microbial strategies with consequences for soil carbon cycling under climate change. *The ISME Journal, 14*(1), 1–9. DOI: 10.1038/s41396-019-0510-0 PMID: 31554911

Meena, R. S., Kumar, S., Datta, R., Lal, R., Vijayakumar, V., Brtnicky, M., Sharma, M., Yadav, G., Jhariya, M., Jangir, C., Pathan, S., Dokulilova, T., Pecina, V., & Marfo, T. D. (2020). Impact of Agrochemicals on Soil Microbiota and Management: A Review. *Land (Basel), 9*(2), 34. https://www.mdpi.com/2073-445X/9/2/34. DOI: 10.3390/land9020034

Mendes, R., Garbeva, P., & Raaijmakers, J. M. (2013). The rhizosphere microbiome: Significance of plant beneficial, plant pathogenic, and human pathogenic microorganisms. *FEMS Microbiology Reviews, 37*(5), 634–663. DOI: 10.1111/1574-6976.12028 PMID: 23790204

Meng, W., Chang, L., Qu, Z., Liu, B., Liu, K., Zhang, Y., & Huang, L. (2024). Dominant Tree Species and Litter Quality Govern Fungal Community Dynamics during Litter Decomposition. *10*(10). DOI: 10.3390/jof10100690

Morella, N. M., Weng, F. C.-H., Joubert, P. M., Metcalf, C. J. E., Lindow, S., & Koskella, B. (2020). Successive passaging of a plant-associated microbiome reveals robust habitat and host genotype-dependent selection. *Proceedings of the National Academy of Sciences of the United States of America, 117*(2), 1148–1159. DOI: 10.1073/pnas.1908600116 PMID: 31806755

Nannipieri, P., Ascher, J., Ceccherini, M., Landi, L., Pietramellara, G., & Renella, G. (2017). Microbial diversity and soil functions. *European Journal of Soil Science, 68*(1), 12–26. DOI: 10.1111/ejss.4_12398

Neuenkamp, L., García de León, D., Hamer, U., Hölzel, N., McGale, E., & Hannula, S. E. (2024). Comprehensive tools for ecological restoration of soils foster sustainable use and resilience of agricultural land. *Communications Biology, 7*(1), 1577. DOI: 10.1038/s42003-024-07275-2 PMID: 39592854

Nguyen, N. H., Song, Z., Bates, S. T., Branco, S., Tedersoo, L., Menke, J., Schilling, J. S., & Kennedy, P. G. (2016). FUNGuild: An open annotation tool for parsing fungal community datasets by ecological guild. *Fungal Ecology, 20*, 241–248. DOI: 10.1016/j.funeco.2015.06.006

Ni, B., Xiao, L., Lin, D., Zhang, T.-L., Zhang, Q., Liu, Y., & Rillig, M. C. (2025). Increasing pesticide diversity impairs soil microbial functions. *Proceedings of the National Academy of Sciences of the United States of America, 122*(2), e2419917122. DOI: 10.1073/pnas.2419917122 PMID: 39786931

Nilsson, R. H., Larsson, K.-H., Taylor, A. F. S., Bengtsson-Palme, J., Jeppesen, T. S., Schigel, D., & Tedersoo, L. (2019). The UNITE database for molecular identification of fungi: Handling dark taxa and parallel taxonomic classifications. *Nucleic Acids Research*, *47*(D1), D259–D264. DOI: 10.1093/nar/gky1022 PMID: 30371820

Nyström, M. (2006). Redundancy and response diversity of functional groups: Implications for the resilience of coral reefs. *Ambio*, *35*(1), 30–35. DOI: 10.1579/0044-7447-35.1.30 PMID: 16615697

O'neill, B., Grossman, J., Tsai, M., Gomes, J. E., Lehmann, J., Peterson, J., Neves, E., & Thies, J. E. (2009). Bacterial community composition in Brazilian Anthrosols and adjacent soils characterized using culturing and molecular identification. *Microbial Ecology*, *58*(1), 23–35. DOI: 10.1007/s00248-009-9515-y PMID: 19381712

Onet, A., Grenni, P., Onet, C., Stoian, V., & Crisan, V. (2025). Forest soil microbiomes: A review of key research from 2003 to 2023. *Forests*, *16*(1), 148. DOI: 10.3390/f16010148

Onyango, L. A., Ngonga, F. A., Karanja, E. N., Kuja, J. O., Boga, H. I., Cowan, D. A., Mwangi, K. W., Maghenda, M. W., Marinho Lebre, P. B. N., & Kambura, A. K. (2023). The soil microbiomes of forest ecosystems in Kenya: Their diversity and environmental drivers. *Scientific Reports*, *13*(1), 7156. DOI: 10.1038/s41598-023-33993-4 PMID: 37130890

Pan, H., Wattiez, R., & Gillan, D. (2024). Soil Metaproteomics for Microbial Community Profiling: Methodologies and Challenges. *Current Microbiology*, *81*(8), 257. DOI: 10.1007/s00284-024-03781-y PMID: 38955825

Parks, D. H., Rinke, C., Chuvochina, M., Chaumeil, P.-A., Woodcroft, B. J., Evans, P. N., Hugenholtz, P., & Tyson, G. W. (2017). Recovery of nearly 8,000 metagenome-assembled genomes substantially expands the tree of life. *Nature Microbiology*, *2*(11), 1533–1542. DOI: 10.1038/s41564-017-0012-7 PMID: 28894102

Parrotta, J. A., & Agnoletti, M. (2011). Traditional forest-related knowledge and climate change. In *Traditional forest-related knowledge: sustaining communities, ecosystems and biocultural diversity* (pp. 491–533). Springer.

Peddle, S. D., Hodgson, R. J., Borrett, R. J., Brachmann, S., Davies, T. C., Erickson, T. E., & Watson, C. D. (2025). Practical applications of soil microbiota to improve ecosystem restoration: Current knowledge and future directions. *Biological Reviews of the Cambridge Philosophical Society*, *100*(1), 1–18. DOI: 10.1111/brv.13124 PMID: 39075839

Peng, Z., Qian, X., Liu, Y., Li, X., Gao, H., An, Y., & Chen, S. (2024). Land conversion to agriculture induces taxonomic homogenization of soil microbial communities globally. *Nature Communications*, *15*(1), 3624. DOI: 10.1038/s41467-024-47348-8 PMID: 38684659

Pezdevšek Malovrh, Š., Bećirović, D., Marić, B., Nedeljković, J., Posavec, S., Petrović, N., & Avdibegović, M. (2019). Contribution of forest stewardship council certification to sustainable forest management of state forests in selected Southeast European countries. *Forests*, *10*(8), 648. DOI: 10.3390/f10080648

Pickles, B. J., Twieg, B. D., O'Neill, G. A., Mohn, W. W., & Simard, S. W. (2015). Local adaptation in migrated interior Douglas-fir seedlings is mediated by ectomycorrhizas and other soil factors. *The New Phytologist*, *207*(3), 858–871. DOI: 10.1111/nph.13360 PMID: 25757098

Pieterse, C. M., Zamioudis, C., Berendsen, R. L., Weller, D. M., Van Wees, S. C., & Bakker, P. A. (2014). Induced systemic resistance by beneficial microbes. *Annual Review of Phytopathology*, *52*(1), 347–375. DOI: 10.1146/annurev-phyto-082712-102340 PMID: 24906124

Poe, M. R., Donatuto, J., & Satterfield, T. (2016). "Sense of Place": Human wellbeing considerations for ecological restoration in Puget Sound. *Coastal Management*, *44*(5), 409–426. DOI: 10.1080/08920753.2016.1208037

Qin, X., Xu, J., An, X., Yang, J., Wang, Y., Dou, M., Wang, M., Huang, J., & Fu, Y. (2024). Insight of endophytic fungi promoting the growth and development of woody plants. *Critical Reviews in Biotechnology*, *44*(1), 78–99. DOI: 10.1080/07388551.2022.2129579 PMID: 36592988

Quince, C., Walker, A. W., Simpson, J. T., Loman, N. J., & Segata, N. (2017). Shotgun metagenomics, from sampling to analysis. *Nature Biotechnology*, *35*(9), 833–844. DOI: 10.1038/nbt.3935 PMID: 28898207

Raimbault, A., Brin, A., Manzi, S., Savoie, J. M., Gandois, L., Oliva, P., Fogliani, O., Roy-Camille, C., Gratacap, L., & Roy, M. (2024). Influence of habitat fragmentation and habitat amount on soil fungi communities in ancient forests. *Landscape Ecology*, *39*(2), 19. DOI: 10.1007/s10980-024-01821-3

Reyes-García, V., Fernández-Llamazares, Á., Aumeeruddy-Thomas, Y., Benyei, P., Bussmann, R. W., Diamond, S. K., & Kosoy, N. (2022). Recognizing Indigenous peoples' and local communities' rights and agency in the post-2020 Biodiversity Agenda. *Ambio*, *51*(1), 84–92. DOI: 10.1007/s13280-021-01561-7 PMID: 34008095

Robinson, J. M., Liddicoat, C., Muñoz-Rojas, M., & Breed, M. F. (2024). Restoring soil biodiversity. *Current Biology, 34*(9), R393–R398. DOI: 10.1016/j.cub.2024.02.035 PMID: 38714171

Robinson, S. J. B., Elias, D. M. O., Goodall, T., Nottingham, A. T., McNamara, N. P., Griffiths, R., . . . Ostle, N. J. (2024). Selective logging impacts on soil microbial communities and functioning in Bornean tropical forest [Original Research]. *Frontiers in microbiology, Volume 15 - 2024*. DOI: 10.3389/fmicb.2024.1447999

Rodrigues, J. L., Pellizari, V. H., Mueller, R., Baek, K., Jesus, E. C., Paula, F. S., & Feigl, B. (2013). Conversion of the Amazon rainforest to agriculture results in biotic homogenization of soil bacterial communities. *Proceedings of the National Academy of Sciences of the United States of America, 110*(3), 988–993. DOI: 10.1073/pnas.1220608110 PMID: 23271810

Rodriguez, R., White, J.Jr, Arnold, A. E., & Redman, R. (2009). Fungal endophytes: Diversity and functional roles. *The New Phytologist, 182*(2), 314–330. DOI: 10.1111/j.1469-8137.2009.02773.x PMID: 19236579

Ruan, Y., Ling, N., Jiang, S., Jing, X., He, J.-S., Shen, Q., & Nan, Z. (2024). Warming and altered precipitation independently and interactively suppress alpine soil microbial growth in a decadal-long experiment. *eLife, 12*, RP89392. DOI: 10.7554/eLife.89392 PMID: 38647539

Segata, N., Izard, J., Waldron, L., Gevers, D., Miropolsky, L., Garrett, W. S., & Huttenhower, C. (2011). Metagenomic biomarker discovery and explanation. *Genome Biology, 12*(6), R60. DOI: 10.1186/gb-2011-12-6-r60 PMID: 21702898

Semenov, M. V., Zhelezova, A. D., Ksenofontova, N. A., Ivanova, E. A., Nikitin, D. A., & Semenov, V. M. (2025). Microbiological indicators for assessing the effects of agricultural practices on soil health: A review. *Agronomy (Basel), 15*(2), 335. DOI: 10.3390/agronomy15020335

Shade, A., Peter, H., Allison, S. D., Baho, D. L., Berga, M., Bürgmann, H., & Martiny, J. B. (2012). Fundamentals of microbial community resistance and resilience. *Frontiers in Microbiology, 3*, 417. DOI: 10.3389/fmicb.2012.00417 PMID: 23267351

Sharpton, T. J. (2014). An introduction to the analysis of shotgun metagenomic data. *Frontiers in Plant Science, 5*, 209. DOI: 10.3389/fpls.2014.00209 PMID: 24982662

Shi, Y., He, Y., Zheng, Y., Liu, X., Wang, S., Xiong, T. e., . . . Nian, F. (2024). Characteristics of the phyllosphere microbial community and its relationship with major aroma precursors during the tobacco maturation process [Original Research]. *Frontiers in plant science, Volume 15 - 2024*. DOI: 10.3389/fpls.2024.1346154

Sibalekile, A., Araya, T., Castillo Hernandez, J., & Kotzé, E. (2025a). Glyphosate-microbial interactions: metagenomic insights and future directions [Review]. *Frontiers in microbiology, Volume 16 - 2025*. DOI: 10.3389/fmicb.2025.1570235

Sibalekile, A., Araya, T., Castillo Hernandez, J., & Kotzé, E. (2025b). Glyphosate-microbial interactions: Metagenomic insights and future directions. *Frontiers in Microbiology*, *16*, 1570235. DOI: 10.3389/fmicb.2025.1570235 PMID: 40485833

Singavarapu, B., Du, J., Beugnon, R., Cesarz, S., Eisenhauer, N., Xue, K., Wang, Y., Bruelheide, H., & Wubet, T. (2023). Functional Potential of Soil Microbial Communities and Their Subcommunities Varies with Tree Mycorrhizal Type and Tree Diversity. *Microbiology Spectrum*, *11*(2), e0457822. DOI: 10.1128/spectrum.04578-22 PMID: 36951585

Smith, C. J., & Osborn, A. M. (2009). Advantages and limitations of quantitative PCR (Q-PCR)-based approaches in microbial ecology. *FEMS Microbiology Ecology*, *67*(1), 6–20. DOI: 10.1111/j.1574-6941.2008.00629.x PMID: 19120456

Song, F., Tian, X., Fan, X., & He, X. (2010). Decomposing ability of filamentous fungi on litter is involved in a subtropical mixed forest. *Mycologia*, *102*(1), 20–26. DOI: 10.3852/09-047 PMID: 20120224

Spooren, J., van Bentum, S., Thomashow, L. S., Pieterse, C. M., Weller, D. M., & Berendsen, R. L. (2024). Plant-driven assembly of disease-suppressive soil microbiomes. *Annual Review of Phytopathology*, •••, 62. PMID: 38857541

Steidinger, B. S., Crowther, T. W., Liang, J., Van Nuland, M. E., Werner, G. D., Reich, P. B., & Picard, N. (2019). Climatic controls of decomposition drive the global biogeography of forest-tree symbioses. *Nature*, *569*(7756), 404–408. DOI: 10.1038/s41586-019-1128-0 PMID: 31092941

Taberlet, P., Bonin, A., Zinger, L., & Coissac, E. (2018). *Environmental DNA: For biodiversity research and monitoring*. Oxford University Press. DOI: 10.1093/oso/9780198767220.001.0001

Taylor, B. N. (2025). Symbiotic nitrogen fixation in trees: Patterns, controls and ecosystem consequences. *Tree Physiology*, *45*(1), tpae159. DOI: 10.1093/treephys/tpae159 PMID: 39658308

Thapa, S., & Prasanna, R. (2018). Prospecting the characteristics and significance of the phyllosphere microbiome. *Annals of Microbiology*, *68*(5), 229–245. DOI: 10.1007/s13213-018-1331-5

Thomsen, P. F., & Willerslev, E. (2015). Environmental DNA–An emerging tool in conservation for monitoring past and present biodiversity. *Biological Conservation*, *183*, 4–18. DOI: 10.1016/j.biocon.2014.11.019

Tláskal, V., Brabcová, V., Větrovský, T., Jomura, M., López-Mondéjar, R., Oliveira Monteiro, L. M., Saraiva, J. P., Human, Z. R., Cajthaml, T. Á., Nunes da Rocha, U., & Baldrian, P. (2021). Complementary Roles of Wood-Inhabiting Fungi and Bacteria Facilitate Deadwood Decomposition. *mSystems*, *6*(1), e01078-20. Advance online publication. DOI: 10.1128/mSystems.01078-20 PMID: 33436515

Toju, H., Peay, K. G., Yamamichi, M., Narisawa, K., Hiruma, K., Naito, K., & Onoda, Y. (2018). Core microbiomes for sustainable agroecosystems. *Nature Plants*, *4*(5), 247–257. DOI: 10.1038/s41477-018-0139-4 PMID: 29725101

Torres, N., Herrera, I., Fajardo, L., & Bustamante, R. O. (2021). Meta-analysis of the impact of plant invasions on soil microbial communities. *BMC Ecology and Evolution*, *21*(1), 172. DOI: 10.1186/s12862-021-01899-2 PMID: 34496752

Torsvik, V., & Øvreås, L. (2002). Microbial diversity and function in soil: From genes to ecosystems. *Current Opinion in Microbiology*, *5*(3), 240–245. DOI: 10.1016/S1369-5274(02)00324-7 PMID: 12057676

Trevelline, B. K., Fontaine, S. S., Hartup, B. K., & Kohl, K. D. (2019). Conservation biology needs a microbial renaissance: a call for the consideration of host-associated microbiota in wildlife management practices. *Proceedings of the Royal Society B*, *286*(1895), 20182448. DOI: 10.1098/rspb.2018.2448

Trivedi, P., Anderson, I. C., & Singh, B. K. (2013). Microbial modulators of soil carbon storage: Integrating genomic and metabolic knowledge for global prediction. *Trends in Microbiology*, *21*(12), 641–651. DOI: 10.1016/j.tim.2013.09.005 PMID: 24139848

Turner, N. J., Ignace, M. B., & Ignace, R. (2000). Traditional ecological knowledge and wisdom of aboriginal peoples in British Columbia. *Ecological Applications*, *10*(5), 1275–1287. DOI: 10.1890/1051-0761(2000)010[1275:TEKAWO]2.0.CO;2

Tyson, G. W., Chapman, J., Hugenholtz, P., Allen, E. E., Ram, R. J., Richardson, P. M., Solovyev, V. V., Rubin, E. M., Rokhsar, D. S., & Banfield, J. F. (2004). Community structure and metabolism through reconstruction of microbial genomes from the environment. *Nature*, *428*(6978), 37–43. DOI: 10.1038/nature02340 PMID: 14961025

Van Der Heijden, M. G., Bardgett, R. D., & Van Straalen, N. M. (2008). The unseen majority: Soil microbes as drivers of plant diversity and productivity in terrestrial ecosystems. *Ecology Letters*, *11*(3), 296–310. DOI: 10.1111/j.1461-0248.2007.01139.x PMID: 18047587

Vasar, M., Davison, J., Sepp, S. K., Mucina, L., Oja, J., Al-Quraishy, S., & Cantero, J. J. (2022). Global soil microbiomes: A new frontline of biome-ecology research. *Global Ecology and Biogeography*, *31*(6), 1120–1132. DOI: 10.1111/geb.13487

Visca, A., Di Gregorio, L., Costanzo, M., Clagnan, E., Nolfi, L., Bernini, R., & Mocali, S. (2025). Microbial bioindicators for monitoring the impact of emerging contaminants on soil health in the European framework. *Sustainability (Basel)*, *17*(3), 1093. DOI: 10.3390/su17031093

Vivian, J., Chazdon, R. L., Catling, A. A., & Lee, D. J. (2025). Early evidences of links between soil microbes and forest restoration through multiple ecosystem metrics. *Frontiers in Forests and Global Change*, *8*, 1540513. DOI: 10.3389/ffgc.2025.1540513

Wang, F., Xiang, L., Sze-Yin Leung, K., Elsner, M., Zhang, Y., Guo, Y., & Ying, G. (2024). Emerging contaminants: A One Health perspective. [In: REVIEW.]. *Innovation (Cambridge (Mass.))*, *5*(4), 100612. DOI: 10.1016/j.xinn.2024.100612 PMID: 38756954

Wang, H., Deng, N., Wu, D., Hu, S., & Kou, M. (2017). Long-term net transformation and quantitative molecular mechanisms of soil nitrogen during natural vegetation recovery of abandoned farmland on the Loess Plateau of China. *The Science of the Total Environment*, *607*, 152–159. DOI: 10.1016/j.scitotenv.2017.07.014 PMID: 28689119

Wang, H., Wang, H., Crowther, T. W., Isobe, K., Reich, P. B., Tateno, R., & Shi, W. (2024). Metagenomic insights into inhibition of soil microbial carbon metabolism by phosphorus limitation during vegetation succession. *ISME Communications*, *4*(1), ycae128. DOI: 10.1093/ismeco/ycae128 PMID: 39507397

Wang, Z., Yun, S., An, Y., Shu, L., Li, S., Sun, K., & Zhang, W. (2025). Effect of fungicides on soil respiration, microbial community, and enzyme activity: A global meta-analysis (1975–2024). *Ecotoxicology and Environmental Safety*, *289*, 117433. DOI: 10.1016/j.ecoenv.2024.117433 PMID: 39637631

Widyati, E., Budiharta, S., Akbar, A., Susilo, A., Kurniawan, A., Sadili, A., & Siregar, M. (2025). Changes in soil-root-organism interactions following tropical forest conversion to tree and oil palm plantations. *Applied Soil Ecology*, *213*, 106253. DOI: 10.1016/j.apsoil.2025.106253

Wu, S., Qu, Z., Chen, D., Wu, H., Caiyin, Q., & Qiao, J. (2024). Deciphering and designing microbial communities by genome-scale metabolic modelling. *Computational and Structural Biotechnology Journal, 23*, 1990–2000. DOI: 10.1016/j.csbj.2024.04.055 PMID: 38765607

Xiong, Q., Yang, J., & Ni, S. (2023). Microbiome-mediated protection against pathogens in woody plants. *International Journal of Molecular Sciences, 24*(22), 16118. DOI: 10.3390/ijms242216118 PMID: 38003306

Xu, M., Fan, L., Li, A., Liu, Q., Yu, G., Wang, S., . . . Zheng, M. (2024). Plant and microbial carbon are important drivers of free-living nitrogen fixation in tropical forest soils: A new discovery of carbon-driven nitrogen input. *Geophysical Research Letters, 51*(20), e2024GL111238.

Yan, F., Hermansen, C., & Norgaard, T. (2025). Effects of microplastics on soil microbial diversity and community structure revealed by meta-analysis. *Agriculture, Ecosystems & Environment, 390*, 109720. https://doi.org/https://doi.org/10.1016/j.agee.2025.109720. DOI: 10.1016/j.agee.2025.109720

Yang, Z., He, G., Yang, Q., Zhang, D., Zhang, Y., Wen, S., Yang, X., Yang, L., & Ji, L. (2024). Nitrogen enrichment stimulates nutrient cycling genes of rhizosphere soil bacteria in the Phoebe bournei young plantations. *Journal of Environmental Management, 371*, 123101. DOI: 10.1016/j.jenvman.2024.123101 PMID: 39486293

Ye, H., Tu, N., Wu, Z., He, S., Zhao, Y., Yue, M., & Hong, M. (2024). Identification of bacteria and fungi responsible for litter decomposition in desert steppes via combined DNA stable isotope probing [Original Research]. *Frontiers in microbiology, Volume 15 - 2024*. DOI: 10.3389/fmicb.2024.1353629

Zhang, L.-M., Hu, H.-W., Shen, J.-P., & He, J.-Z. (2012). Ammonia-oxidizing archaea have more important role than ammonia-oxidizing bacteria in ammonia oxidation of strongly acidic soils. *The ISME Journal, 6*(5), 1032–1045. DOI: 10.1038/ismej.2011.168 PMID: 22134644

Zhang, Y., Cong, J., Sheng, Y., Lu, H., Feng, K., Li, G., & Deng, Y. (2020). Soil microbial distribution pattern and their ecological linkage with plant diversity along a 2500 km forest latitudinal gradient. *Authorea Preprints*.

Zhao, M., Yang, Y., Zhang, H., Li, Q., Zhao, X., Guo, X., Liu, W., & Wan, F. (2024). Asymmetric succession in soil microbial communities enhances the competitive advantage of invasive alien plants. *Microbiome, 12*(1), 265. DOI: 10.1186/s40168-024-01989-5 PMID: 39707566

Chapter 11
Agroforestry and Indigenous Knowledge:
Bridging Biodiversity Conservation and Community Resilience

Arun Kumar
Lovely Professional University, India

Vikanksha Vikanksha
https://orcid.org/0009-0009-0257-4319
Lovely Professional University, India

Jatinder Singh
https://orcid.org/0000-0001-7691-6689
Lovely Professional University, India

ABSTRACT

Agroforestry, the integration of trees and shrubs into agricultural landscapes, has emerged as a critical practice for enhancing biodiversity conservation and fostering community resilience. This paper explores the intersection of agroforestry and indigenous knowledge systems, highlighting how traditional ecological practices can contribute to sustainable land management. Indigenous communities possess valuable insights regarding local biodiversity, ecosystem services, and sustainable agricultural practices that have evolved over generations. By bridging indigenous knowledge with modern agroforestry techniques, this approach not only promotes biodiversity but also strengthens community resilience against climate change and environmental degradation. This study emphasizes the importance of participatory methods in the design and implementation of agroforestry systems, ensuring that the voices of indigenous peoples are central to conservation efforts.

DOI: 10.4018/979-8-3373-3992-4.ch011

1. INTRODUCTION

As the global population continues to grow and expand into previously uninhabited regions, human activities increasingly threaten biodiversity. Civil society advocates, despite the significant global impact of biodiversity loss, are fighting back. However, responses and recoveries are delayed by poverty, poor governance, and insufficient land reform. Such political processes are equally behind the resistance against climate change. Biodiversity is upheld by the rules or norms that govern how humans interact with the other species of a given ecosystem. If there is no normative agreement among people, then biodiversity is loose. This chapter focuses on the role of local legislation and environmental monitors in maintaining biodiversity and indigenous knowledge of agroforestry. Management models differ markedly across socio-ecological contexts. While closer spatial proximity to highly managed lands can be predicted to have more trees by a less extensive local model, environmental monitoring gets broader. The local management model indicates the prior need to provide information on regulators and land insurance that predict trees more broadly than forest proximity or indigenous agriculture with sparse information on the dependent variable. It is argued that the conduct of biodiversity management research needs to be a reciprocal effort involving management actors to access the tacit knowledge often overlooked by social researchers. It is concluded that biodiversity management may be an intelligent entrepreneurial process bridging understanding and practice across social domains rather than a rational political approach founded on the involved knowledge of social scientists (Rae & Blenker, 2024).

Close-up tropical lowland agroforestry in Southwest Java focuses on commercially pioneering perennial row crops arranged into populations accounting symbiotically trees and monocots on row banks. Ornamental monocots limited to the seasonally wet marshlands are captured for this purpose. In this smallholder experiment, public state power contributed to individual risks by preventing the more profitable, upward, and environmentally friendly large-scale production of row crops and the piggery agroindustries reliant on them. In a constrained farm territory, the principled uptake of public goods granted acquisition priority to transitory smallholders. Native agroforestry systems have been developed and adapted in spaces of agrarian difficulty over decades or centuries by local people in South and Southeast Asia (Dahal et al. 2025). Heterogeneity at multiple scales presents conservation scholarships and management challenges for a perennial tree-based crop in transcalant cacao landscape mosaics.

2. UNDERSTANDING AGROFORESTRY

The term agroforestry describes farming systems that include trees, woody shrubs, or palms as part of the near-permanent agricultural system. Trees could be deliberately planted, be retained from a previous crop or pasture, or recruited from wild. These systems also typically include a matrix of non-woody vegetation and numerous species important for animal production. Likewise, both tree and non-tree components of the farming system could be rich in species, although, historically, scholars have focused mainly on trees and tall woody plants. Agroforestry is widely regarded as a sustainable land-use system that simultaneously contributes to environmental conservation, climate mitigation, and household livelihoods (Keprate et al. 2024). Perennial or semi-perennial trees and shrubs, planted together with other crops or pasture, improve the micro-climate and soil quality, provide food and fodder, and act as sinks for carbon as they accumulate biomass. They also tend to stabilize or rehabilitate the landscape by reducing erosion or sedimentation, modulating the water cycle, and reducing vegetation flammability. Smallholder agroforestry can improve food security and farming resilience to climate change by intensifying crop or pasture with less investment, innovating new species or management, diversifying income sources, and restoring degraded land. A system is generally accepted as agroforestry when it combines plots dedicated to permanent cultures of trees with plots for non-woody plants and has some synergy components; that is (Nair et al. 2021), trees should confer at least one sustainability function extending beyond them. Agroforestry could include managed parklands of scattered trees or a forest matrix with few plots with manicured canopies of small, less-diverse woody vegetation. Both forest cover predominantly over cropped pastures represent very low-intensity agroforestry systems, widely regarded as less socioculturally or economically progressive (Figure 1).

Figure 1. Integrated benefits of agroforestry

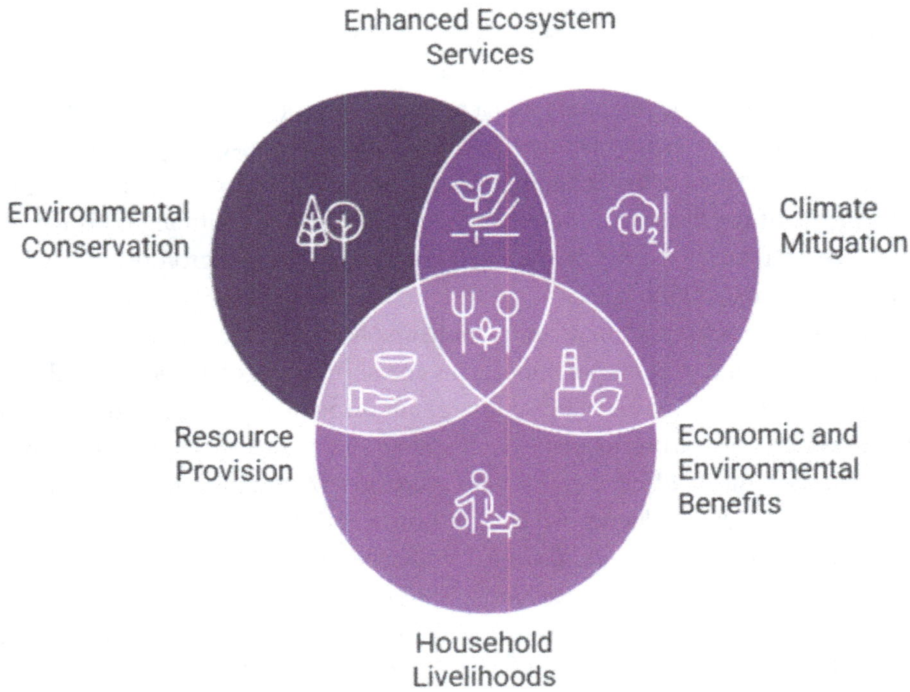

2.1. Definition and Principles

Agroforestry has various definitions, paradigms, approaches, and design models. This complexity is likely to increase rather than decrease in future discussions. As a consequence, definitions of agroforestry have been pluralized as people begin, or continue, discourses using agroforestry as a touchstone term for different ideas. The 'forestry-agriculture interaction model' for conceptualizing various agroforestry designs involves edge effect models inspired by methodologies, but they restrict their definitions of agroforestry to complex systems only (Achmad et al. 2022). The role of perennial plants in agriculture is excluded in their definitions, even though traditional 'forest gardening', shade agriculture, taungya, and other agroforestry systems worldwide create forests, plantations, or woodlots from 'transitional forests'. The tree cover is often little less than 10% in these timber-shedding systems, which is insufficient to qualify as agroforests in these models. However, complex systems should also coexist with simpler systems with limited tree cover: these systems can be called 'agroforestry' as well. There are three paradigms: natural resource and vegetation management paradigms based on management of tree cover, intensification of tree-crop interactions paradigm based on management of agroecological processes, and sustainable development paradigm based on tolerable trade-offs between various values of trees. All three paradigms have been explored and debated by various scholars and proponents from all over the globe for nearly three decades. Agroforestry approaches can be categorized into diversification and intensification approaches (Gómez et al. 2022). Intensification approach incorporates various agroecological technologies for improving tree-crop interactions. Some agroforestry systems with complex designs are often constrained at landscape scale due to common pool tragedies, institutional constraints, practical sequencing of interventions, foregone short-term benefits, and land tenurial constraints. However, diversification approaches with simpler designs which can co-exist with monocultures and low-input offer greater choice and flexibility across biophysical and socio-economic circumstances, and 'forestry-agriculture interaction paradigms' need to be a pluralistic way of thinking on attributes of agroforestry or tree-crop interactions.

2.2. Types of Agroforestry Systems

Agroforestry practices, defined broadly as intersecting spatially or temporally trees with crops and/or livestock in the same agricultural management unit, are long-lived multi-functional systems that provide a wide variety of ecosystem services crucial for sustainable rural development. Based on diverse arrangements, interaction, and management of trees and shrubs with crops and/or livestock, it has been classified into four categories: i.e., alley cropping, silvopasture, parkland, and home gardens (Sauer et al. 2021). System descriptions with emphasis on typology classification criteria will be highlighted in this section. Agroforestry practices based on trees and shrubs purposely planted in alleys already planted with crops in an annual or semi-perennial basis to take advantage of interspecific competition for resources between agroforestry components is referred to as alley cropping. Dry temperate and subtropical areas populated with herbaceous legume crops have been, mostly, the common sites of such practices for soil improvement and property enhancement due to intercropping benefits. Alley cropping practices with fodder shrubs and trees in tropic and subtropic environments for the same properties were, however, less common. Silvopasture systems, which integrate trees and shrubs with forage and livestock in the same management unit, are defined as agroforestry practices with woody perennials purposely interacting with herbaceous perennials in a spatially or temporally overlapping manner to take advantage of interspecific competitiveness for resources or to alleviate environmental stresses (Chappa et al. 2024). This kind of agroforestry system can be designed with varied arrangements and densities of trees, shrubs, forages and livestock based on the required ecosystem services. Although individual components in silvopastures may substantially vary in types and purpose, the four-layer structure composed of a tree and/or shrub layer at the top, a forage layer at the mid, a terrestrial herbaceous layer, and an animal layer at the bottom could be simplified to describe almost all existing silvopasture practices. Pastures populated with tree or shrub species voluntarily emerged are termed as "spontaneous silvopastures," while those purposes planted with economically important cash tree or shrub species are termed as "deliberate silvopastures."

Table 1. Overview of Various Agroforestry Systems, Their Descriptions, Examples, and Benefits

Type of Agroforestry System	Description	Examples	Benefits
Alley Cropping	Crops grown between rows of trees	Corn and walnut trees	Improved soil fertility, reduced erosion
Silvopasture	Integration of trees with livestock	Cattle grazing among pines	Enhanced animal welfare, diversified income
Forest Farming	Cultivation of crops under tree canopy	Mushrooms under oak trees	Shade protection, increased biodiversity
Windbreaks	Rows of trees to protect crops	Trees along field edges	Reduced wind damage, improved microclimate
Homegardens	Diverse plants around homesteads	Vegetables, fruits, and herbs	Food security, nutritional diversity
Taungya System	Crops grown in early stages of forest establishment	Rice and young rubber trees	Soil conservation, sustainable forestry
Riparian Buffers	Vegetation along waterways	Trees and shrubs along rivers	Water quality improvement, habitat creation
Tree Crop Systems	Cultivation of trees for specific products	Cocoa and coffee plantations	Long-term income, carbon sequestration
Shade Coffee Systems	Coffee grown under shade of trees	Arabica coffee under canopy	Biodiversity support, better quality coffee
Intercropping	Growing multiple crops together	Beans with maize	Maximized land use, pest management
Improved Fallows	Crops planted in fallow periods	Legumes in rotation	Soil fertility restoration, weed suppression
Agroforestry Parkland	Trees scattered across pasture lands	Acacia or baobab trees	Livestock shading, forage improvement
Conservation Agroforestry	Agroforestry practices for conservation	Mixed tree and crop systems	Biodiversity enhancement, habitat connectivity
Community Agroforestry	Collective agroforestry practices by communities	Shared forest gardens	Community cohesion, shared resources
Integrated Pest Management (IPM)	Use of agroforestry techniques for pest control	Diverse plantings to attract predators	Reduced pesticide use, enhanced ecosystem health

2.3. Benefits of Agroforestry

Most landscapes worldwide are affected by interaction among climate, land use, and land cover effects on agricultural activities. Sediment production and aerial coverage are the main components of landscape changes. Sediment production conveys off-site environmental effects such as river dam siltation, sediment transport, and sedimentary processes which affect water quality. Increased land cultivation across Ethiopia has resulted in extensive land degradation, soil erosion, sediment transport, and environmental challenges at the river watershed scale. In addition to natural events, it is suggested that human interventions such as tree planting significantly control hydrological implications in the watershed. Agroforestry has been practiced traditionally in many parts of the country for both economic and environmental benefits (Mukhlis et al. 2022). However, the role of agroforestry in sediment production, transport, and environmental management at the watershed scale has not been adequately addressed yet. Semi-structured interviews were administered to key informants, household surveys, focus group discussions, and soil sample collection and analysis. Two-staged sediment source sampling with particle-size analysis was performed. Visualization of the spatiotemporal interaction among climate and land use and land cover change effects on sediment flux on the landscape was conducted. Households with higher knowledge capital usually practice crooked/contorted technology against sedimentation. Rainfall events and agriculture are the main sources of sediment in the watershed and sediment yields show annual and monthly variability across space and time. Application of street tree planting, agroforestry, and conservation technologies are recommended to mitigate soil erosion-causing sediment fluxes at the watershed scale (Gobinath et al. 2024). Alpha diversity indices of soil organic carbon pools or soil microbial communities did not differ significantly with land uses. However, taxonomic composition of soil microbial communities and the size of soil carbon pools were significantly influenced by land use types. Agroforestry showed distinct Bacterial Communalities among other land uses, but not Fungi. Pollarding of trees in agroforestry may affect substrate availability and microbial diversity due to reduced leaf litter inputs. Technologies improved soil organic carbon quality and the fertility status of the soil. Improved soil fertility affects the abundance of SOC pools controlling microbial growth and biodiversity by affecting nutrient and water dynamics in the soil. Agroforestry significantly contributed to soil carbon pools and microbial community structural variation especially the Bacterial Phylum.

3. INDIGENOUS KNOWLEDGE SYSTEMS

Indigenous knowledge systems (IKS) have been applied widely in a variety of systems and cultures to manage a diverse range of landscapes of importance to a variety of stakeholders. Indigenous and local communities have devised cultural practices embedded in cultural and religious values that have maintained species and habitats of biocultural importance through IKS. This topic represents an emerging frontier in the field of biodiversity conservation where a diverse range of biophysical, ethnobotanical, economic, socio-political, and cultural knowledge can be brought to bear on the question of biodiversity management and conservation (Correia et al. 2025). It also highlights the effort of disentangling the knowledge systems relevant to the past, present or future interactions of different communities' cultural practices, beliefs, spatial and temporal domain, and natural resources. These systems have been shaped by the evolution of worldviews, cultural practices, institutions and other elements forming a nested knowledge-practice-belief complex that provide insights into ways of knowing and governing social-ecological systems for contemporary biodiversity management and conservation. In Indonesia, agroforestry has been practiced for centuries as a strategy to balance agricultural expansion and rainforest cover loss. Nevertheless, the challenges for sustainability remain substantial in tropical biodiversity hotspots. Traditional knowledge of indigenous and local communities that has enabled this agroforestry systems to persist for so long needs to be studied and raised awareness of to stimulate innovation and flexibility in response to evolving ecological and socio-political contexts (Sharifian et al. 2022). In Melanesia, inclusive development has stimulated new partnerships to maintain traditional indigenous knowledge with enduring bio-cultural significance that needs to be transcribed or revitalized in a contemporary format. Safeguarding measures are also needed to match the evolving use of indigenous knowledge, including customs-based biocultural protocols, novel partnerships with researchers or citizens, identification and protection of traditional domains of practice and resources through co-custodianship.

3.1. Definition and Importance

Traditionally ecological knowledge (TEK) consists of knowledge, beliefs, practices and skills regarding ecological landscapes that accrued over generations of lived experience of social-ecological systems. TEK addresses a variety of issues at different spatial and temporal scales and it intimately integrates knowledge of biophysical components of the environment with a sense of place and environmental ethics (Albuquerque et al. 2021). TEK operates within overarching cosmological beliefs system that recognizes interconnectedness, change, meaning, education and temporality. TEK systems mesh local environmental knowledge between children and parents and elaborate on what is perceived to count as knowledge. Despite conceptual and methodological advances to render this complex phenomena TEK remains neglected in biodiversity conservation. Indigenous lands overlap priority sites for conservation of biodiversity and ecosystem services, yet indigenous communities do not share equal inclusion in decision making and governance of conservation initiatives. Thus, TEK is mentioned in ethics and scientific approaches but is not fully included in policy frameworks. Models of coexistence grounded in TEK and blending science with indigenous spirituality and morals offer hope for counteracting the degeneration of the planet (Goshu& Ridwan, 2024). The ongoing dissemination of biocultural conservation models across the world demonstrates the urgency for definitive and new arrangements of how to coexist with all the diverse nonhuman species. Agroforestry systems are some of the key biocultural landscapes and focus of the case study. AFS are defined as integrated systems that incorporate woody perennials either simultaneously or sequentially with crops and/or livestock on the same land management unit. AFS can be characterized as complex (many species of trees, crops and/or livestock, with interactions and diversity in structure), diversified (one or a few species of trees and/or crops with low interaction and structured diversity), and simple/plantation-like (few species of trees planted in a spatially delimitated area with mainly agricultural crops) (Moreno-Quintero et al. 2021). AFS are claimed to be potential biodiversity conservation areas and providers of a wide range of ecosystem services such as climate mitigation/adaptation, water, soil, as well as culturally relevant services. AFS are socially sustainable and resilient due to their social and ecological dynamics. Several pressures, however, threaten the AFS. On the one hand, some kinds of AFS decrease or change over time, an issue related to intensifying and converting them into monoculture agriculture. On the other hand, some kinds of AFS are promoted and expanded whilst intensifying agroecological transformation. The change of balance of pressures changes the proposal, implementation and willingness of engagements with AFS.

Table 2. Indigenous Knowledge Systems – Descriptions, Examples, and Associated Benefits

Indigenous Knowledge System	Description	Examples	Benefits
Traditional Ecological Knowledge	Understanding of ecosystems developed over generations	Seasonal migration patterns	Sustainable resource management
Medicinal Plant Knowledge	Use of local plants for health care	Herbal remedies	Cost-effective healthcare
Agricultural Practices	Techniques for sustainable farming	Crop rotation, polyculture	Increased food security
Land Management Practices	Strategies for land use and conservation	Controlled burns, sacred sites	Biodiversity conservation
Water Management Systems	Traditional methods for managing water resources	Rainwater harvesting	Improved water quality and availability
Oral Traditions	Storytelling and knowledge transmission	Myths, legends, and proverbs	Cultural preservation
Spiritual Beliefs	Connection to the land and nature	Ancestral worship	Enhanced community identity
Animal Husbandry Knowledge	Practices related to livestock management	Breeding techniques	Improved livestock health
Climate Adaptation Strategies	Knowledge of adapting to climate change	Crop varieties for drought	Resilience to climate impacts
Resource Sharing Practices	Community-based sharing of resources	Communal hunting and fishing	Enhanced community cooperation
Indigenous Governance Systems	Traditional decision-making structures	Tribal councils	Empowerment and self-determination
Craftsmanship and Art	Skills related to traditional crafts	Weaving, pottery, and carving	Cultural identity and economic opportunity
Fishing Techniques	Sustainable practices for fishing	Traditional nets and traps	Conservation of aquatic ecosystems
Food Preservation Methods	Techniques for storing and preserving food	Drying, smoking, and fermenting	Reduced food waste
Community Health Practices	Holistic approaches to health and well-being	Community gatherings, rituals	Strengthened social bonds

3.2. Cultural Practices and Traditions

Collective knowledge about natural systems including species, ecosystems, and the interactions among these elements is important for human well-being and may be considered traditional ecological knowledge (TEK). Information, techniques, and management practices of plants are culturally transmitted and evolve through time. The accurate understanding of traditional agroforestry systems helps to appreciate the biocultural diversity that has resulted from human–natural systems interactions, and to promote the sustainable utilization of biodiversity (Vallejo-Ramos et al. 2016). In the arid Tehuacán Valley, Mexico, people manage traditionally a variety of woody and nonwoody species in diverse agroforestry systems including multi-crop "milpa", cactus forest, and the combination of both systems. The aim of this study was to explore, for woody species only, the management of culturally important plants in agroforestry systems of the Tehuacán Valley in the context of cultural diversity by (i) analyzing important plant species for non-wood products and their perceived benefits, (ii) describing traditional management methods and land-use practices for cultivated and spontaneous woody species, (iii) exploring the influence of socio-cultural differences on distinct management methods and species, and (iv) assessing public knowledge about agroforestry species. Indigenous knowledge about agroforestry benefits and species. The majority of the respondents described both spontaneous and cultivated woody species in the agroforestry systems around the village (Baliton et al. 2020). Unlike other studies in Latin America, where species from temperate and tropical zones were reported, the species described in the Tehuacán Valley are mainly cacti from the subfamily Opuntioideae, which may be due to the unique climatic conditions. Agroforestry species are utilized in several ways, but the greatest proportion of species is used as food. Two discourse categories were constructed to describe same or closely linked ideas and terms, including (i) food, fiber and fuel and (i) foraging Bees, wild fauna, and erosion control. First category focused on the product of plant use and the second emphasized ecosystems services provided by the species.

3.3. Case Studies of Indigenous Knowledge

Indigenous knowledge (IK) is shown to play a significant role in forest management among the Pokot tribes of Kenya (Guto, 2020). The forests have a general, traditional category to which different, practical, and functional classifications can be ascribed. The understanding of tree species can be very involved and detailed, tapping into accepted systems of classification that draw on a range of knowledge about the plants and the environments in which they grow. Knowledge can be, indeed is, of a multidisciplinary nature as different stakeholders come to share the same tree (e.g., the school, community, etc.); however, these local categories are often not reflected in formal practices of environmental management. In this light, it is argued that IK has important values in the search for new solutions in sustainable forestry practices. Considering a case study from Kenya, it is shown that recycling recycled knowledge is of great importance in the search for new solutions. This opens up ways to include knowledgeable locals and local practices in the processes of environmental management. Such involvement can be viewed as a benign and empowering measure that gives increased voice and agency to those who have been neglected or oppressed. It is likewise argued that such involvement presupposes respectful, productive, and sustained action and dialogue with different partners with scarce knowledge and resources, and at different levels of economy. Accordingly, insights from the literature on bridge-building and continual action and change networks are drawn upon and further dimensions are added regarding the ideals of voice, agency, and relations with outsiders. This study contributes to environmental sociology, environmental geography, and the anthropology of landscapes by exploring a particular case of action-orientated research in environmental management that draws on a resource domain that is at once spatial, local, elusive, and in the changing in a rapid modernity. It also informs southern viewpoints on IK in environmental management and the challenges this raises in standardized systems of classification, reasoning, and management. The study area lies in West Pokot District, Kenya.

4. BIODIVERSITY CONSERVATION

The biodiversity crisis is serious and global. Responses have taken many forms, including additional political and financial resources to protected areas in richer countries (Baynham-Herd et al. 2018). However, in a relationship of power and philosophical differences, this response has either ignored indigenous lands and authorities and/or is creating a new and serious crisis on indigenous lands. Moreover, there is a lack of local, indigenous agency and of explicit attempts to incorporate indigenous knowledge and practice in the post-Humboldt conservation paradigm.

The resilience of some indigenous systems and gradual response to crisis through re-connection are described. It is argued that bridging proposals must be based on co-governance and co-management initiatives, drawing on existing indigenous conservation standards. Collective action by indigenous funds and local authority on contaminated indigenous territory is also described. Also the limitations and difficulties faced by indigenous groups and NGOs in trying to negotiate with states, and the partial and conflicting responses of states to these efforts are sketched. The challenges of creating a post-Humboldt conservation paradigm that is equitable and effective, in a relationship of power full of history and hypocrisy, is daunting (Sułkowski, 2023). Nevertheless, many indigenous groups still have rights to deep reservoirs of biocultural knowledge, culturally important species and habitats, and pre-Hispanic landscapes and territories. Nonetheless damage is being done; habitat loss, pollution and invasion are occurring at unprecedented rates, partly because of large scale commodity agriculture responded to growing global demand for food and biofuels, all at the expense of wild systems and the people who depend on them.

This 'Green Revolution' has involved the disconnection of the Mesoamerican people from their biocultural landscapes (del Castillo et al. 2022). The ability of some indigenous systems to respond to a changing world is illustrated through cultivation and management of formerly submerged, now arid, plains in Mexico, systems which contributed to the resilience of the Mecho outdated. Co-governance and indigenous conservation standards efforts are described and analysed. There is also the need for tools of appreciation, recognition and negotiation that bridge indigenous knowledge and governance with dominant conceptual and practice frameworks.

Table 3. Approaches to Biodiversity Conservation – Descriptions, Examples, and Benefits

Biodiversity Conservation Approach	Description	Examples	Benefits
Protected Areas	Designated regions for habitat protection	National parks, wildlife reserves	Habitat preservation, species protection
Community Conservation	Local communities managing natural resources	Community forests, marine reserves	Empowerment, sustainable livelihoods
Ecological Restoration	Restoring degraded ecosystems	Reforestation projects	Enhanced ecosystem services
Sustainable Land Use Practices	Practices that maintain biodiversity while utilizing land	Agroforestry, permaculture	Soil health, increased productivity
Species Recovery Programs	Initiatives focused on endangered species	Breeding programs, habitat restoration	Preventing extinction, genetic diversity
Conservation Genetics	Study of genetic diversity for conservation	DNA analysis in wildlife	Improved management strategies
Urban Biodiversity Initiatives	Enhancing biodiversity in urban areas	Green roofs, urban parks	Improved air quality, community well-being
Conservation Education	Informing and engaging communities about biodiversity	Workshops, school programs	Increased awareness, behavioral change
Invasive Species Management	Strategies to control non-native species	Eradication programs	Protection of native ecosystems
Ecosystem Services Valuation	Assessing the economic value of ecosystem services	Carbon credits, water purification	Informed decision-making
Policy and Legislation	Laws and regulations aimed at protecting biodiversity	Endangered Species Act	Legal frameworks for protection
Traditional Ecological Knowledge	Indigenous knowledge used in conservation efforts	Sustainable hunting practices	Culturally relevant conservation
Biodiversity Offsets	Compensation for biodiversity loss due to development	Habitat creation elsewhere	Balancing development and conservation

continued on following page

Table 3. Continued

Biodiversity Conservation Approach	Description	Examples	Benefits
International Agreements	Global treaties focused on biodiversity conservation	Convention on Biological Diversity	International cooperation
Research and Monitoring	Scientific studies to track biodiversity changes	Species inventories, biodiversity assessments	Informed policy-making

4.1. Role of Biodiversity in Ecosystems

Ecosystem services are the benefits that people obtain from ecosystems. These include, among others, supporting services (nutrient cycling, soil formation, primary production), provisioning services (food, fresh water, timber, fuel wood), regulating services (climate regulation, flood regulation, water purification), and cultural services (landscape) (Ali & Kamraju, 2023). The role of biodiversity at the community, ecosystem, and landscape levels in underlying most of the services provided by natural ecosystems. Such services are needed to reduce the negative impacts of climatic variability and extremes, or other stressors, on socio-ecological systems' resilience. Community, ecosystem, and landscape levels of biodiversity should be included in designing, implementing, and evaluating strategies aimed at enhancing the provision and management of ecosystem services. Biodiversity is essential for the functioning of ecosystems and its conservation has become a major concern (Rawat & Agarwal, 2015). Most of the ecological literature has focused on the effects of biodiversity on the functioning of single trophic levels in simplistic and artificial experimental systems, mostly in experimental plots. In contrast to what happens at these lower levels, the effects of biodiversity at the community or ecosystem levels, and the parameters used to quantify it, have received little attention. Most ecosystems are the result of the interaction of different biological mechanisms that operate at different temporal and spatial scales, whose roles may vary depending on biogeographical and ecological contexts. Therefore, responses of ecosystem functioning to changes in their composition and structure may also vary depending on such contexts. Nonetheless, there is an overwhelming convergence of evidence from different ecosystems highlighting the importance of biodiversity for the functioning of natural ecosystems, at least to a degree. A proper ecological understanding of how biodiversity controls ecosystems and ecosystem services across gradients of complexity may contribute to identify the biodiversity thresholds beyond which they collapse, essential for designing them for the provision of desirable services.

Meanwhile, the conservation and restoration of biodiversity should be included in management decisions (Hiers et al. 2016).

4.2. Threats to Biodiversity

Changes in land use patterns, largely due to agriculture intensification, have fostered biodiversity loss on a global scale. In the search for higher economic returns, traditional agroforestry systems comprised of a complex variety of indigenous perennial plants have been substituted by simplified monoculture systems. Most people in the world still search for food and income by farming ecologically sustainable agroforestry systems (Raj et al. 2022). However, these systems face an array of social and ecological threats. As subsistence-oriented production systems, traditional agroforestry systems tend to be more vulnerable to external shocks such as migrating diseases and pests. Traditional knowledge practices to cope with climate variability, the suitability of crops under changing climate, or dealing with soil erosion might not be appropriate any longer. On another front, global market and trade forces have rendered indigenous crops vulnerable to exogenous factors leading farmers to restrict the variety of crops planted. Ultimately, increased economic integration and social stratification have rendered the traditional market holders unable to contend with the prevailing institutions in the globalized economy. Recognizing the interconnectedness of these threats is crucial in addressing them effectively, or there is a risk of unintended consequences, such as consolidation of power by landowners or a rise of destructive crop choices. The foci of interdisciplinary research on agroforestry typically depend on scholars' profession and discipline, and vary across regions (Tranchina et al. 2024). Political ecologists concentrate on land tenure models or power relations shaping decision-making processes. Ethnobiologists examine the assemblage of useful plants and structures, needs and management practices. Biodiversity ecologists analyze biodiversity patterns and results of agrobiodiversity conservation actions. Climate ecologists assess agroforestry-driven carbon sequestration potential or climate-mediated production changes. However, it is still unclear whether and how these foci interrelate or whether they are applicable across scales and contexts. Yet, coherent knowledge to root agroforestry in a larger social-ecological landscape is essential for effective biodiversity conservation or climate mitigation studies and for policy-making. For that to happen, first, social-ecological frameworks linking ecological, economic, and sociopolitical driving forces and processes need to be developed as relevant theory exploration. Second, transdisciplinary research to acquire context-specific knowledge using these frameworks and to engage various stakeholders in identifying and addressing pressing socio-environmental issues need to be conducted.

5. LINKING AGROFORESTRY AND INDIGENOUS KNOWLEDGE

There are a range of narratives in the literature regarding the utility of IK in the conservation of forest biodiversity. Many scholars explicate the cultural and material aspects of IK, asserting that knowledge, practice and belief are generally employed in a holistic sense and are important to the health of both people and forests. Such holism is important to note as it has implications for policy and practice in the interface between conservation and IK. A similar holistic approach is evident within the field of agroforestry (Akalibey et al. 2024), which maps a variety of practices onto diverse ecological, social and cultural contexts. If the agricultural landscapes inhabited by indigenous and local communities are to be brought into mainstream notions of agricultural biodiversity, it is essential to understand the commonalities and contributions of different forms of knowledge, and the trade-offs, synergies and tensions within and between them. It is vital to ascertain what supports or inhibits the productivity and biodiversity of agricultural systems in diverse contexts (Figure 2). Understanding how biocultural knowledge systems work is necessary to ascertain their contributions to food sovereignty and safety, ecosystems resilience, sustainable livelihoods and bio-cultural conservation. However, it also points to a need for new partnerships to be forged between conservationists, government actors and local and indigenous knowledge holders, to move beyond monolithic discourses of knowledge towards a more multifaceted understanding of knowledge co-production. Just as there is an urgent need to assess the potential of sulphuric haze mitigation through bioenergy in the West African Monsoon, so too is there a pressing need to explore the adequacy and efficacy of existing agroforestry and IK frameworks in bridging the reshaping, resource-using epistemological divide. The HI-AW framework could enable much better engagement in the above by being both formally broad-based and directly field-based. Through engagement in both, the debate on agroforestry and IK could be fostered between different forms of knowledge and in diverse forums, thereby broadening the debate and linking it up with the world at large.

Figure 2. Integrating indigenous knowledge in conservation and agroforestry

5.1. Synergies Between Practices

The results demonstrate how TEK captures and explains complex ecological interactions and supports the adoption and implementation of sustainable management practices. The key factors enabling these interactions are: (1) prototypic agroforestry systems that can be replicated; (2) appropriate knowledge transmission systems; and (3) wise political decisions and attitudes fostering trust, respect, continuity, integration, and sharing (Awazi, 2024). It is suggested that international biodiversity conservation efforts would benefit from effective mediators and alliance builders with good knowledge of local resources and organisational, education and communication skills. The paper closes with recommendations for action and further research. Empirical framework and case study contexts. In this introductory section the objectives and focus of the paper are outlined, as well as its key concepts, relevance, and research context. The case studies are introduced, a description justified by their representativeness. The next four sections provide the results, while the last presents a brief discussion and concluding remarks. Synergies between conservation and development (livelihood) objectives are threatened by coupled social and ecological processes, leading to a decline in alternative livelihoods and landscape biodiversity (He and Jiao, 2023). Evidence is reviewed from case studies in guava production, where ecological change mediated by agricultural intensification has direct social consequences. Using participatory photography with agriculturalists, the social-ecological dynamics of the guava landscape in Southern Mexico are accounted, arguing for a broader conceptualisation of alternative livelihoods. Agent-based models to simulate incentives that encourage good management of double-edged resources are presented and their implications for research and management. In Indonesia, there is a paper considering how local bioindicators provide important assessment tools for policy and decision making. Indonesia is a mega-biodiverse country with one of the world's highest deforestation rates and is one of 17 global biodiversity hotspots. Local indicators can help meet local need for knowledge of conditions and trends in the environment, which is poorly served by formal systems, particularly in remote areas. But they are rarely incorporated into official systems because they are often subjective and non-standardised.

5.2. Community Involvement and Empowerment

Diverse organizations have made considerable investments in community-based natural resource management (CBNRM) across southern Africa over the last twenty years (Child & Barnes, 2010). CBNRM evaluates the development of community forestry in Tanzania through a public goods framework and asks whether the focus on community forestry has improved the likelihood of sustainability within a highly degraded forest landscape. It looks at the discarding of fifteen years of national forest policy and practices as a possible occasion to change the structures of forest and livelihood governance for the best. The trunk of the Baobab is a keystone tree that gives ecosystem services to an arid region, its loss conferring many costs, but its regeneration requires understanding community biosocial systems and empowering societies to manage, protect and restore acorns. Understanding of local community knowledge and attitudes on forest conservation and the sustainable management of natural resources is crucial (Obradović et al. 2023). Community empowerment strategies using bottom-up approaches that consider local community knowledge, attitudes, needs, wishes, and behaviors have been developed, and community forestry initiatives have been accommodated in a way that promotes local empowerment. The relationship between citizens and governance in natural resource governance has been superbly explored across a multi-tiered governance context using IPA and a simultaneous determination governance rationality model. With a wealth of empirical examples across a variety of situations, policy issues and governance modalities, these contributions explore the relational characteristics of citizens and governance, the socially differentiated impacts of governance on citizen empowerment and disempowerment, and the wider implications for sustainability and conservation. Accompanying the insights are voices that call for publishing research that opens up the kind of possibility envisaged. The disappointment is reflected in the ink used in the checklist, as sustainability studies become compromised by "a huge wave of commercial interest" which reveals disconnections between motives and actions. The perspectives, however, are somewhat at odds with the observations reported by others in this issue that the academic community has proven resilient and innovative.

5.3. Traditional Ecological Knowledge

Governance of biodiversity resources is subject to historical, sociopolitical, economic and cultural conditions. It is important to understand the links between biodiversity and governance in order to use biodiversity to satisfy human needs in a sustainable manner (Savari et al. 2020). Biodiversity governance determines the relative influence of different stewards who have many ways of knowing in which to conduct their governing influence. As such, there is a rapid movement to share and integrate different forms of biodiversity knowledge. These often involve science and local or traditional forms of knowledge. However, understanding of what is meant by these different terms and the assumptions made by many in the movement to integrate these ways of knowing remains poorly understood. Integrating different forms of knowledge is an inherently political endeavour. It is well known that science is not value-free. A considerable literature discusses the suppression and/or elevation of different ways of knowing biodiversity over historical time across the globe. When knowledge systems differ, power differences often accompany them. This might lead to issues of prerogatives over how biodiversity is described, interpreted, represented, made actionable and enforced (Clement et al. 2015). Specifically, arrangements to share knowledge ought to ensure that standard scientific and indigenous categories of biodiversity are upstream and downstream alike and that power inequities between the forms of knowledge are rendered explicit before they are negotiated by mutually acceptable ground rules, procedures and agreements. However, it may not be reasonable to expect parity in the distribution of benefits from a deliberative process of knowledge integration. Some power differences may be inherent in the incommensurability of some ways of knowing, in which case expectations of equal access may be misplaced or misplaced at certain conceptual stages. There may be knowledge clades within which knowledge systems are derived from a shared epistemology and held with acceptable fairness and parity. There may exist amplifying synergies between knowledge clades that support the maintenance of biodiversity prospects for the future.

6. COMMUNITY RESILIENCE

Analyzing the potential of agroforestry for biodiversity and community resilience across ten biogeographical regions is a steep challenge (Jackson et al., 2012). The information needed to do this is often incomplete, incommensurate, and under different CVs. Lack of adequate measures and indicators to evaluate biodiversity and community resilience further complicates matters. Biodiversity indicators are needed at the level of species, assemblages of species, and nested hierarchies of species assemblages. Community resilience indicators are needed for different measurable aspects of community resilience, and across temporal scales that often range from hours to decades. Biodiversity, knowledge and practices, and associated cultural aspects can be difficult to measure. This runs the risk of attempts to quantify biodiversity either missing some really important things or becoming unmanageable in their sheer scope and volume. Inclusive processes are needed to assess all the forms of biodiversity in agroforestry systems, as well as the benefits they provide. Similarly, it needs to be recognized that while some claims and counterclaims about the social impacts of proposed interventions might be difficult to validate, this does not mean they are not based on authentic and genuine beliefs about what is good and bad for people. There is a need to build social impact prediction models to better conceptualize the interactions of different components of the social systems involved in complex agroforestry interventions. This can include aspects such as land/forest tenure, economic structure, pattern of power relations, cultural values, and knowledge systems and practices, and the human, biological, artifact, and financial flows and feedbacks that crisscross these different social system components. Each of these system parts and flows that interact needs to be monitored to predict the probable societal consequences of a proposed agroforestry intervention (Castle et al. 2021). With regard to knowledge systems, such problems include identification of the stakeholders concerned and relevant assemblies of knowledge and practices, mapping of knowledge systems and networks, and estimation of the profile and dynamics of knowledge flows and exchanges. Excluding indigenous knowledge systems or dismissing them as romantic and outdated systems of human–nature relations that cannot cope with modernity is both naively simplistic and completeness-challenged.

6.1. Definition and Importance

Agroforestry is a land-use system that combines agriculture and forestry within the same plot, providing both immediate and long-term benefits (Jose et al. 2021). In Mexico, agroforestry systems have existed for centuries, as Indigenous People have combined crop cultivation with tree management. These systems have varying degrees of complexity; some consist of associated vegetable crops with scattered trees, and others involve the complex management of multiple tree layers with understory crops. Agroforestry systems cover a large geographical area in Mexico, with different levels of intensification, being practiced in rain-fed and irrigated agroecological regions, and allowing a diversity of crops and trees. Agroforestry systems simultaneously promote biodiversity and increases resilience through the broad provision of ecosystem services that broadly benefit the entire landscape (Rolo et al. 2021). Conserving the integrity of biodiversity, ecosystems, and ecological processes across time and space ensures ecosystem functioning, regulating ecosystem services provisioning. Human livelihoods are inseparable from this natural resource base, and the sustainability of biodiversity largely depends on the social and institutional systems and values across cultural groups. Assessment of the vulnerability of agroforestry communities to climate change is key to adaptive resource management. It helps determine the extent to which agroforestry communities are affected by climate variability and how to orient efforts to understand and tackle consequences on human lives and human-nature relationships. Identifying social networks of information exchange about impacts, demands, and management practices supports adaptation to climate-induced shifts. Community resilience measures were adapted to the context of the communities studied, including their environmental and social capital, resources, tensions, individual-community relations, and management priorities. Governance and collaboration with local institutions and outside agents were highlighted as key variables for resilience assessment in other rural contexts. Ultimately, there is a need to understand how agroforestry governance depends on local institutions, community members, and social processes.

6.2. Factors Influencing Resilience

Recent years have seen a more holistic view of the factors influencing resilience. Resilience is defined as a system's capacity to absorb disturbance without changing its basic structure and processes. An interpretation of resilience as a synonym for stability, namely the speed of return to a certain equilibrium was criticized. A focus on feedbacks and the importance of multi-stability in the form of alternative attractors prompted the definition of resilience in terms of the ability to stay within an equilibrium domain. In such a multi-stable system, each attractor has distinct properties that exert different controls on ecosystem functioning. Such differences include changes in species composition and controls on nutrient cycles and systemic risk. There is still a need for bridging analysis and management of concepts such as small-scale processes, feedbacks, and cross-scale interactions in resilience. Ecological economic approaches often address the concern of excessive erosion of biodiversity without knowing what, in terms of data and models (Jean & Mouysset, 2022), and assume polycentric governance structures for the footprint of traditional ecological knowledge on resource use management. By contrast to the manager-side of ecosystem service assessment and economics, alternative actors like policy-makers and emergence-economists underlining a more biophysical view are concerned with the ecological integrity of systems in the knowledge of what diversity loss to its function. Economists and policy-makers have the black box of the ecosystem-services and their uncertainties behind which government-backed R&D are promoted. In avoiding the conflict associated with each view, and the associated cavalier expansion of different ways modelling both actions and emergent ascertains, the farming systems perspective is suggested to be a cost-effective knowledge producer. Agro-ecologists starting from spatial genetic population models are familiar with codifying input-output and linkages between interactions in respect to operator-side feedbacks (Grabowski et al. 2024). Such research uses farm data to estimate coefficient values for reduced extrinsic estimation hinterlines. In super-imposed surplus approaches ideally repeated simulations allow learning of a variable-response capacity, whereas global economic-scale potential of agriculture is addressed in farm/regional approach.

6.3. Role of Agroforestry in Enhancing Resilience

The immense global pressures on shared watersheds can lead to resource depletion, degradation, or complete collapse. Agroforestry systems in forests can create interlinkages between land use and water management (Kumar et al. 2024). Experience in Fairhaven shows how the adoption of commons-based forest management by a resilient rural community enabled citizens to negotiate with commercial water bottlers that would otherwise deplete the aquifer water supply. Complex traditional agroforestry systems can also buffer climatic risks, promote food security, and provide lifeways and commodities. Traditional ecological knowledge (TEK) here does not only refer to the knowledge of indigenous peoples, but it is also present in the landscape as culturally-modified trees and plants, traditional crops and food. Indigenous peoples and local communities have developed complex agroforestry systems as cultural landscapes that conserve biodiversity and are derived from TEK. As a result of historical land use and management, TEK is embedded in biodiversity and has been crucial for in situ conservation of biodiversity and its culture in mountainous regions (Sinthumule, 2023). Further research is needed to document TEK in agroforestry systems around the world and translate its relevance into biodiversity conservation policy. Agroforestry systems broaden the ways in which to think about resilience, given that these systems have been or can be constructed based on TEK. Agroforestry systems sui generis to locally adapted crops, cultivars, and uses conserve forest-inspired agrobiodiversity beyond designated protected areas and also buffer climatic risks due to species functional redundancy. This type of persistence of traditional agroforestry systems with biodiversity conservation measures and congruence with a particular bioclimatic zone has been termed a 'contribution' to resilience (Pancholi et al. 2023). In mangrove coasts, traditional agroforestry systems with different levels of anthropogenic interventions including pristine mangroves also serve as local adaptation to climate change, poverty alleviation mechanisms, and revenue sources. Agroforests of wild and cultivated perennial multipurpose trees and shrubs can promote the conservation of soil restoration species.

7. CHALLENGES AND OPPORTUNITIES

Indigenous Knowledge (IK) systems have accumulated environmental and social monitoring data over generations, which contribute to food security and the sustainable management of natural resources. Rural people uniquely combine traditional knowledge with adopted innovations to manage diverse production systems with changing resources (Adefila et al. 2024). Agri-enviro-social innovations that combine BKSs are however neglected, and strategies to explore BKSs for analysing production systems are under-explored. For instance, Mexican indigenous peoples use vast biocultural knowledge to enhance agro-biodiversity importance in people's connectivity to nature. However, they meet tension of knowledge loss despite the adoption of commercial practices. Central Mexico hosts a unique Rainbow Agroforestry Production System (RAPS) fostering and conserving diverse endemic resources. Rural dwellers innovate by maintaining traditional land uses and cross-managing the introduction of new cropping systems (Romeo et al. 2021). Indigenous actors claim tradition in decision making empowered by a BKS. Their agency is challenged by colonial practices of knowledge separation that study knowledge systems in isolation. Rural people adapt a biodiversity lens integrating observations of pest/disease immunity, flooding/diversity and connectivity via multi-knowledge systems. Multiple practices maintain connectivity though knowledge acquisition and networking with land porch-holders, searching agents and markets. Biocultural innovation for governance and management in participatory approaches is to be explored. Important to analyse the role of documentation, stewardship contracts, involvement of teachers and biocultural brokers for cross-system knowledge integration for biodiversity stewardship capacity. There is an Understand Implementability-Accessibility Limitations approaches to maintain indigenous people's stewardship. Innate adoption potential of the latter is to be studied. Agri-enviro-social innovations that combine BKSs for study production systems are to be explored.

8. FUTURE DIRECTIONS

Past works on TEK highlighted this knowledge's high capacity for biodiversity management and enrichment in regions with neglectful to moderately threatened ecosystems and traditional knowledge. However, it is not yet fully understood how TEK acts in highly industrialized and long-neglected regions with ecosystem overexploitation and knowledge loss. Potential agroforestry systems are available worldwide, with varying degrees of knowledge and biodiverse management (Adamou et al. 2021). Future research could focus on understanding the underrepresented Indigenous perspective in formerly prehispanic periods. Cultivation systems developed by Native American communities improve ecosystem services and serve as an alternative to industrial globally traded commodities. Some biocultural practices should require as much care before implementation as the good intentions validating them. Transmission of native biocultural practices, particularly in top-down conservation programs like biosphere reserves or national parks, should consider Indigenous alterity and resistance. Accepting notions of diversity may also favor their appropriation and commodification by cosmopolitan actors (Elías Jiménez, 2025). A biocultural network of active ESP ecosystems and pools of traditional ecological management knowledge for scholars and Indigenous communities to maintain ties with biodiverse agroforestry systems and improve biodiversity conservation and resilience should be installed. The Indigenous groups implementing or being resettled into territories needing biodiverse cultivation systems, landscapes, or reserve networks must be prioritized. Building from historical precedent may counteract globalization effects. Only benefitting nature-conserving agricultural methods could be legally required under the country's General Law of Ecological Equilibrium and Environmental Protection. However, communities must be suitably compensated due to their prior overexploitation and impoverished state. Failure to legally mandate measures to correct this issue logistically binds in the chain of oblivion the formerly indigenous roles and associated cultural richness owing to the creation of more fallow desert. Harm in this case far exceeded compensation.

9. CONCLUSION

The context of exploring indigenous agroforestry systems as an alternative for the implementation of sustainability initiatives is provided by climate change, deforestation, loss of biodiversity, increase inequality, and poverty aggravation. Historically, indigenous agroforestry systems sustainably provided food, timber, shelter, fuel, biodiversity conservation, and more services. But now they are under stress. Alternatively, the adoption of technical measures to mitigate the impacts of climate change and to sequester carbon may also aggravate the social situation. Climate-smart agroforestry has been introduced to engage traditional agroforestry knowledge as a basis for modern agroforestry systems application in Eastern Africa, whereas traditional perennial crop-based agroforestry systems along with complementary mixed rainfed and wetland agriculture were explored across a broad range of agro-ecological and agricultural intensification gradients in Mesoamerica. In either case, the assertion is made that agroforestry provides multiple opportunities for reconciling biodiversity conservation with the need of local communities to adapt to a changing world provided the broader socio-economic context is accounted for. Biodiversity is a critical component of ecosystem functionality and service delivery. Agricultural intensification and climate change are however new threats to biodiversity, and their global prevalence is likely to rapidly deteriorate the adaptive capacity of current agroecosystems and the ecosystem services delivered thereby. This drives research need to explore transcending resilience approaches that may also readily bring biodiversity to agricultural and climate adaptive responses. The evolutionary path of existent agricultural biodiversity across a given region in relation to the geo-physical reality provides the basis for investigating the long-lasting 'agroecological memory' therein. Exploration of the evolution of the globally valuable and provisionally inscribed key-heritage, traditional, and indigenous perennial crop-based agroforestry systems across various regions is discussed as a means to highlight the relevance of agroecological memory and their future priority for biodiversity, restoration, and management and regimes for low/spread energy approaches to reconciling agricultural and climate adaptation with biodiversity conservation. There is an urgent need to co-design and co-develop biodiversity-centric agroforestry alternatives and sustainable development pathways for rural communities that are affected by rapid socio-economic and environmental change and which have experienced a dwindling of traditional indigenous agroforestry systems and the biodiversity provision therein. Such initiatives may be stimulated and reinforced by a range of sites- and context-specific indigenous knowledge approaches applicable to contemporary social-ecological challenges.

10. REFERENCES

Achmad, B., Sanudin, , Siarudin, M., Widiyanto, A., Diniyati, D., Sudomo, A., Hani, A., Fauziyah, E., Suhaendah, E., Widyaningsih, T. S., Handayani, W., Maharani, D., Suhartono, , Palmolina, M., Swestiani, D., Budi Santoso Sulistiadi, H., Winara, A., Nur, Y. H., Diana, M., & Ruswandi, A. (2022). Traditional subsistence farming of smallholder agroforestry systems in Indonesia: A review. *Sustainability (Basel), 14*(14), 8631. DOI: 10.3390/su14148631

Adamou, R., Ibrahim, B., Bonkaney, A. L., Seyni, A. A., & Idrissa, M. (2021). *Niger-Land, climate, energy, agriculture and development: A study in the Sudano-Sahel Initiative for Regional Development.* Jobs, and Food Security.

Adefila, A. O., Ajayi, O. O., Toromade, A. S., & Sam-Bulya, N. J. (2024). Integrating traditional knowledge with modern agricultural practices: A sociocultural framework for sustainable development. *Journal of Sustainable Agriculture and Development.*

Akalibey, S., Hlaváčková, P., Schneider, J., Fialová, J., Darkwah, S., & Ahenkan, A. (2024). Integrating indigenous knowledge and culture in sustainable forest management via global environmental policies. *Journal of Forest Science, 70*(6), 265–280. DOI: 10.17221/20/2024-JFS

Albuquerque, U. P., Ludwig, D., Feitosa, I. S., De Moura, J. M. B., Gonçalves, P. H. S., Da Silva, R. H., & Ferreira Júnior, W. S. (2021). Integrating traditional ecological knowledge into academic research at local and global scales. *Regional Environmental Change, 21*(2), 1–11. DOI: 10.1007/s10113-021-01774-2

Ali, M. A., & Kamraju, M. (2023). Ecosystem services. In *Natural resources and society: understanding the complex relationship between humans and the environment* (pp. 51–63). Springer Nature Switzerland. DOI: 10.1007/978-3-031-46720-2_4

Awazi, N. P. (2024, June). The Future of Agroforestry Systems for Sustainable Livelihoods: Policy and Governance Dimensions. In *Turbulence & Energy Laboratory Annual Conference* (pp. 187-208). Cham: Springer Nature Switzerland.

Baliton, R., Landicho, L., Cabahug, R. E., Laruan, K., & Rodriguez, R.ROSELYN F. PAELMOROBERTO G. VISCOARNOLD KARL A. CASTILLO. (2020). Ecological services of agroforestry systems in selected upland farming communities in the Philippines. *Biodiversitas (Surakarta), 21*(2). Advance online publication. DOI: 10.13057/biodiv/d210237

Baynham-Herd, Z., Amano, T., Sutherland, W. J., & Donald, P. F. (2018). Governance explains variation in national responses to the biodiversity crisis. *Environmental Conservation, 45*(4), 407–418. DOI: 10.1017/S037689291700056X

Castle, S. E., Miller, D. C., Ordonez, P. J., Baylis, K., & Hughes, K. (2021). The impacts of agroforestry interventions on agricultural productivity, ecosystem services, and human well-being in low-and middle-income countries: A systematic review. *Campbell Systematic Reviews*, *17*(2), e1167. DOI: 10.1002/cl2.1167 PMID: 37131923

Chappa, L. R., Nungula, E. Z., Makwinja, Y. H., Ranjan, S., Sow, S., Alnemari, A. M., & Gitari, H. I. (2024). Outlooks on major agroforestry systems. *Agroforestry*, 21-48.

Child, B., & Barnes, G. (2010). The conceptual evolution and practice of community-based natural resource management in southern Africa: Past, present and future. *Environmental Conservation*, *37*(3), 283–295. DOI: 10.1017/S0376892910000512

Clement, S., Moore, S. A., & Lockwood, M. (2015). Authority, responsibility and process in Australian biodiversity policy. *Environmental and Planning Law Journal*, *32*(2), 93–114.

Correia, J. E., Piaguage Lucitante, J., Weiss, L., Narváez, N., Lucitante, L., Biaguaje, A., & Esbach, M. (2025). Stimulating reciprocity: How human–plant relations support Indigenous cultural revitalization and stewardship in the Ecuadorian Amazon. *People and Nature*, *7*(5), 1151–1170. DOI: 10.1002/pan3.70019

Dahal, R. P., Kafle, S., & Khanal, K., KC, S., Subedi, M., Ojha, P., & Din, H. U. (. (2025). Agroforestry and traditional knowledge: Lessons from indigenous practices in South Asian Countries: A review. *Archives of Agriculture and Environmental Science*, *10*(1), 189–196. DOI: 10.26832/24566632.2025.1001027

del Castillo, R. F., & Rivera-García, R. (2022). Biophysical and Biocultural Upheavals in Mesoamerica, a Conservation Perspective: Mountains, Maize-Milpa, and Globalization. *Frontiers in Forests and Global Change*, *5*, 763009. DOI: 10.3389/ffgc.2022.763009

Elías Jiménez, A. G. (2025). Multicultural commodification in New York City's culinary field: Resistance among Mexican Indigenous chefs. *Food, Culture, & Society*, *28*(1), 252–266. DOI: 10.1080/15528014.2023.2279803

Gobinath, R., Ekambaram, G., Prakash, P., Priya, K., & Marella, V. S. (2024). Soil Bioengineering Practices for Sustainable Ecosystem Restoration in Landslide-Affected Areas. In *Sustainability of Natural Resources* (pp. 41–60). CRC Press. DOI: 10.1201/9781003303237-3

Gómez, M. U., Bueno, A. L., León, A. C., Uribe Bernal, J. I., & Hernández Aguirre, S. A. (2022). Traditional agroforestry systems: A methodological proposal for its analysis, intervention, and development. *Agroforestry Systems, 96*(3), 1–13. DOI: 10.1007/s10457-021-00692-w

Goshu, B. S., & Ridwan, M. (2024). Earth: The Nexus of Science, Spirituality, and Societal Strife. *Lakhomi Journal Scientific Journal of Culture, 5*(1), 34–50.

Grabowski, P., Slater, D., Gichohi-Wainaina, W., Kihara, J., Chikowo, R., Mwangwela, A., & Bekunda, M. (2024). Research agenda for holistically assessing agricultural strategies for human micronutrient deficiencies in east and southern Africa. *Agricultural Systems, 220*, 104094. DOI: 10.1016/j.agsy.2024.104094

Guto, R. (2020). A meta-analytical review of the role of indigenous knowledge on environmental conservation and climate change in Kenya. *Regional Journal of Information and Knowledge Management, 5*(2), 65–84. DOI: 10.70759/8zg90b28

He, S., & Jiao, W. (2023). Conservation-compatible livelihoods: An approach to rural development in protected areas of developing countries. *Environmental Development, 45*, 100797. DOI: 10.1016/j.envdev.2022.100797

Hiers, J. K., Jackson, S. T., Hobbs, R. J., Bernhardt, E. S., & Valentine, L. E. (2016). The precision problem in conservation and restoration. *Trends in Ecology & Evolution, 31*(11), 820–830. DOI: 10.1016/j.tree.2016.08.001 PMID: 27622815

Jackson, L. E., Pulleman, M. M., Brussaard, L., Bawa, K. S., Brown, G. G., Cardoso, I. M., & Van Noordwijk, M. (2012). Social-ecological and regional adaptation of agrobiodiversity management across a global set of research regions. *Global Environmental Change, 22*(3), 623–639. DOI: 10.1016/j.gloenvcha.2012.05.002

Jean, S., & Mouysset, L. (2022). Bioeconomic models for terrestrial social–ecological system management: A review. *International Review of Environmental and Resource Economics, 16*(1), 43–92. DOI: 10.1561/101.00000131

Jose, S., Garrett, H. E. G., Gold, M. A., Lassoie, J. P., Buck, L. E., & Current, D. (2021). Agroforestry as an integrated, multifunctional land use management strategy. *North American Agroforestry*, 1-25.

Keprate, A., Bhardwaj, D. R., Sharma, P., Verma, K., Abbas, G., Sharma, V., & Janju, S. (2024). Climate resilient agroforestry systems for sustainable land use and livelihood. In *Transforming agricultural management for a sustainable future: climate change and machine learning perspectives* (pp. 141–161). Springer Nature Switzerland. DOI: 10.1007/978-3-031-63430-7_7

Kumar, D., Yadav, R. S., & Kadam, D. M. (2024). Agroforestry Practices: A Possible Way to Achieve Land Degradation Neutrality. *Agroforestry Solutions for Climate Change and Environmental Restoration*, 373-391.

Moreno-Quintero, R., Córdoba, D., & Acevedo, R. (2021). Decolonizing local planning through new social cartography: Making Black geographies visible in a plantation context in Colombia. *Third World Thematics: A TWQ Journal, 6*(4-6), 225-249.

Mukhlis, I., Rizaludin, M. S., & Hidayah, I. (2022). Understanding socio-economic and environmental impacts of agroforestry on rural communities. *Forests, 13*(4), 556. DOI: 10.3390/f13040556

Nair, P. R., Kumar, B. M., Nair, V. D., Nair, P. R., Kumar, B. M., & Nair, V. D. (2021). Multipurpose trees (MPTs) and other agroforestry species. *An Introduction to Agroforestry: Four Decades of Scientific Developments*, 281-351.

Obradović, S., Stojanović, V., & Milić, D. (2023). The importance of understanding local community attitudes and perceptions regarding nature conservation. *Wetlands, 43*(1), 2. DOI: 10.1007/s13157-022-01652-5

Pancholi, R., Yadav, R., Gupta, H., Vasure, N., Choudhary, S., Singh, M. N., & Rastogi, M. (2023). The role of agroforestry systems in enhancing climate resilience and sustainability-a review. *Int. J. Environ. Clim. Change, 13*(11), 4342–4353. DOI: 10.9734/ijecc/2023/v13i113615

Rae, D., & Blenker, P. (2024). The role of collective intelligence and collective agency in enterprising communities. *Journal of Enterprising Communities: People and Places in the Global Economy, 18*(1), 7–28. DOI: 10.1108/JEC-09-2022-0141

Raj, A., Jhariya, M. K., Banerjee, A., Meena, R. S., Nema, S., Khan, N., & Pradhan, G. (2022). Agroforestry a model for ecological sustainability. In *Natural resources conservation and advances for sustainability* (pp. 289–307). Elsevier. DOI: 10.1016/B978-0-12-822976-7.00002-8

Rawat, U. S., & Agarwal, N. K. (2015). Biodiversity: Concept, threats and conservation. *Environment Conservation Journal, 16*(3), 19–28. DOI: 10.36953/ECJ.2015.16303

Rolo, V., Roces-Diaz, J. V., Torralba, M., Kay, S., Fagerholm, N., Aviron, S., & Moreno, G. (2021). Mixtures of forest and agroforestry alleviate trade-offs between ecosystem services in European rural landscapes. *Ecosystem Services, 50*, 101318. DOI: 10.1016/j.ecoser.2021.101318

Romeo, R., Manuelli, S., Geringer, M., & Barchiesi, V. (2021). *Mountain farming systems–Seeds for the future: Sustainable agricultural practices for resilient mountain livelihoods*. Food & Agriculture Org.

Sauer, T. J., Dold, C., Ashworth, A. J., Nieman, C. C., Hernandez-Ramirez, G., Philipp, D., & Chendev, Y. G. (2021). Agroforestry practices for soil conservation and resilient agriculture. *Agroforestry and ecosystem services*, 19-48.

Savari, M., Eskandari Damaneh, H., & Eskandari Damaneh, H. (2020). Factors influencing local people's participation in sustainable forest management. *Arabian Journal of Geosciences*, *13*(13), 513. DOI: 10.1007/s12517-020-05519-z

Sharifian, A., Fernández-Llamazares, Á., Wario, H. T., Molnár, Z., & Cabeza, M. (2022). Dynamics of pastoral traditional ecological knowledge: A global state-of-the-art review. *Ecology and Society*, *27*(1), 14. DOI: 10.5751/ES-12918-270114

Sinthumule, N. I. (2023). Traditional ecological knowledge and its role in biodiversity conservation: A systematic review. *Frontiers in Environmental Science*, *11*, 1164900. DOI: 10.3389/fenvs.2023.1164900

Sułkowski, Ł. (2023). *Managing the digital university: paradigms, leadership, and organization*. Taylor & Francis. DOI: 10.4324/9781003366409

Tranchina, M., Reubens, B., Frey, M., Mele, M., & Mantino, A. (2024). What challenges impede the adoption of agroforestry practices? A global perspective through a systematic literature review. *Agroforestry Systems*, *98*(6), 1817–1837. DOI: 10.1007/s10457-024-00993-w

Vallejo-Ramos, M., Moreno-Calles, A. I., & Casas, A. (2016). TEK and biodiversity management in agroforestry systems of different socio-ecological contexts of the Tehuacán Valley. *Journal of Ethnobiology and Ethnomedicine*, *12*(1), 1–15. DOI: 10.1186/s13002-016-0102-2 PMID: 27449507

Chapter 12
Community Knowledge, Traditional Leadership, and Forest Conservation in Ghana

Seth Tweneboah
https://orcid.org/0000-0003-2144-7760
University of Education, Winneba, Ghana

Vincent Assanful
University of Cape Coast, Ghana

ABSTRACT

This study sought to explore the role of traditional leaders in forest conservation in Ghana as part of broader global sustainability efforts. Drawing on oral traditions from Edwenase in the Amansie Central District, media reports of the Dormaa Anti-Galamsey operation, and relevant scholarly works, the paper demonstrates the ambivalent role of chiefs as both custodians and exploiters of biodiversity. While the Kobiri River and sacred forests of Edwenase highlight the enduring significance of taboos, rituals, and ancestral lands, the unchecked spread of galamsey reveals the erosion of cultural authority and the monetization of communal resources. By contrast, the Dormaa case illustrates how chiefs can mobilize customary legitimacy and collaborate with state institutions to enforce environmental laws, with strong public support. The study concludes that sustainable forest governance requires bridging traditional authority with statutory structures to protect biodiversity, preserve cultural heritage, and strengthen Ghana's environmental policy framework.

DOI: 10.4018/979-8-3373-3992-4.ch012

INTRODUCTION

In Ghana, illegal small-scale mining commonly called *galamsey* has gradually left deep and visible scars across the country's ecological and human landscape. The country's once-clear rivers now run thick with mud and mercury residue. Farmlands are stripped of topsoil and rendered infertile. Importantly, communities report rising cases of skin diseases, respiratory problems, and suspected heavy-metal poisoning linked to contaminated water sources (Agyemang & Oppong, 2025; Mensah et al., 2025; Ofori et al., 2024; Obiri et al., 2016). Cocoa farms, food crops, and forest livelihoods have been destroyed, forcing families to abandon ancestral lands in search of survival elsewhere. It is against this backdrop of ecological devastation and human vulnerability that the struggle against galamsey must be understood, not merely as an environmental crisis, but as a public health emergency and a threat to cultural survival.

On Monday, 15 September 2025, the Ghanaian pressure group *#FixTheCountry* issued a stark warning that the nation risks sliding into what it described as a "Mad Max wasteland" if *galamsey* is not tackled with urgency (MyJoyOnline, 2025). In a statement released on X (formerly Twitter), the group emphasized that the fight against *galamsey* should go beyond political talk and turn into a real, community-driven effort. The group indicated that their investigations suggest that certain traditional authorities are not only aware of but, in some cases, complicit in the perpetuation of illegal mining activities (MyJoyOnline, 2025). Consequently, the *#FixTheCountry* group called for a radical rethinking of the state's strategy. They suggested the need for the government to hold chiefs accountable and assign them a frontline role in the anti-*galamsey* campaign.

Such calls, to be sure, are not entirely new. Over the past decade, well-meaning Ghanaians ranging from environmental activists, religious leaders and academics to political actors have repeatedly highlighted the need to involve traditional authorities in the fight against *galamsey* (Wunpini, 2022; Dzakpata, 2021). These appeals stem from a recognition that chiefs historically functioned as custodians of the land and mediators between communities and natural resources. In the contemporary context, however, their authority has often been sidelined by statutory regulations or, worse still, compromised by allegations of complicity in illegal mining. The *#FixTheCountry* intervention, thus, reopens an important conversation about how traditional leadership can be meaningfully integrated into formal conservation and governance strategies to safeguard biodiversity and address the growing devastation of *galamsey*.

In this chapter, we present traditional leaders as a missing dimension in Ghana's forest conservation efforts. The objective of this chapter is to demonstrate the influence of traditional leaders in community-driven biodiversity protection and cultural preservation. In Ghana, forests are not only ecological treasures but also cultural heritage sites, holding spiritual significance for many communities. Due to their secular and spiritual importance, forests are governed by a range of legal, institutional, and policy frameworks, some of which originated from the colonial era of the 1920s. This, to be sure, demonstrates how the protection of forests and reserves has long been considered vital. Despite the various state-led initiatives aimed at conserving these natural resources, deforestation and illegal mining activities (*galamsey*), continue to frustrate the conservation of forest reserves and water bodies across the country. The recent spate of mining in forest reserves and its politicisation has generated diverse media and academic discussions. Some have proposed the need for newer conservation policies, practices, and strategies, as well as tougher sanctions for violators to protect and promote the sustainable use of forest and wildlife resources (Akonnor, 2025; GhanaWeb, 2025, 2024). The John Mahama-led government's inauguration of forest guards, officially christened as Blue Water Guards, demonstrates its commitment to protecting the country's environment, particularly water bodies and forest reserves, from illegal mining and pollution (Ali, 2025; Quansah, 2025).

It is important, however, to situate *galamsey* within wider political–economic structures that extend beyond local communities. Increasing evidence suggests that contemporary *galamsey* is sustained by transnational criminal networks, elite capture, and political patronage systems that provide financial backing, legal protection, and access to heavy mining equipment. Powerful actors, often operating across national borders, finance operations, launder proceeds, and shield local miners through connections to political and security elites. This convergence of criminal capital and political influence has significantly weakened customary authority, rendering traditional sanctions ineffective and exposing chiefs to coercion, co-optation, or marginalisation. In this context, indigenous cultural systems are not inherently obsolete. Instead, they are overwhelmed by asymmetric power relations in which globalised extractive interests systematically undermine local governance structures.

Despite this and other efforts, significant concerns persist. As the opening concerns of the *#FixTheCountry* movement have garnered attention, very little has been paid to the indispensability of local actors, particularly traditional authorities defined broadly to include chiefs, queenmother, clan heads, and community elders. Indeed, some have noted that the surge in *galamsey* activities exposes gaps in Ghana's governance system (Sempah, 2025). Critics point to accusations of traditional authorities with complicit behaviour (Kuzor, 2025). Some of them are accused of being actively involved in the encroachment on protected lands. Yet, the underrepresentation of the chief in matters of environmental preservation is particularly worrying. First, traditionally, chiefs are held as those who hold custodial rights over community lands. Traditional leaders act as representatives of their ancestors, the supposed owners of the land on which the state is built. Second, as sacral ancestral representative, they combine both political and spiritual powers, and notwithstanding the influence of modern forces on the office of the chief, for example, he remains the single most influential figure in his community (Busia, 1968). Third, chiefs have historically played significant roles in environmental governance through the enforcement of customary laws and the protection of sacred groves (Osei & Asantewa, 2025; Ormsby, 2012; Awuah-Nyamekye, 2012, 2009; Busia, 1968; Kimble, 1963). These groves, often untouched by modern deforestation, evidence indigenous conservation practices.

The goal of this chapter, then, is to reflect on how the fight against *galamsey* in Ghana presents an opportunity to reassess the role of traditional leaders in forest conservation. We demonstrate how the state's crackdown on illegal mining has met resistance in some quarters due to socio-economic dependencies on *galamsey* as a livelihood option. Nevertheless, we demonstrate, traditional leaders remain influential in community mobilization and the enforcement of local norms. We argue that an effective strategy for forest conservation and anti-*galamsey* efforts must prioritize collaboration between the state and traditional authorities. By leveraging indigenous knowledge systems and community trust, chiefs can be reempowered to act as custodians of forest reserves, enforcing local sanctions against illegal mining while promoting sustainable land use practices. The chapter, thus, explores how traditional leadership can be integrated into formal conservation strategies to address both biodiversity loss and the threat of *galamsey*. It also examines the role of chiefs in mediating local environmental practices, the challenges they face in resisting illegal mining pressures, and the potential for state collaboration to enhance community-based conservation efforts. Through qualitative analysis of a case study from Edwenase in the Ashanti region of Ghana and the media reportage of Dormaahene-led Anti-*Galamsey* operation, the chapter further provides insights into how indigenous governance can complement state-led initiatives for sustainable development and environmental protection. This integration, it posits, is not only crucial for sustainable resource management but also for reinforcing cultural heritage and combating environmental degradation in postcolonial African contexts.

THEORETICAL FRAMEWORK AND REVIEW OF LITERATURE

This section situates the conversation within the frameworks of Indigenous knowledge systems and postcolonial ecology. It provides vital lenses for examining the role of traditional leaders in forest conservation and resistance to *galamsey*.

Indigenous knowledge systems

Earlier efforts to develop effective climate strategies often overlooked the importance of Indigenous knowledge systems. Indigenous knowledge systems here refer to the accumulated wisdom, practices, and beliefs developed by communities over generations to interact with their environment (Berkes, Colding, & Folke, 2000). For years, Indigenous knowledge and modern approaches were viewed as two distinct fields, with the latter deemed as more scientific and rational. Recent studies have paid attention to how integrating Indigenous knowledge with Western science can strengthen climate policy, presenting this as an attempt at bridging the two worlds (Indigenous Climate Hub, 2024). Indeed, the role of Indigenous knowledge systems has been recognised in a wide range of arenas, including popular, institutional and academic circles. A July 2024 article by the UNDP's Climate Promise titled *"Indigenous knowledge is crucial in the fight against climate change — here's why"* observes that Indigenous Peoples hold profound ecological knowledge and sustainable practices, such as agroforestry, rotating "forest gardens," and traditional fire management. These practices, the article notes, enhance climate mitigation, adaptation, and resilience. Despite managing vast carbon-rich ecosystems and biodiversity hotspots, they are often marginalized in policy and financing. The article argues that achieving the Paris Agreement goals hinges on fully recognizing their rights, integrating their knowledge into climate strategies, and providing direct access to climate finance.

Earlier studies have shown that understanding Indigenous knowledge helps planners use flexible, creative methods and avoid harming local ecology or culture in development projects. A systematic review by Ijatuyi and colleagues (2025) explores the integration of indigenous and scientific knowledge systems across fields such as environmental conservation, agriculture, healthcare, and education. The authors identify key benefits, including more holistic and sustainable solutions, as well as challenges like power imbalances, epistemological differences, and the need for equitable collaboration, intellectual property protections, and culturally sensitive frameworks. Other studies have also lamented how Indigenous knowledge, despite its usefulness, has remained elusive (McGregor, 2021). In her 2021 article, Deborah McGregor examines the persistent shortcomings in efforts to integrate Indigenous knowledge into Canada's environmental governance. She argues that the prevailing paradigm treats Indigenous knowledge as a set of extractable data, stripped from its cultural context. This approach, she argues, undermines both the integrity of Indigenous knowledge and the societies that sustain it. True recognition, McGregor (2021) contends, demands treating Indigenous knowledge not merely as information but as a lived system, embedded in Indigenous knowledge systems. Proper inclusion requires acknowledging these systems and supporting Indigenous self-determination.

McGregor (2021) examines how Indigenous knowledge systems can be integrated into contemporary environmental conservation strategies, highlighting their value in promoting sustainable resource management. Blending traditional ecological practices with modern scientific approaches, she notes, enhances biodiversity protection and fosters culturally grounded environmental stewardship.

In Ghana, Indigenous knowledge systems manifest in the form of sacred groves, taboos, seasonal restrictions on resource use, and community-based sanctions (Murray, G., & Agyare, 2018; Awuah-Nyamekye, 2012, 2009; Robinson & Sasu, 2013; Sarfo-Mensah & Oduro, 2010). These practices are not only ecological management strategies but also cultural expressions of identity and spirituality (Campbell, 2005). Chiefs, elders, and traditional councils serve as custodians of these knowledge systems, reinforcing the interconnectedness between people and nature. Within the context of forest conservation, Indigenous knowledge systems emphasize the significance of traditional leaders in maintaining ecological balance through culturally embedded practices. Studies on sacred groves in Ghana, for instance, have shown that they harbor higher biodiversity levels compared to state-managed reserves, largely because of effective traditional enforcement mechanisms (Awuah-Nyamekye, 2014).

The overall significance of the above is that they show how traditional leaders use culturally embedded practices such as taboos, sacred groves, and community sanctions to regulate the environment. By grounding the chapter in Indigenous knowledge systems, we demonstrate that chiefs are not only political figures but also custodians of ecological wisdom, making them vital to biodiversity protection and the fight against *galamsey*.

Postcolonial Ecology

This chapter also employs postcolonial ecology as a guiding framework. Postcolonial ecologies here refer to the study of how colonial histories, power relations, and cultural encounters shape human-environment interactions in formerly colonized societies. It highlights the ways ecological practices, knowledge systems, and environmental struggles are entangled with legacies of colonial exploitation and contemporary forms of resistance. In doing so, postcolonial ecologies critique the dominance of Western conservation models that often marginalize indigenous practices (Huggan & Tiffin, 2015). Postcolonial ecologies thereby emphasize the need to decolonize environmental governance by recognizing and integrating local knowledge systems into national and global frameworks. In *Postcolonial Ecocriticism: Literature, Animals, Environment*, Graham Huggan and Helen Tiffin (2010) explore how literature engages with issues of the environment, animal life, and ecological justice from a postcolonial perspective. They emphasize the interconnectedness of colonial histories, environmental exploitation, and cultural representation, showing how literature can both critique and reimagine human-nature relationships. Their study ultimately calls for more inclusive and ethically responsible ways of thinking about ecology that account for power, history, and global inequality. In Ghana, colonial legacies continue to shape land tenure and conservation policies, sometimes undermining the authority of chiefs and the relevance of customary practices (Amanor, 2006). Postcolonial ecology thus provides a critical lens for analyzing the tensions between state-led anti-*galamsey* interventions and indigenous governance systems. Previous scholarship has argued that empowering traditional leaders through recognition of their ecological authority can serve as a corrective to historically imposed conservation strategies. By embedding Indigenous knowledge within formal policy, conservation becomes not only environmentally sustainable but also culturally resonant and socially legitimate.

Like Indigenous knowledge systems, the postcolonial ecological scholarship is relevant because it critiques the dominance of Western conservation models that sideline indigenous practices. The above studies allow us to situate the role of traditional leaders within a broader historical context of colonial legacies that weakened customary authority. By using these works, the current chapter positions Ghanaian traditional leaders as key actors whose authority, if recognized and integrated into state-led governance, could decolonize conservation strategies and make forest management both effective and culturally legitimate. Both Indigenous knowledge and postcolonial ecological thinking show that protecting Ghana's forests cannot be done by the government alone. Rather traditional leaders must also play a key role in tackling biodiversity loss and illegal mining (*galamsey*).

METHODOLOGY

The chapter largely draws on findings from a phenomenological and an ethnographic investigation in Edwenase, to provide a compelling example of the overlap between tradition, ecology, and resource exploitation. The primary empirical material was generated from field data collected at Edwenase, a town in the Amansie Central District of the Ashanti region. Edwenase, like many Akan societies, is known for its forest conservation practices and the contestations surrounding land use.

A trained research assistant, a native of the area, was assigned to gather relevant data through oral accounts, informal conversations, personal lived experiences and observation of local practices. This method was employed to mitigate the authors' absence while still ensuring that community voices were documented authentically. Our research assistant was guided by semi-structured prompts, but we encouraged flexibility to allow community members to shape the narratives shared. In addition to field data, we also draw on secondary sources, including archival materials, policy documents, scholarly works and media reportage on Akan customs and legal norms, forest governance, and *galamsey* in Ghana. These sources provide a broader context for situating the Edwenase case within national and global debates on environmental sustainability.

A purposive sampling strategy was employed to focus on key stakeholders, particularly traditional authorities, elders, and community members directly engaged in forest conservation practices. The research assistant targeted individuals identified by his father and some local chiefs and community elders as knowledgeable about forest-related customs, as well as ordinary community members who had witnessed changes in forest use over time. This approach ensured a diversity of perspectives while centering the authority of traditional custodians. Data collection techniques include oral narratives and storytelling as oral narratives were collected from elders and traditional leaders, focusing on myths, taboos, and rituals surrounding forest use. This informs the Akan cultural modes of knowledge transmission and adds depth to understanding the spiritual significance of biodiversity. The chapter also relied on observational data as the research assistant documented local conservation practices, rituals, and, as the images below show, provided pictorial evidence of forest degradation. We also used informal interviews, whereby conversations with community members, including farmers whose farmlands have been destroyed through illegal mining activities and youth, provided insight into tensions between traditional conservation norms and pressures from illegal mining (*galamsey*).

Ethical protocols were strictly observed. Informed consent was sought from participants through verbal agreement, as is culturally appropriate in Edwenase. The anonymity of participants was maintained, except in cases where community leaders preferred to be identified in recognition of their public role. The involvement of a research assistant also required oversight to ensure accurate representation and accountability in the data collection process.

To contextualize local findings within national debates, the study also analyzed media reports of significant events, most notably the Anti-Galamsey operation led by Osaagyefo Oseadeeyo Agyemang Badu II, the Omanhene of Dormaa Traditional Area. Reports from major Ghanaian outlets (including *Graphic Online*, *Citi Newsroom*, *ModernGhana*, and *GhanaWeb*) were examined alongside viral social media commentary. This triangulation provided insights into public perceptions of chiefly interventions, the symbolic power of traditional authority, and the possibilities for collaboration between traditional authorities and the state in combating illegal mining. Media sources were treated not only as factual reports but also as texts reflecting the moral imagination and expectations of Ghanaian society regarding chieftaincy and conservation.

The study was further grounded in academic literature on chieftaincy, land governance, and indigenous conservation strategies in Ghana. These sources allowed the study to frame the Dormaa and Edwenase cases within wider debates on the ambivalence of traditional authority, its capacity both to protect and to endanger natural resources.

We analyzed our data using an inductive thematic approach. Narratives from Edwenase, events from Dormaa, and media and scholarly materials were coded for recurring themes such as custodianship, ritual erosion, complicity of chiefs, community trust, and state–traditional collaboration. This comparative method enabled the study to highlight both divergences and convergences. The multi-sourced methodology reflects the complex and layered nature of forest conservation in Ghana. Oral community narratives reveal lived experiences and cultural practices, media reports capture real-time interventions and public discourse, and scholarly sources provide historical and theoretical grounding. Taken together, these data sources ensure a holistic understanding of the role of chiefs as both protectors and potential exploiters of biodiversity, as well as the urgent need for state–traditional partnerships in forest governance.

THE STATE AND AUTHORITY OVER LAND, FORESTS, AND MINERAL RIGHTS

The problem of regulating small-scale mining and protecting forest resources in Ghana today cannot be fully understood without examining the colonial roots of land and mineral legislation in the Gold Coast. It is observable that from the late nineteenth century, British colonial authorities introduced a series of ordinances that fundamentally reconfigured the authority of chiefs, redefined rights over land and minerals, and centralized regulatory power in the hands of the colonial state. Indeed, European interest in the colonised land and its resources led to a series of legal and physical aggression against local authorities. A 1877 correspondence from the Colonial Officer acknowledged that the chiefs were the rightful owners of the land in the Protectorate and would likely resist any attempt to change ownership, even if it was claimed to be in their best interest (Hemming, 1877). As Kimble (1963) writes, "if Europeans were to come in to develop mining successfully, the Government would have to take over the land, since the tribes would never alienate it willingly" (Kimble, 1963:16). Colonial laws and policies, accordingly, laid the groundwork for contemporary tensions over "*galamsey*," forest conservation, and the contest between state and customary authority in natural resource governance.

First, colonial legislation on land, forest, and mineral rights raised the question of sovereignty. To be sure, the starting point of colonial intervention in resource control was the contested question of land ownership. As just seen above, prior to colonization, land was vested in stools, clans, and families under the custodianship of chiefs, who held it in trust for their people. Rights to land use were defined within customary frameworks and enforced by local authorities. There is an abundance of evidence that, with the expansion of British influence after the Bond of 1844 and the Anglo-Asante wars, the colonial state increasingly sought to undermine this autonomy. In particular, the introduction of the Lands Bill of 1897 was the earliest attempt by the colonial administration to vest all "waste" or "unoccupied" land in the Crown, a legislation that sparked strong opposition from the Aborigines' Rights Protection Society (ARPS). Chiefs and educated elites argued that the bill threatened the very foundation of the Akan political and social order. Though the bill was withdrawn, the controversy revealed the colonial ambition to redefine land tenure by carving out state control over "unused" lands, thereby restricting the authority of chiefs and undermining customary rights. This marked the beginning of a dual system of land law, viz, customary and statutory, that persists to this day.

The second concern can be raised in relation to minerals and the centralisation of control. Here, it is pertinent to indicate that the discovery of gold in commercial quantities heightened the colonial state's interest in asserting control over mineral resources. The Concessions Ordinance of 1900 stands out. This Ordinance is largely justified on the basis that it aimed to bring order to the way land was being given out, as the uncontrolled granting of land was hurting farming and creating many land disputes. Thus, it was justified as seeking to regulate the granting of mining rights by chiefs, requiring colonial approval before any concession could take effect. In essence, however, this Ordinance sought to transfer decision-making power from chiefs to colonial officials, while ensuring that mining interests, particularly those of European companies, were protected under British law.

The Mining Ordinance of 1905, also referred to as the Gold Mining Products Protection Ordinance (CAP. 149) of 1905, further entrenched state dominance by introducing licensing regimes that limited indigenous participation in mining. While foreign companies gained access to large concessions, African small-scale miners, now labelled as "*galamsey*" operators, were gradually pushed to the margins, criminalized, or left unregulated. By insisting that all concessions and licenses required colonial sanction, the law redefined minerals as a resource held by the state rather than local communities. This framework has endured into Ghana's postcolonial mineral legislation.

Alongside mining, colonial authorities were also concerned with forest resources (Oduro, Marfo, Agyeman, & Gyan, 2011). The Forestry Ordinance of 1927 is particularly significant. It introduced the concept of "reserves," demarcating forest lands under the control of the colonial state, ostensibly for scientific forestry and environmental protection. Section 3 of this Ordinance states that: "The [Government] may constitute any forest-land or waste-land [or any land suitable for afforestation] which is the property of Government, or over which the Government has proprietary rights, or to the whole or any part of the forest-produce of which the Government is entitled, a reserved forest in the manner hereinafter provided." Chiefs and communities were often sidelined in this process, with their customary rights to hunt, farm, or gather forest products either restricted or made subject to colonial permission. By vesting forests in the state, the ordinance not only centralized environmental governance but also reinforced the principle that natural resources were subject to statutory, not customary, authority. This shift echoes in Ghana's contemporary struggles with illegal logging, mining in forest reserves, and debates over who has the final authority to sanction or prohibit resource exploitation.

These colonial laws collectively reshaped authority in profound ways. Chiefs lost significant autonomy over land and mineral rights, as their role shifted from custodians and negotiators of concessions to local enforcers of colonial law. The dualism of customary rights and statutory frameworks created persistent ambiguities, often exploited by both state officials and mining entrepreneurs. As already argued, the exclusion of indigenous miners from formal mining regimes, combined with the privileging of foreign concessionaires, helped create the category of informal mining now labeled *galamsey*. What was once a community-sanctioned form of resource use became criminalized under statutory law. Similarly, the alienation of communities from forests under the 1927 ordinance created continuing resentment and limited local incentives for conservation.

The legacy of the aforementioned colonial interventions remains vital to appreciating Ghana's current struggles with *galamsey*. As already noted, today's regulatory frameworks, which vest all minerals in the state and require licensing through centralized bureaucracies, are direct continuations of colonial ordinances. At the same time, chiefs and communities continue to assert customary authority over land and resources, leading to frequent clashes of jurisdiction. Thus, the colonial reshaping of authority over land, forests, and minerals not only disrupted indigenous governance structures but also planted the seeds of the dual legal and institutional tensions that define contemporary resource governance in Ghana.

In the postcolonial era, Ghanaian governments have inherited and reproduced this framework, often relying on centralised bureaucracies, militarised enforcement, and statutory regulation, while sidelining chiefs in matters of resource governance. The last few years have been particularly germane. The Nana Addo Dankwa Akufo-Addo's government (2017–2024) exemplified this state-centred approach. Faced with mounting public outrage over environmental destruction, shortly after taking office in 2017, Akufo-Addo declared a moratorium on small-scale mining and set up a military-police Joint Task Force (Operation Vanguard) and an inter-ministerial committee to suppress illegal mining (Awuah, 2017). The strategy emphasized arrests, destruction/seizure of machinery and mobile taskforce raids (Botchwey and Crawford, 2025). Later operations, including the high-profile

Operation Halt, extended military and naval enforcement to polluted rivers, where excavators and dredging machines were destroyed and arrests made (Gyesi, 2024; Albrecht, Aubyn and Edu-Afful, 2021; Lartey, 2018). These actions echoed the colonial state's practice of policing resource access through force, privileging statutory authority over customary processes. The administration used L.I.s and other subordinate instruments to regulate mining in sensitive zones; some instruments (and later actions to revoke or amend them) became politically contested because of their implications for mining in forest reserves (Hawkson, 2024). For instance, the government introduced the Environmental Protection (Mining in Forest Reserves) (Revocation) Instrument, 2024, to revoke the Environmental Protection (Mining in Forest Reserves) Regulations, 2022 (L.I. 2462). This new L.I. sought to make any form of mining in forest reserves illegal. On the one hand, the government claimed to be tightening regulation; on the other, critics argued that some instruments opened space for exploitation, undermining forest conservation. Public protests in 2024, for example, forced the state to withdraw or revise aspects of these measures. Yet throughout, traditional leaders who are historically the custodians of these lands and forests were treated less as partners in governance and more as potential obstacles to enforcement.

The neglect of chiefs was particularly evident in prosecutions and community engagement. Although the government sought faster prosecutions and collaborated with agencies like the Minerals Commission and Environmental Protection Agency, chiefs were rarely given meaningful roles in enforcement or monitoring (Aikins, 2024). Instead, state forces targeted artisanal miners, often displacing local livelihoods without dismantling the more entrenched networks that supplied machinery and exported gold. The outcome mirrored the colonial dynamic: centralised state enforcement displaced customary authority while leaving the deeper structures of extraction intact.

In a similar vein, President John Dramani Mahama's early initiatives since returning to power in 2025 suggest a reorientation, though still within the same statist logic. His administration has shifted from visible crackdowns to structural regulation, focusing on the supply chains and markets that sustain *galamsey*. New measures include a permit regime for excavator imports, the tracking of heavy machinery, and the establishment of the Ghana Gold Board to oversee gold transactions (Addae, 2025; Kokutse, 2025; MyJoyOnline, 2025; Graphic Online, 2025). The creation of a cross-agency taskforce with accountability tools such as body cameras signals an attempt to address corruption within enforcement itself. Initial reports of equipment seizures and forest reserve recoveries suggest early results.

These new models notwithstanding, chiefs remain marginal. The Mahama government's emphasis on controlling machinery imports and monitoring gold flows bypasses the authority of traditional leaders, who continue to exercise customary jurisdiction over land allocation and community livelihoods. While the state portrays its strategy as technocratic and systemic, its reliance on administrative instruments and cross-agency enforcement reflects the legacy of colonial centralisation. Chiefs are once again positioned at the periphery of governance, despite their embeddedness in the very communities where galamsey thrives (Kanarku, 2025).

From the actions of both Akufo-Addo and Mahama, one thing becomes clear: Ghana's resource governance still struggles with a profound problem, that is, the central government tightly controls land, forests, and minerals, while the role of chiefs is often sidelined. Military crackdowns under Akufo-Addo and regulatory innovations under Mahama may differ in form, but both rest on the assumption that statutory institutions, not customary authorities, are the legitimate custodians of the nation's resources. This marginalization of chiefs, this chapter maintains, helps to explain why *galamsey* persists. It is our averment that without the involvement of traditional leaders in enforcement and regulation, government crackdowns risk repeating colonial-era mistakes of showing power without solving the real conflicts between state laws, local customs, and people's livelihoods.

EDWENASE, TRADITIONAL BELIEF AND FOREST GOVERNANCE

Like other Akan societies, ancestral lands in Edwenase are strongly linked to spiritual identity and stewardship responsibilities (Berry, 2008). The forest and river are not simply economic assets but sacred spaces embodying communal heritage. Though a relatively little-known town in the Amansie Central District of the Ashanti region, the town of Edwenase is an integral place in Asante history, particularly owing to the supposed role the Kobiri River deity played. The deity's strength was once evident in the protection it was believed to have provided for the Asante people. In Edwenase historical narrative, it is said that the Kobiri River played a decisive role in defeating the enemies of the Asante as it fought on behalf of the people. Even in the contemporary era, especially during the 1970s, the Kobiri Deity was still regarded as potent, serving about six communities through the guidance of a traditional priestess. Over time, however, the spread of Christianity and other modernising and socioeconomic changes have reduced the belief in the inherent force of this riverine deity, weakening its influence.

Edwenase oral tradition, as recounted by Nana Kojo Antwi, an elderly informant, has it that the town was founded by Nana Yaw Anim who sought refuge in the mountains. The place grew to become home to various groups whose arrival created tensions over leadership and land ownership. These disputes drew the intervention of the Kobiri River deity, whose presence has since played a central role in shaping the community's ecological and cultural identity. The Kobiri River deity is revered as a guardian of the community and its forest. Rituals, including libation, animal sacrifices, and taboos, govern human interaction with the environment. Like other Akan riverine deities, menstruating women are forbidden from approaching the river, sexual activity is prohibited near the river, and bathing with soap in the river is tabooed. These practices, embedded within Indigenous knowledge systems, serve as ecological safeguards that maintain water purity, regulate human activity, and protect biodiversity (Gumo et al., 2012; Twumasi-Ankrah, 2023). The deity's power is further illustrated in oral narratives: when people believed to have desecrated the Kobiri River through acts of defilement suffered mass deaths, earning the deity the epithet *Kum Aduosia* ("killer of sixty"). This myth reinforces ecological ethics by illustrating the dire consequences of disregarding taboos, aligning with arguments that ecological ethics ascribe moral value to natural entities. The ritual norms tied to the Kobiri River deity demonstrate how spiritual governance reinforces ecological conservation. Such practices reecho the sacredness of the forest surrounding the river and embody the Indigenous knowledge systems that regulate human–environment interaction.

In recent times, however, tensions have emerged as portions of the forest have been sold to *galamsey* operators. Despite the resistance of the current chief, according to elderly informants, land transactions driven by competing claims undermine communal ecological stewardship.

Figure 1 A section of the forest where the shrine is located

Figure 2 Current state of the Kobiri River

Figure 3 A picture of one of the machines used to wash the gold

The figure above shows a section of the forest where the shrine of Kobiri; a River Diety at Edwensi can be found.

The case of Edwenase highlights the dual pressures facing traditional leaders: on one hand, they act as custodians of ecological and cultural heritage; on the other, they contend with internal disputes and external pressures from illegal mining. This underscores the need for collaborative governance where state authorities engage chiefs and shrine custodians in forest conservation and anti-galamsey strategies.

THE AMBIVALENCE OF TRADITIONAL LEADERSHIP AND FOREST CONSERVATION

On September 15, 2025, the Omanhene (Paramount Chief) of Dormaa Traditional Area, Osaagyefo Oseadeeyo Dr. Agyemang Badu II, went viral across Ghana after a video showed him personally leading a bold crackdown on illegal mining in his territory. The move was not a symbolic gesture, but a hands-on intervention in which the Omanhene, dressed in ordinary cloth and surrounded by a delegation of his sub-chiefs, descended on suspected galamsey sites in the communities of Dormaa-Akwamu, Akontanim, and Kobedi. Media reports indicate that the operation, which was conducted in collaboration with the Dormaa Anti-Galamsey Task Force, resulted in the arrest of three individuals suspected of engaging in illegal mining (Abisa, 2025; GhanaWeb, 2025). Alongside the arrests, the reports suggest, the task force confiscated mining equipment, including water-pumping machines and other tools used to dredge riverbeds and degrade forest lands. In a move that underscored both the seriousness and procedural legitimacy of the crackdown, all suspects were promptly handed over to the Ghana Police Service for further investigation and prosecution (Abisa, 2025; GhanaWeb, 2025).

Following the operation, Osaagyefo Oseadeeyo Dr. Agyemang Badu II issued a stern warning to his sub-chiefs. He insisted that any chief found to have leased stool lands to miners would face destoolment. He called for the reclamation of lands that had already been wrongfully allocated for illegal mining and reiterated that the environment, especially rivers and forests, must be safeguarded for future generations. Video footage of the Dormaahene's direct involvement quickly went viral on social media, sparking widespread public debate. Many Ghanaians praised his bold action, contrasting it with what they perceived as the state's sluggish response to the menace of galamsey. Social media commentators hailed him as a courageous leader willing to protect stool lands and rivers at the risk of personal exposure. His actions were framed as evidence that chiefs, when they choose to act, can be more effective than politicians and state agencies in curbing environmental destruction.

We maintain that by handing the suspects over to the police rather than taking extrajudicial measures, the Dormaahene also demonstrated a model of state-traditional authority collaboration in addressing environmental crimes (GhanaWeb, 2025). The Dormaa incident has since become a touchstone in debates on the role of chiefs in forest conservation and the fight against galamsey. Yet, as we shall demonstrate below, while it reinforced the perception of traditional leaders as protectors of the land, it also draws attention to the ambivalence of chieftaincy as an institution since not all traditional leaders act decisively against illegal mining, and some are complicit in the very practices the Dormaahene denounced.

ANTI-GALAMSEY OPERATION

The public reaction amplified the event's significance. Video and social-media posts circulated widely, with many Ghanaians praising the Dormaahene's visible leadership and contrasting it with what they perceive as uneven state enforcement of galamsey. Commentators framed the operation as evidence that traditional authorities because of their local legitimacy and immediate reach can be effective first responders to environmental crimes, particularly where state capacity is thin or enforcement is politicized. The high level of popular support for the Dormaa sweep demonstrates how a traditional leader's moral authority can translate quickly into community mobilization and reputational pressure on would-be offenders (ModernGhana, 2025).

Set against this dramatic example of chieftaincy in action, the Edwenase case presented earlier offers a more ambivalent picture. Edwenase's Kobiri River and its surrounding forest are saturated with ritual value and longstanding taboos: shrine rites, libation protocols, and strict behavioural proscriptions. As noted, the restrictions around menstruation, prohibitions on bathing with soap at the riverbank, and limits on farming in sacred groves, have historically functioned as local mechanisms of environmental protection. The ethnographic material gathered for this study shows how these practices are not merely symbolic but are embedded in everyday governance. They serve as controlling access, regulating resource use, and inculcating an ecological ethic across generations. Scholars of Ghanaian sacred groves and indigenous conservation, we have also seen, confirm that such cultural regimes often preserve pockets of biodiversity and act as informal protected areas within otherwise degraded landscapes balance (Ntiamoa-Baidu, 2008).

Yet despite that dense web of ritual protection, the Kobiri River and its enclave have been extensively degraded by galamsey activity. The Edwenase narrative reveals why: contested lineage claims, competing interpretations of ancestral ownership, and the monetization of stool/lineage claims have enabled some actors to sell or lease land parcels to miners. In other words, the customary rules that once protected the grove have been undermined from within the community's own property politics. Where claims to ancestral title are fragmented or disputed, custodial authority weakens; when lucrative mineral rents enter the equation, spiritual sanctions alone are often insufficient to deter exploitation. This pattern (sacred protection collapsing under intensified commercial pressure) mirrors findings across Ghana that sacred groves conserve biodiversity only so long as customary authority remains relatively unified and insulated from market incentives (Ubink & Amanor, 2008).

The situations in Dormaa and Edwenase pinpoint significant consequential dynamics about the ambivalence of chiefs and the prospects for state-traditional collaboration. First, there is the performance of authority and the fragmentation of authority. The Dormaa case illustrates how a (relatively) united traditional leadership can perform enforcement visibly and redirect suspects into statutory channels. When this happens, it reinforces both customary norms and the rule of law. The situation in Edwenase, on the other hand, shows the opposite. It points to situations where a fragmented authority (multiple competing lineage claims, some actors selling land contrary to the chief's position), customary protections are hollowed out, and conservation fails. Second, we also see that there are legitimacy and enforcement pathways. The Dormaa operation combined legitimacy (the chief's moral authority) with procedure (police handover). That combination matters: it avoids vigilantism, signals due process, and links customary policing to state adjudication. In Edwenase, the absence of such procedural linkages — and the prevalence of opaque land transactions — created a permissive environment for galamseyers to operate with impunity. Third, political-economic pressures serve as the decisive variable. Across both cases, it is clear that spiritual belief and customary law are powerful but not determinative. When the political economy of land changes (i.e., when gold values and outsider demand rise), incentives can override taboo and tradition. This is consistent with scholarship that emphasizes the embeddedness of illegality in local political economies and the limits of coercive or ritual sanctions when economic rents are large (Adombila, 2024).

The above linked narratives provide analytical leverage in three ways. First, they empirically demonstrate the chapter's central thesis of chiefly ambivalence: traditional leaders can be frontline protectors (Dormaa) or, through either complicity or loss of control, enablers of destruction (Edwenase). Second, the comparison supplies a practical template for policy: Dormaa's model suggests that formalizing joint task forces, clarifying land transactions, and building a clear handover procedure to statutory enforcement can harness chiefs' local reach while limiting rogue transactions that undercut conservation. Third, the Edwenase experience underscores that conservation strategies which rely solely on ritual authority will falter unless accompanied by transparent land-management institutions, safeguards against fraudulent transfers of stool/ancestral lands, and economic alternatives that reduce incentives for illicit extraction.

IMPLICATIONS AND POLICY-RELEVANT RECOMMENDATIONS

The juxtaposition of Dormaa's forceful anti-galamsey intervention with Edwenase's experience of ritual erosion and forest loss reveals profound implications for environmental governance, the authority of chiefs, and the prospects for state–traditional collaboration in Ghana. The whole episode reveals traditional authorities as double-edged swords. In particular, the Dormaa case illustrates how traditional authorities, when acting in concert with their councils and aligning with state law enforcement, can become effective first responders to environmental crime. Their legitimacy is immediate and entrenched in community trust. Yet the Edwenase case shows the other edge of the sword: where lineage disputes, commodification of stool lands, or intra-family rivalries undermine chiefly authority, chiefs can become unable—or at times unwilling—to enforce customary taboos. Historically, this duality has been well noted in Ghanaian land politics, where some chiefly authorities have defended communal land rights while others have commoditized and alienated land in ways that undermine both tradition and sustainability (Ubink, 2008; Amanor, 2010; Kimble, 1963). From the above discussion, we also get a sense of fragility of ritual conservation. The Edwenase sacred grove and Kobiri river illustrate the limits of relying solely on ritual authority and taboo to safeguard biodiversity. Earlier studies highlight how sacred groves served as de facto protected areas, preserving biodiversity amidst agricultural expansion (Ntiamoa-Baidu, 1995; Campbell, 2009, Shumway, 2011). However, as shown in Edwenase, ritual systems collapse under the combined weight of contested custodianship and market-driven incentives from galamsey. This underpins a crucial implication: cultural institutions are necessary but insufficient for conservation in a monetized political economy. In addition, we also observe the centrality of legitimacy and due process. Dormaa's handover of suspects to the Ghana Police contrasts sharply with vigilante-style enforcement seen elsewhere. By emphasizing due process, the Dormaahene avoided accusations of abuse of power while reinforcing state legitimacy. For policy, this suggests that conservation efforts underpinned by both customary legitimacy and statutory legality are more sustainable than efforts relying on one domain alone (Boafo-Arthur, 2003; Lund, 2016).

Policy-Relevant Recommendations

The contrasting experiences of Dormaa and Edwenase highlight that Ghana's environmental governance cannot succeed through state policy or cultural authority alone. What is required is a careful weaving together of both structures into a hybrid framework that leverages the legitimacy of chiefs and the enforcement powers of the state, while simultaneously addressing the economic drivers of illegal mining. From the above insights, some interconnected policy recommendations emerge. First, we propose the need to formalize state-traditional task force. The Dormaahene's decisive operation demonstrates the potential of state-traditional authority task forces as effective guardians of natural resources. When traditional leaders visibly act in defense of rivers and forests, their authority resonates more strongly with local populations than the distant reach of state institutions. Yet to avoid arbitrariness, we propose that such interventions should be institutionalized through district-level agreements that outline clear procedures for raids, arrests, and prosecution. This ensures that traditional leaders who act against galamsey are not left vulnerable to accusations of vigilantism, while also embedding their actions within the state's legal framework. Such collaboration, we hope, would provide a template for other mining-prone areas, making traditional leaders reliable partners rather than unpredictable actors.

Second, we also propose the necessity of strengthening customary land governance mechanisms. The case of Edwenase shows that the collapse of ritual authority is often rooted in weak customary land governance. Traditional leaders, especially chiefs and family heads sometimes allocate sacred lands for mining in pursuit of personal gain or under pressure from factions in succession disputes. The earlier instances of the Asantehene destooling chiefs accused of supporting illegal mining show the reality of traditional leaders' involvement in environmental degradation. To close this loophole, we propose that customary land governance must be strengthened through transparency and accountability mechanisms. This could include strengthening the ongoing digitization of stool and family land records to prevent clandestine leasing, mandatory consultation with local councils before granting concessions, and oversight mechanisms at the traditional council level. These measures not only protect sacred groves and rivers from illicit leasing but also restore the moral credibility of traditional leaders, who are too often seen as complicit in environmental degradation.

Third, we also propose reinforcing the protection of sacred sites through statutory recognition. Indeed, sacred sites such as the Kobiri river and forest require statutory recognition as part of Ghana's conservation regime. While rituals and taboos have historically preserved these sites, as we have also seen, their authority weakens in the face of monetized incentives from galamsey. Integrating sacred groves into national frameworks such as the Community Resource Management Areas (CREMAs), or gazetting them as community-managed reserves, would add a protective legal layer while still respecting cultural traditions. It is hoped that such recognition ensures that violations are not treated merely as cultural infractions but as statutory crimes subject to state sanction. This hybrid approach strengthens both the symbolic and legal weight of conservation.

Fourth, we propose the need to provide livelihood alternatives to reduce galamsey incentives. We believe that the persistence of galamsey also highlights that conservation cannot be achieved without addressing the livelihood needs of local populations. All cases of galamsey in Ghana reveal that poverty and limited employment opportunities create fertile ground for collaboration between locals and illegal miners. The Edwenase case, as elsewhere, shows that without viable alternatives, even strong taboos or state prohibitions will eventually give way to economic desperation. Policy must therefore integrate biodiversity protection with livelihood diversification, through investment in agroforestry, ecotourism, sustainable farming practices, and small-scale green enterprises. These alternatives not only reduce the allure of galamsey but also align environmental conservation with tangible community benefits.

Finally, there is a need for continuous civic education that frames conservation not only as an economic or legal issue but also as a matter of cultural survival and intergenerational justice. Chiefs, faith leaders, schools, and civil society actors can and actually do play a pivotal role in reminding communities that rivers and forests are not merely resources but sacred inheritances. Such moral reframing can help revive communal pride in stewardship traditions, while reinforcing the state's conservation agenda with grassroots legitimacy.

Collectively, these recommendations call for a layered governance framework in which chiefs and the state serve as joint custodians of Ghana's forests and rivers. The Dormaa case illustrates the potential of traditional leaders who honor their sacred duty, while the Edwenase case warns of the risks that arise when chieftaincy falters under internal conflict and economic strain. Reconciling these contrasting realities demands legal innovation, cultural renewal, and economic creativity—essential ingredients for protecting biodiversity against the threat of *galamsey*.

CONCLUSION

In this chapter, we have examined the role of traditional leaders in forest conservation in Ghana by situating our analysis within the context of biodiversity loss and the pervasive threat of *galamsey*. The chapter relies heavily on oral accounts from Edwenase,a town in the Amansie Central District of the Ashanti region. media reports of the Dormaahene-led Anti-Galamsey operation, and other galamsey-related activities as well as relevant scholarly literature to demonstrate the ambivalent position of traditional leaders as both custodians and potential exploiters of the environment.

We have positioned the case of Edwenase as illustrating the fragility of conservation when rooted solely in ritual authority and traditional taboos. Despite the persistence of cultural beliefs, contested lineage claims and the monetization of stool lands in most communities in Ghana have eroded the authority of traditional leaders and created space for the unchecked destruction of river bodies and surrounding forests. This suggests the limitations of relying exclusively on cultural norms in a context where economic incentives from *galamsey* often outweigh ritual obligations. Conversely, we have also used the Dormaa operation to show that traditional leaders, when acting decisively and in collaboration with state institutions, can play a transformative role in environmental governance. By personally leading a raid, arresting offenders, and handing them over to the police, Osaagyefo Oseadeeyo Dr. Agyemang Badu II modeled a form of hybrid governance that aligns customary legitimacy with statutory enforcement. The chapter has shown that the strong public support for his actions reveals the enduring moral authority of traditional leaders in Ghana and the untapped potential of their leadership in addressing ecological crises.

The comparative analysis of these cases affirms that the future of forest conservation in Ghana lies in bridging cultural authority with state structures. Traditional leaders, the chapter demonstrates, cannot be assumed to be automatic guardians of biodiversity. Their effectiveness depends on accountability mechanisms, community trust, and collaboration with formal institutions. Similarly, the state cannot rely solely on policing and legislation, which often lack grassroots legitimacy. We have proposed that a model of co-governance hinged on transparency, statutory recognition and protection of sacred sites, and livelihood alternatives for local communities emerges as the most viable pathway toward sustainable conservation. Ultimately, the chapter maintains that protecting forests in Ghana is not simply an ecological imperative but also a cultural, moral, and political one. Reported and unreported cases in the media have shown that traditional leaders embody the ambivalence of this struggle. They may either deepen the crisis through complicity or inspire its resolution through decisive stewardship. Recognizing and institutionalizing their positive contributions, while constraining their potential for exploitation, is vital to advancing Ghana's environmental sustainability agenda. In this way, the chapter concludes, the integration of indigenous knowledge, traditional authority, and state policy becomes not only a strategy for forest conservation but also a vital component of global sustainability efforts.

REFERENCES

Abisa, C. A. (2025, September 15). Dormaahene leads Anti-Galamsey Task Force, arrests 3 persons in major crackdown. *3News*. https://3news.com/news/dormaahene-leads-anti-galamsey-task-force-arrests-3-persons-in-major-crackdown

Addae, J. O. (2025, May 25). Finance Minister inaugurates Gold Board tribunal. *Graphic Online*. https://www.graphic.com.gh/news/general-news/finance-minister-inaugurates-gold-board-tribunal.html

Agyemang, S., & Oppong, P. (2025). Health impacts of illegal mining in Ghana: An urgent public health crisis. *African Journal of Current Medical Research*, *8*(1). https://doi.org/10.31191/afrijcmr.v8i1.158

Aikins, E. R. (2024, September 24). Ghana must stop galamsey before it sinks the country. *ISS Today*. https://issafrica.org/iss-today/ghana-must-stop-galamsey-before-it-sinks-the-country

Akonnor, E. (2025, June 20). Galamsey fight: 'Withdraw military and use suicide drones to monitor, kill galamseyers' — St. Sark urges gov't. *ModernGhana*. https://www.modernghana.com/news/1410291/galamsey-fight-withdraw-military-and-use-suicide.html#google_vignette

Albrecht, P., Aubyn, F., & Edu-Afful, F. (2021). *Halt and Vanguard: Two military operations in Ghana and their consequences* (DIIS Policy Brief No. December 2021). Danish Institute for International Studies. https://www.diis.dk/node/25270

Ali, M. (2025, July 21). Second batch of Blue Water Guard trainees to join anti-galamsey patrols this week. *Graphic Online*. https://www.graphic.com.gh/news/general-news/second-batch-of-blue-water-guard-trainees-to-join-anti-galamsey-patrols-this-week.html

Amanor, K. S. (2006). Customary land, mobile labor and alienation in the Eastern Region of Ghana. In Kuba, R., & Lentz, C. (Eds.), *Land and the politics of belonging in West Africa* (pp. 137–159). Brill.

Awuah, I. F. J. (2025, August 6). Chief arrested over galamsey. *DailyGuide Network*. https://dailyguidenetwork.com/chief-arrested-over-galamsey/

Awuah-Nyamekye, S. (2009). Salvaging nature: The Akan religio-cultural perspective. *Worldviews: Global Religions, Culture, and Ecology*, *13*(3), 251–282.

Awuah-Nyamekye, S. (2012). Belief in sasa: Its implications for flora and fauna conservation in Ghana. *Nature and Culture*, *7*(1), 1–15.

Awuah-Nyamekye, S. (2014). *Managing the environmental crisis in Ghana: The role of African traditional religion and culture with special reference to the Bono people of the Brong Ahafo region*. Cambridge Scholars Publishing.

Baah, E. (2024, August 28). Asantehene destools 3 chiefs over galamsey. *Graphic Online*. https://www.graphic.com.gh/news/general-news/ghana-news-asantehene-destools-3-chiefs-over-galamsey.html

Berkes, F., Colding, J., & Folke, C. (2000). Rediscovery of traditional ecological knowledge as adaptive management. *Ecological Applications*, *10*(5), 1251–1262.

Boafo-Arthur, K. (2003). Chieftaincy in Ghana: Challenges and prospects in the 21st century. *African and Asian Studies*, *2*(2), 125–153.

Boateng, J. A. (2025, August 20). Barekese youth petition Otumfuo to probe chief over alleged galamsey involvement. *ModernGhana*. https://www.modernghana.com/news/1425742/barekese-youth-petition-otumfuo-to-probe-chief.html

Botchwey, G., & Crawford, G. (2025). Small-scale mining, environmental destruction and democracy capture: The Government of Ghana's failed "war on galamsey.". *Democratization*, *•••*, 1–22.

Busia, K.A. (1968). *The Position of the Chief in the Modern Political System of Ashanti: A Study of the Influence of Contemporary Social Changes on Ashanti Political Institutions*. London: Frank Cass and Co., Ltd.

Campbell, M. (2005). Sacred groves for forest conservation in Ghana's coastal savannas: Assessing ecological and social dimensions. *Singapore Journal of Tropical Geography*, *26*(2), 151–169.

Documentation and Information Section-UCC. (2023, September 18). Support fight against Galamsey — Prof. Kwarteng to Chiefs. *University of Cape Coast*. https://ucc.edu.gh/news/support-fight-against-galamsey-%E2%80%94-prof-kwarteng-chiefs

Duodu, S. (2024, October 21). CPP proposes community-led approach to combat galamsey. *Graphic Online*. https://www.graphic.com.gh/news/politics/ghana-news-cpp-proposes-community-led-approach-to-combat-galamsey.html

Frimpong Manso, A. (2025, September 14). E/R: Akoase Chief accused of aiding galamsey, misusing GH¢560,000 COCOBOD cash. *MyRepublicOnline*. https://myrepubliconline.com/e-r-akoase-chief-accused-of-aiding-galamsey-misusing-gh%c2%a2560000-cocobod-cash/

GhanaWeb. (2024, October 10). Shoot to kill: Joe Wise reinforces bold stance in tackling galamsey menace. https://www.ghanaweb.com/GhanaHomePage/NewsArchive/Shoot-to-kill-Joe-Wise-reinforces-bold-stance-in-tackling-galamsey-menace-1954695

GhanaWeb. (2025, April 2). I'll use shoot-to-kill policy to end galamsey in a month – Nablah. https://www.ghanaweb.com/GhanaHomePage/NewsArchive/I-ll-use-shoot-to-kill-policy-to-end-galamsey-in-a-month-Nablah-1978073

GhanaWeb. (2025, September 16). Three arrested in Dormaahene-led anti-galamsey operation. https://www.ghanaweb.com/GhanaHomePage/NewsArchive/Three-arrested-in-Dormaahene-led-anti-galamsey-operation-2000874

Graphic Online. (2025, June 6). Ghana bans importation of excavators in bid to curb illegal mining. *Graphic Online*. https://www.graphic.com.gh/news/general-news/ghana-bans-importation-of-excavators-in-bid-to-curb-illegal-mining.html

Gumo, S., Gisege, S. O., Raballah, E., & Ouma, C. (2012). Communicating African spirituality through ecology: Challenges and prospects for the 21st century. *Religions*, *3*(2), 523–543.

Gyesi, Z. K. (2024, October 9). Government intensifies "Operation Halt" to combat illegal small-scale mining. *Graphic Online*. https://www.graphic.com.gh/news/general-news/government-intensifies-operation-halt-to-combat-illegal-small-scale-mining.html

Hawkson, E. E. (2024, October 14). Govt presents L.I. revoking mining in forest reserves to Parliament. *Daily Graphic*. https://www.graphic.com.gh/news/general-news/ghana-news-govt-initiates-forest-mining-law-revocation.html

Hemming, A. W. L. (1877, September 29). Dispatch no. 143 [Dispatch]. *Colonial Office: Gold Coast correspondence, CO 96/121*. Ghana National Archives (GNA), Kew, United Kingdom.

Henyoh, M. K. (2025, April 21). Involve chiefs to stop galamsey – govt urged. *News Ghana*. https://www.newsghana.com.gh/involve-chiefs-to-stop-galamsey-govt-urged/

Hilson, G., & Maconachie, R. (2020). For the environment: An assessment of recent military intervention in informal gold mining communities in Ghana. *Land Use Policy*, *96*, 104706.

Huggan, G., & Tiffin, H. (2010). *Postcolonial ecocriticism: Literature, animals, environment* (2nd ed.). Routledge.

Indigenous Climate Hub. (2024, August 1). Bridging two worlds: Integrating Indigenous knowledge and Western science in climate policy. *Indigenous Climate Hub.* https://indigenousclimatehub.ca/2024/08/bridging-two-worlds-integrating-indigenous-knowledge-and-western-science-in-climate-policy/

Ijatuyi, E. J., Lamm, A., Yessoufou, K., Suinyuy, T., & Patrick, H. O. (2025). Integration of indigenous knowledge with scientific knowledge: A systematic review. *Environmental Science & Policy, 170*, 104119.

Kanarku, N. N. (2025, March 23). Eastern Regional House of Chiefs urge greater role in galamsey fight. *CitiNewsroom.* https://citinewsroom.com/2025/03/eastern-regional-house-of-chiefs-urge-greater-role-in-galamsey-fight/#google_vignette

Kimble, D. (1963). *A political history of Ghana: The rise of Gold Coast nationalism, 1850–1928.* Clarendon Press.

Kokutse, F. (2025, April). Ghana prohibits foreigners from trading gold in the country starting May 1. *AP News.* https://apnews.com/article/ghana-gold-board-mining-857b3be91b8a9e649de6097b23cabcb2

Kuzor, A. (2025, June 5). Corrupt chiefs, politicians, others fuelling galamsey in Ghana – Report. *MyJoyOnline.* https://www.myjoyonline.com/corrupt-chiefs-politicians-others-fuelling-galamsey-in-ghana-report/

Lartey, N. L. (2018, September 29). Destroy equipment of illegal miners – Minister orders Operation Vanguard. *Citi Newsroom.* https://citinewsroom.com/2018/09/destroy-equipment-illegal-miners-minister-orders-operation-vanguard/

Lund, C., & Eilenberg, M. (Eds.). (2017). *Rule and rupture: State formation through the production of property and citizenship.* Wiley-Blackwell.

McGregor, D. (2021). Indigenous knowledge systems in environmental governance in Canada. *KULA: Knowledge Creation, Dissemination, and Preservation Studies, 5*(1). Advance online publication.

Mensah, J. P., Oduro-Kwarteng, S., Miezah, K., Boakye, P., & Koomson, B. (2025). Health risks from persistent heavy metal contamination in crops and water at an abandoned naturally revegetated galamsey site in Ghana. *Scientific Reports, 15*(1), 34377. https://doi.org/10.1038/s41598-025-17149-0

Murray, G., & Agyare, A. (2018). Religion and perceptions of community-based conservation in Ghana, West Africa. *PLoS One, 13*(4), e0195498. PMID: 29621348

MyJoyOnline. (2025, June 5). Government to ban importation of excavators without permit under "No Permit, No Excavator" policy. *MyJoyOnline*. https://www.myjoyonline.com/government-to-ban-importation-of-excavators-without-permit-under-no-permit-no-excavator-policy/

Ntiamoa-Baidu, Y. (2008). Indigenous beliefs and biodiversity conservation: The effectiveness of sacred groves, taboos and totems in Ghana for habitat and species conservation. *Journal for the Study of Religion, Nature and Culture*, *2*(3), 309–326.

Obiri, S., Yeboah, P. O., Osae, S., Adu-Kumi, S., Cobbina, S. J., Armah, F. A., Ason, B., Antwi, E., & Quansah, R. (2016). Human health risk assessment of artisanal miners exposed to toxic chemicals in water and sediments in the Prestea Huni Valley District of Ghana. *International Journal of Environmental Research and Public Health*, *13*(1). https://doi.org/10.3390/ijerph13010139

Oduro, K. A., Marfo, E., Agyeman, V. K., & Gyan, K. (2011). One hundred years of forestry in Ghana: A review of policy and regulatory discourses on timber legality. *Ghana Journal of Foresty*, *27*(3), 15–32.

Ormsby, A. (2012). Cultural and conservation values of sacred forests in Ghana. In G. Pungetti, G. Oviedo, & D. Hooke (Eds.), *Sacred species and sites: Advances in biocultural conservation* (pp. 335–350). Cambridge University Press.

Ofori, S. A., Dwomoh, J., Yeboah, E. O., Aggrey, M. L., Nti, S., Philip, A., & Asante, C. (2024). A ecological study of galamsey activities in Ghana and their physiological toxicity. *Asian Journal of Toxicology, Environmental, and Occupational Health*, *2*(1), 40–57. https://doi.org/10.61511/ajteoh.v2i1.2024.395

Osei, B. K., & Asantewa, J. (2025). Traditional leadership, indigenous knowledge and environmental conservation in Ghana: A content analysis. *Open Journal of Social Sciences*, *13*(6), 136–156.

Quansah, I.-T. (2025, March 17). 400 guards trained under 'Blue Water' initiative to protect river bodies from galamsey. *MyJoyOnline*. https://www.myjoyonline.com/400-guards-trained-under-blue-water-initiative-to-protect-river-bodies-from-galamsey/#google_vignette

Reuter, A. M. (Adombila, M. A.). (2024, October 7). Ghana's wildcat gold mining booms, poisoning people and nature. *Reuters*. https://www.reuters.com/markets/commodities/ghanas-wildcat-gold-mining-booms-poisoning-people-nature-2024-10-07/

Robinson, L. W., & Sasu, K. A. (2013). The role of values in a community-based Conservation initiative in northern Ghana. *Environmental Values*, *22*(5), 647–664.

Sarfo-Mensah, P., & Oduro, W. (2010). Changes in beliefs and perceptions about the natural environment in the forest-savanna transitional zone of Ghana: The influence of religion. *Journal of African Studies and Development*, *2*(7), 176–184.

Sempah, D. B. (2025, March 4). Leadership failure, environmental rape, and environmental terrorism: Unraveling Ghana's galamsey crisis and its multifaceted threats. *ModernGhana*. https://www.modernghana.com/news/1383466/leadership-failure-environmental-rape-and-enviro.html

Stone, G. (1920). The mining laws of the West African colonies and protectorates. *Journal of Comparative Legislation and International Law*, *2*(3), 259–266.

Tawiah, O. (2024, October 23). Paramount Chief of Twifo Atti-Morkwa sued again for alleged involvement in galamsey. *MyJoyOnline*. https://www.myjoyonline.com/paramount-chief-of-twifo-atti-morkwa-sued-again-for-alleged-involvement-in-galamsey/

Twumasi-Ankrah, E. (2023). Akan worldview on human-earth relations for environmental conservation in Ghana. *The International Journal of Humanities & Social Studies*, *11*(6). Advance online publication.

Ubink, J. M., & Amanor, K. S. (Eds.). (2008). *Contesting land and custom in Ghana: State, chief and the citizen*. Amsterdam University Press.

United Nations Development Programme. (2024, July 31). Indigenous knowledge is crucial in the fight against climate change—here's why. *Climate Promise*. https://climatepromise.undp.org/news-and-stories/indigenous-knowledge-crucial-fight-against-climate-change-heres-why

Ward, D. F. (2024, January 26). Galamsey fight needs effective collaboration among stakeholders – chiefs. *Ghana News Agency*. https://gna.org.gh/2024/01/galamsey-fights-needs-effective-collaboration-among-stakeholders-chiefs/

Wunpini, H. Y. (2022, October 11). *Involve chiefs in galamsey fight – Osabarima Otu Darko appeals to government*. https://www.graphic.com.gh/news/general-news/involve-chiefs-in-galamsey-fight-osabarima-otu-darko-appeals-to-government.html

Yeboah, I. (2011, November 25). Okyenhene's accuser fined 72 sheep. *ModernGhana*. https://www.modernghana.com/news/363148/okyenhenes-accuser-fined-72-sheep.html

KEY TERMS AND DEFINITIONS

Community-based conservation: emphasizes participatory, locally driven approaches to protecting ecosystems by aligning conservation strategies with cultural values and community livelihoods.

Environmental sustainability: denotes the responsible use and protection of natural resources to ensure ecological balance and the well-being of present and future generations.

Galamsey: is the local Ghanaian term for illegal small-scale mining, often linked to severe environmental degradation and the destruction of water bodies and forests.

Indigenous knowledge: encompasses locally grounded, intergenerational wisdom and practices that guide ecological stewardship, cultural survival, and community governance.

Postcolonial ecology: examines how colonial histories, global capitalism, and local power dynamics shape contemporary environmental struggles and responses.

Chapter 13
Spatial Strategies for Biodiversity Conservation and Forest Management

Yasir Ansari
https://orcid.org/0009-0007-2277-8154
Independent Researcher, Pakistan

Munir Ahmad
https://orcid.org/0000-0003-4836-6151
Survey of Pakistan, Pakistan

ABSTRACT

This chapter explores the transformative role of geospatial technologies in forest and biodiversity management. By leveraging tools such as GIS, remote sensing, spatial modeling, and UAVs, conservation efforts have become more data-driven, responsive, and precise. The integration of participatory mapping and indigenous knowledge enhances data quality and supports ethical, inclusive governance models. Geospatial platforms enable real-time monitoring, adaptive management, and early warning systems for environmental threats. While these advances offer immense promise, challenges such as data fragmentation, capacity gaps, and ethical concerns remain. The chapter concludes with recommendations for scaling innovations through inclusive policies, open data, and capacity building to ensure equitable and effective conservation outcomes.

DOI: 10.4018/979-8-3373-3992-4.ch013

INTRODUCTION

Loss of biodiversity and forest degradation are two of the most burning environmental problems nowadays (UNEP-WCMC, 2024). More than 80 percent of land-based species exist in the forests, and forests are considered to be pivotal ecosystems in the regulation of climate, water cycles, and carbon storage (BirdLife International, 2024). However, ecosystem integrity is becoming increasingly affected by the locally and globally accelerating rates of deforestation, land conversion, habitat fragmentation, and climate change (Forest Declaration, 2024). Unsustainable land use, illegal logging, agricultural growth, and infrastructure construction are other unsustainable activities that are continually encroaching on biodiversity hotspots (Sodhi et al., 2010). Such trends are not only threatening the survival of the species but are also undermining that which sustains human well-being, especially for those communities that are dependent on forests (Friedman et al., 2022; Hein et al., 2018; UNEP-WCMC, 2024).

Spatial thinking and geospatial technologies have become an unavoidable weapon in the war on biodiversity loss and forest degradation (World Resources Institute, 2025). High spatial and temporal resolution mapping, modeling, and monitoring of the Earth's surface have enabled conservationists, forest managers, and policymakers to make evidence-based decisions. Geographic Information Systems (GIS), remote sensing, and spatial modeling tools help in giving critical information on the distribution of habitation, forest cover change, hotspots of deforestation, and ecological connectivity (World Resources Institute, 2025). With the help of these technologies, it is possible to move towards proactive approaches to conservation and management of resources, becoming more accurate in their interventions, efficient management of resources, and more accountable.

This chapter elaborates on how spatial strategies could support the advancement of biodiversity and the governance of forests. It examines how spatial data and geospatial technology have been used to design protected areas, support community-based forest management, and track our progress on conservation. The chapter presents a practical understanding of the power of spatial approaches with realistic behaviors and understandings of the ability of stakeholders to solve the complicated environment using data in practical ways. It also comments on critical implementation issues such as the existence of data gaps, discrepancies in scale, and the domestication of local knowledge within the spatial analyses.

GEOSPATIAL TECHNOLOGIES IN FOREST AND BIODIVERSITY MANAGEMENT

Forest governance and biodiversity conservation landscapes have improved with technology integration of advanced geospatial technologies. The current environmental issues, such as illegal logging, species extinction, fire outbreaks, and habitat fragmentation, require accurate, precise, real-time, and scalable tools. GIS, satellite-remote sensing, spatial modeling platforms, and unmanned aerial vehicles (UAVs) are all technologies that have become the fundamental tools in both planning and implementing and tracking conservation plans. These tools are the data and analytics foundation for the policy and management decisions that are made. Enclosing this technical core are three related fields:

Institutional Arrangements

Governance structures - which range from national forest departments to local institutions of communities - translate the spatial intelligence into coordinated decision-making. Open data policies, inter-agency data sharing, and transparent monitoring systems help to improve accountability and support adaptive management.

Regulatory and Policies

Spatial information is helping guide policy formation and implementation, such as the designation of protected areas, support of conservation laws, and fulfillment of international biodiversity commitments (e.g., CBD, SDG 15). The policy layer translates the spatial evidence into decision-making processes and strategic planning for regulatory action and operational planning, guaranteeing that the data-informed information can be turned into meaningful conservation actions.

Socio-Economic and Community Related Dimensions

The outer layer is a reflection of the social and economic systems that influence and are influenced by spatial strategies. Sustainability is inextricably linked to the attributes of long-term socio-economic sustainability and sustained benefits that support livelihoods, foundational values of community participation, and the sharing of benefits and benefits equitably. Spatial tools contribute to socio-economic analysis in mapping ecosystem services, vulnerability assessment, or identifying priority zones of inclusive interventions.

Geographic information systems (GIS)

GIS is the main tipping point for integrating and viewing spatial data (Longley et al., 2015). It allows the user to map, compare, and interpret a wide variety of data sets-including elevation models, forest inventory data, and sightings of species, as well as anthropogenic pressures. The conservation professionals apply the GIS to test land use patterns, carry out site selection of the conserved sites, map ecological corridors, as well as simulate the effects of infrastructure development.

Looking at this more in terms of implementation, GIS software programs such as ArcGIS, QGIS, and GRASS GIS offer powerful spatial analysis capabilities, such as buffer zoning, suitability modeling, hotspot mapping, and multi-criteria decision analysis (MCDA) (Longley et al., 2015). Advanced mapping of suitability based on GIS played an important role in exploring possible wildlife corridors of the snow leopard in the Western Himalayas, where terrain ruggedness, abundance of prey items, and human distribution data formed the core information. Few applications, such as the ones above, depict the relevance of GIS in planning and conservation design at the level of the landscape.

Remote Sensing

Remote sensing, mainly by satellite, is characterized by great spatial extent and time continuity, which makes it inevitable when detecting changes in the cover of forests, deforestation, and monitoring the health of the vegetation (de Bem et al., 2020; Hansen, Potapov, Moore, Hancher, et al., 2013; Pettorelli et al., 2014; Torres et al., 2021). NASA Landsat series and other public applications, in particular, ESA Sentinel-1 and Sentinel-2, provide multispectral imagery that allows calculating and analyzing vegetation indices (e.g., NDVI, EVI), canopy density, land surface temperature (European Space Agency, 2024).

At a smaller scale, more immediate activities, commercial assets are used instead, such as PlanetScope and WorldView satellites by Maxar, which have sub-meter resolution data, which is used in fine-scale habitat evaluation and logging concession compliance monitoring. This has played a critical role in the illegal deforestation tracking in real-time in the Amazon Basin (Asner et al., 2012) Where enforcement interventions are made to take place through platforms such as Global Forest Watch (World Resources Institute, 2025). New LiDAR (Light Detection and Ranging) and Synthetic Aperture Radar (SAR) technologies are increasing the possibilities to peek beneath cloud cover and through dense canopy cover-important in monitoring tropical forests, estimating the carbon stock in forests, and remote sensing of topography in cloud-covered mountains or cloud-prone areas (Asner et al., 2012).

Spatial Modeling

The use of spatial modeling acts as a predictive and prioritization method that translates raw spatial information into useful knowledge. These models are based on environmental, topographic, and socio-economic factors to model habitat suitability, species distributions, areas of fire hazards, and climate change resulting effects. The most popular platforms and tools are: MaxEnt of species distribution modeling (SDM), Circuitscape and Linkage Mapper connectivity and corridor modeling, InVEST to value ecosystem services and to prioritize their conservation, and Wildfire spread by FlamMap and FARSITE.

As in the Western Ghats of India, researchers have modelled the shifting habitat of endemic amphibians under various climate regimes and used MaxEnt. In Australia, there is a mapping of fire-prone areas employing real-time fuel moisture and spatial fire weather forecasts to assist in pre-emptive action by the forest services. Conservation planning with these tools has the possibility of scenario-planning, where a decision-maker can see alternative futures and appropriately allocate resources.

Unmanned Aerial Vehicles (UAVs)

The use of drones has increased in pace to become an accurate tool in local and responsive surveillance. Bearing RGB, multispectral, thermal, or LiDAR payloads, drones are capable of creating high-resolution aerial imagery, 3D models, and canopy structure information in inaccessible satellite or ground locations. Applications include: Checking the success (or failure) of reforestation and the health of the plantation, detection of gaps and illegal encroachments in the canopy, use of thermal imaging or automatic detection of animals in estimating wildlife populations, and enabling participatory mapping of forests with local communities.

In Borneo, thermal-equipped drones deployed to survey orangutan nesting sites have significantly reduced disruptions to carrying out a population survey or survey of the orangutans. UAVs have been used to survey the location of charcoal production that was largely inaccessible due to safety concerns and inaccessible regions in East Africa. Besides data collection, drones can assist in emergency assessment after the disaster, and one example is the vegetation analysis in mangrove forests of cyclone devastation, and another instance is Fire damage assessment in Mediterranean pine grasslands (Hardin et al., 2019).

Converging Technologies

The real strength of geospatial technologies can be seen where they are assimilated at the top and scale levels. The GIS system can allow the addition of remote sensing-derived vegetation change maps, drone-created orthomosaics, and outputs of spatial models into the same decision-support environment (Longley et al., 2015). More and more, AI and machine learning are being applied to handle and categorize terabyte-scale datasets of satellite and drone imagery and autonomously spot unlawful conduct or environmental transformations. Spatial data sharing and scalable processing are carried out with the help of cloud-based platforms, like Google Earth Engine, Microsoft Planetary Computer, or AWS Open Data (Google Earth Engine, 2024). Such tools also democratize access to high-end geospatial analytics and enable even low-resource institutions to engage in conservation planning using global datasets and real-time updates.

Conclusively, geospatial technologies not only open the door to mapping, but they also open the door to the multi-layered, intelligent method of observation, analysis, and action of complex ecological patterns and processes. They can be irreplaceable to the forest and biodiversity managers who aim to move into predictive and adaptive styles of conservation that are no longer responsive (Phalan et al., 2011). Adapting to particular situations and combining them with robust institutional structures, the technologies become transformative means to achieve the sustainability and resilience of forest ecosystems. Table 1 provides an overview of geospatial technologies, whereas Figure 1 presents a visual interpretation of the technology stack for spatially enabled forest and biodiversity management.

Table 1. Overview of Geospatial Technologies for Forest and Biodiversity Management

Technology	Key Functions	Tools / Platforms	Example Use Cases
Geographic Information Systems (GIS)	Data integration, spatial analysis, and multi-layer mapping	ArcGIS, QGIS, GRASS GIS	Corridor planning for snow leopards in the Himalayas: protected area gap analysis
Remote Sensing (Satellite-based)	Land cover change detection, vegetation health monitoring, fire mapping	Landsat, Sentinel-2, PlanetScope, NDVI, LiDAR, SAR	Deforestation alerts in the Amazon; forest health monitoring in African drylands
Spatial Modeling	Predictive modeling, scenario analysis, risk mapping	MaxEnt, InVEST, Circuitscape, Linkage Mapper, FlamMap	Species distribution under climate change in India; fire risk zones in Australia
Unmanned Aerial Vehicles (UAVs / Drones)	High-resolution mapping, localized monitoring, and participatory mapping	DJI, Parrot, Pix4D, Agisoft Metashape	Orangutan nest detection in Borneo; damage assessment in mangrove forests after cyclones
Integrated Platforms / Cloud-based Tools	Scalable analysis, big data processing, and real-time monitoring	Google Earth Engine, Microsoft Planetary Computer, AWS Earth Data	Large-scale forest monitoring, change detection automation, and democratizing access to spatial analytics

Figure 1. Technology Stack for Spatially Enabled Forest and Biodiversity Management

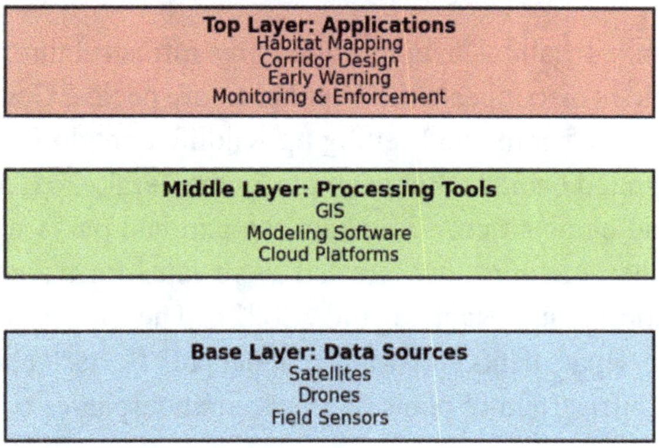

GEOSPATIAL STRATEGIES FOR FOREST AND BIODIVERSITY MANAGEMENT

The contemporary conservation and forest management planning are based on geospatial strategies. Using spatial data, decision-makers can be able to 'maneuver the complicated terrain of ecological priorities, land-use requirements, and community needs. The strategies are not limited to the task of mapping out a situation-they allow dynamic, scenario-based planning which underpins proactive, adaptive management.

Identifying Gaps in Protected Area Networks

Gap analysis, which refers to the estimation of existing designated areas in comparison to biodiversity patterns and ecosystem representation, is one of the fundamental geospatial application strategies. This assists in going through ecological areas that are underrepresented or defenseless. In Madagascar, scientists applied GIS, species distribution, and map out where biodiversity was high but was outside the network of protected areas. Due to this, new reserves were to be introduced to cater to vital lemur habitats (Smith et al., 1997). Application is possible at a national level, where national conservation agencies may employ this strategy in revising or expanding existing protected areas to manage protected areas compatible with the national biodiversity strategies and the Convention on Biological Diversity (CBD) targets.

Designing Ecological Corridors to Enhance Connectivity

The subdivision of habitat neighborhoods by infrastructure, agriculture, and the extension of cities disconnects the population of species. Connectivity models in space would lead to finding and setting up wildlife corridors to reconnect these numerous disintegrated habitats. For example, in the Terai Arc Landscape of Nepal and India, tiger and elephant corridors between national parks were mapped with spatial models. The corridors thus created were later formalized and built into national conservation plans (Rautela et al., 2022) . The model of corridors can be applied by wildlife departments and urban planners in efforts to combine ecological connectivity with infrastructure plans (e.g., green underpasses or bridges).

Zoning for Sustainable Forest Use and Conflict Mitigation

Geospatial products are employed in the designation of zoning plans in forest landscapes to enhance conservation and sustainable utilization, and community access. For example, Tanzania Miombo woodlands, Satellite imagery, along with participatory GIS, were used to identify conservation areas, harvest sustainability areas, and buffer areas to lessen the conflict of resources (Chitayat et al., 2024). The forest management plans can be spatially optimized in order to control logging, grazing, and collection of NTFP (non-timber forest products) and to conserve sensitive habitats.

Prioritizing Species-Specific Conservation Areas

Spatial models (e.g., MaxEnt, InVEST) allow planners to test the habitat suitability of endangered or endemic species and find priority areas on which to focus specific conservation efforts. For example, in the Deosai National Park of Pakistan, climate and topographic data were used to model the habitat where the Himalayan brown bear would live (Fida et al., 2024). This results in a future strategy on how to protect habitats and prevent poaching. Such modeling may be applied by wildlife departments and other non-governmental organizations, or NGOs, to prioritize reintroduction efforts, population monitoring, and the selection of captive breeding locations.

Planning Reforestation and Restoration Efforts

Spatial analysis assists restoration projects by defining the eroded territory, the quality of soils, and their proximity to either seed deposition or the water sources. For example, in Brazil, spatial prioritization was used to determine which areas of the Atlantic Forest should be replenished with trees to achieve as much beneficial biodiversity and carbon sequestration as possible (Giorgi et al., 2014). This information can help governments and climate finance institutions to plan carbon offset initiatives and tree planting programs to have the greatest ecological and climate outcomes.

Early Warning and Fire Risk Mapping

As climate change makes the occurrences of forest fires more frequent, geospatial analysis of the areas at risk of fires can assist in implementing a warning system and planning strategies on how to combat fires. In Malaysia, GIS and remote sensing-based hazard modeling have been used in peat swamp forests to map fire risk zones for early warning (Mahmud et al., 2009). Forest protection agencies can be able to apply the monitoring resources more strategically and create fire lines or community-based preparedness measures in the high-risk areas.

Integrating Local Knowledge with Spatial Planning

Indigenous knowledge is such that it can highly incorporate the implementation of community-based conservation in the spatial structures, so that the conservation process is locally sensitive and is constructed in local realities. In the Philippines, Participatory GIS was used to map sacred forests and areas of biodiversity-rich hunting grounds in collaboration with indigenous communities, with the support of NGOs. These lands were then included in local plans of conservation (Rambaldi et al., 2002). Participatory mapping methods can be implemented by local governments as well as NGOs to help ensure bottom-up participatory conservation planning.

In all these scenarios, geospatial strategies can provide an integrative, scalable, decision-supportive framework to forest and biodiversity management. Spatial tools would help in having a clear, accurate, and forward-looking perspective in their planning of preserved places, restoring landscapes, dealing with risks, and empowering communities. Notably, the employment of these tools is most efficient in conjunction with ground-based data, involvement of stakeholders, and adaptive management that molds to align with changing ecological and socio-economic conditions.

COMMUNITY-BASED AND PARTICIPATORY FOREST MANAGEMENT

The nature of maintained forests has increasingly realized that in addition to scientific evidence or the regulation provided at the top, success in the sustained conservation of forests also relies on vigorous involvement of local communities and indigenous people who are usually stakeholders, often affected by the most, and the traditional custodians of the forested land (Agrawal & Ostrom, 2001; Charnley & Poe, 2007). Use of community-based forest management (CBFM) is a paradigm change as it has faced exclusionary conservation strategies with a more inclusive, rights-based, and socially centered approach. Spatial tools have played a critical role in achieving this transition because they are used to promote local participation, increase transparency, and incorporate indigenous knowledge systems in planning and decision-making exercises.

The Role of Spatial Tools in Community Engagement

Geospatial technologies that were originally the property of technical specialists are today more widely used and conform to participatory use. Our communities can now see what they have, track the changes, and communicate effectively using user-friendly GIS systems, mobile data collection applications, and mapping via drones (IFAD, 2009; Rambaldi et al., 2006). Such tools democratize spatial knowledge, opening up a fully participatory relationship between communities and policy-making processes.

In particular, open-source software, such as QGIS, and mobile applications, such as OpenDataKit (ODK), KoboToolbox, and Maptionnaire, enable local actors to gather geotagged data about the use of forests, their material resources, and environmental risks. The instruments are being incorporated in forest co-management, whereby the communities have the power to capture their usual land territories, sacred spaces, and areas of medicinal plants, and fauna habitats using maps.

Participatory Mapping and Integration of Indigenous Knowledge

Participatory mapping is a ground-up, collaborative, spatial documentation of the landscape through data produced by communities, oral histories, and traditional ecological knowledge. In addition to validating local experiences, this process improves the granularity and cultural sensitivity of spatial datasets in areas where there is either no official or out-of-date data. For example, participatory GIS (PGIS) has also been implemented in the Philippines with indigenous associations mapping ancestral territories in forested uplands (Rambaldi et al., 2002). One of the aims was to lobby the government to give the territories legal status, under which they would have the right to use the Indigenous Peoples Rights Act.

In Peru, the Ashaninka communities used both satellite images and photographic flights by drones over the region, together with oral testimony, to map traditional hunting ranges, water points, and degradation potential due to illegal logging, the raw material of community forest monitoring agreements (Paneque-Gálvez et al., 2014). In the northern regions of Pakistan, community resources mapping in Gilgit-Baltistan has been able to define grazing land, fuel wood, and forest regeneration areas, which are subsequently incorporated into the local conservation plans with the assistance of the Forest Department (Ali et al., 2021). That these are most applicable through participatory measures in enhancing the quality of spatial data, as exemplified, is an eye-opener in restoring confidence and emotional possession that can enhance social integrity and sustainability.

Community-Managed Forests

The community forestry program in Nepal is one of the largest programs of the kind in Nepal, with more than 22,000 Community Forest User Groups (CFUGs) (Pokharel et al., 2007). GPS mapping and GIS-based forest inventories have become one of the many approaches taken by CFUGs to oversee the extraction of timber, to surveil illegal cutting, and to ensure local bylaws. In most places under the community tenure, there has been an expansion of forest cover as well as the provision of income to rural households.

In Tanzania, the locally managed forest reserves have allowed the Joint Forest Management (JFM) and Community-Based Forest Management (CBFM) to empower the local communities to acquire some responsibility in the control of forest reserves (Blomley et al., 2008). The spatial planning devices assisted in zoning the management regionalities, which allowed superior fire control, prohibition of charcoal production, and control of wildlife. In Mexico, indigenous ejido groups run sustainable GIS and aerial imagery to manage concessions of forests in such states as Oaxaca and Durango (Bray et al., 2005). They carry out forest inventory, set out harvest rotations, and preserve areas of biodiversity conservation, and all this is mapped and monitored by local technicians with technical assistance. Indonesia has a Social Forestry Program that has employed participatory mapping to make customary forests visible and enforce land tenure rights as well as indigenous rights by communities (Myers et al., 2017). Harnessing of drone-based surveys along with traditional land knowledge has enhanced accountability and national acknowledgement of more than 5 million hectares of community-managed forests.

When community-based and participatory forest management are well supported with handy spatial gadgets, the management gives rise to better governance of the forest, more effective conservation, and equitable sharing of resources. Such strategies not only harmonize scientific and traditional knowledge systems but also promote local agency and environmental justice. Notably, these strategies are only successful when the capacity to execute them at a local level is developed, when the technology that is available is accessible, and when there is institutional support to facilitate the achievement of meaningful results with respect to the attainment of spatial planning processes by the communities. With the changing international policy on conservation focusing more on human rights-based conservation (e.g., under the UNFCCC and CBD), integrating a participatory geospatial approach to forest planning is no longer a best practice, but a requirement in achieving inclusive and sustainable futures in forests. Table 2 endorses some of the initiatives related to community-based forest management using spatial tools.

Table 2. Community-Based Forest Management Using Spatial Tools

Country	Initiative / Region	Spatial Tools Used	Key Outcomes
Nepal	Community Forestry User Groups (CFUGs)	GPS, GIS forest inventories	Forest regeneration, local bylaw enforcement, and increased community income
Philippines	Indigenous Peoples' Ancestral Domain Mapping	Participatory GIS, oral mapping	Legal recognition of ancestral lands under IPRA law
Peru	Asháninka community mapping	Drones, satellite imagery, and local knowledge	Community forest monitoring, land-use planning, and threat identification
Tanzania	Participatory Forest Management (PFM)	GIS zoning, fire risk mapping	Regulated resource use, better fire control, and wildlife monitoring
Mexico	Ejido Forest Enterprises (Oaxaca, Durango)	Aerial imagery, GIS, forest inventory tools	Sustainable timber management, biodiversity zone protection
Indonesia	Social Forestry Program (Hutan Adat)	Drones, PGIS, spatial land claims	Customary forest recognition, improved transparency, and national mapping inclusion
Pakistan	Gilgit-Baltistan Resource Mapping	GPS, sketch mapping, participatory GIS	Demarcated grazing/fuelwood zones, integrated with Forest Department planning

MONITORING, EVALUATION, AND ADAPTIVE MANAGEMENT

Adaptive management and long-term monitoring are the key tenets of proper biodiversity safeguarding and forest governance. Ecosystems being dynamic, susceptible to the impacts of natural and human activities, monitoring and dynamic decision-making are needed, since this is always an ongoing process. These geospatial technologies, such as remote sensing, GIS, and analytical data, have transformed the monitoring, interpretation, and action of forest cover change, biodiversity trends, and ecological stressors.

Geospatial Tools for Long-Term Monitoring

Geospatial tools enable repeatable and systematic observation of the very large and frequently inaccessible forested areas. Satellite sources of remote sensing data can be categorized under Landsat, Sentinel, and MODIS, which are multispectral sample data that help accomplish forest loss, fragmentation, degradation, and regrowth over the years (Drusch & others, 2012; Huete et al., 2002; Roy et al., 2014). Satellite measurements obtained through the Normalized Difference Vegetation Index (NDVI) and Enhanced Vegetation Index (EVI) are used to monitor vegetation growth and vitality.

In the case of biodiversity, camera traps, acoustic sensors, and GPS-collared animal data can be used to build models of species movement and habitat usage with the aid of GIS-based tools (Kays & others, 2020; Sugai et al., 2019). Real-time, geotagged records are also obtained by citizen science data (e.g., iNaturalist, eBird), which add to the enrichment of monitoring activities (Aristeidou et al., 2021; Campbell et al., 2023; Sullivan & others, 2014). Such data allows stakeholders to know when there is a shift in species population, nesting loci, or migration pathways-which is very important in measuring the effect of conservation.

Feedback Loops and Adaptive Strategies

Development of feedback loops through adaptive management is one of the most effective uses of geospatial tools. Relevant conservation priorities and policies about forest policies can be re-examined promptly based on such insights as new data (monitoring of deforestation alerts, encroachment, and surprise changes in biodiversity) (Holling, 1978; Walters, 1986). As an example, in case satellite data shows an increase in illicit logging within a particular protected area, enforcement agencies can be able to change patrol schedules and redistribute resources. Similarly, as habitat models suggest a decline of habitat suitability towards a flagship species because of climate change, managers can review their zoning policies, improve upon corridors, or adopt an assisted migration technique.

Such spatially-based feedback loops underpin iterative decision-making, whereby decision-making is made continuously informed, tested, and refined based on constantly updated spatial intelligence. Future land-use impacts can be simulated using scenario modeling, via platforms such as InVEST, Marxan, or CLUE, providing a glimpse of future trade-offs in conservation or restoration benefits in alternative intervention futures (Ball et al., 2009; Sharp et al., 2020; Verburg et al., 2002).

Remote Sensing and Early Warning Systems

The creation of a forest and biodiversity early warning system fits in the framework of remote sensing. Fire monitoring satellites (e.g., NASA FIRMS) or drought stress monitoring thermal sensors, or because they monitor the land in real-time (e.g., Global Forest Watch), allow decision-makers to respond immediately when ecological limits are crossed (Hansen, Potapov, Moore, & others, 2013; Justice et al., 2002). As an example, where forest fires are more common, fire risk zoning maps overlain with meteorological information could be deployed in advance to steer community preparedness, resource deployment, and fire suppression readiness. An example is the remote sensors in the tropics through cloud-penetrating radar (e.g., Sentinel-1) that provides steady monitoring in spite of monsoon seasons and limits information lapses and response slacks.

Equally, deforestation early warning systems, such as those used in the Amazon basin, use high-frequency images in order to highlight the loss of tree cover within days, and this brings about enforcement mechanisms or mobilization of the people in order to limit the extent of damage, destruction, or losses. Adding continuous geospatial monitoring to the adaptive-management concepts allows the forest managers and conservationists to shift towards proactive stewardship instead of responding to the changes. Such tools not only shed light upon where and how the change is taking place, but have the power to facilitate evidence-based adjustments, which would increase the ecological resilience and sustainability. Table 3 presents an overview of widely used geospatial tools and the key ecological indicators they help track. To effectively link monitoring data with actionable strategies, a geospatially enabled adaptive management cycle is essential. Figure 2 illustrates a typical workflow for integrating diverse spatial data sources into continuous monitoring, scenario modeling, and iterative policy refinement.

Figure 2. Workflow of Geospatially-Enabled Monitoring and Adaptive Forest and Biodiversity Management

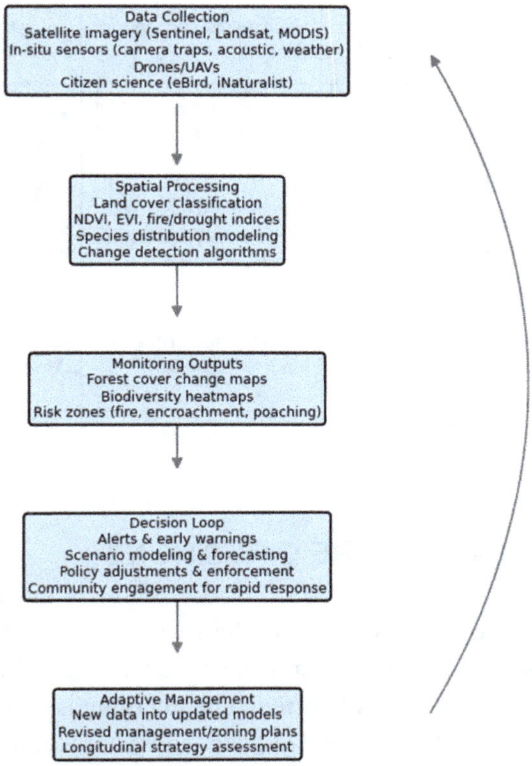

Table 3. Key Geospatial Tools and Indicators for Monitoring Forest and Biodiversity Systems

Tool/Technology	Monitoring Focus	Key Indicators	Example Applications
Remote Sensing (e.g., Landsat, Sentinel-2)	Forest cover change, deforestation, degradation	NDVI, canopy density, forest loss rate	Forest degradation mapping in the Congo Basin
GIS-Based Spatial Analysis	Zoning, habitat fragmentation, and corridor identification	Patch size, edge density, proximity to threats	Protected area planning in India using corridor analysis
Drones / UAVs	High-resolution monitoring, illegal activity detection	Tree height, biomass estimation, and visual change detection	Monitoring illegal logging in the Amazon
Change Detection Algorithms	Temporal analysis of land cover	Annual deforestation rate, reforestation gains	Myanmar forest recovery post-community intervention

continued on following page

Table 3. Continued

Tool/Technology	Monitoring Focus	Key Indicators	Example Applications
Species Distribution Modeling (SDM)	Priority conservation areas, migration tracking	Habitat suitability, species richness	Snow leopard corridor mapping in Central Asia
Early Warning Systems (EWS)	Fire, drought, flood, or pest outbreak prediction	Fire risk index, drought severity, pest alert thresholds	MODIS-based fire alerts in Indonesia's peatlands
Community Mapping Tools (e.g., GPS, ODK, participatory GIS)	Local monitoring and reporting	Presence/absence, land-use claims, culturally significant zones	Community-led Forest tracking in Nepal
Biodiversity Information Platforms (e.g., GBIF, iNaturalist)	Species sightings, ecological data	Occurrence records, observer frequency	Bird migration trends in East Africa

CHALLENGES AND FUTURE DIRECTIONS

Although geospatial technologies have the capacity to potentially transform the forest and biodiversity management sector, their overall effective implementation and use are constrained by a number of current and emerging issues. These difficulties range from technical and infrastructural restrictions as well as ethical and governance-related issues, but in particular, the data-sparse and marginalized territories.

Data Availability, Scale Mismatches, and Technical Capacity Gaps

The first challenge of the first order is the disparity in capacity and accessibility of quality spatial information. Most areas of the Global South, especially, do not have sufficient investment in monitoring infrastructure and open data policies, resulting in unreliable and up-to-date geospatial datasets being unavailable (Fritz et al., 2009; McCall & Minang, 2005). Although satellite imagery might be available (e.g., Landsat missions or the Sentinel series), very special skills and resources are needed to convert raw imagery into decision-ready information (Wulder et al., 2012).

Besides, mixes in scale can pose problems to planning processes, often as in the resolution of spatial data and decision scale (Cash et al., 2005). As an example, microhabitats that support populations of endemic species (such as forest inventories at the national level) or high-resolution data collected with drones (planning of large areas) may be out of reach. Major technical issues revolve around the balancing of macro and micro perceptions. The other key bottleneck is technical capacity. Even in most of the developing countries, the forestry department and other conservation agencies do not have skilled labor, up-to-date GIS-based infrastructure, and analytical facilities (Dhungana et al., 2024). Capacity-building projects tend to be skewed to the project and short-term, and there is little institutionalization of spatial planning processes.

Ethical Issues and Inclusivity in Spatial Decision-Making

Spatial tools, though very powerful, are not necessarily neutral. They are subjective in terms of their priorities, assumptions, and biases of their creators and users, and tend to enhance already existing social and spatial inequalities. Moral issues are brought up by the fact that spatial information is gathered or utilized without the involvement or approval of the impacted communities, or that analytical frameworks do not take into account the social, cultural, and political aspects of space. As an example, mapping processes that ignore indigenous land-use knowledge may marginalize traditional custodianship systems, displace customary rights, and unintentionally support land grabbing or resource struggles (Chambers, 2006).

Social justice in spatial planning must acknowledge that space is not a physical phenomenon and is a socio-political phenomenon that exists in relation to power. When spatial analyses are biased towards economic or technical efficiency to the neglect of community well-being, they will validate unequal access to land, infrastructure, and services. The disproportionate allocation of mapping resources, including access to high-resolution data, analytical functions, or technical skills, might further cement the differences between urban and rural, or technologically advanced and resource-limited communities (Cowie et al., 2022).

Equity and inclusivity are also new issues in algorithmic modeling, remote sensing, and spatial machine learning. Models that have been mostly trained on global or generalized data might miss the local ecological and socio-economic context and give output that is not reflective of the realities of the community. As an example, species distribution models can be more focused on environmental parameters but fail to consider cultural and livelihood aspects of landscapes. Conservation or development plans grounded on such models can therefore serve to marginalize or evict local people instead of empowering them.

Spatial decision-making inclusivity can hence not be achieved by token representation only, but a real redistribution of epistemic power is needed. Women and indigenous people, and other marginalized groups should be brought on board not just as sources of data, but as co-producers of spatial knowledge. Community-based mapping, participatory GIS (PGIS), and citizen science projects have demonstrated a great potential for democratizing spatial planning. But in most instances, such tools are only used in a mere form, with little bearing on the ultimate decisions or implementation of the policy (Mccall & Dunn, 2012). Notable inclusion involves the incorporation of participatory mechanisms in the entire planning process, such as data collection and analysis, decision-making, and sharing of benefits.

To better understand the landscape of barriers and potential enablers in applying geospatial strategies to forest and biodiversity management, Figure 3 presents a comparative visualization.

Figure 3. Barriers and Enablers in the Application of Geospatial Strategies for Forest and Biodiversity Management

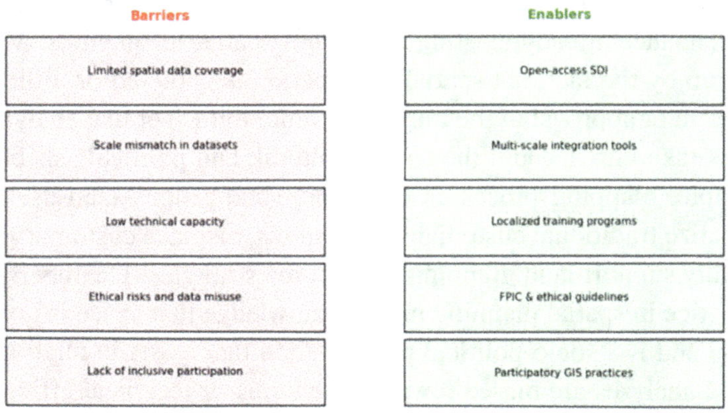

Recommendations for Policy, Capacity Building, and Research

To realize the full potential of spatial strategies in forest and biodiversity governance, what is suggested in terms of directions should be taken is as follows:

- Governments ought to invest in open-access spatial information infrastructures and interoperability across agencies. The planning of integrated land-use and biodiversity can receive a backbone through National Spatial Data Infrastructures (NSDIs).
- Technical education, GIS training centers, and technical capacity building of in-service forest managers and conservation personnel need long-term funding. Skills emphasized should be in the area of remote sensing interpretation, spatial modeling, and participatory mapping.
- There has to be a policy around Free, Prior, and Informed Consent (FPIC) of spatial data collection within indigenous lands (Corbera & Schroeder, 2011).

Geospatial practitioners should establish ethical codes to protect the rights of the community, just as is the case with biomedical research.

- This requires transdisciplinary research that can fuse ecology, technology, and social sciences. These are advancements in low-cost sensors, AI-enabled habitat modeling, and participatory mobile data collection that put experts and communities on equal footing.
- Last, spatial data must not be bound to an academic or technical realm. It requires mechanisms to translate geospatial understanding into actionable policy, which requires the availability of clear visualization tools and decision dashboards, and scenario simulation, available to policymakers.

The future of spatial strategies in such areas of work as biodiversity conservation and forest management can be shifted towards a more leveled and data-driven, stronger model with the recognition of these issues and their timely resolution.

CONCLUSION

The currently used geospatial technologies have been adopted in the management of forests and biodiversity, and have fundamentally changed the way the perception of ecosystem is perceived, controlled, and managed. These tools, which include GIS, remote sensing, spatial modeling, and UAVs, have become essential in the planning of conservation and implementation. These tools allow viewing and interpreting complicated ecological patterns on numerous spatial and temporal scales, both in near-real-time observations of deforestation and in advanced projections of the consequences of climate change on community composition.

What is clear about this discussion is not merely about the individual strength of any individual tool, but the power to change when there is an interplay between these technologies and a participatory process, and a community-based strategy. The addition of local and indigenous knowledge by using participatory mapping is enriching the quality of spatial data and instills the aspect of ethical stewardship in conservation activities. Through spatial platforms and mobile applications, community-based forest management has demonstrated wonderful success in numerous locations, lending even additional power to the significance of inclusive governance and local agency in the accomplishment of ecological results.

Geospatial tools have changed monitoring, evaluation, and adaptive management as well. Satellite imagery and ground observations relating to conservation over time can be used to reassess the effectiveness of the given conservation plans in terms of providing feedback to improve them over time. Once remote-sensing dependent early warning systems are in place, forest fires as well as illegal logging, and biodiversity threats can be met very objectively and rapidly, and this has further favored more proactive and resilient management of the ecosystems involved. There are, however, limitations to this geospatial revolution. Other long-standing issues like data fragmentation, scale mismatch, inconsistent technical capabilities, and ethical issues with respect to data ownership and representation still prevail as obstacles to optimal implementation. They need to address the gaps in terms of their investment in open data infrastructures, training opportunities, equality in policy frameworks, and interdisciplinary research programmes.

The future of conserving forests and biodiversity is in the increase of technological innovations in a manner that is not only fair and context sensitive but also ethically grounded. Simultaneously, in the next ten years geospatial intelligence, artificial intelligence, and environmental sensing will progressively be able to monitor ecosystems, deforestation, and habitat fragmentation in near real-time. New technologies will include AI-based biodiversity models, blockchain-based forest product traceability, and drone-based ecological survey tools that will provide greater accuracy and transparency in conservation initiatives than ever before. The key to reaching their full potential will however lie in bridging the perennial divide between the latest geospatial science to cutting-edge geospatial science and the realities of communities on the ground, especially in the Global South where data access, infrastructure and governance facilities are disproportionately uneven.

In the policy arena, the new frameworks will focus on data sovereignty, cross-border ecosystem management, and ethical use of AI in environmental management. It is anticipated that governments and international bodies should encourage the open geospatial data standards, interoperability and participatory spatial governance so that technological benefits can be fairly distributed. Meanwhile, geospatial verification systems will become more central to climate adaptation processes and financing mechanisms centering around carbon markets and nature-based solutions to recalculate environmental performance and social co-benefits. The ecological challenges of the next ten years will be even greater with increasingly pressing changes of climate change, invasive species and land-use changes requiring dynamic, predictive and community-based spatial models. Planning conservation will have to be informed by ecological forecasting, indigenous ecological knowledge and scenario based spatial modeling in order to predict and control the biodiversity loss in the changing climatic conditions.

To fulfill these aspirations, a multi-stakeholder ecosystem, which includes governments, research centers, indigenous people, civil society, and individual technology creators, will have to be developed. It is only with the principles of inclusive collaboration and shared governance and the element of ethical design that geospatial technologies can truly be of service to people and planet, making conservation proactive, equitable, and resilient to the future.

REFERENCES

Agrawal, A., & Ostrom, E. (2001). Collective action, property rights, and decentralization in resource use in India and Nepal. *Politics & Society, 29*(4). Advance online publication. DOI: 10.1177/0032329201029004002

Ali, S., Wang, D., Hussain, T., Lu, X., & Nurunnabi, M. (2021). Forest resource management: An empirical study in northern pakistan. *Sustainability (Switzerland), 13*(16). Advance online publication. DOI: 10.3390/su13168752

Aristeidou, M., Herodotou, C., Ballard, H. L., Higgins, L., Johnson, R. F., Miller, A. E., Young, A. N., & Robinson, L. D. (2021). How do young community and citizen science volunteers support scientific research on biodiversity? The case of inaturalist. *Diversity, 13*(7). Advance online publication. DOI: 10.3390/d13070318

Asner, G. P., Mascaro, J., Anderson, C., Knapp, D. E., Martin, R. E., Kennedy-Bowdoin, T., van Breugel, M., Davies, S., Hall, J. S., Muller-Landau, H. C., & Potvin, C. (2012). High-resolution mapping of forest carbon stocks in the Colombian Amazon. *Biogeosciences, 9*(7), 2683–2696. DOI: 10.5194/bg-9-2683-2012

Ball, I. R., Possingham, H. P., & Watts, M. E. (2009). Marxan and relatives: Software for spatial conservation prioritisation. *Spatial Conservation Prioritisation: Quantitative Methods and Computational Tools*, 185–195.

BirdLife International. (2024). *Over half of forest within Key Biodiversity Areas identified for forest species no longer has high integrity*. https://datazone.birdlife.org/articles/over-half-of-forest-within-key-biodiversity-areas-identified-for-forest-species-no-longer-has-high-integrity

Blomley, T., Pfliegner, K., Isango, J., Zahabu, E., Ahrends, A., & Burgess, N. (2008). Seeing the wood for the trees: An assessment of the impact of participatory forest management on forest condition in Tanzania. *Oryx, 42*(3). Advance online publication. DOI: 10.1017/S0030605308071433

Bray, D. B., Merino-Pérez, L., & Barry, D. (2005). Community managed in the strong sense of the phrase: The community forest enterprises of Mexico. In *The Community Forests of Mexico*. Managing for Sustainable Landscapes. DOI: 10.7560/706378-003

Campbell, C. J., Barve, V., Belitz, M. W., Doby, J. R., White, E., Seltzer, C., Di Cecco, G., Hurlbert, A. H., & Guralnick, R. (2023). Identifying the identifiers: How iNaturalist facilitates collaborative, research-relevant data generation and why it matters for biodiversity science. *Bioscience, 73*(7). Advance online publication. DOI: 10.1093/biosci/biad051

Cash, D., Clark, W. C., Alcock, F., Dickson, N., Eckley, N., & J ger, J. (2005). Salience, Credibility, Legitimacy and Boundaries: Linking Research, Assessment and Decision Making. *SSRN Electronic Journal*. https://doi.org/DOI: 10.2139/ssrn.372280

Chambers, R. (2006). Participatory Mapping and Geographic Information Systems: Whose Map? Who is Empowered and Who Disempowered? Who Gains and Who Loses? *The Electronic Journal on Information Systems in Developing Countries*, *25*(1). Advance online publication. DOI: 10.1002/j.1681-4835.2006.tb00163.x

Charnley, S., & Poe, M. R. (2007). Community forestry in theory and practice: Where are we now? In *Annual Review of Anthropology* (Vol. 36). https://doi.org/DOI: 10.1146/annurev.anthro.35.081705.123143

Chitayat, A. B., Lewis, M., Anyelwisye, M., Laizer, G. S., Stewart, F. A., Wich, S. A., & Piel, A. K. (2024). Development of spatial models and maps for tree species diversity and biomass in a miombo ecosystem, western Tanzania. *Applied Vegetation Science*, *27*(4), e70002. DOI: 10.1111/AVSC.70002

Corbera, E., & Schroeder, H. (2011). Governing and implementing REDD+. *Environmental Science & Policy*, *14*(2). Advance online publication. DOI: 10.1016/j.envsci.2010.11.002

Cowie, R. H., Bouchet, P., & Fontaine, B. (2022). The Sixth Mass Extinction: Fact, fiction or speculation? *Biological Reviews of the Cambridge Philosophical Society*, *97*(2), 640–663. DOI: 10.1111/BRV.12816

de Bem, P. P., de Carvalho, O. A., Guimarães, R. F., & Gomes, R. A. T. (2020). Change detection of deforestation in the brazilian amazon using landsat data and convolutional neural networks. *Remote Sensing*, *12*(6). Advance online publication. DOI: 10.3390/rs12060901

Dhungana, N., Lee, C. H., Khadka, C., Adhikari, S., Pudasaini, N., & Ghimire, P. (2024). Evaluating Community Forest User Groups (CFUGs)' Performance in Managing Community Forests: A Case Study in Central Nepal. *Sustainability (Switzerland)*, *16*(11), 4471. DOI: 10.3390/SU16114471/S1

Drusch, M. (2012). Sentinel-2: ESA's Optical High-Resolution Mission for GMES Operational Services. *Remote Sensing of Environment*, *120*, 25–36.

European Space Agency. (2024). *Copernicus Sentinel Missions*. https://sentinels.copernicus.eu

Fida, T., Mohammadi, A., Almasieh, K., Bosso, L., Ud Din, S., Shamas, U., Nawaz, M. A., & Kabir, M. (2024). Species distribution modelling and landscape connectivity as tools to inform management and conservation for the critically endangered Himalayan brown bear (Ursus arctos isabellinus) in the Deosai National Park, Pakistan. *Frontiers in Ecology and Evolution*, *12*, 1477480. DOI: 10.3389/FEVO.2024.1477480

Forest Declaration. (2024). *Conserve and restore forests and other terrestrial ecosystems*. https://dashboard.forestdeclaration.org/article/conserve-and-restore-forests-and-other-terrestrial-ecosystems

Friedman, K., Bridgewater, P., Agostini, V., Agardy, T., Arico, S., Biermann, F., Brown, K., Cresswell, I. D., Ellis, E. C., Failler, P., Kim, R. E., Pratt, C., Rice, J., Rivera, V. S., & Teneva, L. (2022). The CBD Post-2020 biodiversity framework: People's place within the rest of nature. *People and Nature*, *4*(6). Advance online publication. DOI: 10.1002/pan3.10403

Fritz, S., McCallum, I., Schill, C., Perger, C., Grillmayer, R., Achard, F., Kraxner, F., & Obersteiner, M. (2009). Geo-wiki.org: The use of crowdsourcing to improve global land cover. In *Remote Sensing* (Vol. 1, Issue 3). DOI: 10.3390/rs1030345

Giorgi, A. P., Rovzar, C., Davis, K. S., Fuller, T., Buermann, W., Saatchi, S., Smith, T. B., Silveira, L. F., & Gillespie, T. W. (2014). Spatial conservation planning framework for assessing conservation opportunities in the Atlantic Forest of Brazil. *Applied Geography (Sevenoaks, England)*, *53*. Advance online publication. DOI: 10.1016/j.apgeog.2014.06.013

Google Earth Engine. (2024). *Google Earth Engine: Data Catalog*. https://earthengine.google.com/datasets/

Hansen, M. C., Potapov, P. V., Moore, R., Hancher, M., Turubanova, S. A., Tyukavina, A., Thau, D., Stehman, S. V., Goetz, S. J., Loveland, T. R., Kommareddy, A., Egorov, A., Chini, L., Justice, C. O., & Townshend, J. R. G. (2013). High-resolution global maps of 21st-century forest cover change. *Science*, *342*(6160), 850–853. DOI: 10.1126/science.1244693

Hansen, M. C., Potapov, P. V., & Moore, R. (2013). High-resolution global maps of 21st-century forest cover change. *Science*, *342*(6160), 850–853.

Hardin, P. J., Lulla, V., Jensen, R. R., & Jensen, J. R. (2019). Small Unmanned Aerial Systems (sUAS) for environmental remote sensing: Challenges and opportunities revisited. *GIScience & Remote Sensing*, *56*(2). Advance online publication. DOI: 10.1080/15481603.2018.1510088

Hein, J., Guarin, A., Frommé, E., & Pauw, P. (2018). Deforestation and the Paris climate agreement: An assessment of REDD + in the national climate action plans. In *Forest Policy and Economics* (Vol. 90). DOI: 10.1016/j.forpol.2018.01.005

Holling, C. S. (1978). *Adaptive Environmental Assessment and Management*. John Wiley & Sons.

Huete, A. R., Didan, K., Miura, T., Rodriguez, E. P., Gao, X., & Ferreira, L. G. (2002). Overview of the radiometric and biophysical performance of the MODIS vegetation indices. *Remote Sensing of Environment*, *83*(1–2), 195–213.

IFAD. (2009). Good Practices in Participatory Mapping: A Review Prepared for the International Fund for Agricultural Development (IFAD). In *International Fund for Agricultural Development*.

Justice, C. O., Giglio, L., & Korontzi, S. (2002). The MODIS fire products. *Remote Sensing of Environment*, *83*(1–2), 244–262.

Kays, R. (2020). Does hunting or hiking affect wildlife communities in protected areas? *Ecological Applications*, *30*(7), e02133.

Longley, P. A., Goodchild, M. F., Maguire, D. J., & Rhind, D. W. (2015). *Geographic Information Systems and Science* (5th ed.). Wiley.

Mahmud, A., Setiawan, I., Mansor, S., Shariff, A., Pradhan, B., & Nuruddin, A. (2009). Utilization of geoinformation tools for the development of forest fire hazard mapping system: Example of Pekan fire, Malaysia. *Central European Journal of Geosciences*, *1*(4). Advance online publication. DOI: 10.2478/v10085-009-0032-5

Mccall, M. K., & Dunn, C. E. (2012). Geo-information tools for participatory spatial planning: Fulfilling the criteria for "good" governance? *Geoforum*, *43*(1). Advance online publication. DOI: 10.1016/j.geoforum.2011.07.007

McCall, M. K., & Minang, P. A. (2005). Assessing participatory GIS for community-based natural resource management: Claiming community forests in Cameroon. In *Geographical Journal* (Vol. 171, Issue 4). DOI: 10.1111/j.1475-4959.2005.00173.x

Myers, R., Intarini, D., Sirait, M. T., & Maryudi, A. (2017). Claiming the forest: Inclusions and exclusions under Indonesia's 'new' forest policies on customary forests. *Land Use Policy*, *66*. Advance online publication. DOI: 10.1016/j.landusepol.2017.04.039

Paneque-Gálvez, J., McCall, M. K., Napoletano, B. M., Wich, S. A., & Koh, L. P. (2014). Small drones for community-based forest monitoring: An assessment of their feasibility and potential in tropical areas. *Forests*, *5*(6). Advance online publication. DOI: 10.3390/f5061481

Pettorelli, N., Laurance, W. F., O'Brien, T. G., Wegmann, M., Nagendra, H., & Turner, W. (2014). Satellite remote sensing for applied ecologists: Opportunities and challenges. *Journal of Applied Ecology*, *51*(4), 839–848. DOI: 10.1111/1365-2664.12261

Phalan, B., Balmford, A., Green, R. E., & Scharlemann, J. P. W. (2011). Minimizing the harm to biodiversity of producing more food globally. *Food Policy*, *36*, S62–S71. DOI: 10.1016/j.foodpol.2010.11.008

Pokharel, B. K., Branney, P., Nurse, M., & Malla, Y. B. (2007). Community Forestry : Conserving Forests, Sustaining Livelihoods and Strengthening Democracy. *Journal of Forest and Livelihood*, *6*(2).

Rambaldi, G., Callosa-Tarr, J., & others. (2002). *Participatory 3-Dimensional Modelling: guiding principles and applications.*

Rambaldi, G., Chambers, R., McCall, M., & Fox, J. (2006). Practical ethics for PGIS practitioners, facilitators, technology intermediaries and researchers. *Participatory Learning and Action*, •••, 54.

Rautela, N., Shanu, S., Agarwal, A., Bhattacharya, S., & Roy, A. (2022). Geospatial modelling of overlapping habitats for identification of tiger corridor networks in the Terai Arc landscape of India. *Geocarto International*, *37*(27). Advance online publication. DOI: 10.1080/10106049.2022.2095444

Roy, D. P., Wulder, M. A., & Loveland, T. R. (2014). Landsat-8: Science and product vision for terrestrial global change research. *Remote Sensing of Environment*, *145*, 154–172.

Sharp, R., Tallis, H., Ricketts, T., & Associates. (2020). *InVEST 3.9.0 User's Guide*.

Smith, A. P., Horning, N., & Moore, D. (1997). Regional biodiversity planning and lemur conservation with GIS in western Madagascar. *Conservation Biology*, *11*(2). Advance online publication. DOI: 10.1046/j.1523-1739.1997.95283.x

Sodhi, N. S., Posa, M. R. C., Lee, T. M., Bickford, D., Koh, L. P., & Brook, B. W. (2010). The state and conservation of Southeast Asian biodiversity. *Biodiversity and Conservation*, *19*(2). Advance online publication. DOI: 10.1007/s10531-009-9607-5

Sugai, L. S. M., Silva, T. S. F., Ribeiro, M. A.Jr, & Llusia, D. (2019). Terrestrial passive acoustic monitoring: Review and perspectives. *Bioscience*, *69*(1), 15–25.

Sullivan, B. L. (2014). The eBird enterprise: An integrated approach to development and application of citizen science. *Biological Conservation*, *169*, 31–40.

Torres, D. L., Turnes, J. N., Vega, P. J. S., Feitosa, R. Q., Silva, D. E., Marcato, J.Junior, & Almeida, C. (2021). Deforestation detection with fully convolutional networks in the amazon forest from landsat-8 and sentinel-2 images. *Remote Sensing*, *13*(24). Advance online publication. DOI: 10.3390/rs13245084

UNEP-WCMC. (2024). *Earth's biodiversity depends on the world's forests*. https://www.unep-wcmc.org/en/news/earths-biodiversity-depends-on-the-worlds-forests

Verburg, P. H., Soepboer, W., Veldkamp, A., Limpiada, R., Espaldon, V., & Mastura, S. S. A. (2002). Modeling the spatial dynamics of regional land use: The CLUE-S model. *Environmental Management*, *30*(3), 391–405.

Walters, C. (1986). *Adaptive Management of Renewable Resources*. Macmillan.

World Resources Institute. (2025). *Global Forest Watch*. https://www.globalforestwatch.org

Wulder, M. A., Masek, J. G., Cohen, W. B., Loveland, T. R., & Woodcock, C. E. (2012). Opening the archive: How free data has enabled the science and monitoring promise of Landsat. *Remote Sensing of Environment*, *122*. Advance online publication. DOI: 10.1016/j.rse.2012.01.010

KEY TERMS AND DEFINITIONS

Biodiversity: refers to the variety and variability of life on Earth, encompassing diversity within species, between species, and within ecosystems. It includes all living organisms and the ecological complexes they form.

Forest degradation: is the decline in a forest's ability to provide ecosystem services, such as carbon storage, biodiversity habitat, and soil and water protection, due to human or natural disturbances. It does not necessarily mean a complete loss of forest cover, but rather reflects a decline in forest quality.

Geospatial data: is data containing information about the geographic location and characteristics of natural or human-made features on the Earth's surface. This data is typically presented as coordinates, maps, or spatial layers and is used for analysis, decision making, and visualization.

Geographic information system (GIS): is a computing system for collecting, storing, managing, analyzing, and visualizing geospatial data. GIS integrates spatial and non-spatial information to support mapping, spatial analysis, modeling, and decision making in various fields.

Remote sensing: is the science and technology of obtaining information about objects or areas at a distance, typically using sensors mounted on satellites, aircraft, drones, or ground-based systems. It records electromagnetic radiation reflected or emitted by the Earth's surface to analyze land cover, environmental changes, and other phenomena.

Unmanned aerial vehicles (UAVs): commonly referred to as drones, are aerial vehicles controlled without a pilot on board. Equipped with cameras, sensors, or other payloads, UAVs are used for aerial photography, remote sensing, mapping, environmental monitoring, and data collection in hard-to-reach areas.

Chapter 14
Revolutionizing Forest Monitoring:
The Role of Drones and Remote Sensing

Nitin Goyal
https://orcid.org/0000-0002-0043-5472
Lovely Professional University, India

Shyam Sunder Agrawal
https://orcid.org/0009-0004-3648-0584
Lovely Professional University, India

Amanjot Singh Syan
Canadore Stanford International College, Canada

ABSTRACT

This chapter explored how remote sensing and drone technologies transformed forest monitoring and management practices between 2015 to 2025. The objective was to assess their role in evaluating forest health, biodiversity, and carbon storage. A key research gap addressed was the limited understanding of how integrated platforms, combining satellites, drones, LiDAR, and vegetation indices, could offer more precise and scalable forest data. Using a conceptual framework and the Systematic Literature Review (SLR) technique for literature review, the study analyzed tools such as Landsat, Sentinel, MODIS, NDVI, EVI, and LiDAR to examine their applications in tracking deforestation, estimating biomass, and detecting illegal logging. The findings showed that merging satellite imagery with drone and ground-based sensors significantly improved the accuracy and frequency of forest assessments.

DOI: 10.4018/979-8-3373-3992-4.ch014

Integration with GIS and machine learning further enhanced decision-making capabilities. Multi-source remote sensing enhances forest conservation, climate monitoring, policy development.

1 INTRODUCTION

1.1 Background of the Study

1.1.1 Overview of Forest Management

Forest ecosystems - the most essential elements in an environment- offer environmental services such as carbon sink, biodiversity habitat as well as water cycling regulation (Ahmad et al., 2025). These important systems; however, face adverse challenges which complicate their management and conservation. With increase in human destruction from forest and the changing climate, the calls for effective management have never been so urgent. Below are a few points (Dasipah et al., 2024).

Deforestation

Indeed, deforestation remains one of the most serious global problems. Most of this clearing occurs in tropical areas, where forests are felled in great quantities to make room for agricultural expansion, urbanization, and logging (de Almeida et al., 2025). Millions of hectares of forests are lost each year, directly affecting one of the most valuable resources-not only hastening climate change (Doornbos et al., 2024). All of these can happen because forests provide important carbon sink functions, taking carbon dioxide from the atmosphere. The moment they are destroyed, however, not only is this capacity diminished, but the carbon stored in the trees is released (Gonzalez et al., 2016), thus contributing to global warming. In addition, deforestation is not merely a carbon emission. It can affect local climates and result in reduced rainfall and even increase the vulnerability of indigenous communities to all forms of subsistence-dependence on foods, shelters, and cultures that have close ties with forests (Harle et al., 2024).

Forest Degradation

Forest degradation is a gradual decline in the quality of the structure and function of a forest. This does not always mean total elimination of trees, but it can severely damage the ability of a forest to support biodiversity, regulate water, and sequester carbon (Lame, 2019). Degradation is often the result of unsustainable practices such as illegal logging, overgrazing, and poor land management (Li & Wang, 2025). It can bring soil erosion, reduce water retention, and lower forest productivity. Degraded forests often appear to be "forests" as seen from the strata, but their ecological functions are largely absent, and they have reduced resilience to disturbances like droughts, fires or pest outbreaks (Mulatu et al., 2017). This somehow widely spread degradation poses a serious problem for the remaining forest managers, who need to balance restoration on one side against continued pressure from land-use demands on the other side (Machireddy, 2023).

Biodiversity Loss

The forests themselves are biodiversity havens, habitats for terrestrial species that account for more than 80 percent of their total diversity. Degraded through either deforestation or degradation, forests impact on biodiversity in cascade effects (Neves & de Aquino, 2024). The loss of habitat threatens the majority of species with extinction and disrupts ecosystem services. In addition to their natural implications, they have deep effects on human societies since forests are the most critical sources of food, medicine, and materials (Pádua et al., 2017). Deforestation reduces the resilience of forests against pests, diseases, or climate change. Therefore, effective forest management needs to consider the preservation of biodiversity and restoration of degraded ecosystems for maintaining the stability and functioning of forest landscapes (Pandey & Arellano, 2022).

The Need for Sustainable Forest Management

Sustainable forest management (SFM) presents great challenges in the complex interplay of how forests operate as ecosystems. In summary, SFM embodies the vital principle of the richness of biodiversity reservation, as it strives to equate environmental sustainability and the efficiency of socioeconomic forest use (Raj et al., 2024). So, forest resources ought to be consumed such that they fulfil the current generation's wants without compromising the future generation's ability to do the same. Requires a holistic approach of conservation, restoration, and use-stewardship of forest resources well-grounded in science and enabling governance: how such practices can be implemented at the local level and at the same time scaled (Rane & Choudhary, 2023). This is the challenge in implementing SFM, considering it can be locally relevant and scaled to the global forestry conservation goal. The heart of that requirement is innovative technologies that will describe with good accuracy their conditions in real time and will guide management decisions. This is where remote sensing technologies, including satellite-based systems, are playing an increasingly critical role (Stone, 2012).

1.1.2 Emergence of Remote Sensing Technologies in Forest Management

Everything regarding the new technologies, efficiency, and monitoring requirements has taken on new dimensions in forest management. Remote sensing technology collects data from a distance, mainly satellite and airborne systems such as drones or aircraft (Thakur et al., 2021). These technologies have made it possible to monitor large-scale, difficult-to-access forest areas in high frequency and high resolution, thereby providing new insights into assessing forest condition, biodiversity, and carbon sequestration (Vélez et al., 2023).

In the last 20 years, this knowledge has rapidly evolved, from simple aerial photographs to state-of-the-art satellites and drone systems capable of acquiring multispectral and hyperspectral data (Wavrek et al., 2023). As an example, satellite remote sensing data provides global coverage and long-term monitoring, which can be useful to track relatively slow changes over a larger scale, such as regional deforestation and degradation processes (Zakari et al., 2024). Conversely, drones provide comparatively high-resolution images and data for capturing canopy architecture, biodiversity, and forest management interventions.

The advances in technology facilitate the collection of real-time, large-scale, and low-labour input forest data. With new technologies that enhance data collection, conserving forests and forest management have become more cost-effective and less risky than traditional methods of fieldwork (Zhang et al., 2016). Remote sensing has thus become an indispensable tool for sustainable forest management and conservation.

Figure 1. Trends in remote sensing technologies usage in forestry (2010 to 2025)

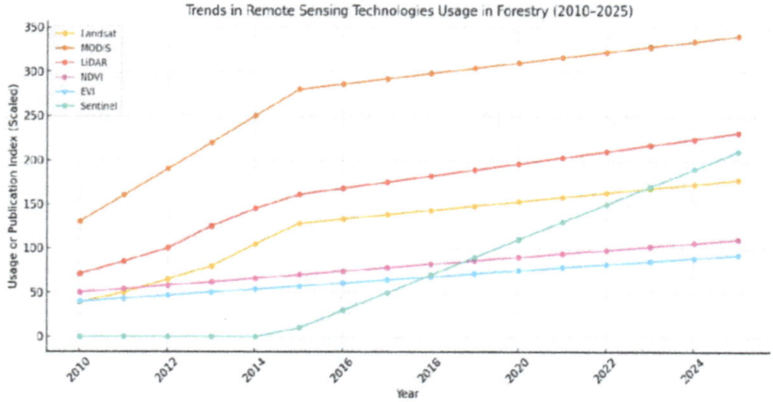

Source: Self-created

Figure 1 clearly illustrates the growth patterns of six major technologies: Landsat, MODIS, LiDAR, NDVI, EVI, and Sentinel. From the data, MODIS consistently held the highest usage or publication index over the years, showing steady growth. Sentinel, although starting from a low point in 2015, experienced a sharp and consistent rise, indicating its increasing popularity in forestry applications. LiDAR also showed a positive upward trend, reflecting its growing role in high-precision forest mapping. NDVI and EVI followed moderate upward trends, suggesting their steady use for vegetation monitoring. Landsat showed a gradual but stable increase in usage. Overall, the graph highlights a general rise in the adoption of remote sensing tools in forestry between 2015 and 2025, with Sentinel and LiDAR gaining significant momentum. These trends suggest a growing reliance on diverse remote sensing technologies for better forest management and environmental monitoring.

1.1.3 Technological Revolution.:

Technological advancement in forest monitoring through the Integration of Drones, Satellites, and Ground-Based Systematic approaches results in strong consideration by remote sensing tools like satellites, drones, and land observatories that are dealt with smartly (Pandey & Arellano, 2022). Each of these has unique advantages, and the combined application gives a more complete picture of forest ecosystems than any one tool could do on its own.

Satellites

Satellite-based systems such as Landsat, Sentinel, and MODIS provide continuous monitoring of the ecosystem of forests on a large scale, which is in fact the lifeblood of global forest monitoring. Such satellites measure multispectral bands capturing forest cover, vegetation health, and biomass-related information (Sabir et al., 2024). Natural NDVI and Enhanced NDVI are important vegetation indices that are generally used to assess forest health and detect changes in forest cover and value carbon pooling. The key satellite capability is large-scale forest monitoring with consistent annual/daily quality between years and decades, which is irreplaceable in monitoring the change in long-term deforestation, forest degradation, and the climate (Ogunbamowo et al., 2024).

Drones

While aircraft can provide a broad view over a major area and/or state, drones can collect high-resolution local data to short distances for specific forest surveillance. Drones equipped with cameras, LiDAR, or multispectral sensors have been examined in depth for analysis of forest-canopy structure, tree health, and biodiversity assessment on a much finer level (Doornbos et al., 2024). For instance, they allow fire damage assessment on forests, timber recovery after disturbances, and wildlife tracking in the difficult accessibility of forest areas. Therefore, drones are an easier and quicker way to carry out such studies compared to satellite or ground-based methods, which would be much less efficient or too pricey (Haq et al., 2024).

Ground-Based Sensors

The tools used on the ground, such as LiDAR (Light Detection and Ranging), handheld sensors, and other geospatial technologies, also play an essential role in complementing the information provided from airborne sensors and satellites (Sofia et al., 2022). Here, these technologies create very accurate field-level data about parameters associated with the forest, including tree height, density of canopy, soil quality, and structure of the forest stands. LiDAR is an imaging technology that captures 3D information about forest canopy structure and topography at a very high resolution, which allows forest managers to assess the health of forests and biomass estimation quite accurately (Pandey & Arellano, 2022).

By integrating satellite data for large-scale monitoring, drone imagery for high-resolution monitoring, and ground sensors for accurate measurements, forest managers would have a richer and deeper understanding of forest conditions that would allow for wiser decision-making (Mulatu et al., 2017).

1.1.4 Importance of Forest Health: Assessing Forest Health, Biodiversity, and Carbon Sequestration Through Technology

The viability of an environment depends on the health of its ecosystem, and forests are surely the most reliable witnesses. The forests are major actors in nurturing biodiversity, in carbon sequestration, and in the maintenance of air and water quality. Because of this, forest health assessment finds its way into both forest management and global environmental conservation efforts (Machireddy, 2023).

These remote sensing techniques have thus become invaluable to forest health assessment, providing suitable real-time information from a variety of parameters (Mishra & Rai, 2021) For example, satellite-based vegetation indices such as NDVI and EVI detect forest stress caused by drought, pests, or disease. Drones and ground sensors also give more detailed information regarding forest conditions like tree canopy density, soil properties, and health conditions of trees (Nandasena et al., 2023).

Aside from improving forest health, remote sensing also serves to provide information regarding biodiversity assessments. These would include the estimation of forest structure, canopy cover, and the health of vegetation; with these instruments, ecologists can determine wildlife habitats and monitor species population changes and dynamics in biodiversity over time (Neves & de Aquino, 2024).

In addition, forests act as some of the largest carbon pools on earth, and one of the essential functions forests play in carbon sequestration is a major contributory factor in alleviating the impacts of climate change (Rane & Choudhary, 2023). Using remote sensing technologies helps make reliable estimates of forest biomass and carbon stocks based on height, density, and volume of trees, thereby enabling quantification of the role of forests in the global carbon cycle and evaluation of any reforestation or afforestation activities (Silva et al., 2022).

All in all, with remote-sensing technology integration, forest managers have access to a full toolbox for monitoring forest health, conserving biodiversity, and ensuring that forests continue to function, providing crucial ecosystem services for the survival of the planet.

1.2 Research Gap

Systematic Literature Review (SLR) is a structured research methodology that helps identify existing knowledge, analyze trends, and uncover research gaps by critically evaluating and synthesizing previous studies in a comprehensive and unbiased manner, Table 1 summarizes key gaps in forest monitoring, highlighting limitations of traditional methods, poor data integration, and insufficient real-time capabilities. These gaps emphasize the need for multi-source remote sensing and advanced technologies for accurate, scalable, and timely forest management.

Table 1. Research Gap

Research Focus	Identified Gap	Justification for Bridging the Gap	Citations
Existing Gaps in Forest Monitoring: Limitations of Traditional Methods	Traditional forest monitoring methods, such as manual surveys and fieldwork, are time-consuming, expensive, and often provide limited spatial and temporal coverage.	Remote sensing technologies can overcome the limitations of traditional methods by providing faster, more scalable, and more comprehensive data for large areas.	Cocchia, A. (2014), Dasipah et al., (2024)
Challenges with Data Integration: Satellite, Drone, and Ground-Based Data	There is a lack of effective integration and synchronization of data from satellite imagery, drones, and ground-based sensors, which hinders the development of a unified forest monitoring system.	Integrating multiple data sources allows for a more accurate and holistic view of forest conditions, improving management practices, and enabling better decision-making.	Ehrlich-Sommer et al., (2024), Jiménez López & Mulero-Pázmány, (2019)
Need for High-Resolution and High-Frequency Observations	Current remote sensing methods often face trade-offs between resolution and temporal frequency, with the need for more frequent and precise monitoring to detect short-term forest changes like pests, diseases, or climate impacts.	High-resolution, high-frequency observations are crucial for detecting subtle, short-term changes in forest ecosystems and allowing for more responsive management strategies.	Lame, (2019), Li& Wang, (2025), Messinger et al., (2016)
Role of GIS and Machine Learning in Forest Monitoring	There are challenges in effectively integrating remote sensing data with geospatial technologies (GIS) and machine learning algorithms, especially for large-scale, real-time forest monitoring.	The integration of GIS and machine learning enables advanced analyses of complex forest data, improving predictive modeling, decision-making, and the ability to address specific forest management challenges.	Mishra & Rai, (2021), Mulatu et al., (2017)
Integration of Satellite and Drone Data for Forest Monitoring	Limited research on the fusion of satellite and drone data for comprehensive forest monitoring. Most studies focus on either satellite or drone technology separately.	Integrating both technologies can provide more detailed and accurate assessments of forest health, combining large-scale satellite data with high-resolution drone imagery.	Ogunbamowo et al., (2024), Pandey & Arellano, (2022)
Real-time Forest Health Monitoring Using Remote Sensing	A gap exists in real-time forest health monitoring using integrated systems of UAVs, satellites, and ground sensors, particularly for high-frequency updates and dynamic forest conditions.	Real-time monitoring is crucial for early detection of forest health issues such as pest outbreaks or diseases, allowing for timely interventions.	Quamar et al., (2023), Raj et al., (2024)

continued on following page

Table 1. Continued

Research Focus	Identified Gap	Justification for Bridging the Gap	Citations
Role of Remote Sensing in Biodiversity Conservation	Insufficient integration of remote sensing technologies with biodiversity conservation models to assess habitat fragmentation and species distribution in forests.	A more integrated approach would enhance conservation efforts by providing more accurate data on species habitats and biodiversity hotspots, improving decision-making for protection.	Sabir et al., (2024), Sanchez-Azofeifa et al., (2017)
Carbon Sequestration and Forest Carbon Stocks Estimation	A need for more accurate and consistent methods of carbon sequestration estimation that combine satellite, drone, and ground-based data for forest carbon monitoring.	Accurately estimating carbon stocks is crucial for evaluating the role of forests in mitigating climate change and supporting reforestation or afforestation initiatives.	Mishra et al., (2024), Maurya et al., (2022)
Advancement in LiDAR and Ground-Based Remote Sensing Tools	Limited exploration of integrating LiDAR and handheld sensors with UAV and satellite data to provide more accurate field-level measurements of forest parameters.	LiDAR and ground sensors offer precise data on forest structure and soil quality, which can complement aerial and satellite data for improved forest management and health assessments.	Sofia et al., (2022), Zakari et al., (2024)
Impact of Climate Change on Forest Health through Remote Sensing	Current remote sensing methods cannot often accurately capture the long-term impacts of climate change on forest ecosystems. There is also insufficient integration of data to predict future climate impacts.	Effective forest management in the face of climate change requires the ability to predict and monitor the effects of climate shifts on forest health, which can be achieved by integrating multiple remote sensing data sources.	Sabir et al., (2024), Oliveira et al., (2024)

Source: Created by authors

1.3 Research Objectives

- Examine the role of satellite technologies (Landsat, Sentinel, MODIS) for assessing forest health and their integration with drone and ground-based data.
- Investigate the applications of drones and UAVs for high-resolution, localized forest data collection to enhance forest monitoring.
- Examine the use of ground-based tools like LiDAR to complement satellite and drone data for accurate field-level forest measurements.
- Explore how remote sensing technologies address key forest management challenges, such as deforestation, biodiversity loss, and climate change mitigation.

- Assess the integration of GIS and machine learning with remote sensing data to improve forest management decision-making and climate change responses.

1.4 Conceptual Framework

The conceptual framework of monitoring and managing forest health includes sophisticated technologies for examining, assessing, and alleviating environmental concerns (Figure 2). Satellite technologies include Landsat, Sentinel, and MODIS. These provide detail and resolution in remote sensing data, which are then cross-referenced with GIS and machine learning models for complex analysis and pattern detection in forest ecosystems (Wulder et al., 2008). Using UAVs and terrestrial devices, including LiDAR and handheld non-contact techniques, adds another layer of data collection that provides localized, precise metrics. This technology allows the adequate monitoring of important phenomena such as deforestation, biodiversity loss, and climate change impact (Aarif KO et al., 2025). With this multi-source information, decision-making is greatly enhanced, resulting in more timely, informed, and efficient management policies (Ali et al., 2025). This strategy benefits not only in the preservation of forest health but also has very significant implications in policy formulation with a view toward more focused measures for climate change mitigation and management of forests that is sustainable (Ahmad, 2024).

Figure 2. Conceptual Framework

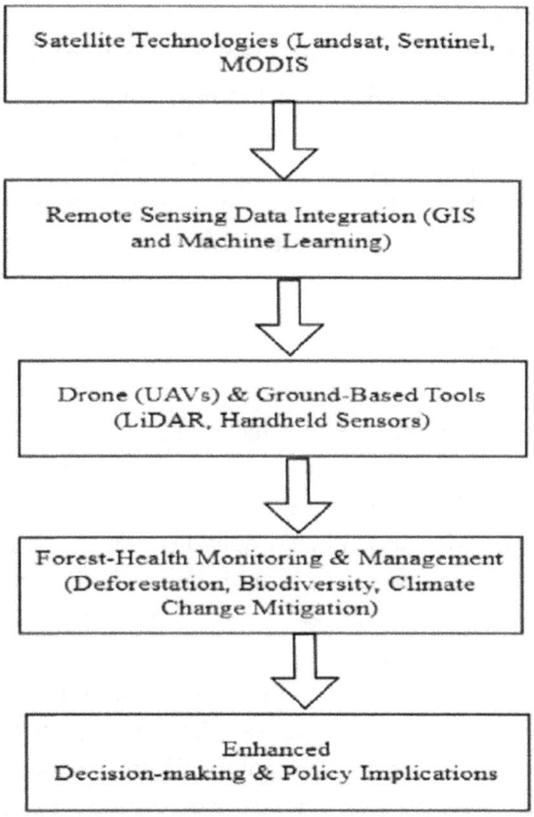

Source: created by authors

2. REVIEW OF LITERATURE

2.1 New Avenues for Assessing Forest Health

Remote sensing technology has actually found its way into presenting a lot of other methods to monitor and assess the health of forests-so far overcoming a more complete definition of conventional approaches to monitoring forests. The technologies allow monitoring of the forest condition on large spatial scales with high resolution and near real-time dataset on forest dynamics (Anderson, 2024).

- Integration of Remote Sensing

Remote sensing provides a non-intrusive method of large-scale monitoring at high temporal frequencies, overcoming the inherent limitations of ground survey. Satellite, unmanned aerial vehicles (UAVs), and LiDAR sensors all provide complementary data sets that further bring into view the complete panorama of health in the forest (Guimarães et al., 2020). Such integration permits better and more efficient tracking of forest health over time, so even remote and inhospitable areas can be accessed for more accurate and timely perception of changes in forest cover, biodiversity, and overall ecosystem health. All this serves to enhance the ability to combine so many diverse sources of data, which will in turn advance forest management and inform strategic decision making in sustainable aspects of conservation (Ecke et al., 2022).

- Key Metrics of Forest Health

With the emergence of remote sensing technologies, assessment of various forest health problems-from vegetation health to forest canopy structure, biodiversity, and carbon holding capacity-have gained prestigious applications. Using NDVI and EVI has helped monitor leaf density and chlorophyll content to quantify forest vitality (Mouafik et al., 2024). Remote sensing could also be used to analyze forest canopy structures in detail as this knowledge is also important for biodiversity conservation since it gives information about forest composition and structural complexity (Quamar et al., 2023). UAVs and other forms of remote sensing tools, for example, are employed to keep a continuous and conscientious tab on species distribution, forest type, and structural diversity-three dynamic areas concerning conservation. Apart and concurrent use of these techniques involves biomass estimation and carbon stock applications that are two important determinants in understanding the role of forests in a changing climate and the global carbon cycle (Narayanappa, 2024).

2.2 Role of Drone and Remote Sensing Technology

Drones or UAVs have set their foot in the field of forestry due to their capacity to obtain high resolution, localized data even finer than that provided by satellite data. Therefore, combined with other remote sensing technologies, it is a good technique behind the scenes to know what is really going on in the forest (Oliveira, 2024).

- Unmanned Aerial Vehicle

Unmanned Aerial Vehicles (UAVs) are also called drones, and they are increasingly becoming the most important instruments in forest monitoring and management. Their capability in high-resolution imaging and data gathering both from the ground and from the air gives them an edge in all aspects of traditional forest monitoring

(Tang & Shao, 2015). UAVs have a use for determining forest canopy structure and can help very effectively in the analysis of vegetation health, tree density, and composition of forests. These attributes afford a precision that is often unattainable through ground-based or satellite surveys-technologies in areas that are usually difficult or remote in access. Non-invasive wildlife tracking using UAVs is still one of the most complete applications seen in the concept of UAV and uses characteristics for wildlife monitoring (Victor et al., 2024). It enables population and behaviour tracking of animals in their natural habitats. Drones are now also being used more for damage assessments following fires, where they can scan wildfire-affected areas relatively quickly and relay such important data, with corresponding urgency for recovery and management planning. UAVs have the cost-effective flexibility to collect high-quality data, supporting their ultimate role as indispensable tools in the monitoring and conservation efforts of forest ecosystems (Anderson, 2024).

- Remote Sensing Technologies

Drones are able to capture imaging or 3D models of forest canopies to provide information concerning the vegetation health and forest structure with remarkable precision. These models are used to analyze the vegetation health and structure of the forest using much more high-resolution canopy structure data (Cocchia, 2014). UAVs are also very useful in wildlife monitoring, allowing non-invasive tracking of animal populations within their natural environments. Drones collect timely and accurately information regarding areas that wildfires have affected post-fire during recovery and management planning. Besides, these excel at collecting localized data, particularly in smaller, more specific forested areas where satellite imagery may not provide the required resolution-allowing much more fine-tuned, targeted assessments of forest health as a result (Ehrlich-Sommer, 2024).

- Impact of Climate Change

Rapid changes in climate bring new changes to the structure and function of forest ecosystems. The health of forests elevates their biodiversity and carbon stores. Monitoring these changes is important for ensuring sustainable management of forests to be able to continue to provide vital ecosystem services (Harle et al., 2024).

- Climate Change and Forest Dynamics:

Climate change currently acts on forest ecosystems, with alterations in patterns of temperature and precipitation that deeply impact health, distribution, and growth. Increasing temperatures and altered patterns of rainfall worsen forests' susceptibility to outbreaks by pests and diseases, which lead to reduced biodiversity and forest productivity (Lame, 2019). In addition, climate change is altering species compositions as some adapt to the new climate conditions while others struggle to survive, therefore disrupting ecosystem equilibrium (Li & Wang, 2025). It adds the scale of the challenge due to the frequency and intensity of forest fires and drought to which they have been subjected and along with impacts from other areas, brings extensive destruction to forests, disrupts ecosystems, and makes forests less able to sequester carbon, which increases the effects of climate change (Messinger et al., 2016).

- Remote Monitoring in Climate Change Mitigation:

Continuous supply of relevant data for the monitoring and accounting of carbon stocks is one of the roles of remote sensing for the understanding of forests' roles in mitigating climate change through carbon sequestration over time (Mouafik et al., 2024). Monitoring deforestation and the successful reforestation interventions will be very significant contributions of satellites and UAVs. They will also yield a very crucial insight into the changes of forest cover and thus help evaluate climate change mitigation strategies more efficiently (de Almeida et al., 2020). In addition, remote sensing goes a long way in assessing forest vulnerability for this by detecting precursory signs of stress, such as leafy deaths or tree mortality due to drought, giving room for timely intervention in respect of the health of these ecosystems to be resilient in the context of climate change.

2.3 The Role of Satellite

Satellite-based systems are the major platforms to provide global, continuous monitoring in the world with regard to forest ecosystems (Lame, 2019). Their wide longitudinal and collective datable information about the condition, cover, and dynamics of forests gives them a large scope in managing forests at very large scales or in scientific research (Ahmad, 2024).

- Satellite Based Systems

The terrestrial satellites, namely Landsat, Sentinel, and MODIS, are indispensable components in forest monitoring dynamics and health. The temporal series of Landsat satellites thus becomes critical, both historic and present data upon which the track record of deforestation and degradation may be made over very long periods-with values that may go up to several decades (Choudhry & O'Kelly, 2018). Satellite series Sentinel greatly improve forest health observations: such data with high resolution and no less differentiated between optical and radar imagery is, at that time, given much more detailed readings on canopy structure and forest vegetation dynamics (Ecke et al., 2022). On the other spectrum, MODIS provides a very broad, worldwide eye on the phenomena of forest cover change and vegetation change over time within the context of large-scale assessments of tree health, carbon storage, and global environment monitoring (Jiménez López & Mulero-Pázmány, 2019). Complementing each other, these satellite systems together can give a large picture of forests from local to global scales.

- Data Types

The multispectral, hyperspectral, and radiometric data are key components for the comprehensive evaluation of forests and forest health. Multispectral data provides valuable information at a broader scale about the forest ecosystem; it is primarily used for mapping forest cover, monitoring vegetation health, and estimating forest biomass (Maurya et al., 2022). Hyperspectral imagery provides detailed analysis, enabling differentiation of forest types and the detection of subtle alterations in species composition and health, particularly significant for biodiversity tracking (Tereşneu et al., 2016). Radiometric measurements collected from satellite sensors reflect very well the forest canopy reflectance, thereby helping characterize the important properties of forests such as density or leaf area index, strengthening the forest structure and health evaluation further. The presence of such variety in data adds much detail to the knowledge of forest dynamics and aids informed forest management decisions (Zakari et al., 2024).

2.4 Some Key Indices

- We define NDVI as the normalized difference vegetation index. This index has been widely used as a measure of vegetation health that conveys a higher amount of green biomass and photosynthesizing activities (Torresan et al., 2017).
- EVI (Enhanced Vegetation Index): EVI is an extension of NDVI because it reduces atmospheric disturbance. It is more suited to dense forests or areas experiencing high atmospheric variability (Mohan et al., 2021).
- Early Uses for Forest Monitoring: Both NDVI and EVI have been important in characterizing forest dynamics, early signs of stress, and a long-term assessment of vegetation health (Espíndola & Ebecken, 2023)

2.5 Advancements in Forestry Practices

The decision-making capability of forest management has greatly hinged upon the integration of remote sensing technology with aerial and spatial observation analysis techniques for advanced analysis such as GIS and machine learning algorithms (Torresan, 2017). These transformed practices in forestry have advanced the industry from very tedious and manual processes to modern data-rich and efficient processes (Vélez et al., 2023).

- Technological Integration

It then became possible to use GIS and machine learning data to integrate remote sensing, announcing improvements in forest management practices. The analysis and visualization of spatial data takes place within a GIS; for instance, an analysis could help identify forest management zones, biodiversity hotspots, and at-risk areas (Zhang et al., 2016). If such an analysis is then combined with the machine-automated processing of large volumes of remote sensing data, these could result in improved accuracy in the prediction of the health and dynamics of forests (Ashutosh & Roy, 2021). All this integrates and enhances the process for decision-making, allowing informed judgments to be made by the forest managers regarding resource allocation, conservation strategies, and adaptation to climate change, thus furthering more sustainable and effective forest management (Doornbos et al., 2024).

- Recent Advancements

The transfer of management from traditional and labour-intensive methods to data-driven, technologically advanced practices has completely transformed forest management in speed and improved accuracy of assessments made on forests (Camarretta et al., 2020). Particularly, through advances that remote sensing technologies brought forth, biodiversity monitoring has been able to realize real-time tracking of information on species' abundance within forests, forest composition, as well as the effects of disturbances on ecosystems (Gonzalez et al., 2016). These technologies also enabled more sustainable resource use patterns because of information provided about growth rates and biomass production, as well as harvest patterns within forests, because they will lead to decisions made on informed and responsible bases (Lame, 2019). Besides that, through the combination of remote sensing, GIS, and machine-learned technological methods, further effective climate change adaptation strategies have been developed with demonstrations such as monitoring carbon sequestration improvement and reforestation offer improvement, thus reducing the adverse impact of climate change on forest ecosystems (Rane & Choudhary, 2023).

3. METHODOLOGY

A systematic literature review technique was used in this chapter which reveals several critical research gaps related to forest monitoring as summarized in Table 1. This methodology helps to get various insights related to the conceptual framework. According to Nightingale (2009), the literature review is a significant method for formulating research objectives, and it helps to find research gaps and It also provides insights related to research design and segregation of research articles, research reports, chapters, and so on. Boolean operators have been used for extracting articles. The following steps have been taken for the Systematic Literature Review, which is mentioned in Figure 3;

3.1 Search Strategy

In this section, Boolean Operators have been used to find suitable articles for the study, and the following keywords have been used for a search like "forest management" OR "smart forestry" AND "remote sensing" OR "satellite technology" AND "forest monitoring,", "remote sensing technology in forest" etc. Boolean operators were used from academic sources like Google Scholar, Scopus, and IEEE Xplore. Top top-cited research articles have been considered for this study from various sources.

3.2 Inclusion

Relevant studies have been included from 2015 to 2025 focusing on the role of drones, satellite technology, and remote sensing and how these technologies help forest monitoring for which we have applied a filter from the search outcomes of 3120 articles in the initial step and then we found that 1670 articles have been excluded due to other languages and the remaining 1450 articles were extracted based on title, abstract. After this, 1200 articles are similar kinds of keywords, but these articles were not directly aligned with our study because these articles were related to satellite technology in other fields, and the role of drones in other fields so these studies we have excluded these articles. Then we further filtered out and applied the domain categorization method from the remaining 250 articles, at last after using this method and exclusion of duplicates, we finally included 55 articles based on top citations and quality research.

3.3 Data Extraction

Research articles have been extracted that helped to formulate the research objectives, research design, and identification of variables like deforestation rates, remote sensing technologies, monitoring of forests, LiDAR, GIS, and the role of technology (Van et al., 2012).

3.4 Assessment

These articles were carefully evaluated based on their applicability to the progressive application of remote sensing technologies, i.e., drones, in forest monitoring and management. All studies were analyzed for their impact on enhancing the knowledge of the effects caused by drones and satellite technology, i.e., LiDAR, UAVs, and other remote sensing tools, on monitoring forest health, deforestation monitoring, and biodiversity assessment (Xiao & Watson, 2019). The quality of these studies has been evaluated based on their research strategy, data sources, and findings, which enabled a comprehensive representation of technological developments as well as practical implementations (Sonti, 2015). Data integration with GIS, machine learning-based predictive analysis, and remote sensing applications in climate change are some identified major themes from these studies. This review also provides critical reviews of some of the challenges that affect the technologies, such as data accuracy, scalability, and cost-effectiveness (Banu et al., 2016). Top-cited articles/research papers were considered, which have been summarised in Table 2. This systematic review, comprising 55 articles, provides a basis for the understanding of how drones and remote sensing could transform forest monitoring and management practices across the globe. Figure 3 depicts the information related to article extractions.

Table 2. Top-cited articles included in the review as per citations

Name of the Journal	Number of Documents	Area of Focus	Citations
Journal of Sensors	1	Precision Agriculture, Smart Sensor Technologies (Aarif KO et al., 2025)	18
Smart Cities and Sustainable Manufacturing	1	Environmental Sustainability, Advanced Sensors, Monitoring Systems (Ahmad et al., 2025)	654
Telecommunication Systems	1	Disaster Management, Smart Remote Sensing Network (Ahmad, 2024)	4400
Babylonian Journal of Internet of Things	1	Sustainable Forest Management, IoT, AI, Remote Sensing (Ali et al., 2025)	649
Journal of the Indian Society of Remote Sensing	1	Nationwide Forest Cover Mapping, Conservation (Ashutosh & Roy, 2021)	179

continued on following page

Table 2. Continued

Name of the Journal	Number of Documents	Area of Focus	Citations
Journal of Environmental Science and Engineering B	1	Drone Usage in Forestry (Banu et al., 2016)	697
American Journal of Agricultural and Forest	1	Remote Sensing, GIS in Tropical Forest Management (Beckline et al., 2017)	55
New Forests	1	Forest Restoration, Remote Sensing (Camarretta et al., 2020)	64
IEEE International Conference on Artificial Intelligence	1	Autonomous Drone Technology, Agriculture (Dasipah et al., 2024)	15
Restoration Ecology	1	Forest Restoration, Remote Sensing Approaches (de Almeida et al., 2020)	264
Journal of Applied Ecology	1	Tropical Forest Restoration, Remote Sensing (de Almeida et al., 2025)	998
International Journal of Environmental Monitoring	1	AI-Enhanced Remote Sensing, IOT Networks (Divine et al., 2024)	156
IEEE Access	1	Drone Technologies, Systematic Literature Review (Doornbos et al., 2024)	1121
Remote Sensing	1	UAV-Based Forest Health Monitoring (Ecke et al., 2022)	151
Sensors	1	Smart Forestry, Digital Transformation (Ehrlich-Sommer et al., 2024)	356
Revista Caribeña de Ciencias Sociales	1	Sustainable Forest Management, Remote Sensing (Espíndola & Ebecken, 2023)	864
Sensors	1	AI, UAVs, Conservation, Remote Sensing (Gonzalez et al., 2016)	54
Remote Sensing	1	Forestry Remote Sensing, UAV Data Processing (Guimarães et al., 2020)	667
IEEE Internet of Things Journal	1	IoT, AI, Remote Sensing for Forest Conservation (Haq et al., 2024)	464
Current Applied Materials	1	UAVs, Land Productivity, Vegetation Trends (Harle et al., 2024)	54
Drones	1	UAVs in Protected Areas, Conservation (Jiménez López & & Mulero-Pázmány, 2019)	69
Science in China Series E	1	Forest Resource Management, GIS-Based Decision Support (Li & Zhao, 2006)	881
Scientific Reports	1	UAV Remote Sensing, Pest Monitoring (Li & Wang, 2025)	1641

continued on following page

Table 2. Continued

Name of the Journal	Number of Documents	Area of Focus	Citations
Environmental Science and Engineering Journal	1	Natural Resource Management, Remote Sensing (Machireddy, 2023)	6348
Remote Sensing	1	UAVs for Biomass and Amazon Forest Structure (Messinger et al., 2016)	2016
Journal of the Indian Society of Remote Sensing	1	UAVs for Natural Resource Management (Mishra & Rai, 2021))	2216
Journal of Science Research International (JSRI)	1	AI, Machine Learning Applications in Forestry (Mishra et al., 2024)	7894
Remote Sensing	1	UAV-supported Forest Regeneration (Mohan et al., 2024)	255
Geographies	1	Remote Sensing Technologies, Argane Forest Monitoring (Mouafik et al., 2024)	654
Fire	1	UAV-Driven Fire Monitoring, Wildfire Detection (Muksimova et al., 2024)	1648
Remote Sensing	1	Biodiversity Monitoring, Tropical Forests (Mulatu et al., 2017)	664
Springer International Publishing	1	Remote Sensing for Forest Management in Developing Countries (Nandasena et al., 2023)	1244
Remote Sensing in Earth Systems Sciences	1	UAV, IoT for Agricultural Field Monitoring (Narayanappa et al., 2024)	267
Springer Nature Switzerland	1	Urban Forest Management, Remote Sensing (Neves & de Aquino, 2024)	5462
African Journal of Science and Technology	1	Forestry Sustainability, UAV Technologies (Ogunbamowo et al., 2024)	421
The forestry chronicle	1	The role of LiDAR in sustainable forest management (Wulder et al., 2008)	545
Springer Nature	1	Remote Sensing Monitoring of Forests (Sabir et al., 2024)	674

Source: Table created by authors

Figure 3. Flow diagram of Systematic Literature Review

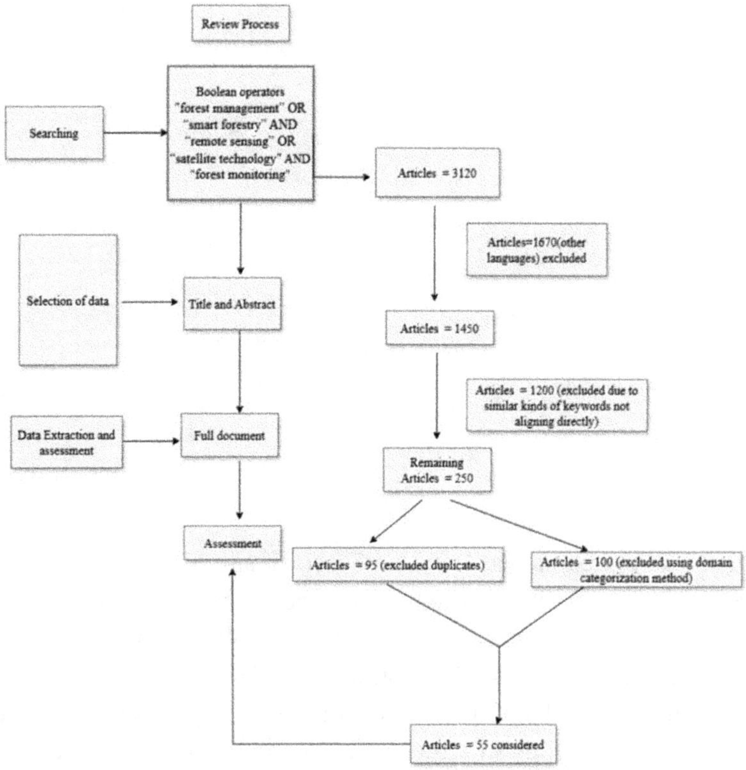

Source: Figure created by authors

4. FINDINGS OF THE STUDY

4.1 Forest Health Monitoring

Effectiveness of Satellite, Drone, and Ground-Based Remote Sensing Technologies, which have been confirmed by the SLR. This study has employed a systematic literature review from 2015 to 2025. Several studies, which have been enumerated in Figure 4, have been published over this Period.

Figure 4. Number of Studies published 2015-2025

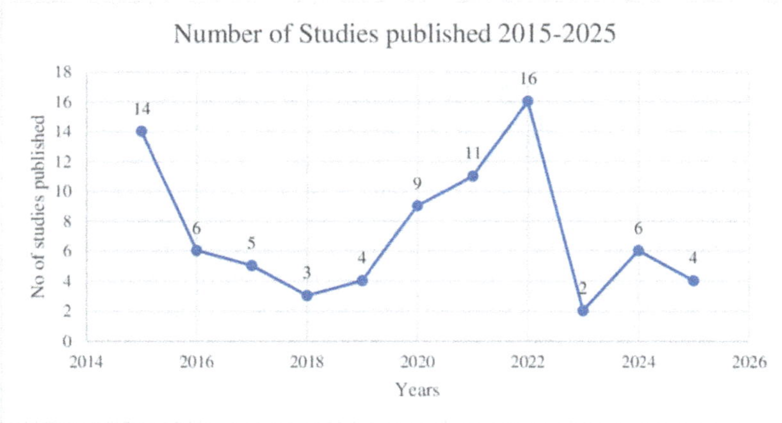

Source: Figure created by authors

Satellite technologies such as Landsat, Sentinel, and MODIS play a vital role in large-scale and long-term monitoring of forest health by providing comprehensive data on forest cover, vegetation vitality, biomass estimation, and carbon stocks across the globe (Ecke et al., 2022; Choudhry & O'Kelly, 2018). Complementing these, drone technologies (UAVs) enable localized and high-resolution observations essential for analyzing canopy structures, forest composition, and vegetation health, particularly in remote or inaccessible areas where satellite coverage is limited (Banu et al., 2016; Aarif KO et al., 2025). Additionally, ground-based tools such as LiDAR and advanced field instruments offer highly precise measurements of parameters like tree height, canopy density, and soil conditions, enriching satellite and aerial data with detailed field-level insights (Mishra et al., 2024). Together, these integrated technologies provide a comprehensive, multi-scale framework for assessing and managing forest ecosystems with both accuracy and spatial depth.

4.2 Technological Synergies: Integration of Satellite, Drone, and Ground-Based Data

Satellite, UAV, and ground technologies act in concert to provide a holistic understanding of forest health monitoring. The integration of satellite, aerial, and terrestrial perspectives helps obtain a detailed understanding of forest phenomena on varying spatial scales (Muksimova et al., 2024).

Satellite data provides broad, regional observations; drones provide localized, high-resolution visual imagery, and ground data serve to ground-truth results (Pádua et al., 2017). This union delineates the broad scope of identifying forest stress, changes in vegetation, and variations in biodiversity over time.

4.3 Data Integration: GIS and Machine Learning used for Improved Decision-Making

Combining GIS, machine learning, and remote sensing data enhances decision-making regarding forests. These three sources projected models to visualize spatial patterns, model processes in the forest, and indicate the areas that are most prone to degradation and loss of biodiversity based on the information that comes from satellites, aircraft, and ground surveys (Ramani et al., 2025).

Big data holds promise for revealing fine-scale changes in forests that can predict stress from pests, diseases, and climate, often only detectable early within the processes. Consequently, immediate attention will be required in planning for such forests to facilitate proactive, effective conservation and management of forest lands (Thakur et al., 2021).

4.4 Forest Management Challenges

4.4.1 Deforestation

Satellite and unmanned aerial vehicle remote sensing technology is one of the most advanced to monitor deforestation trends while providing real-time information for forest cover change. This is essential for stopping illegal logging and assessing the success of conservation efforts (Xiao & Watson, 2019).

4.4.2 Loss of Biodiversity

Monitoring biodiversity through UAVs and remote sensing technology is possible through the collection of data regarding species distribution, forest structure, and habitat quality. It is critical in protecting diverse ecosystems and in the maintenance of biodiversity in forests (Sanchez-Azofeifa et al., 2017).

4.4.3 Climate Change Mitigation

Although essential in assessing carbon stocks in the forests and monitoring reforestation or afforestation, remote sensing technologies contribute to mitigating climate change. Satellite images, aerial surveys by drones, and LiDAR can be immensely beneficial for the monitoring of carbon sequestration and the quantification of the impacts of forest management on climate change (Anderson, 2024).

What remote sensing techniques measure applications in forests to assess the carbon they hold as well to measure the afforested area and reconstituted areas for mitigating the effects of climate. Satellite images, aerial surveys, and LiDAR can be vital for measuring the carbon sequestered in forests and quantifying how well the management of forests has changed the carbon balance for climate change (Anderson 2024).

4.5 Integration of GIS and Machine Learning for Climate Change Responses

Various applications of GIS and remote sensing data improve the observation and analysis of forest systems at present and their vulnerability to climate change. GIS allows for the analysis and deployment of extensive spatial datasets to identify areas that are high-risk zones and implements interventions within those specific areas. (Li & Zhao 2006)

Machine learning constructs that exploit remote sensing and GIS data enable effective climate change adaptation decisions. It emphasizes the prediction of forest health deterioration for informed decision-making by forest managers in terms of useful interventions to counter adverse climate-change-induced impacts, namely enhancing carbon sequestration or reforestation efforts (Cocchia 2014).

5. DISCUSSION OF THE STUDY

5.1 Technical Limitations: Resolution, Coverage, and Integration

Monitoring forest health through remote sensing faces persistent challenges related to data resolution, coverage, and integration. Satellite systems such as Landsat, Sentinel, and MODIS provide valuable broad-scale data for long-term observation but lack the spatial resolution needed to capture detailed forest attributes like canopy structure or species composition (Doornbos et al., 2024; Kubiak et al., 2024; Messinger et al., 2016). This limitation often results in the under-detection of localized disturbances such as pest infestations or small-scale deforestation. Drone technologies, though capable of generating high-resolution imagery, fall short in coverage, making them unsuitable for large-scale forest assessments (Quamar et al., 2023; Maurya et al., 2022). Meanwhile, LiDAR and other ground-based sensors deliver highly precise measurements but are constrained by high costs and limited scalability across vast forested landscapes (Beckline et al., 2017; Divine et al., 2024; Shanthalakshmi et al., 2025). Furthermore, integrating data from satellites, drones, and ground instruments remains technically complex due to differences in data formats, spatial alignment, and temporal frequency, posing a significant barrier to creating unified forest health assessments (Ecke et al., 2022; Jiménez López & Mulero-Pázmány, 2019).

5.2 Policy and Management Challenges: Standardization, Data Accessibility, and Policy Integration

Remote sensing technologies are excellent tools, but they are not always properly utilized in existing forest management systems due to policy and management problems. One of the key problems is that there are no standard protocols to gather and analyze remote sensing data (Camarretta et al., 2020). Forest managers and policymakers frequently struggle to ensure that remote sensing data are gathered and analyzed in a manner that is compatible with national or international forest management standards. In addition, data sharing and access remain a difficult challenge. Although satellite data are free, accessing high-resolution drone and ground data is frequently restricted by ownership or operation regulations (Thakur et al., 2021). The integration of remote sensing technology into forest management policy is a real challenge at the local, national, and international levels. Shortcomings in implementing an appropriate plan for remote sensing data use and integration into the decision-making process mean that the advantages accrued from their application for improved forest management would only be partially realized. Areas in which new policy guidelines and frameworks have to be developed revolve around the use of remote sensing technology and coordination among data sources across various regions to get the expected benefits from the technologies (Zakari et al., 2024).

5.3 Climate Change Impact: The Challenge of Measuring and Predicting Effects

In assessing climate change impacts on forests, remote sensing technologies can provide useful evidence, but some weaknesses prevent the certification and forecasting of these effects. Forests exist as ever-changing systems, and often with complex and counterintuitive responses to climate change when susceptible (Sabir et al., 2024). It is not always possible or desirable to obtain all the required information through remote sensing methods. For example, any changes in temperature and rainfall affecting the growth and composition of forests may have been too slow or subtle for satellite sensors operating at present to register immediately. So, while remote sensing can track general trends regarding forest health, such as deforestation and biomass loss, its ability to record specific impacts such as the slow loss of certain species or increased susceptibility to disease and pest impacts-which are often early warning indicators of climate change-may not be observable in any immediate future by satellite or drone imaging (Sofia et al., 2022).

The remote sensing technology's capacity to speculate on the behaviour of different forest types under climate change remains very much in its infancy. Machine learning and predictive models have a bright future, but capacity requirements are immense, and data are not necessarily available or accurate enough to fully generate climate impacts in areas with low monitoring or where the dynamics of forest change are poorly characterized (Zhang et al., 2016). Furthermore, the dynamic nature of climate-related hazards, such as increased frequency of fires in the forest or droughts, calls for periodic and frequent measurements to detect gradually changing forest conditions over time. Drones can provide real-time information on specific points in space, but satellite data might not always be available within a rapid-response timeframe (Divine et al., 2024). It is necessary to integrate the strengths of all of these technologies to develop a more efficient, real-time monitoring system capable of predicting and responding to the impacts of climate change on forests, but this is still under development (Victor et al., 2024).

5.4 Economic and logistical barriers

In spite of drones and remote sensing technology being immensely transforming and useful tools for forest monitoring, economic and logistical barriers sometimes pose problems in the way of adoption. These encompass high asset and maintenance costs, heavy data processing requirements, and specialized technical know-how. Perhaps, the Indian Forest Survey's 2019 project in Madhya Pradesh could be cited: Here, drones bearing multispectral sensors were flown over teak and sal forests to detect illegal logging and estimate biomass (Zakari et al., 2024). Although the project made an excellent case in favor of drone imagery, greatly raising detection accuracy and shortening response times in comparison with traditional patrols, a universal rollout was never on the cards due to a lack of funds, dependence on skilled personnel, and the absence of means to process huge volumes of imagery within governmental facilities in a time-bound manner. In this case, it teaches us that even successful pilots cannot scale in the absence of investment in capacity development, data infrastructure, and sustainable financing models. By acknowledging these practical limitations, the chapter provides a more nuanced and practical framework for practitioners and policymakers seeking to integrate drone and remote sensing technologies into forest management (de Almeida et al., 2025).

6. CONCLUSION

In this chapter, SLR techniques have been applied from 2015 to 2025. The integration of remote sensing technologies from satellites and drones to terrestrial tools has transformed forest health monitoring and forestry operations (Banu et al., 2016). The real-time provision of high-resolution, comprehensive information enables the use of remote sensing technologies in a manner that supports unparalleled capability for tracking forest ecosystems, tracking forest cover loss, evaluating biodiversity, and quantifying carbon stock, all of which are vital in the pursuit of sustainable forest management and mitigation against climate change. Synergizing the use of satellite, UAV, and ground data has introduced new windows into how forest challenges may be overcome and how the dynamics of the forest are better understood, particularly in distant or intractable areas (Doornbos et al., 2024).

However, there are still issues to be addressed, the economic and logistical challenges emerge as a key area for future policy and investment, alongside the technical and data integration issues already discussed, including limitations in the resolution of the data, complexities in integrating heterogeneous data sources, and the need to improve policy frameworks and standardized approaches (Harle et al., 2024). The technological limitations of current remote sensing technologies, as well as the limitations of data availability and integration, highlight the need to continuously develop and enhance forest monitoring systems. Furthermore, the effects on forests due to climate change may require timely, accurate, and sustained monitoring, elements that remote sensing technologies provide only if their present shortcomings are resolved (Kubiak et al., 2024).

The implications of these findings are highly applicable to policymakers, researchers, and forest managers. One of the significant takeaways is that remote sensing technologies can produce better conservation plans and thus improve decision-making concerning forest management and give support to the policies on climate change mitigation (Mishra & Rai, 2021). Reinforcing the integration of remote sensing information of geospatial technologies, such as GIS, and, in particular, machine learning, can make forest management a much anticipatory and information-based process by conserving it in a better way and using it sustainably (Mohan et al., 2021). In the days ahead, these technologies will have to overcome technical, policy, and especially integration hurdles to realize their potential, and then the forest management strategy will be more resilient and adaptive against changes related to climate change (Ali et al., 2025).

7. FUTURE SCOPE OF STUDY

Continuous advancement of remote sensing is fundamental for the future of forest monitoring and management. Remote sensing could improve data resolution, frequency, and coverage to give more accurate assessments of forest health (Aarif KO et al., 2025). New technologies such as AI, machine learning, and predictive analytics will greatly assist decision-making by allowing real-time processing of large datasets while giving insights into forest processes, climate change, and biodiversity conservation (Dasipah et al., 2024). However, all these technologies should be supported by good policy frameworks that ensure accessibility and integration, especially in deforestation and degradation hotspots. Another urgent need that has arisen is global research on the application of these technologies in developing countries (Harle et al., 2024). Forest management in the future could greatly benefit from improved climate change mitigation and sustainable practices through technological innovations, enabling policies, and global uptake (Mishra et al., 2024).

REFERENCES

Aarif, K. O. M., Alam, A., & Hotak, Y. (2025). Smart Sensor Technologies Shaping the Future of Precision Agriculture: Recent Advances and Future Outlooks. *Journal of Sensors*, *2025*(1), 2460098.

Ahmad, I., Zulfiqar, A., Shabbir, M., Khan, J. S., Çalışkan, H., & Hong, H. (2025). Advanced sensors, monitoring, and control systems for environmental sustainability. In *Smart Cities and Sustainable Manufacturing* (pp. 47–55). Elsevier.

Ahmad, R. (2024). Smart remote sensing network for disaster management: An overview. *Telecommunication Systems*, *87*(1), 213–237.

Ali, G., Mijwil, M. M., Adamopoulos, I., & Ayad, J. (2025). Leveraging the Internet of Things, Remote Sensing, and Artificial Intelligence for Sustainable Forest Management. *Babylonian Journal of Internet of Things*, *2025*, 1–65.

Anderson, K. (2024). Evaluation Of Forest Canopy Characteristics Using Drone-Based Spectral Data Analysis.

Ashutosh, S., & Roy, P. S. (2021). Three decades of nationwide forest cover mapping using Indian remote sensing satellite data: A success story of monitoring forests for conservation in India. *Photonirvachak (Dehra Dun)*, *49*(1), 61–70.

Banu, T. P., Borlea, G. F., & Banu, C. (2016). The use of drones in forestry. *Journal of Environmental Science and Engineering B*, *5*(11), 557–562.

Beckline, M., Yujun, S., Yvette, B., John, A. B., Mor-Achankap, B., Saeed, S., & Paul, C. (2017). Perspectives of remote sensing and GIS applications in tropical forest management. *Journal of Agriculture and Forestry (New York, N.Y.)*, *5*(3), 33–39.

Camarretta, N., Harrison, P. A., Bailey, T., Potts, B., Lucieer, A., Davidson, N., & Hunt, M. (2020). Monitoring forest structure to guide adaptive management of forest restoration: A review of remote sensing approaches. *New Forests*, *51*(4), 573–596.

Choudhry, H., & O'Kelly, G. (2018). Precision forestry: a revolution in the woods. *Basic materials, paper & forest products*.

Cocchia, A. (2014). Smart and digital city: A systematic literature review. *Smart city: How to create public and economic value with high technology in urban space*, 13-43.

Dasipah, E., Akbari, G. F., Syafei, N. S., Turnip, A., Le Hoa, N., & Sitompul, E. (2024). Autonomous drone technology based random forest classifier for revolutionizing agriculture. In *2024 IEEE International Conference on Artificial Intelligence and Mechatronics Systems (AIMS)* (pp. 1-5). IEEE.

de Almeida, D. R., Stark, S. C., Valbuena, R., Broadbent, E. N., Silva, T. S., de Resende, A. F., & Brancalion, P. H. (2020). A new era in forest restoration monitoring. *Restoration Ecology*, *28*(1), 8–11.

de Almeida, D. R., Vedovato, L. B., Fuza, M., Molin, P., Cassol, H., Resende, A. F., & Brancalion, P. H. (2025). Remote sensing approaches to monitor tropical forest restoration: Current methods and future possibilities. *Journal of Applied Ecology*, *62*(2), 188–206.

Divine, I. L. O. H., Aguma, C. P., & Olagunju, A. O. (2024). Integrating AI enhanced remote sensing technologies with IOT networks for precision environmental monitoring and predicative ecosystem management.

Doornbos, J., Bennin, K. E., Babur, Ö., & Valente, J. (2024). Drone technologies: A tertiary systematic literature review on a decade of improvements. *IEEE Access : Practical Innovations, Open Solutions*, *12*, 23220–23239.

Ecke, S., Dempewolf, J., Frey, J., Schwaller, A., Endres, E., Klemmt, H. J., & Seifert, T. (2022). UAV-based forest health monitoring: A systematic review. *Remote Sensing*, *14*(13), 3205.

Ehrlich-Sommer, F., Hoenigsberger, F., Gollob, C., Nothdurft, A., Stampfer, K., & Holzinger, A. (2024). Sensors for digital transformation in smart forestry. *Sensors (Basel)*, *24*(3), 798.

Espíndola, R. P., & Ebecken, N. F. F. (2023). Advances in remote sensing for sustainable forest management: Monitoring and protecting natural resources. *Revista Caribeña de Ciencias Sociales*, *12*(4), 1605–1617.

Gonzalez, L. F., Montes, G. A., Puig, E., Johnson, S., Mengersen, K., & Gaston, K. J. (2016). Unmanned aerial vehicles (UAVs) and artificial intelligence revolutionizing wildlife monitoring and conservation. *Sensors (Basel)*, *16*(1), 97.

Guimarães, N., Pádua, L., Marques, P., Silva, N., Peres, E., & Sousa, J. J. (2020). Forestry remote sensing from unmanned aerial vehicles: A review focusing on the data, processing and potentialities. *Remote Sensing*, *12*(6), 1046.

Haq, B., Jamshed, M. A., Ali, K., Kasi, B., Arshad, S., Kasi, M. K., & Ur-Rehman, M. (2024). Tech-driven Forest conservation: Combating deforestation with internet of things, artificial intelligence, and remote sensing. *IEEE Internet of Things Journal*.

Harle, S., Bhagat, A., & Dash, A. K. (2024). Remote Sensing Revolution: Mapping Land Productivity and Vegetation Trends with Unmanned Aerial Vehicles (UAVs). *Current Applied Materials*, *3*(1), e070224226752.

Jiménez López, J., & Mulero-Pázmány, M. (2019). Drones for conservation in protected areas: Present and future. *Drones (Basel), 3*(1), 10.

Kubiak, K., Spiralski, M., Pompeu, J., Levavasseur, V., & Wawer, R. (2024). *Advances in remote sensing for monitoring soil conditions in forest ecosystems: techniques, challenges, and applications*. Transactions on Aerospace Research.

Lame, G. (2019). Systematic literature reviews: An introduction. In *Proceedings of the design society: international conference on engineering design* (Vol. 1, No. 1, pp. 1633-1642). Cambridge University Press.

Li, F., & Zhao, Y. (2006). Development of a GIS-based decision-support system of forest resource management. *Science in China. Series E, Technological Sciences, 49*, 76–85.

Li, X., & Wang, A. (2025). Forest pest monitoring and early warning using UAV remote sensing and computer vision techniques. *Scientific Reports, 15*(1), 401.

Machireddy, S. R. (2023). Natural resource management using remote sensing and geographic information systems. *Environ. Sci. Eng, 2*(2), 73–82.

Maurya, N. K., Tripathi, A. K., Chauhan, A., Pandey, P. C., & Lamine, S. (2022). Recent advancement and role of drones in forest monitoring: Research and practices. *Advances in Remote Sensing for Forest Monitoring*, 221-254.

Messinger, M., Asner, G. P., & Silman, M. (2016). Rapid assessments of Amazon forest structure and biomass using small unmanned aerial systems. *Remote Sensing, 8*(8), 615.

Mishra, P. K., & Rai, A. (2021). Role of unmanned aerial systems for natural resource management. *Photonirvachak (Dehra Dun), 49*(3), 671–679.

Mishra, R. K., Mishra, D., & Agarwal, R. (2024). Artificial intelligence and machine learning applications in forestry. *Journal of Science Research International (JSRI). ISSN, 2456*, 6365.

Mohan, M., Richardson, G., Gopan, G., Aghai, M. M., Bajaj, S., Galgamuwa, G. P., & Cardil, A. (2021). UAV-supported forest regeneration: Current trends, challenges and implications. *Remote Sensing, 13*(13), 2596.

Mouafik, M., Chakhchar, A., Fouad, M., & El Aboudi, A. (2024). Remote Sensing Technologies for Monitoring Argane Forest Stands: A Comprehensive Review. *Geographies, 4*(3), 441–461.

Muksimova, S., Umirzakova, S., Mardieva, S., Abdullaev, M., & Cho, Y. I. (2024). Revolutionizing Wildfire Detection Through UAV-Driven Fire Monitoring with a Transformer-Based Approach. *Fire (2571-6255), 7*(12).

Mulatu, K. A., Mora, B., Kooistra, L., & Herold, M. (2017). Biodiversity monitoring in changing tropical forests: A review of approaches and new opportunities. *Remote Sensing, 9*(10), 1059.

Nandasena, W. D. K. V., Brabyn, L., & Serrao-Neumann, S. (2023). Using remote sensing for sustainable forest management in developing countries. In *The Palgrave Handbook of Global Sustainability* (pp. 487–508). Springer International Publishing.

Narayanappa, G. B. C., Abbas, S. H., Annamalai, L., Meenakshi, R., Singh, M., Yadav, T. N., & Kumar, A. R. (2024). Revolutionizing UAV: Experimental Evaluation of IoT-Enabled Unmanned Aerial Vehicle-Based Agricultural Field Monitoring Using Remote Sensing Strategy. *Remote Sensing in Earth Systems Sciences, 7*(4), 411–425.

Neves, F. R., & de Aquino, A. C. B. (2024). Bringing the Green into the Screen: Remote Sensing and Smart Green Governance in Urban Forest Management. In *Mobile Crowdsensing and Remote Sensing in Smart Cities* (pp. 51–71). Springer Nature Switzerland.

Nightingale, A. (2009). A guide to systematic literature reviews. *Surgery (Oxford), 27*(9), 381–384.

Ogunbamowo, P. O., Olaniyi, M. B., Ariwoola, O. S., & Popoola, O. O. (2024). Digital Innovations for Sustainable Forest Management: A Review of Strategies and Applications in the Nigerian Context. In *e-Proceedings of the Faculty of Agriculture International Conference* (pp. 124-129).

Oliveira, A. H. M., Matricardi, E. A., de Aragão, L. E. O. E. C., Felix, I. M., Chaves, J. H., Magliano, M. M., & Martorano, L. G. (2024). Assessing Forest Degradation Through Remote Sensing in the Brazilian Amazon: Implications and Perspectives for Sustainable Forest Management. *Remote Sensing, 16*(23), 4557.

Pádua, L., Vanko, J., Hruška, J., Adão, T., Sousa, J. J., Peres, E., & Morais, R. (2017). UAS, sensors, and data processing in agroforestry: A review towards practical applications. *International Journal of Remote Sensing, 38*(8-10), 2349–2391.

Pandey, P. C., & Arellano, P. (2022). Introduction to Forest Monitoring Using Advanced Remote Sensing Technology–An Editorial Message. *Advances in Remote Sensing for Forest Monitoring*, 1-9.

Pandey, P. C., & Arellano, P. (Eds.). (2022). *Advances in remote sensing for forest monitoring*. John Wiley & Sons.

Quamar, M. M., Al-Ramadan, B., Khan, K., Shafiullah, M., & El Ferik, S. (2023). Advancements and applications of drone-integrated geographic information system technology—A review. *Remote Sensing, 15*(20), 5039.

Raj, E. F. I., Manimala, K., Rani, E. F. I., & Darwin, S. (2024). Environmental Monitoring and Conservation Using Drones. In *Drone Applications for Industry 5.0* (pp. 358–380). IGI Global.

Ramani, D. R., Kalaiarasan, K., & Tangade, S. (2025). Remote Monitoring Advancements: A New Approach to Biodiversity Conservation. *Environmental Monitoring Using Artificial Intelligence*, 61-69.

Rane, N. L., & Choudhary, S. P. (2023). Remote sensing (RS), UAV/drones, and machine learning (ML) as powerful techniques for precision agriculture: Effective applications in agriculture. *International Research Journal of Modernization in Engineering Technology and Science, 5*(4), 4375–4392.

Sabir, R. M., Mehmood, K., Sarwar, A., Safdar, M., Muhammad, N. E., Gul, N., & Akram, H. M. B. (2024). Remote sensing and precision agriculture: A sustainable future. In *Transforming Agricultural Management for a Sustainable Future: Climate Change and Machine Learning Perspectives* (pp. 75–103). Springer Nature Switzerland.

Sanchez-Azofeifa, A., Antonio Guzmán, J., Campos, C. A., Castro, S., Garcia-Millan, V., Nightingale, J., & Rankine, C. (2017). Twenty-first century remote sensing technologies are revolutionizing the study of tropical forests. *Biotropica, 49*(5), 604–619.

Shanthalakshmi, M., Jeevasree, M., Kavitha, R., Madhumathi, V., Mythreye, S., & Yusra, A. N. (2025). Forestry Management with AI and Drone Technology–Digital Forestry. *Computer Vision in Smart Agriculture and Crop Management*, 259-281.

Silva, A. A., Machado, C. C., Gomes, R. R. M., Schettini, B. L. S., Minette, L. J., Nunes, Í. L., & Villanova, P. H. (2022). Forest extraction management with the indicator of overall efficiency of forest machines (OEFM). *Revista Árvore, 46*, e4618.

Sofia, S., Maetzke, F. G., Crescimanno, M., Coticchio, A., La Mela Veca, D. S., & Galati, A. (2022). The efficiency of LiDAR HMLS scanning in monitoring forest structure parameters: Implications for sustainable forest management. *EuroMed Journal of Business, 17*(3), 350–373.

Sonti, S. H. (2015). Application of geographic information system (GIS) in forest management. *Journal of Geography & Natural Disasters, 5*(3), 1000145.

Stone, K. B. (2012). Four decades of lean: A systematic literature review. *International Journal of Lean Six Sigma, 3*(2), 112–132.

Tang, L., & Shao, G. (2015). Drone remote sensing for forestry research and practices. *Journal of Forestry Research*, *26*, 791–797.

Tereşneu, C. C., Clinciu, I., Vasilescu, M. M., & Biali, G. (2016). Using the GIS tools for a Sustainable Forest Management. [EEMJ]. *Environmental Engineering and Management Journal*, *15*(2).

Thakur, T. K., Patel, D. K., Thakur, A., Kumar, A., Bijalwan, A., Bhat, J. A., & Kumar, A. (2021). Biomass production assessment in a protected area of dry tropical forest ecosystem of India: A field to satellite observation approach. *Frontiers in Environmental Science*, *9*, 757976.

Torresan, C., Berton, A., Carotenuto, F., Di Gennaro, S. F., Gioli, B., Matese, A., & Wallace, L. (2017). Forestry applications of UAVs in Europe: A review. *International Journal of Remote Sensing*, *38*(8-10), 2427–2447.

Van Dinter, R., Tekinerdogan, B., & Catal, C. (2021). Automation of systematic literature reviews: A systematic literature review. *Information and Software Technology*, *136*, 106589.

Vélez, S., Martínez-Peña, R., & Castrillo, D. (2023). Beyond vegetation: A review unveiling additional insights into agriculture and forestry through the application of vegetation indices. *J*, *6*(3), 421-436.

Victor, N., Maddikunta, P. K. R., Mary, D. R. K., Murugan, R., Chengoden, R., Gadekallu, T. R., & Paek, J. (2024). Remote Sensing for Agriculture in the Era of Industry 5.0–A survey. *IEEE Journal of Selected Topics in Applied Earth Observations and Remote Sensing*.

Wavrek, M. T., Carr, E., Jean-Philippe, S., & McKinney, M. L. (2023). Drone remote sensing in urban forest management: A case study. *Urban Forestry & Urban Greening*, *86*, 127978.

Wulder, M. A., Bater, C. W., Coops, N. C., Hilker, T., & White, J. C. (2008). The role of LiDAR in sustainable forest management. *Forestry Chronicle*, *84*(6), 807–826.

Xiao, Y., & Watson, M. (2019). Guidance on conducting a systematic literature review. *Journal of Planning Education and Research*, *39*(1), 93–112.

Zakari, R. Y., Shafik, W., Kalinaki, K., & Iheaturu, C. J. (2024). 15 Internet Things (IoFT) of Forestry Technologies and Applications in Forest Management. *Advanced IoT Technologies and Applications in the Industry 4.0 Digital Economy*, 275.

Zhang, J., Hu, J., Lian, J., Fan, Z., Ouyang, X., & Ye, W. (2016). Seeing the forest from drones: Testing the potential of lightweight drones as a tool for long-term forest monitoring. *Biological Conservation*, *198*, 60–69.

KEY TERMS AND DEFINITIONS

EVI: An advanced vegetation index that provides clearer insights into dense forest canopies by minimizing atmospheric interference, improving accuracy in monitoring forest vitality.

GIS: Geographic Information System (GIS) was used to store, analyze, and visualize forest data collected from satellites, drones, and ground surveys.

LiDAR: A laser-based remote sensing technology that creates detailed 3D maps of forest structure, tree height, and biomass, helping researchers understand forest composition and carbon storage.

MODIS: A powerful satellite sensor providing continuous, large-scale forest observations that help track changes in forest cover, fires, and vegetation health over time.

NDVI: A vegetation health indicator used in remote sensing to assess forest density, stress, and greenness, enabling early detection of degradation or deforestation.

UAVs (Drones): Highly flexible flying platforms that capture high-resolution images and real-time data, revolutionizing forest monitoring through precise mapping, rapid assessment, and cost-effective surveillance.

Chapter 15

Ecological Modelling and Biodiversity Conservation Health Impact:

Sustainability is a Global Responsibility

Wasswa Shafik
https://orcid.org/0000-0002-9320-3186
Dig Connectivity Research Laboratory (DCRLab), Kampala, Uganda & School of Digital Science, Universiti Brunei Darussalam, Gadong, Brunei

ABSTRACT

This chapter examines the crucial role of ecological modeling in biodiversity conservation and its associated health implications. It discusses how various modelling approaches, including species distribution models, agent-based simulations, and systems dynamics, are employed to forecast ecological responses to anthropogenic pressures such as habitat fragmentation, climate change, and pollution. The chapter emphasizes the connection between biodiversity loss and public health, highlighting the increased risks of zoonotic diseases, reduced ecosystem services, and deterioration in mental health. Integrating ecological and epidemiological data enables the identification of conservation priorities, assessment of intervention effectiveness, and support for informed decision-making for sustainable ecosystem management. The chapter advocates for interdisciplinary frameworks that combine ecological modeling, remote sensing, and health analytics to inform policy and promote resilience in socio-ecological systems.

DOI: 10.4018/979-8-3373-3992-4.ch015

INTRODUCTION

Over the past few decades, ecological modeling has made significant contributions to the discipline of ecology. Ecological models have proven invaluable in hypothesis testing and parameter estimation, prediction and forecasting, and evaluating management decisions. With morphological or dynamic structures and parameters, to a certain extent, ecological models can simulate biological systems, provide a quantitative and accurate understanding of these systems, and also uncover great potential in promoting global biodiversity and public health (Chiaverini et al., 2022). Ecological models can represent ecological theories whose parameters are poorly quantified or provide reliable management strategies for local communities with scarce species population data. The applications of ecological models cover, but are certainly not limited to, fish, forest, grassland, coral reef, and urban ecosystems. These ecological models are determined by the research goals and available data, as well as the complexities and characteristics of the problems (Diniz-Filho et al., 2024; Simpson et al., 2022).

Ecological modeling attempts to link ecological theory with the changes observed in ecosystems by quantifying common observations, such as species-area relationships, dominance-diversity patterns, or species associations. As such, models play a key role in the integration of processes and data to predict, understand, and manage complex biological systems. Such models fall into two broad, interdependent categories that range from descriptive biophysical to predictive ecological ones (GIRARD et al., 2023). Both approaches depend on empirical data for their formulation, while confronting the limits of data availability, the errors in their measurements, and the uncertainties concerning the application domain. Ultimately, model predictions require careful validation against new data, updated knowledge, and an improved mental picture of biological systems in the minds of the interested stakeholders. Otherwise, careful application of ecological conceptual models to address questions about how ecosystems are structured and function makes the knowledge of ecological theories rather sterile (Drechsler et al., 2022; Gillingham et al., 2024).

Descriptive models attempt to represent the systems of interest using logical and mathematical structures without making specific predictions about system behavior. Instead, they identify new or recurring patterns, fundamental aspects, or emergent properties that explain apparent biological phenomena. Such models clarify observations and speed up predictions, mainly by providing evidence to support or reject ecological hypotheses, guiding experimental design, and collecting data to statistically and ecologically demonstrate or interpret observed system configurations (Maharjan et al., 2022). Given the conceptual nature of these models, they are commonly discussed at a general level, paying attention to boundaries rather than precise details about their demonstration, interpretation, and potential simple predictions. Their results are mostly presented in probabilistic terms to reflect the realistic variability and unpredictability of study systems. Descriptive models belong to a top-down approach in which general patterns are conveyed via a mixture of model structures, expert knowledge, hypotheses, and database information (Huang et al., 2020). Modeling ecological processes has often played a key role in these recent advances in ecology and its links to health. While conservation, and especially public education, helping to achieve conservation aims, can be improved when society is appreciative of these values, the case for their excessive use could also be an inherently weak one (Grabowski et al., 2023). In practice, morbidity and loss of life, and functioning are prevented or mitigated by understanding the abiotic and biotic environment and our impact on it. However, narrower conservation objectives may dilute the science. An example of potential thinking on biodiversity conservation is from a pilot study on mapping the global ecosystem services of nature. Each biome was evaluated for seventeen services and its trends forecast. A running total plot based upon an analysis to guide conservation of biodiversity globally included direct values and indirect values significantly higher (Carrara & Roig-Juñent, 2022).

To further strengthen the conceptual foundation and methodological rigor of this chapter, it is important to situate the discussion within established scholarship in ecological and conservation modeling. Foundational literature such as Burgman's work on Population Viability Analysis (PVA), which provides critical frameworks for assessing species extinction risk and guiding conservation decision-making, and Peterson's seminal contributions to Species Distribution Models (SDMs), which underpin contemporary approaches to predicting species' ecological niches and responses to environmental change, offer essential theoretical grounding for the techniques and perspectives explored here (GIRARD et al., 2023). Integrating these seminal sources not only acknowledges the intellectual traditions that have shaped modern conservation analytics but also enhances the chapter's academic depth and credibility. In addition, to demonstrate the practical relevance and policy impact of ecological modeling, this chapter incorporates a case study in which model-based insights directly informed conservation policy and management actions. This real-world example illustrates how analytical tools transition from theoretical constructs to applied decision-support systems, highlighting the vital role of modeling in shaping evidence-based biodiversity governance and environmental sustainability strategies (Diniz-Filho et al., 2024; Simpson et al., 2022).

VITAL NEXUS: BIODIVERSITY, ECOSYSTEM SERVICES, AND PUBLIC HEALTH

Ecological modeling is the practice of creating mathematical descriptions or analogs of natural phenomena or systems, whether these are in the form of mathematical equations, computer simulations, or tangible models using physical media. While a purely abstract model may exist solely in the mind of its creator, in practice, most models used in ecology are implemented within various media and software programs to numerically simulate the behavior of ecosystems under some combination of environmental or evolutionary drivers. These systems can also involve humans and the impact of climate change on relevant portions of the Earth system. Examples of the outputs from these models include spatially defined maps of species distributions, as well as graphical and analytical visualizations of model projections as extracted from model simulations. Ecological modeling has had a considerable impact on both basic and applied ecological research, in part because ecological rules are typically expressed as complex interwoven processes across heterogeneous landscapes, which may reflect the activity of disparate biological hierarchies or 'functioning units.' Similarly, models can also estimate connections linking ecosystem services, climate change, and public health. Depending upon the research question addressed, features in ecological models will possess any combination of properties that are spatial, conditional, local, autoregressive, or hierarchical, among others (Grabowski et al., 2023). Consequently, the added value for 'stand-alone' ecological models is relatively limited, as classical ecology had already utilized a range of empirical and quantitative tools to address these questions on biodiversity comparison, as presented in Figure 1.

Figure 1. Biodiversity Comparison, (a) High and (b) Low diversity

There are several types of ecological models. The simplest type is the conceptual model that is used in education as well as in common language. It is a kind of verbal representation of relationships. The second type is the visual model that can be demonstrated by diagrammatic presentations that comprise a suite of boxes, showing particular components and links between these components. A visual model is less straightforward in formulating mathematical functions than more detailed and quantitative models. Mathematical models in ecology consist of theoretical models, process-based models, spatial models, and stochastic models. The oldest and still valuable is the theoretical model that does not include data from field measurements. Its primary goal is to understand essential mechanisms that are responsible for observed phenomena and generate new hypotheses. The process-based model incorporates processes such as growth, dispersal, competition, predation, and succession, which operate within a system, organism, or ecological community (Huang et al., 2020). While theoretical models are usually temporally more stable than data, it is the opposite for the process-based models. A spatial model is primarily designed for simulating variations in space.

Applications in Environmental Science

Ecological modeling in the field of environmental science allows the analysis of process interactions in complex ecological settings. Human activities are linked to profound changes in ecological systems. A notable example is the increase in homogeneity and the fragmentation of landscapes, negatively impacting species that require environmental conditions or large habitat patches. On the other hand, landscapes provide valuable and diverse services, such as lumber, firewood, and non-timber resources, mainly collected for subsistence purposes. Long-established practices can also create diverse habitats. Agro-silvo-pastoral systems have been implicated in the nurturing of total burrow-dependent mammal diversity and the conservation of open habitat species. Ecological models can spatially predict the relationship between species of particular conservation interest and a set of multidimensional stressors (Russell-Bennett et al., 2024). Conservation planning has been extensively used in spatial analyses in ecology with the objective of representing a set of biological and conservation targets. Agent-based modeling approaches may identify motivations fueling land use decisions, whether promotional or limiting entities, policy support, or stakeholder biases. Ecological econometric model structures may be formed to weigh the effectiveness of subsidies and other incentives on land use pattern changes in order to direct development towards multi-objective international and local benefits. The simulations of management and political actions on dynamics may help develop targeted measures to contrast infection status decline (European Commission, 2020; Zhong et al., 2024).

THE ECOLOGICAL MODELLER'S TOOLKIT: FRAMEWORKS FOR ASSESSMENT AND PREDICTION

Ecology has a long tradition in the estimation of the size of populations. These estimates can be based on numerical accounts, distance or trap-recapture sampling, sampling of tracks and signs, analysis of vocalizations, and many other approaches. Most of the techniques employed in the sampling of populations, to a greater or lesser extent, try to correct the so-called "n0 non-availability bias" that is related to the fact that it is much easier to count or see those individuals that are more abundant (Grabowski et al., 2023). Although the advances of the last fifty years have been tremendous, many of the techniques that are frequently used, such as distance sampling or simple capture-recapture models, were developed in the 1960s and 1970s. Although they perfectly fulfill their purpose when the assumptions made do apply, a new generation of biologists and ecologists has developed sophisticated tools that permit data gathering with an enormous reduction in costs. The development of these new techniques is based on advances in computer power, the increasing access to different technological tools, the development of electronics, and the computer programming that has been performed mostly, but not exclusively, by social demand aiming to improve the management of game populations (Carrara & Roig-Juñent, 2022). The emergence of several new techniques that yield, as a result, the so-called "virtual censuses" has changed the subject radically compared to 20 years ago. Along these pages, we review some of these techniques and point to their potential applications in the field of biodiversity conservation. A review of camera-trap-based capture-recapture models and their applications is well beyond the scope of this study, but we make a brief introduction and provide some basis to understand the literature on the design and analysis of camera-trap studies (Drechsler, 2020).

Role of Ecological Modelling in Biodiversity

In the past two centuries, ecologists relied on statistical analyses of field data to interpret these patterns and to test the biological hypotheses that explain ecological phenomena. The main advances in understanding have been based on first-principles models of processes such as energy flow and resource and nutrient cycling; these first-principles models have built our contemporary synthesis of ecological knowledge. The debate as to whether or not the field data clashed with the first-principles understanding has continued in some topics, for example, with regard to the relative importance of resource supply and herbivory on plant productivity, but in reality, of course, the two sources of knowledge can complement each other (Rodríguez et al., 2007). Despite their continuing importance, separate first-principles models are seldom possible for populations, as such separate models only work with certain assumptions that make perfect sense when addressing individual sub-systems of individual organisms, such as enzyme kinetics, for example. Even more complex models for whole populations that rely on first principles of growth, emergence, and individual behaviors are available, such as cellular automata, individual-based models, and dynamical systems of particle models (Drechsler et al., 2007).

Frameworks for Biodiversity Assessment

It is virtually impossible to make site comparisons for the eight to fifteen million species that inhabit the Earth. Sampling will almost certainly be biased by the vagaries of the investigator's location, time, and technique. Ecological models offer a solution to this situation by providing a framework to guide data collection and assemble known information. By using principles of data processing, models can transform myriad locally derived data into information at the scale of a relative few summary parameters and hypothesis tests, providing the most efficient link between microscale data and global-scale problems (Drechsler & Wätzold, 2007). Graduated levels of biodiversity infer the proper scales of investigation and set standards by which successful or unsuccessful data collection can be evaluated. In this way, ecological modeling can guide hierarchical investigations aimed at understanding hierarchically structured problems. Biodiversity measures are not easily made, however, and consequently, measures inadvertently trigger packaging and can result in inappropriate biological assessments. Some data requirements can best be formulated as questions, and doing this often makes the available methods or available data less well suited to answer the question since they were not designed to answer it when first developed. Such inappropriate elements can be reduced in number by formally asking questions before measuring anything, and selecting the best combination of method and data that can be expected to provide the most reliable answer to the question (Schöttker & Wätzold, 2022).

Conservation Strategies

Ecological data is mostly fine-grained and has high spatial and temporal resolution, but it is irregularly distributed. Consequently, the point estimates or predictions from ecological models may not address most of the pressing challenges, but predictive modeling can effectively address many of these through preprocessing. There are two stages in the model-building process: examining the underlying ecological model so that the data and predictions comply with ecological knowledge, and using the ecological model to explain the summarized data quantitatively (Zhu et al., 2023). The statistical approaches that directly estimate population parameters not only address these two challenges, but they also address the ecological-level challenges of uncertainty about generalization, uncertainty about population parameters, and uncertainty about relationships. Incorporating spatial models into applied Bayesian models over spatially organized data should be standard for applied ecologists. Different integration methods can be applied across the combination of predictions in modern ecology, including stacking, Bayesian model averaging, boosting, and bagging. These stochastic integration methods make the most efficient use of the available data to produce accurate predictions (Huang et al., 2020; Maharjan et al., 2022). Predictive mean matching mimics the real target values of the response distribution, even when the values are outside the range of the available data. Predictive modeling not only predicts a desired attribute but also identifies the attributes that are most influential in this task. Boosting is a type of variable selection and should always be an option for prediction. Predictive models in ecology encompass two major branches of modeling: species distribution models, treating species occurrence data, and the distribution of biotic and abiotic traits in the environment, and trait prediction models, which address the relationship between traits and the environment (Lv et al., 2019).

Case Studies in Conservation Efforts

Islands are key resources for conservation and are frequently used as test beds for conservation strategies. For example, one island was chosen as the location for a conservation breeding program for endangered species in part because of the isolation of the island. In some cases, islands have unique ecological properties, making them easier to model. Many locally derived species richness models have been created on islands, including a detailed and widely used model for a specific archipelago (Tytar, Sobolenko, et al., 2015). Island models add a significant dimension of complexity to the task of extrapolating global patterns from global models. In addition to island ecological models, other, more specific models address multiple issues of conservation concern. The most common ecological models for conservation efforts are species richness models. There are a number of types of species richness models, which include locally derived models, model islands, hotspots, and global models—both vicariant and ecological. The general idea of using hotspots and other forms of species richness models is to identify a small area that contributes the largest number of species to a larger region (Tytar, Mezhzherin, et al., 2015). That area, whatever its size, is then protected. The methodologies for these different types of species models lead to several criticisms because the criteria to identify significant clusters of species vary so widely. Another reason for criticism is that many models were created without due consideration of how they would be used to derive priorities for conservation action. The use of ecological models, however, mitigates the impacts of the two main criticisms of species-based approaches because, by working to model 'autecological conditions' instead of just species, they eliminate the need for conservation professionals to predefine species clusters and provide a framework for many types of conservation actions to be developed as management strategies emanate relatively directly from predictive models (Drechsler et al., 2022).

Modeling Infectious Diseases

An understanding of the ecology of infectious diseases is ultimately founded on the principles of ecology itself: patterns of disease, such as those observed in time-space clusters, point to mechanisms which can, in turn, help to validate ecological and larger-scale theories. Molecular and epidemiological evidence, together with multilayer networks, now provide a means for the identification of a wide range of event sequences and geospatial features, including who inflicts disease and who gets sick; whereas four-dimensional macroecological models, with breeding and feeding habitats and wind migration, are predicting upcoming disease waves (Chiaverini et al., 2022). Microbial evolution and population dynamics, along with high-throughput omics and satellite remote sensing, can elucidate drug resistance patterns and host behavior. As such, ontologies can help to build a framework capable of resolving discrepancies in genomic and network analysis. Furthermore, more thoughtful, comprehensive spatial statistics can lead either to a better understanding of truly new, alarming epidemic signals or, be it the likely outcome in many cases, the recognition of known, non-threatening endemics. Beyond acute infections, the updated view of the gastrointestinal microbiome and surrounding environment can be further improved by metagenomics, whereas the transmission of food and waterborne infectious outbreaks lends itself to triangulating satellite environmental data with surveillance records and econometric modeling (Diniz-Filho et al., 2024).

Ecological Factors in Public Health Crises

Public health crises are often ecological and include the population regulation of socially relevant species, such as pests, vectors, and parasites. Integrative approaches may be required to break down the disciplinary silos and generate understanding beyond the sum of contributions from traditional biomedical sciences like pharmacology, immunology, molecular biology, and epidemiology. Key interdisciplinary contributions in public health crises are advanced conceptually, using mathematical models, and supported with data (Gillingham et al., 2024). The ecological models presented include the community ecology of the human body, the infection biology of vector-borne diseases, and the unplanned perturbations of human-driven evolutionary pressures. We emphasize the importance of understanding ecological niches in public health, including the abiotic, biotic, behavioral, and anthropogenic factors that collectively determine the ecology of diseases. Public health crises are often at least partially the result of evolutionary processes (GIRARD et al., 2023). Conversely, they offer opportunities for scientific breakthroughs that the biological and medical sciences advance in parallel. Responding to major public health challenges in ecosystems inherently involves interdisciplinary applications and contributions from experts in evolution, ecology, genetics, evolutionary physiology, sociobiology, and bioethics, among others. Ongoing and future public health challenges include defense against emerging infectious diseases and opportunities that arise from innovative medicine, including the management of manipulating mutations in the human population (Tytar, Mezhzherin, et al., 2015). Interdisciplinary applications in action are emerging as biologists are increasingly integrated with other experts in public health, including clinicians, social workers, health decision-makers, and non-experts who contribute to healthcare, as presented in Table 1.

Table 1. Types of Ecological Models and Their Applications in Biodiversity Conservation

Model Type	Key Features	Application in Conservation	Data Required	Limitations	Reference
Species Distribution	Predicts species occurrence	Habitat suitability mapping	Species presence data, climate, topography	Limited by data quality and scale	(Grabowski et al., 2023)
Population Viability	Projects extinction risk	Endangered species management	Demographic, life-history, environmental variability	Sensitive to demographic and environmental uncertainty	(Carrara & Roig-Juñent, 2022)
Agent-Based	Models individual behaviors	Land-use impact assessment	Behavioral rules, spatial data	Computationally intensive	(Drechsler, 2020)
Ecosystem Dynamics	Simulates interactions between species	Assessing ecosystem stability	Biotic/abiotic interactions, trophic links	Requires detailed interaction matrices	(Rodríguez et al., 2007)
Metapopulation	Considers spatial population structures	Habitat connectivity analysis	Patch data, dispersal rates	Oversimplification of habitat dynamics	(Drechsler et al., 2007)
Dynamic Global Vegetation	Simulates vegetation-climate interaction	Climate-biodiversity impact	Climate, soil, vegetation parameters	Complex parameterization	(Drechsler & Wätzold, 2007)

continued on following page

Table 1. Continued

Model Type	Key Features	Application in Conservation	Data Required	Limitations	Reference
Habitat Suitability	Evaluates suitable areas for survival	Reserve design and species reintroduction	Environmental variables, species preferences	Ignores species interactions	(Schöttker & Wätzold, 2022)
Landscape Models	Represents habitat fragmentation	Urban and agricultural planning	GIS data, land-use change	May lack species-specific detail	(Zhu et al., 2023)
Food Web Models	Represents trophic relationships	Predicting collapse scenarios	Species interaction data	Often data-limited, static assumptions	(Lv et al., 2019)
Niche Modelling	Defines ecological niche	Predict range shifts under climate change	Bioclimatic variables, species data	May ignore biotic interactions	(Maharjan et al., 2022)
Spatially Explicit	Accounts for spatial patterns	Corridor design, dispersal pathways	Landscape metrics, species distribution	Requires high-resolution spatial data	(Tytar, Sobolenko, et al., 2015)
Bioclimatic Envelope	Defines species' climatic tolerances	Assessing climate impacts on distribution	Climate envelopes, species limits	Assumes static species-climate relationships	(Drechsler et al., 2022)
Individual-Based	Simulates traits and behavior of individuals	Genetic diversity maintenance	Trait data, mutation rates	High computational demand	(GIRARD et al., 2023)
Hybrid Models	Combines multiple modelling approaches	Comprehensive policy support	Mixed datasets	Integration complexity	(Simpson et al., 2022)

Spatial and temporal heterogeneity in the environment and ecology make it difficult to predict biological patterns at regional and temporal scales. In addition to this variation and partly because of it, there is strong ambivalence about some of the fundamental aspects of large-scale ecological patterns. Rather than regarding this as evidence of impending chaos or as an excuse for making little or no progress, ecologists are now developing an ever-increasing toolkit of relevant theoretical and practical modeling techniques (Chiaverini et al., 2022). Decision-makers are recognizing the potential implications of these ecological models and are applying them in many realms of environmental management and conservation. We concentrate on the use of modeling approaches for the management and conservation of biodiversity at the regional and temporal scales where ecosystem and public health linkages are particularly relevant. We identify some of the challenges and some direct and indirect opportunities for ecologists and modelers to reduce uncertainty and improve decision-making for biodiversity conservation policy and damage limitation, while simultaneously alerting society to the wider benefits in and around our regions and living organisms as presented in Figure 2 (Diniz-Filho et al., 2024).

Figure 2. Organism illustrations and their ecosystem management, (a) a robust ecological system encompasses a diverse array of flora and fauna[1] Every species possesses a distinct ecological function (photo by Nora Sherwood). (b) The process of deforestation has the potential to augment the populations of mosquitoes, which in turn can facilitate the transmission of diseases to human beings[2] (Credit: Oregon State University), and (c) The growing of native shrubs along roadways by farms has the potential to enhance biodiversity[3] (Credit: Alayne Blickle).

Ecologists must now operate within a new biodiversity context: resignation to a major human-induced extinction event, acceptance that present-day local and regional diversity is a direct consequence of past human activity, analogies with aesthetic values even for much of the 'natural' world, widespread fears of future discontinuities in ecosystem processes upon which we are fundamentally dependent, and recognition that most people now live in and often depend upon highly modified and homogeneous 'cultural' landscapes (Gillingham et al., 2024). Is 'managing for diversity' still an appropriate and useful biological conservation goal? Should ecologists still choose to identify and conserve multiple 'natural' and representative sample analogues of most biomes for future generations? Can we afford to focus on endangered species at the expense of less vulnerable ones? Even the responses to these questions have value: they have the potential to define what many believe to be society's required ethical relationship with the living world in which we evolved (Simpson et al., 2022).

Public Health And Ecological Modelling

Recent availability of enormous and diverse ecological information has resulted in the burgeoning of ecological modeling, and the prospects for further growth appear tremendous. These models could also prove to be useful tools for assessing future pandemic risks and scheduling interventions. I offer some reflections about the use of ecological models in assisting public health decisions and demonstrate how one class of ecological models, spatially explicit transmission dynamic models, can be trained to help derive practical advice for pandemic disease control for wildlife species (Adhikari et al., 2019). Such a goal of ecological modeling work will provide a direct link to significant human health issues while also providing insights into a series of important theoretical questions about maternal effect models that are difficult to assess directly. Modern ecological studies generate data at a rate that would have been unfathomable several years ago. Much of this data has been used to parameterize ecological models, leading to the bulk of current ecological models being data-driven (GIRARD et al., 2023; Tytar, Mezhzherin, et al., 2015).

Linkages Between Biodiversity and Public Health

The link between biodiversity and public health has been the focus of considerable research in recent years, and there is now a substantial body of work documenting the multiple ways in which this linkage operates both positively and negatively. For instance, those living in close proximity to biodiverse ecosystems obtain a range of health benefits from doing so, which are in part derived from the diverse range of food, clean water, and medicines, as well as the reduced stress provided by the local ecosystems. This study builds on and extends prior work by emphasizing the interrelationships between ecological modeling, conservation, and public health (Maharjan et al., 2022). First, within this section, the biodiversity provided by the diverse characteristics of ecosystems at different spatial scales is briefly summarized as background. However, certain forms of contact can also lead to the spread of diseases, and the populations affected can be among the poorest in the world. Second, the same biodiversity can offer both conservation and public health benefits, thereby making co-management of the land by the two sectors a potentially attractive option. Hence, as argued in the conclusion, the development of eco-epidemiological modeling offers clear promise in conserving biodiversity, public health, and global ecosystems (Lv et al., 2019).

Disease Control and Management

A model can be defined as a system of equations or rules that describe the relationships among different components. In predictive modeling, the aim is to forecast data outside the original dataset in which the model was developed. Predictive modeling applications focus on one of three types: binary classification or regression, multiclass classification, or multi-step predictive modeling. In the latter, the model is used to predict several future events (Grabowski et al., 2023). Predictive disease models can be of importance to control infectious diseases, as they provide insights into the potential spread given certain ecological and climatic conditions. Macroecology and disease ecology have not yet contributed as much to predictive ecological forecasting about ecological dynamics, such as disease distribution and spread, as time-series models have done. Guidelines and best practices to construct these categorical models have been provided; single and multi-species studies were performed to develop the necessary background information through species distribution models, expert knowledge, and field data collection in risk assessment and risk mapping, showing that the predicted disease distribution as an output from species distribution models could be used as predictors for risk-based surveillance (Carrara & Roig-Juñent, 2022).

Impact of Environmental Changes on Health

The changes in land use and the environment from the natural state, by the increase in risk factors, allergens, injuries, diseases, and by modifying the interactions with the natural microorganisms that produce disease, provoke a set of benefits and damages for human health. In the same way, the changes in human actions as a result of public health policy cause damages or benefits that are transferred to the environment (Drechsler, 2020). Among a large number of increasing associations of non-structured form, the following are set out: pedagogical natural reserves, green zones of the communities surrounding the teaching centers or hospitals, natural garden zones as anti-depressants, clear urbanism of the surrounding area, and provision of the house with a world of confined shared space potentially dangerous. Other mechanisms arise by which the benefits of the maintenance of the environment can be transferred to human health and quality of life, through the reduction of pathologies generated by the contamination of water, reduction of epidemic diseases by regulation of health, increase in host factors, remote depuration of recycled water, etc (Martínez-Ramos et al., 2016). In turn, the management of parks produces public health benefits derived from the promotion of physical activity, either in the form of body movement or mental stimulation, with psycho-physical effects that can provoke the reduction of noise and air pollution, replace sedentary activities with other activities that have health benefits, and allow the introduction of the most disadvantaged layers into society. The number, however, is less in the new social equipment related to the new emerging occupational profiles (Rodríguez et al., 2007).

FROM THEORY TO PRACTICE: IMPLEMENTING MODELS IN CONSERVATION AND HEALTH POLICY

Ecological modeling has historically been a useful aid for assessing potential outcomes of political or policy actions in the fields of wildlife management and epidemiology. The role of ecologists in policy making, however, has remained one of direction, that is, of predicting ecological outcomes of policies, while acknowledging that the very decision-making processes, the personal and institutional frameworks and considerations that should lead to the most commonly accepted 'number one priority': i.e., sustainable landscapes, were beyond their sphere of competence (Grabowski et al., 2023). With a number of highly sophisticated, mechanistic simulation models intended for solving complex problems of landscape sustainability, we suggest that this hierarchical structure has been unbalanced. The predictions from model implementation are now in advance of the methods to support the myriad decisions by personal, public, and institutional parties that need to be made. In this manuscript, we explore the reasons for this unbalance, and how ecological modeling, an immensely successful scientific approach, can 'rebalance' and return to the fundamental principles of prediction and knowledge synthesis, playing a fundamental role in creating some of the know-how that may help decision makers, personal or institutional, to adjust their needs and decisions while, in the longer perspective, making a transition towards sustainability (Carrara & Roig-Juñent, 2022).

In particular, ecological modelers in landscape complex problems, such as fragmentation and loss of habitats and connectivity, are facing enormous problems in the validation of their predictive results. It is thus clearer that, in order to provide practical support for decision making, the modeler should concentrate on creating tools that address specific problems, with the ultimate goals of negotiating at different levels, assisting the formulation of effective actions, and improving the effectiveness of final policies. These goals are met in the ecological modeling endeavors that propose appropriate, problem-specific indicators and indexes as policy and, in general, decision-making devices (Drechsler, 2020). More recently, agencies requiring an analytical tool to facilitate decisions have sought the capabilities of simulation models. Bioeconomic and population models employed predictive sound research in response to human needs and, to some extent, contributed to demonstrating the importance of ecosystem services to the public. The fact that the results were solidly grounded in complex ecological processes led to the interest of a first group of governmental agencies that were responsible for the management of wildlife resources, which accepted the proposition of 'ecosystem engineering,' relying on biologists and ecologists to provide valid, rigorous predictions of ecological outcomes (Martínez-Ramos et al., 2016).

Frameworks for Decision Support

Ecological models are not built solely for the advancement of knowledge. As a matter of fact, an ecological model today is even more of a representative framework for evidence than a model built prior to the 21st century. The unique ability of well-constructed shared representations to offer multiple snapshots of the same ecosystem, different parts of the same process, synthetic and self-consistent antagonistic explanations from different groups of scientists and stakeholders will undoubtedly increase the role of ecological models in the classification phase of decision making, setting the stage for the design and evaluation of political decisions (Rodríguez et al., 2007). Scientists agree that ecosystems and the services they underpin benefit from decision processes that are built on the best available evidence. Even though in many instances ecological models are not explicitly incorporated by decision makers into their decision and negotiation processes, the preference they exhibit in selecting informed solutions suggests that they have a more subtle influence on them. Again, it is reminded that whenever models are utilized for guided decision making, the entire model is never truly utilized. In some instances, evidence may be employed in the decision without a formal reference to the shared model (Drechsler et al., 2007). This should not be regarded as a diminishing impact of ecological models. One of the first elements to appreciate about environmental decision-making is that many of its alternative strategies are defined and evaluated in relatively general and not necessarily scientific terms. It is therefore the judgment of the future of the ecosystem services that guides the argument, giving the evidence incarnated in the model a more subtle, less direct, but still crucial role. Therefore, even in these instances, ecological modeling can have a role in shaping a satisfactory solution. For those social actors who indeed rely on specific scientific evidence, and not just expert judgment, for taking a position, ecological models can provide the necessary evidence (Schöttker & Wätzold, 2022).

Stakeholder Engagement

Scientists and conservation practitioners recognize the need to engage stakeholders and use results that can be directly applied to a specific ecological problem or issue. Developers of eco-epidemiological models also seek to incorporate knowledge and feedback from field researchers, wildlife managers, health officials, or others in formulating these models and their underlying assumptions and uncertainties. Some have direct experience in leveraging ecological or landscape variables for such applications (Zhu et al., 2023). There is a demonstrated need to design models to address specific questions, meet the practical needs of stakeholders and decision-makers, and secure funding, especially with respect to designing policy rules or uncovering pathways to sell such information in new markets to new customers that implicitly value it for health or conservation purposes. Most importantly, all current models produced results that could inform actual management issues. These researchers engaged with real problems, and models have been influential with respect to disease or damage control strategies (Lv et al., 2019). However, most members of the International Society for Ecology and Culture, or the members of the International Congress on Ecological Economics, can profit from the current models and their experience. The meetings offer a unique opportunity for researchers to present preliminary results, receive critical feedback, and generate interest, not just from other academics, but from potential future partners or stakeholders. In the organizers' view, the wider scientific community, regulators, and the key stakeholders shaping conservation and health policy require a considered stakeholder interaction by these particular researchers in order to pursue research paths ultimately more likely to be financed by stakeholders or applied to informed decision-making by society, managers, and political decision-makers (Maharjan et al., 2022).

Regulatory Implications

In the language of the EC White Paper, a 'regulatory' approach to conservation techniques as emergent properties, stable behavior, sustainability, or success can help to restore or sustain natural systems. The main value of ecological information for regulatory action would be: i. To better justify or predict the impact of management and other human activities on populations of species, habitats, and ecosystems that regulatory action is meant to protect. ii. To assess whether targets for regulatory action have been met, particularly when impacts of proposed activities are uncertain, cumulative, and cross-media, affect threatened and endangered species, or involve risks to public health (Chiaverini et al., 2022; Diniz-Filho et al., 2024). The European Community has been especially active in legislating in areas of potential importance to biodiversity and public health, as presented in Table 2.

Table 2. Linkages Between Biodiversity, Ecosystem Services, and Health Outcomes

Biodiversity Element	Ecosystem Service Provided	Health Benefit	Risk from Biodiversity Loss	Modelling Approach Used	Reference
Forest cover	Air and water purification	Reduced respiratory illness	Increased air pollution, waterborne diseases	Spatial and remote sensing models	(Rissman & Gillon, 2017)
Pollinators	Crop pollination	Food security, nutrition	Malnutrition, lower crop yield	Species distribution, agent-based models	(Chiaverini et al., 2022)
Wetlands	Water filtration, disease control	Reduced waterborne and vector diseases	Flooding, cholera, malaria spread	Hydrological and ecosystem dynamics models	(Diniz-Filho et al., 2024)
Coral reefs	Coastal protection, fish nurseries	Food supply, protection from storms	Malnutrition, coastal vulnerability	Marine ecosystem models	(Opoku, 2019; Singh et al., 2024)
Soil microorganisms	Nutrient cycling, pathogen control	Crop health, disease resistance	Soil degradation, food contamination	Soil-plant interaction models	(Wickberg, 2023)
Forest mammals	Zoonotic disease regulation	Zoonosis suppression (e.g., Ebola hosts)	Disease outbreaks	Population viability, agent-based models	(Grabowski et al., 2023)
Birds	Pest control, seed dispersal	Agricultural stability, mental health benefits	Pest infestations, ecological imbalance	Avian movement and landscape models	(Carrara & Roig-Juñent, 2022)
Amphibians	Insect population control	Reduced vector-borne disease risks	Increase in mosquitoes, disease transmission	Species-climate interaction models	(Drechsler, 2020)
Genetic diversity	Disease resistance in crops/livestock	Agricultural resilience	Epidemics, food insecurity	Genetic algorithm-based models	(Martínez-Ramos et al., 2016)
Freshwater fish	Protein source, ecosystem regulation	Dietary health	Protein deficiency, ecosystem collapse	Riverine system and aquatic species models	(Rodríguez et al., 2007)
Urban green spaces	Recreation, temperature regulation	Mental health, reduced heat stress	Stress disorders, heat-related illness	Urban ecosystem and land use models	(Drechsler & Wätzold, 2007)
Large predators	Regulation of prey populations	Ecological stability, zoonotic balance	Prey overpopulation, pathogen proliferation	Food web and trophic cascade models	(Schöttker & Wätzold, 2022)

continued on following page

Table 2. Continued

Biodiversity Element	Ecosystem Service Provided	Health Benefit	Risk from Biodiversity Loss	Modelling Approach Used	Reference
Insects (e.g., ants)	Decomposition, soil aeration	Healthy soils, pathogen breakdown	Soil compaction, pathogen persistence	Micro-ecosystem and decomposition models	(Huang et al., 2020)
Seagrass beds	Water quality, nursery habitat	Fisheries support, toxin filtering	Poor water quality, fishery decline	Coastal dynamic models	(Lv et al., 2019)

THE TECHNOLOGICAL FRONTIER: BIG DATA, AI, AND REMOTE SENSING

Future technological advances will result in a more general application of ecological models at multiple spatial and temporal scales. Among the significant technological advances are the development of data fusion from multiple sources, the sequencing of complete genomes, the use of rapid ecosystems, and complete single-cell analysis of microorganisms, improved satellite remote sensing capabilities through the breakdown of spectral signature components, and the development of solid-state array detection of hybridization on a DNA biochip platform (Rissman & Gillon, 2017; Chiaverini et al., 2022). Biological meaning is derived from emergent robust patterns that evolve from these runs. Bacteria are presently the only organisms whose details of the genome, protogenomic toolbox, and their ecosystem-defined roles are close to complete. This high level of completeness and genome information has model applicability from species to larger ecosystem levels. Model parameterization has benefited from the parallel existence and enabling method development for the more mature field of metagenome parameterization. Despite this, many gaps remain at present. The mean cell area detection limit in the flux towers is insufficient to permit reliable detection of small ecosystem genera and higher levels of diversity. Modeling complexity has led to thermodynamically scalar indicators that simplify the number of molecular classes required to perform the individual reactions (Diniz-Filho et al., 2024; Opoku, 2019).

Big Data and Machine Learning

Ecologists and medical doctors dealing with questions such as which species will harbor the next disease likely to jump to humans have another commonality.

Both are facing a flood of data produced by recent technical developments. The two types of data are too different to be managed satisfactorily without solid underlying models. The common need for such models and their improvement calls for a guiding discourse. Some of the most important questions of our time, such as the supporting role of diverse ecosystems in coping with recent environmental stress, are directly related to biodiversity (Chiaverini et al., 2022). Ecological models should predict how changes in species distributions and community composition will impact ecosystem functioning, both under today's and under future climatically changed conditions. We are entering, or are actually deeply within, the age of big data and machine learning. What happens to ecological modeling when datasets become huge? An unprecedented change in the types and the scope of ecological models, ranging from entirely data-driven to simulation models, is the answer. Big data and machine learning have become popular buzzwords (Diniz-Filho et al., 2024). Whenever such terms circulate, the question arises whether this is a genuinely new development or just an instance of inflated news bubbles around past research that was long forgotten. The developments leading ecologists from the 1990s to big data were most certainly continuous, marked by increasing computer power along with the collection of ever larger data series. Machine learning's rather recent buzz is related to the development of much better learning algorithms. These need two conditions: huge datasets and superior deep learning architectures. With substantial time lags between their disciplines, vision and Freshwater Collins are publishing ecological research that fulfills both conditions (Gillingham et al., 2024).

Remote Sensing Technologies

Remote sensing refers to the collection and measurement of data at a distance from the Earth using instruments operating on various platforms, which traditionally encompass aircraft and satellites. This remote sensing approach offers average spatial coverage and temporal resolution. There are increasing variations in the type of data captured, coverage, and monitoring, which typically involve acquisitions of time series of optical information as images. Satellite sensors are becoming increasingly sophisticated and yield data of higher resolution (Adhikari et al., 2019). In the past few decades, it has been possible to acquire global images with spatial resolutions for the optical domain at a scale of 30 m through a constellation of satellites. These days, more and more high-resolution optical images can be found in the public domain, notably through a growing group of open and free programs. High-resolution images, notably for certain sensors, are found in the commercial domain, but they have the impediment of high economic costs. They allow mapping large areas at the scale necessary (GIRARD et al., 2023). Moreover, composite hierarchical texture and spatially adjacent principal component analysis with high-resolution texture indices have proven reliable in discriminating exotic gardens containing alien species from macrophyte species that co-exist in a seminatural lake ecosystem. Furthermore, object-oriented capabilities allow the very detailed mapping of small groups of trees, including individual trees. In terms of scale, a criterion of sampling for landscape ecology or operational exploitation of the remote sensing resources lies in the relevant structure of the monitored ecosystems (Grabowski et al., 2023). Broadly, a rule of thumb is that the scale should be commensurate with the physiological and morphological dimensions of the main targets and phenomena. However, one difficult problem concerning the use of aerial systems in sampling natural ecosystems is that image acquisition height and scanning angle result in data acquisition over heterogeneous sample windows. The angular effects may strongly affect the spectral response of the observed targets, which could blur relationships detected in terrestrial conditions, introducing an additional source of difficulty in performing comparisons of ground-based with image-based observations. It is far from clear whether this difficulty could be mitigated by adopting different band ratios and newly proposed methods of correction for the sensor-related atmospheric effects (Carrara & Roig-Juñent, 2022).

Software Tools for Modelling

Ecological modeling is a key tool in both conservation biology and public health, with the potential to influence practices with significant global impacts. Several recent papers have reviewed software tools for modeling in conservation biology, but they have had very little focus on public health. In this chapter, we provide an overview of some of the most widely used software tools for ecological modeling used in both conservation biology and public health, focusing mainly on their history, application, and interoperability. Although some of the software tools reviewed can be used in both areas, we focus mainly on ecological examples for conservation and infectious disease examples for public health (Martínez-Ramos et al., 2016; Rodríguez et al., 2007). In the span of a few decades, numerous software have been developed and/or employed in ecological studies. From classic general statistics software like Minitab and Statistica to more specific software, such as SPSS, PAST, Dinamica, Vensim, NetLogo, QGIS, Habitat-Habitat++, MARXAN, Aegis, RAMAS, PVA Tools, and ArcGIS, among numerous others, many possibilities focus on different statistical and modeling purposes. Ultimately, some of this software is very well specialized in a given type of analysis, while others are more generalist. Alongside this specialized software, there is a trend to use some programming languages like R (Drechsler et al., 2007).

FUTURE DIRECTIONS AND CONCLUSION

Ecological models serve many purposes in the communication between disciplines and between scientists, managers, and policymakers at diverse scales. The preceding case studies demonstrate the breadth of scales over which models and modeling tools can be usefully applied at the interface between ecological research and problems in conservation and public health. The models used here contrast particularly in the finer-scale detail and issues addressed in some of the test-case studies. Messages from population models need to be integrated with those from landscape models, and vice versa, concerning pattern and process (Schöttker & Wätzold, 2022). Those messages need to be coordinated across a range of spatial scales. In addition, all the models provide guidance on what data should be collected, how it should be monitored, and where, improving the interoperability of the core data that form a base map for ecological systems, itself a notable contribution to proactive and problem-based ecology and applied biology. One fast-emerging need is to enable the assimilation of information from different models and different disciplinary backgrounds more efficiently in adaptive management phases (Zhu et al., 2023). The communication technologies required to do this are not so well researched or developed as some of the earlier needs outlined here. Currently, all the models here could be adapted to take into account different kinds of information, on different scales, and for different stakeholders. The problem is that current computing technologies are not yet well adapted to model and resolve the cross-scale and cross-disciplinary issues that emerge from such use. For example, in some cases, there are static links between current models, but as that process becomes more dynamic, a more dynamic architecture is needed to represent them and the choices that emerge (Huang et al., 2020).

Emerging Trends and Innovations

There has been increasing concern and interest in biodiversity over the last few decades, with ongoing research, policy making, and reform helping to provide the necessary toolkits for implementing conservation. Biodiversity research and policy interventions have flourished under the Ecological Modelling umbrella. Thus, as in section seven, increasing computational power and capacity required for ecological modelling applications during the last century created an enabling environment for the development of numerous ecoinformatics, machine learning, and statistical modelling techniques to advance the state of biodiversity science (Lv et al., 2019). Machine learning and deep learning models are increasingly being deployed to identify, monitor, and conserve plant and animal species, estimate Community Level Indicator criteria, and successfully predict biogeographical ranges across different spatial temperature regions. GIS and remote sensing technology using drones and other platforms have transformed both spatial ecology and large-scale ex-ante biodiversity policy-making (Maharjan et al., 2022). Rapid prototype water sampling technologies are enhancing the power of species distribution models. Individual-based models and game theory have a long history of application to conservation planning. Recent work used agent-based systems to investigate the potential impact of green agricultural policies, the impact of urban design strategies on driving biodiversity, bat colony success, the impact of landscape fragmentation on survival rates of roe deer fawns, as well as habitat and land use change trends (Drechsler, 2020; Tytar, Sobolenko, et al., 2015).

Collaborative Research Initiatives

A collaborative research program in West Africa is using ecological models to understand the relationship between deforestation, biodiversity loss, ecosystem function, and public health. This program is examining the relationships between landscape changes in an area in Guinea and their effects on the temporal and spatial abundance of arthropod vector species. The use of remote sensing, cryptographic tools, mathematical models, and mapped landscape patterns allows for targeted identification activities. Multi-disciplinary analysis clarifies the vector-host relationships (Rissman & Gillon, 2017). Molecular epidemiology is used to determine whether highlighted areas promote the transmission of human and potentially zoonotic pathogens. The program builds the capacity of local students and scientists and uses the results to educate community members and the national decision-making centers for a new rational management approach to the land. A collaboration is being established between scientists worldwide to model conservation in an unmanaged protected area in Namibia. Despite the proclamation of many relatively large-scale wildlife conservation areas in southern Africa, most have received little management (Chiaverini et al., 2022). These areas might be considered more conserved as they are not interfered with and hence are considered to have a greater potential for maintaining ecosystem processes over human-managed sites. To date, however, there has been little systematic scientific evaluation of the ecological effects of milestones in wildlife-bearing capacity. With the emerging threats of global change, including large-scale climate change, the potential resilience of large arid and remote protected areas is receiving much-needed attention. It is, therefore, a strategic question as to whether 'do-nothing' conservation policies can allow widespread arid terrestrial ecosystems to survive in a useful and dynamic state. Initial investigations have been initiated in Namibia to master the deterministic environmental drivers, which are co-determined by species distribution rules (Diniz-Filho et al., 2024; Opoku, 2019).

Interdisciplinary Approaches

While the above examples are a testament to the growing role of ecological models in practical conservation, the links between purely theoretical approaches based on statistical or mathematical tools, the development of technological devices, land-use planning, and public health are still embryonic and represent an interesting goal for future research, together with a further development of purely theoretical models that better represent ecological realities at large spatial scales and with large numbers of interacting species. The need to address global environmental issues demanded the development of sophisticated statistical tools that are used in large-scale analyses of complex ecological relationships (Chiaverini et al., 2022). However, most public health applications have been implemented at local scales, far from global processes that are ultimately driving the observed ecological changes. The non-existence of a map that relates scales in ecology to those in epidemiology has kept the advancement of public health ecological studies clipped. This situation poses both good and bad news to the field. The good news is that decision-makers are increasingly aware of the absolute necessity to have environmental scientists engaged in public health issues and, consequently, funds are being obtained to conduct this type of research (GIRARD et al., 2023). This has a very benevolent positive impact on ecology as a whole, with plenty of opportunities to engage in relevant applied research. The bad news is that because of the stove-piping that still characterizes multi- and interdisciplinary fields of study, results from public health don't often make it back to funding international agencies and our colleagues remain 'unaware' of both the contributions of ecological modeling in tackling public health challenging issues and the research social relevance, potentially stranding critical initiatives (Huang et al., 2020; Zhu et al., 2023).

CONCLUSION

We began by arguing that ecological modelling has an important capacity to characterize and predict the behavior of ecological systems and the disposition and nature of the ecological benefits offered to humankind by these systems. In exploring the depths of scientific understanding of these systems, we have also unveiled a range of uncertainties and limitations associated with a tendency in ecological modelling to bypass or circumvent difficult problems and questions - a scientific weakness that restricts progress on substantive problems and a criticizable state of affairs that invites a sustained injection of applied modelers with a science of adequate complexity and detail. We would argue not only for an increased role for applied biomathematicians and statisticians, but also naturalists, ecologists and others who tend, more empirically, to ground their theories and predictions in a multi-dimensional understanding of the correlation between taxon biology and ecology. In keeping with the purpose of this book, applied macroecology is defined by thoughtfully crafting ecological models that describe and manage the interactions between taxon carriers and the natural resources and processes that provide products like recreational landscapes, environmental education, science, biodiversity conservation, ecosystem functions and processes, and public health. In a certain sense there is nothing new about this combination of elements - biologists, conservation planners and education professionals often coalesce essentially to this end.

REFERENCES

Adhikari, D., Tiwary, R., Singh, P. P., Upadhaya, K., Singh, B., Haridasan, K. E., Bhatt, B. B., Chettri, A., & Barik, S. K. (2019). Ecological niche modeling as a cumulative environmental impact assessment tool for biodiversity assessment and conservation planning: A case study of critically endangered plant Lagerstroemia minuticarpa in the Indian Eastern Himalaya. *Journal of Environmental Management, 243*. Advance online publication. DOI: 10.1016/j.jenvman.2019.05.036

Carrara, R., & Roig-Juñent, S. A. (2022). Maps of potential biodiversity: When the tools for regional conservation planning clash with species ecological niches. *Biodiversity and Conservation, 31*(2). Advance online publication. DOI: 10.1007/s10531-022-02355-3

Chiaverini, L., Macdonald, D. W., Bothwell, H. M., Hearn, A. J., Cheyne, S. M., Haidir, I., & Hunter, L. T. B., Kaszta, Macdonald, E. A., Ross, J., & Cushman, S. A. (2022). Multi-scale, multivariate community models improve designation of biodiversity hotspots in the Sunda Islands. *Animal Conservation, 25*(5). Advance online publication. DOI: 10.1111/acv.12771

Diniz-Filho, J. A. F., Bini, L. M., Targueta, C. P., Telles, M. P. de C., Jardim, L., Machado, K. B., Nabout, J. C., Nunes, R., Vieira, L. C. G., & Soares, T. N. (2024). Environmental DNA and biodiversity patterns: a call for a community phylogenetics approach. In *Perspectives in Ecology and Conservation* (Vol. 22, Issue 1). DOI: 10.1016/j.pecon.2024.01.006

Drechsler, M. (2020). Ecological-Economic Modelling for Biodiversity Conservation. In *Ecological-Economic Modelling for Biodiversity Conservation*. DOI: 10.1017/9781108662963

Drechsler, M., Grimm, V., Mysiak, J., & Wätzold, F. (2007). Differences and similarities between ecological and economic models for biodiversity conservation. *Ecological Economics, 62*(2). Advance online publication. DOI: 10.1016/j.ecolecon.2006.03.026

Drechsler, M., & Wätzold, F. (2007). Ecological-economic modelling for the sustainable use and conservation of biodiversity. In *Ecological Economics* (Vol. 62, Issue 2). DOI: 10.1016/j.ecolecon.2006.05.004

Drechsler, M., Wätzold, F., & Grimm, V. (2022). The hitchhiker's guide to generic ecological-economic modelling of land-use-based biodiversity conservation policies. In *Ecological Modelling* (Vol. 465). DOI: 10.1016/j.ecolmodel.2021.109861

European Commission. (2020). Food 2030 pathways for action : research and innovation policy as a driver for sustainable, healthy and inclusive food systems. In *Publications Office of the ....*

Gillingham, P. K., Britton, J. R., Jones, G., Miller-Rushing, A., Stafford, R., & Slater, H. (2024). Climate change adaptation for biodiversity in protected areas: An overview of actions. *Biological Conservation, 289*. Advance online publication. DOI: 10.1016/j.biocon.2023.110375

GIRARD, L., WEZEL, A., & ROBIN, J. (2023). Drying out fish ponds, for an entire growth season, as an agroecological practice: Maintaining primary producers for fish production and biodiversity conservation. *Open Research Europe, 3*. Advance online publication. DOI: 10.12688/openreseurope.16363.1

Grabowski, Z., Fairbairn, A. J., Teixeira, L. H., Micklewright, J., Fakirova, E., Adeleke, E., Meyer, S. T., Traidl-Hoffmann, C., Schloter, M., & Helmreich, B. (2023). Cosmopolitan conservation: the multi-scalar contributions of urban green infrastructure to biodiversity protection. In *Biodiversity and Conservation* (Vol. 32, Issue 11). DOI: 10.1007/s10531-023-02614-x

Huang, Z., Bai, Y., Alatalo, J. M., & Yang, Z. (2020). Mapping biodiversity conservation priorities for protected areas: A case study in Xishuangbanna Tropical Area, China. *Biological Conservation, 249*. Advance online publication. DOI: 10.1016/j.biocon.2020.108741

Lv, Z., Yang, J., Wielstra, B., Wei, J., Xu, F., & Si, Y. (2019). Prioritizing green spaces for biodiversity conservation in Beijing based on habitat network connectivity. *Sustainability (Switzerland), 11*(7). Advance online publication. DOI: 10.3390/su11072042

Maharjan, K. L., Gonzalvo, C., & Aala, W. (2022). Dynamics of Environmental Conservation Agriculture (ECA) Utilization among Fujioka Farmers in Japan with High Biodiversity Conservation Awareness but Low ECA Interest. *Sustainability (Switzerland), 14*(9). Advance online publication. DOI: 10.3390/su14095296

Martínez-Ramos, M., Ortiz-Rodríguez, I. A., Piñero, D., Dirzo, R., & Sarukhán, J. (2016). Anthropogenic disturbances jeopardize biodiversity conservation within tropical rainforest reserves. *Proceedings of the National Academy of Sciences of the United States of America, 113*(19). Advance online publication. DOI: 10.1073/pnas.1602893113

Opoku, A. (2019). Biodiversity and the built environment: Implications for the Sustainable Development Goals (SDGs). *Resources, Conservation and Recycling, 141*. Advance online publication. DOI: 10.1016/j.resconrec.2018.10.011

Rissman, A. R., & Gillon, S. (2017). Where are Ecology and Biodiversity in Social–Ecological Systems Research? A Review of Research Methods and Applied Recommendations. In *Conservation Letters* (Vol. 10, Issue 1). DOI: 10.1111/conl.12250

Rodríguez, J. P., Brotons, L., Bustamante, J., & Seoane, J. (2007). The application of predictive modelling of species distribution to biodiversity conservation. In *Diversity and Distributions* (Vol. 13, Issue 3). DOI: 10.1111/j.1472-4642.2007.00356.x

Russell-Bennett, R., Polonsky, M. J., & Fisk, R. P. (2024). SDG commentary: Services that sustainably manage resources for all humans-the regenerative service economy framework. *Journal of Services Marketing*, *38*(2). Advance online publication. DOI: 10.1108/JSM-09-2023-0327

Schöttker, O., & Wätzold, F. (2022). Climate Change and the Cost-Effective Governance Mode for Biodiversity Conservation. *Environmental and Resource Economics*, *82*(2). Advance online publication. DOI: 10.1007/s10640-022-00684-z

Simpson, K. H., de Vries, F. P., Dallimer, M., Armsworth, P. R., & Hanley, N. (2022). Ecological and economic implications of alternative metrics in biodiversity offset markets. *Conservation Biology*, *36*(5). Advance online publication. DOI: 10.1111/cobi.13906

Singh, R., Shafik, W., Crowther, D., & Kumar, V. (Eds.). (2024). *Transforming Healthcare Sector Through Artificial Intelligence and Environmental Sustainability* (1st ed., Vol. 1). Springer Nature Singapore. DOI: 10.1007/978-981-97-9555-0

Tytar, V., Mezhzherin, S., & Sobolenko, L. (2015). Using ecological niche modeling for biodiversity conservation guidance in the Western Podillya (Ukraine): Amphibians. *Vestnik Zoologii*, *49*(2). Advance online publication. DOI: 10.1515/vzoo-2015-0015

Tytar, V., Sobolenko, L., Nekrasova, O., & Mezhzherin, S. (2015). Using Ecological Niche Modeling for Biodiversity Conservation Guidance in the Western Podillya (Ukraine): Reptiles. *Vestnik Zoologii*, *49*(6). Advance online publication. DOI: 10.1515/vzoo-2015-0065

Wickberg, A. (2023). Environing media and cultural techniques: From the history of agriculture to AI-driven smart farming. *International Journal of Cultural Studies*, *26*(4). Advance online publication. DOI: 10.1177/13678779221144762

Zhong, M., Umar, M., Mirza, N., & Safi, A. (2024). Mineral resource Optimization: The Nexus of sustainability, digital transformation, and green finance in OECD economies. *Resources Policy*, *90*. Advance online publication. DOI: 10.1016/j.resourpol.2024.104829

Zhu, G., Giam, X., Armsworth, P. R., Cho, S. H., & Papeş, M. (2023). Biodiversity conservation adaptation to climate change: Protecting the actors or the stage. *Ecological Applications*, *33*(2). Advance online publication. DOI: 10.1002/eap.2765

KEY TERMS AND DEFINITIONS

Biodiversity Conservation: involves protecting and managing the variety of life forms on Earth to ensure ecosystem stability and resilience.

Climate Change: refers to long-term alterations in global or regional climate patterns primarily caused by human activities such as greenhouse gas emissions.

Ecological Management: refers to the strategic use of scientific principles and practices to maintain and restore the health, productivity, and sustainability of ecosystems.

Life Under Water: focuses on conserving and sustainably using oceans, seas, and marine resources for ecological balance and human prosperity.

Radiation Effect: describes the impact of ionizing or non-ionizing radiation on living organisms, materials, and the environment.

Sustainable Development Goals (SDGs): are a set of 17 global objectives established by the United Nations to promote social, economic, and environmental sustainability by 2030.

ENDNOTES

1 https://www.worldwildlife.org/pages/what-is-biodiversity

2 https://e360.yale.edu/features/how_forest_loss_is_leading_to_a_rise_in_human_disease_malaria_zika_climate_change

3 https://www.who.int/teams/global-malaria-programme/reports/world-malaria-report-2020

About the Contributors

Islam Md. Shafiqul received his PhD and he was a postdoctoral fellow at the London School of Economics and Political Science in London (LSE), UK. With more than thirty years of experience, his professional endeavors encompass the education and development sectors, with a particular emphasis on Sustainable Waste Management, Organic Farming, Climate Change Adaptation, Disaster Risk Reduction, Sustainable Development, and the enhancement of social-ecological resilience in communities that are vulnerable to coastal disasters and environmental threats. Dr. Islam currently holds the position of Associate Professor at the International University of Business Agriculture and Technology. Prior to his current position, he was affiliated with the United International University, State University of Bangladesh, and the University of Liberal Arts Bangladesh. Dr. Islam has authored and published more than 60 articles and titles in international peer-reviewed journals, in addition to contributing to numerous books and book chapters with publishers such as Taylor and Francis, Apple Academic Press, Rutledge, Springer Nature, IGI Global, Green Leaf and Nova Science Publishers. His upcoming publication, Waste Management Case Studies from Developing Countries, is set to be released by Springer Nature. He has also taken on roles as an editor, reviewer, and author for a variety of books, articles, and book chapters with renowned publishers. In 2016, Dr. Islam made contributions to the UN Sustainable Development Report.

***

Arun Agrawal is an Assistant Professor in the Department of Computer Science & Engineering at the Institute of Technology & Management, Gwalior. He is currently pursuing a Ph.D. in Computer Science and Engineering, holds an M.Tech. in Information Technology from AAI-Deemed University, Allahabad (2006), an M.Sc. in Computer Science from Jiwaji University, Gwalior (2002), and a B.Sc. in Electronics from P.G.V. College, Gwalior. With over 17 years of teaching and industry experience, Arun has been serving as an Assistant Professor at ITM, Gwalior since July 2021. His prior experience includes positions at ITM University, ITM Gwalior, and other engineering colleges, along with more than three years of industry experience at companies like Createch Devices and Simoco Telecommunications. His research interests lie in Vehicular Ad-hoc Networks, Digital Image Processing, and Machine Learning. Arun has contributed significantly to his field with 12 publications in international journals, 8 in international/ national conferences, and 5 chapters in edited books.

Shyam Sunder Agrawal, working as Assistant Professor, Mittal School of Business, Lovely Professional University, Phagwara, Punjab, India. He has been associated with the education industry since 2015. He has presented papers at various reputed IIM conferences and published several research papers in UGC CARE, SCOPUS, and ABDC Journals. He has earned his Ph.D. in management from Devi Ahilya Vishwavidyalaya, Indore. His fields of interest are Behavioral Finance, Sustainable Finance, Capital Market, Banking, and Insurance. He has published and granted 2 patents from Govt of India. He received the International Case Study Award in 2022. He is a reviewer for Emerald, Springer, and peer-reviewed journals. He is also a financial advisor at Life Insurance Corporation of India.

Munir Ahmad is a seasoned professional in the realm of Spatial Data Infrastructure (SDI), Geo-Information Productions, Information Systems, and Information Governance, boasting over 25 years of dedicated experience in the field. With a PhD in Computer Science, Dr. Ahmad's expertise spans Spatial Data Production, Management, Processing, Analysis, Visualization, and Quality Control. Throughout his career, Dr. Ahmad has been deeply involved in the development and deployment of SDI systems specially in the context of Pakistan, leveraging his proficiency in Spatial Database Design, Web, Mobile & Desktop GIS, and Geo Web Services Architecture. His contributions to Volunteered Geographic Information (VGI) and Open Source Geoportal & Metadata Portal have significantly enriched the geospatial community. As a trainer and researcher, Dr. Ahmad has authored over 50 publications, advancing the industry's knowledge base and fostering innovation in Geo-Tech, Data Governance, and Information Infrastructure, and Emerging Technologies. His commitment to Research and Development (R&D) is evident in his role as a dedicated educator and mentor in the field.

Ferdous Ahmed, is the Associate Professor of Environmental Science and Sustainability in the Department of Environmental science at IUBAT since 2018. He received his BSc and MSc degrees in Botany from the University of Rajshahi, Bangladesh. He also obtained his PhD in Climate Change Adaptation Policy and

Food Security from the University of Malaya Malaysia with the Bright Spark Scholarship in 2016. He also did a Postdoc in Climate Change Mitigation from the same university in Malaysia. He is the Academic Editor in PLOS CLIMATE. However, He has more than 30 ISI/SCOPUS journal papers including sole book on Climate Change Adaptation in Springer Nature in 2023. However, he comes from the Chapainawanganj District, the Capital of Mangoes in Bangladesh.

Yasir Ansari, born in 1989 in Hyderabad, Sindh, Pakistan, is a dedicated Sub Engineer with a passion for preserving and celebrating the rich cultural heritage of Sindh. With over 13 years of experience in the Culture Tourism Antiquities and Archives Department, Ansari has been instrumental in various projects aimed at conserving and promoting the province's historical legacy. Ansari's educational background in Civil Engineering and General History has equipped him with a unique blend of skills, enabling him to approach heritage preservation with a holistic perspective. His associate degree in Civil Engineering has provided him with a solid foundation in construction and restoration, while his master's degree in General History has deepened his understanding of the cultural and historical significance of Sindh's heritage sites. Beyond his professional endeavors, Ansari is an avid enthusiast of cultural heritage, often visiting heritage sites and collecting data on Sindh's cultural legacy. His ultimate goal is to safeguard this legacy for future generations, ensuring that the province's historical treasures remain a source of pride and inspiration for years to come.

Vincent Assanful is an Associate Professor of Religion and Politics in the Department of Religion and Human Values, University of Cape Coast, Ghana. He holds a Phd in Religion and Human Values. His PhD was on the role of the Indigenous Assin Religion in the practices of inheritance and succession of the Assin. He has published extensively in the area of religion and politics, religious pluralism, chieftaincy and implantation of Christianity and many more. He lectures in African Traditional Religion, Comparative Religion, Religion and Chieftaincy, Nature and Goal of African Traditional Religion at both the undergraduate and graduate levels in the Department of Religion and Human Values.

Aruna Bajpai is an accomplished academic in the field of Computer Science & Engineering, Currently serving as an Assistant Professor at the Institute of Technology & Management, Gwalior. With a strong educational background including M.Tech. in Computer Science, Ms Bajpai is currently pursuing Ph.D. Her career spans over two decades, in academia since 2008 . Over the past 16 years, She has progressed from a Lecturer to his current role as Assistant Professor, demonstrating expertise in areas such as Machine Learning,Image Processing.

Ms. Bajpai's academic contributions include numerous patents, book chapters, publications in international journals and conferences, as well as presentations at IEEE international conferences. Her ongoing research and teaching efforts continue to shape the field of computer science and nurture the next generation of technology professionals.

Niamjit Das is a Research Fellow at the Center for Natural Resource Studies in Dhaka. He completed his B.Sc. and M.Sc. in Forestry from Shahjalal University of Science & Technology, Bangladesh. With more than 12 years of research experience, he has led and contributed to national and international projects, authored fifteen peer reviewed articles, several book chapters, and presented widely at conferences. He has also written for newspapers and magazines. His expertise includes forest and community ecology, biodiversity conservation, and climate change adaptation, applying computational tools and interdisciplinary approaches to advance understanding of forest dynamics and ecosystem function.

Manjunatha Reddy G. R. is a Botanist with over 21 years of academic and research experience. He holds an M.Sc., M.Phil., and Ph.D. in Botany, specializing in in vitro regeneration and arbuscular mycorrhizal (AM) fungi interactions in Withania somnifera. Currently serving as Faculty in the Department of Botany, Bangalore University, he has published extensively with Springer, IGI Global, CRC Press, and other reputed publishers. His research focuses on plant biotechnology, AM fungi, and the integration of AI and omics technologies for environmental sustainability. He also mentors students globally through the Planting Science program (BSA & NSF, USA).

Laila Badriyatul Habibah is a master's student in the Biology Education program at Universitas Negeri Malang (UM) with a scholarship from the Indonesian Education Fund Management Agency (LPDP) in 2023. Previously, she completed her bachelor's degree in the Biology Education program at the Faculty of Mathematics and Natural Sciences (FMIPA) at Universitas Negeri Malang (UM) through a BIDIKMISI scholarship. Her professional experience includes an internship at the International Affairs Office at UM in 2021, research assistance focused on animal reproductive systems, and teaching assistance for several courses focused on animal biodiversity, animal development, and protists over a period of more than three years, she has also been a speaker on several occasions regarding the ethics of animal dissection in laboratories at biology department events. Laila is interested in improving the quality of biology education by developing learning models or strategies integrated with artificial intelligence (AI), and she actively writes as a form of contribution to research and educational innovation.

Md. Sazid Hasan graduated in Business Administration from the IUBAT Business School, majoring in Human Resource Management in 2026. He is currently working as an HR Intern at Debugger Studio, a leading IT firm, where he is developing practical experience in recruitment, organizational development and workflow optimization. Sazid has participated in several academic research projects focused on waste management, tourism, sustainable finance and blockchain. With a keen interest in technology and innovation, he strives to integrate modern tools into HR processes for better efficiency. His leadership skills and passion for continuous learning make him a promising future professional in strategic human resource management. Though he is a Busjiess graduate but he bears great courage for the sustainable development prioritizing forest biodiversity and ecosystem. However, he has completed a course title "Environmental Science and Sustainability" at IUBAT during his BBA study. So far he has published two ISI/SCOPUS ranked journal papers that reflects the concerns for environmental degradation.

Gan Liu is a PhD student at the School of Communication, Universiti Sains Malaysia. His research interest is on Environmental Communication and society.

Mohamad Saifudin Mohamad Saleh is an Associate Professor at the School of Communication, Universiti Sains Malaysia (USM). He earned his Ph.D. in Environmental Communication from the Faculty of Sustainability, Leuphana University of Lüneburg, Germany, in 2016. His expertise lies in environmental and sustainability communication, with numerous publications in academic journals, book chapters, and newspaper articles on these topics. He has recently completed research projects on Islamic environmental communication, sustainable university branding, and sustainable lifestyles in Malaysia.

Ananya Pandey is a Research Scholar in School of Law at Christ University, Bangalore. She has her research interests in the field of Constitutional Law, Environment Law, Human Rights, Food Security, Intellectual Property Law and Jurisprudence.

Wasswa Shafik is a researcher and author specializing in artificial intelligence, sustainability, and inclusive education. He holds a PhD in Computer Science from Universiti Brunei Darussalam and has published extensively on AI-driven solutions for health, agriculture, education, and environmental management. Dr. Wasswa has authored and edited several books with leading publishers, including Springer, Elsevier, CRC Press, and Cambridge Scholars Publishing. His work bridges technology and social good, focusing on innovation for marginalized communities, sustainable development, and responsible AI adoption. He actively contributes to global discussions on digital transformation and equitable futures.

Jagveer Singh is an accomplished academic and chemistry professor with a strong background in organic chemistry. His educational and professional journey reflects a deep commitment to the field of science and education. Education 1989: Completed 12th grade from the Madhya Pradesh Board. 1992: Earned a B.Sc. degree from Guru Ghasidas University, Bilaspur, Chhattisgarh. 1994: Completed an M.Sc. in Organic Chemistry from Guru Ghasidas University, Bilaspur, Chhattisgarh. 1996: Received an M.Phil. in Organic Chemistry from the same institution. Professional Career Since July 1999: Serving as a Professor of Chemistry at the Institute of Technology and Management (ITM), Gwalior, Madhya Pradesh. In this role, he has contributed significantly to educating future scientists and engineers, fostering a strong foundation in chemistry. Expertise and Contribution With over two decades of experience in teaching organic chemistry, Jagveer Singh is recognized for his dedication to students and his field. His academic path, spanning both bachelor's and master's level studies, underscores his expertise in organic chemistry. His long tenure at ITM Gwalior reflects his passion for teaching and nurturing young minds.

Seth Tweneboah (PhD) is a senior lecturer at the Centre for African Studies of the University of Education, Winneba, Ghana. He received his PhD in Religious Studies (with specialization in law and politics) from the Victoria University of Wellington, New Zealand. Seth has an MPhil from the University of Ghana, Legon and an MA from the Florida International University, Miami. His research focuses on the religious human rights and the religion-law interrelationships in Africa. He is the author of Religion, Law, Politics and the State in Africa: Applying Legal Pluralism in Ghana.

Anurag Yadav Associate Professor, Department of Microbiology C.P. College of Agriculture SardarKrushinagar Dantiwada Agricultural University, Gujarat, India Email: anuragyadav123@gmail.com Dr. Anurag Yadav is currently serving as Associate Professor in the Department of Microbiology at SardarKrushinagar Dantiwada Agricultural University (SDAU), Gujarat, India. He holds a Ph.D. in Microbiology and has over two decades of teaching, research, and extension experience in soil microbiology, microbial biotechnology, and bioresource management. His research interests include metagenomics of microbial communities, bioprospecting of plant growth-promoting microbes, organic farming bioinputs, and the application of nanotechnology and artificial intelligence in sustainable agriculture. Dr. Yadav has guided several postgraduate students and published numerous research and review articles in reputed national and international journals. He has also contributed book chapters to various academic publishers and actively participates in editorial and review roles for scientific journals. Dr. Yadav has a special interest in writing scholarly content with interdisciplinary scope, particularly at the interface of microbiology, bioinformatics, and agrotechnology. He is passionate about translating microbial research into field-level applications for climate-resilient and resource-efficient agriculture. His academic and research contributions have been recognized through institutional awards and invitations to contribute to conferences and edited volumes.

Kusum Yadav Professor, Department of Biochemistry University of Lucknow, Lucknow, Uttar Pradesh, India Dr. Kusum Yadav is currently serving as Professor in the Department of Biochemistry at the University of Lucknow, Uttar Pradesh, India. She holds a Ph.D. in Biochemistry and has established a strong academic and research portfolio in the areas of molecular biology, enzymology, and plant biochemistry. Her scholarly trajectory reflects a sustained commitment to excellence in both teaching and research. Dr. Yadav's research interests are centered on biochemical signaling pathways, plant trait-associated molecular markers, and molecular responses to environmental stimuli. She has published extensively in reputed peer-reviewed journals and has contributed book chapters in the fields of plant biochemistry and biotechnology. In addition to her academic pursuits, Dr. Yadav is actively engaged in interdisciplinary research collaborations, integrating biochemistry with microbiological, agricultural, and environmental sciences. She is deeply committed to mentoring emerging researchers and promoting the adoption of advanced methodologies, including omics technologies and computational biology, in both research and pedagogy.

Ester Novi Kurnia Zebua is a young lecturer in the Biology Education Study Program at Universitas Nias (UNIAS), Indonesia. He completed his bachelor's degree in biology education from Universitas Nias Raya (UNIRAYA) and his master's degree from the Department of Biology at Universitas Negeri Malang (UM). He is a passionate researcher and writer who explores innovative topics in the field of education. His research focuses on biology, biology and science education, literacy, thinking skills, artificial intelligence (AI) for education, biodiversity, pedagogy, 21st-century skills, and ethnobiology to enhance student learning experiences and engagement. He is also actively providing valuable insights to educators, researchers, and students, as well as promoting digital literacy and environmental awareness through problem-based, project-based, and community-based approaches.

Nofamataro Zebua is a young writer and researcher. He graduated with a bachelor's degree in biology education from the Faculty of Teacher Training and Education, Nias Raya University. He completed his master's degree in education at the Faculty of Mathematics and Natural Sciences, Malang State University, through a scholarship from the Indonesian Republic's Education Fund Management

Institution (LPDP). His research focus and interests include biology education, biology, biodiversity studies, ecology, botany, learning models and media, learning evaluation and assessment, literacy, and AI for education.

Index

E

F

G

H

I

L

M

N

P

R

S

T

U

W

www.ingramcontent.com/pod-product-compliance
Lightning Source LLC
LaVergne TN
LVHW062338300926
849161LV00007B/1745